Into the Final Frontier:

The Human Exploration of Space

Bernard McNamara
New Mexico State University

HARCOURT COLLEGE PUBLISHERS

Ft Worth • Philadelphia • San Diego • New York • Orlando • Houston • San Antonio
Toronto • Montreal • London • Sydney • Tokyo

Vice President/Publisher	Emily Barrosse
Acquisitions Editor	Kelley Tyner
Marketing Strategist	Kathleen Sharp
Developmental Editor	Jennifer Pine
Production Manager	Susan Shipe
Art Director	Caroline McGowan
Cover Credit	Apollo 8 Earthrise View, NASA or National Aeronautics and Space Administration

McNamara, Bernard

Into the Final Frontier: The Human Exploration of Space
ISBN: 0-03-032016-X

Library of Congress Cataloging-in-Publication Data

McNamara, Bernard.
  Into the final frontier: The Human Exploration of Space/Bernard McNamara.
    p. cm.
  Includes bibliographical references and index.
  ISBN 0-03-032016-X
    1. Manned space flight—History.   2. Astronautics—History.   I. Title.

TL873 .M36 2000
629.45'009—dc21   00-063112

Address for domestic orders:
Harcourt College Publishers
6277 Sea Harbor Drive, Orlando, FL 32887-6777
1-800-782-4479
e-mail collegesales@harcourt.com

Address for international orders:
International Customer Service, Harcourt, Inc.
6277 Sea Harbor Drive, Orlando, FL 32887-6777
(407) 345-3800
Fax (407) 345-4060
e-mail hbintl@harcourt.com

Address for editorial correspondence:
Harcourt College Publishers
Public Ledger Building Suite 1250,
150 S. Independence Mall West, Philadelphia, PA 19106-3412

Web Site Address:
http://www.harcourtcollege.com

Printed in the United States of America
0 1 2 1 2 3 4 5 6 7 8   066   10 9 8 7 6 5 4 3 2 1

Preface

One of the greatest accomplishments of the twentieth century is the human advance into space. For the first time ever, travel beyond the Earth is more than just the subject of adventerous science fiction tales, it is a reality. The purpose of this textbook, *Into the Final Frontier: The Human Exploration of Space* is to trace the development of spaceflight from the late 1800s to the present time and then to discuss the advantages and disadvantages of establishing human outposts in space and on nearby worlds. The scientific results of human spaceflight and the cultural, military, and political environment in which these achievements were made are also examined.

After reading this book a person should (1) be able to recognize the critical obstacles that inhibited the development of space flight and how they were overcome, (2) be informed about the individual sacrifices made to advance this field, (3) be aware of the primary scientific results produced by spaceflight, and (4) be able to discuss the advantages and disadvantages of proposals to build the International Space Station, Moon Base, and a Mars outpost.

This book serves as the foundation for a Junior-level nonscience majors course at New Mexico State University (NMSU). It fulfills a student requirement in the area of Viewing a Wider World. It also satisfies the desire of a significant number of students who wish to become informed about the origin of the world's major space programs and the prospects for future international collaborations in space. I was motivated to write this book when I discovered that although a multitude of technical books exist on spaceflight, few of them discuss this subject from a comprehensive viewpoint. It is relatively easy to find books about the pioneers of space travel, the origin of NASA, the Soviet space program, Mission to Mars, etc., but I was unable to find a single text that dealt with all of these subjects. The gathering and synthesis of this material into this book took several years to complete.

Unique Features of This Book

This book has three unique features. The first is its *scope*. A considerable amount of text is devoted to the Soviet, American, and international space programs. Nevertheless, the book is not an anthology of space missions. It emphasizes the historical context in which missions were undertaken and the human, technological, political, and scientific side of spaceflight. A technical field does not evolve in a vacuum but is often propelled by the convergence of seemingly unrelated events. If circumstances unfold such that a field simultaneously satisfies several human needs, rapid progress in it is possible. Human space flight serves as an excellent example of this process.

A second unique feature of this book is that it discusses an *area of contemporary interest*. The human race is now actively engaged in the journey into space. Progress on the International Space Station is widely reported in the news media. Unlike material covered in many university science courses, all of the students in this class will have been eyewitnesses to many of the events discussed in this book. The text also introduces the rational for some of NASA's major future efforts.

The third unique feature of this book is that it employs a *reader friendly format*. The text is designed to be readable, making frequent use of first-person quotes. Over 540 photographs and drawings have been used to illustrate the text. The chapters are relatively short, making them easy to digest. Technical details that might interrupt the story line have been avoided and explanations of equipment are provided in Focus Boxes. The instructor has great flexibility in tailoring the level of math and physics to the class.

Supplement for This Book

An optional supplement supports this book. The Student Guide is a two part aid to understanding the text. The first half consists of a series of different length exercises that parallel the book's discussion. Many of these activities are short and can be completed in class. Others are longer and more suitable as homework. Some of the assignments rely on an understanding of physics, but most emphasize the historical development of spaceflight. The variety of exercises is purposeful and was done to allow the instructor to teach this class from a variety of perspectives. Despite the different levels of math found in these exercises, they follow the format developed under two of my prior NSF grants on the construction of innovative instructional materials. To engage the student as quickly as possible, a minimal amount of text is included at the beginning of each exercise. It is assumed that the instructor will provide the required background material. To make the activity more inviting, an open format is employed. Each question is placed inside a box for easy identification and the size of the box provides the student with a guide to the length of the expected response. An empty box serves as a visual reminder for the student to ask the instructor for assistance. This process allows learning to take place more efficiently by identifying areas where the student's understanding is incomplete.

The second half is a set of *student notes*. Notetaking is a skill that takes time to develop. Some instructors feel this is an individual student matter and leave them to develop their own conventions. Other instructors provide extensive written notes that students personalize by adding side comments. Finally, some instructors use the initial class sessions to discuss various notetaking techniques. Dialogue notetaking, for example, is a method educational research suggests produces superior learning.

The supplemental student notes provided here employ a hybrid approach I have found to be effective. The left-hand portion of each page consists of a picture or a short statement about the material discussed in a chapter while the student writes lecture notes on the right-hand side of the page. The pictures and phrases serve as visual markers to help the students organize and clarify their writings. The pictures also serve as a locator between their notes and the material in the textbook.

The student guide is offered as an optional course ancillary to give the instructor the greatest flexibility possible in adapting this course to a variety of curriculum objectives, class sizes, and student backgrounds.

Author Acknowledgments

I would like to thank the following people for their advice and assistance in completing this book, Thomas Harrison, Mary French, Christopher Burnham, Eugene Cunnar, Abbey Osborne, and Eugenia Conway (New Mexico State University), Steve Danford (University of North Carolina), James Ryan (University of New Hampshire), Kevin Marvel (American Astronomical Society), Thomas Hockey (University of Northern Iowa), Mark Miksic (Queens College), George McCluskey (Lehigh University), and Timothy Barker (Wheaton College).

I would also like to thank Catherine Lewis and Kate Igoe of the National Air and Space Museum, Dave Dooling, Pamela McCaskill, and Jim Oberg for information about copyright policies. James Harford kindly supplied copies of several of the photographs in the book. Gloria Sanchez of the Johnson Space Center Media Center was of enormous help in collecting the NASA photographs. Mott Linn of the Robert H. Goddard Library of Clark University, Irene Willhite of the U.S. Space and Rocket Center, and Nick Vlasov of theWashington D.C. Bureau of the Russian Information Agency Novosti provided valuable assistance.

Support for this book was provided through a grant from the New Mexico Space Grant Consortium and through a sabbatical leave from NMSU. The Griffith Observatory gave permission to reproduce the chapter on the Mercury 13 which was first published in the Griffith Observer. Finally I would like to thank my wife Brenda for her assistance in all phases of this project.

Contents

Introduction to Section I
The Pioneers of Manned Space Flight

A movement of such sweep and momentum must necessarily be sustained and powered by the minds and energies of many men, and by the dreams of the whole human race. Nevertheless, it often finds supreme expression in the lives and works of a special few, who in some unusual way are caught up in the idea and become the key achievers of it. They are the men who, by virtue of personality, genius, inspiration or opportunity, become the selected instruments of history.

—Harry Guggenheim in *Men of Space*

An imaginary road exists which leads into space. This road is marked by guideposts that have been placed there by past explorers who have taken the first tentative steps into this vast and unknown realm on our behalf. Today, the appearance of this road has changed so dramatically that it bears little resemblance to its earlier state. Initially narrow and treacherous, progress along it was slowed by numerous barriers. Dangerous curves and formidable obstacles were present. Several side trails existed, which if followed, led to dead ends. Each generation of travelers has accepted the challenge of straightening this road, filling in its potholes, and discovering ways of making it safer. Technology has been employed to overcome nature's barriers and to improve the reliability of the space vehicles used by each group of succeeding adventurers.

Progress has come at great personal and material costs. Over the years people have shouldered these costs through numerous acts of courage and nations have expended huge sums of wealth. Today we have come to realize that these types of sacrifices are not limited to a single era. The conquest of a lim-itless space requires an endless commitment. As each generation makes its contribution, the burden is passed on to others who extend the road's boundary further into the heavens.

The Early Space Pioneers

There are so many individuals who have made notable contributions toward the goal of space travel that one book cannot do justice to them all. Three individuals, widely recognized as pioneers in this field, have been selected to serve as examples: Konstantin Tsiolkovsky, Robert Goddard, and Hermann Oberth. As we will see, these men possessed extraordinary vision, perseverance, and a consuming desire to unlock the secrets of spaceflight. They accepted formidable challenges and the harsh reality that they would most likely not live long enough to see their dreams fulfilled.

Konstantin Tsiolkovsky was an unlikely explorer. A self-educated, partially deaf school teacher who grew up in rural Russia, he lived in near poverty for most of his life. His studies on space flight were so important that many people would later claim the road to space ran through Kaluga, the small Russian town in which he conducted his research.

Robert Goddard, the second pioneer, was a shy man whose passion for rocketry grew into a lifelong desire to build a vehicle capable of traveling to Mars. Ridiculed across America as the "Moon man," he constructed the world's first liquid-fueled rocket. Like many other visionaries, Goddard had to deal with the criticism of others who considered his ideas to be too futuristic.

Hermann Oberth, our third pioneer, was frustrated in his rocket research by a lack of support from the academic establishment. Although he foresaw the great potential of the rocket, he could not convince the educated elite that such a device was worthy of serious consideration. His tireless efforts to make space travel a reality would later inspire others in Germany to build the world's first spaceship. Unlike his fellow pioneers, Oberth did live long enough to see the partial fulfillment of his dream when Neil Armstrong set foot on the dusty surface of the Moon in 1969.

Although these three pioneers were from different countries and possessed different educational backgrounds, they

Figure Intro. I-1. The road to space would first lead to earth orbit and then to the moon. A moon trip was a frequent topic of science fiction novels in the early 1900s.

Focus Box: Three of the Early Pioneers of Space Travel

Konstantin Tsiolkovsky   **Robert Goddard**   **Hermann Oberth**

all experienced childhood events that led them to dedicate their lives to the conquest of space. Some of their inspiration came from science fiction writers such as Jules Verne and H.G. Wells. Tsiolkovsky, Goddard, and Oberth were all inspired by stories about space travel and viewed the technological shortcomings inherent in these tales as the starting place for their work. Due to the massive costs associated with rocket research, the progress each space pioneer was able to make was limited, but their approaches to the problems that space flight posed were remarkably similar.

Tsiolkovsky, Goddard, and Oberth lived during the early 1900s and to a small degree were familiar with each other's work. However, their individual personalities, politics, and physical locations dictated that little communication took place among them. Oberth viewed Tsiolkovsky and Goddard as giants in the field of rocketry and eagerly sought to discuss his research with them. He hoped to establish collaborations that

would lead to quicker results. Goddard, on the other hand, saw rocketry as his exclusive field of research. He politely replied to correspondence from Oberth, but did not encourage or welcome it. Tsiolkovsky desired recognition for his work, but was afraid others would claim credit for the results he had labored to produce. Following the collapse of Czarist Russia, he was elevated to the stature of a national hero. The communist government, anxious to elevate the contributions of the common man, praised his discoveries while minimizing the contributions of others. Tsiolkovsky was perhaps the most visionary of the pioneers, foreseeing uses of space that we have yet to exploit.

Unhappily, Tsiolkovsky, Goddard, and Oberth found that the completion of the road to space was far beyond their abilities. The problems were too numerous and the required resources were well beyond what they could afford. The realization that they would have to leave the completion of their work to others was hard to accept.

1   Konstantin Tsiolkovsky (1857–1935)

The Earth is the cradle of mankind, but you cannot live in the cradle forever.

—Konstantin Tsiolkovsky

After a long period of stability, the Russian political system underwent radical changes in the latter half of the nineteenth century. Following the death of Nicolas I, his son Alexander II was crowned Tzar on February 19, 1855. The Russian people welcomed Alexander to the throne because Nicolas had ruled the land harshly and was intolerant of dissent. He forbade the creation of political parties and employed a powerful police force to maintain order. Contrary to his father's opinion, Alexander believed that the country needed to modernize. The humiliating defeat of Russia in the Crimean War in 1856 provided him the excuse to implement reforms. The justice system was overhauled, censorship was moderated, universities were granted autonomy, elected assemblies at the local and provincial level were permitted, and in 1861 the Russian system of serfdom was abolished.

Despite these bold initiatives, life outside the major cities remained difficult. Poverty was widespread and travel was virtually impossible after heavy rains as the roads dissolved into rivers of mud. In prior times, people had lived out their lives near the place of their birth and held few expectations beyond the simple life they had come to know. Now, people in the cities began to demand greater freedom and people in the countryside began to view the government as the source of their problems. It is a curious thing that the space age came into existence in this unstable political environment.

The Birth of Konstantin Tsiolkovsky

Konstantin Tsiolkovsky was born on September 5, 1857, in the small Russian town of Izhevskoye, located about 100 miles southwest of Moscow. His father, Eduardovich Ignatievich was an immigrant from Poland and was an educated man. He taught natural science and philosophy, but a personality difference with the school administrator led to his discharge. The only work he could find was that of a forester. His meager wages kept the family in poverty. Fortunately, Konstantin's mother, Maria Ivanovna, was a warm and caring woman who was able to endure this situation and still lavish love and attention on her family. Her spirit filled their household with a joy that financial hardships could not diminish.

Figure 1-1. Konstantin Tsiolkovsky at the age of 37.

The Pressure to Excel

Up to the age of nine, Konstantin enjoyed a normal childhood. He did well in school and enjoyed making friends. One winter day following a sledding outing, he caught a cold. This was hardly unusual, but the illness persisted and developed into a respiratory infection. Despite the efforts of his mother, it progressed into scarlet fever. He recovered, but not unscathed, as the disease left him nearly deaf.

At this time, deafness was a serious impairment, especially in rural Russia. No social services were available and individuals were left to cope the best they could. At school, Konstantin became the object of cruel jokes by his classmates. To spare him further pain, his parents decided to keep him at home. As a result, Konstantin was forced to educate himself using the few books his father could purchase or borrow. His mother helped him, and through their combined efforts he made slow but steady progress in his education. Nevertheless he longed to escape the physical bondage his deafness imposed. In his autobiography, Tsiolkovsky wrote about his feelings during this period, "I loved reading pas-

Figure 1-2. Jules Verne (shown here at age 40) served as an inspirational guide to Tsiolkovsky through his science fiction stories about space travel.

sionately, and I read whatever books were available. . . . I indeed liked to immerse myself in dreams. . . . In my dreams I was taking high leaps, climbing up poles or ropes like a cat. Absolute lack of gravity was my dream."

The decision to take Konstantin out of school had social as well as educational consequences. He formed few close childhood friends and became increasingly embarrassed by his deafness. To compensate for these feelings of inadequacy, a strong desire grew within him to accomplish some larger-than-life feat, something that was beyond the ability of the average person. As he would later write: "This handicap estranged me from people and prompted me to read, concentrate and dream to keep from boredom. Hurt pride sought satisfaction. There was a desire to do something big, heroic." This self-imposed pressure plus the inherent isolation of rural Russia turned his childhood years into the darkest period of his life. However, some good arose from these difficult circumstances. He developed a mental toughness and survivor's spirit that would prove essential to the ultimate success of his rocket research.

By the age of sixteen, Konstantin's interests had turned to natural science, but he soon exhausted the locally available books about the subject. His father and mother decided to send him to Moscow where he would have access to the libraries of that great city. The family could not afford to give him much money, but they did what they could. On the day of his departure, one brother gave him his shoes, another a belt, and his father presented him with his favorite heavy scarf.

Moscow was an imposing city for a boy from the countryside. Magnificent buildings lined the streets and crowds of people were everywhere. Despite the swirl of activity around him, Konstantin remained focused and quickly immersed himself in his self-education. He spent long hours at the library and attended free lectures in astronomy, chemistry, and physics. During these talks he used a homemade ear trumpet to amplify the speaker's voice.

In the course of these studies, Konstantin became interested in spaceflight, but found very little written material about the subject. Frustrated again by events beyond his control, he wrote: "Books on the subject of flight were practically nonexistent. Therefore, it was necessary for me to think independently and, often, in the wrong direction."

The lack of scholarly material paradoxically had a positive effect, it forced him to develop a faith in his own ability to solve complex mathematical problems and increased his own self-confidence. These traits proved crucial in his later work since most of his research was performed in an isolated setting without the benefit of collaborators.

Despite his tiring study schedule, Konstantin found some time to read science fiction stories. He particularly enjoyed the space adventures written by popular figures such as Jules Verne. They inspired him to think about the role that rockets might play in space travel. He wrote:

> For a long time I thought of the rocket as everybody else did—just as a means of diversion and of petty everyday uses. I do not remember exactly what prompted me to make calculations of its motions. Probably the first seeds of the idea were sown by that great fantastic author Jules Verne—he directed my thoughts along certain channels, then came a desire, and after that, the work of the mind.

The Flying Machine

One night while deep in study in a Moscow library, Konstantin thought he had discovered a way of making space travel possible. He visualized using the centrifugal force caused by two inverted pendulums to propel a craft off the Earth and into space. Centrifugal force causes an object traveling along a curved path to move in an outward direction. An inverted pendulum would, therefore, cause the craft to travel upward. After considering this idea for a few hours he realized that it would not work, but the exhilaration he initially felt about this possible discovery had a lifelong impact on him. He later wrote:

> I was so excited that I could not sleep the whole night, and instead spent it wandering through the streets of Moscow and thinking about the great consequences of my discovery. Toward morning, I was convinced of the fallacy of my invention. I still remember that night, and even now fifty years later, I sometimes dream about rising in my machine.

While studying in Moscow, Konstantin received some financial support from the philosopher-librarian Nikolai Fyodorov, an illegitimate son of Prince Pyotr Gagarin. In a twist of fate, Yuri Gagarin, one of Fyodorov's descendants, would become the first human to orbit the Earth.

Fyodorov believed in young Konstantin's abilities and helped organize a study program for him. From the age of sixteen to nineteen, Konstantin worked hard concentrating on the subjects of mathematics and physics. He devoured books on algebra, differential and integral calculus, analytical geometry, and spherical trigonometry. By the time his studies were complete, he had essentially finished an entire secondary school curriculum and a considerable part of a university program.

Upon completing these studies, Tsiolkovsky took and passed the czarist teaching exam in 1880. He was awarded a certificate as a "People's School Teacher" and received an appointment to teach arithmetic and geometry in the Borovsk district school, located sixty miles from Moscow. He resolved not to become just a teacher, but a "good teacher" and to not allow his deafness to defeat him. He later explained,

> Occasionally I would call on a student seventeen or eighteen years of age, have him or her stand near me at my left ear, and I would listen that way to their answers. The class would laugh good-naturedly.... I did not stint on interesting experiments, so that we had real "performances"; a part of my salary went for these experiments.... The students liked me very much for my fairness, good grades, and tirelessness in explanations.

In the evenings Tsiolkovsky worked on the theory of balloons and dirigibles. These efforts marked the beginning of his theoretical investigations of flight. He sent scientific papers to the Society for Physics and Chemistry in St. Petersburg for review and received a reply from Dimitri Mendeleyev, the inventor of the periodic table. The letter gave an assessment of his work and ended by inviting him to join the society. Tsiolkovsky felt honored by this invitation, but was forced to decline because the high membership fee would place too much of a burden on his meager income.

Newton's Third Law

As early as 1883 Tsiolkovsky realized that Newton's law of action-reaction provided the easiest way to propel an object in space. Although this principle had been discovered almost 200 years earlier by the great English physicist Isaac Newton, he could not find that it had been applied to rocket flight.

Newton's Third Law states that for every action, there is an equal and opposite reaction; that is, by ejecting mass in one direction, motion in the opposite direction results. Tsiolkovsky recognized the importance of this law to spaceflight and used it to describe how a craft could travel in space. He also recognized that motion in any direction would be possible if rockets could be attached at several places on the spacecraft. In his paper, "Free Space," written in 1883, he wrote:

Figure 1-3. Dimitri I. Mendeleyev, the discoverer of the periodic table of the elements.

> Consider a cask filled with a highly compressed gas. If we open one of its taps, the gas will escape through it in a continuous flow, the elasticity of the gas pushing the cask itself. The result will be a continuous change in the motion of the cask. With a sufficient number of valves (6), we may control the exit of the gas so that the motion of the cask will depend entirely on the desires of the one controlling the valves. ... As a general rule uniform motion along a curved line or rectilinear nonuniform motion in free space involves the continuous loss of matter.

In 1892, at the age of thirty-five, Tsiolkovsky was promoted to a teaching position at the diocesan school at Kaluga, a fairly large city located 100 miles southwest of Moscow (see Figure 1-4). Once settled in this new city, he continued to devote his free evenings to the study of flight. From time to time, Konstantin purchased materials to determine experimentally

Figure 1-4. Kaluga around the time that Tsiolkovsky lived there.

Figure 1-5. Tsiolkovsky's home in Kaluga became a pilgrimage site for Soviet rocket engineers.

Figure 1-6. Konstantin Tsiolkovsky (right), and an assistant (left) in Tsiolkovsky's home laboratory in Kaluga.

some of the values he needed in his calculations. He constructed the first wind tunnel in Russia to measure the amount of air resistance on various shaped objects. These purchases, though modest in size, placed a severe burden on the family's income. It became clear to him that outside financial help would be needed to continue this work, but where could such funds be found? He later wrote in a short autobiographical sketch of his embarrassment at receiving a small amount of money from an individual who considered it a charitable donation. Tsiolkovsky felt embarrassed by the gift, but accepted it out of necessity. Lack of funds would be a recurring problem, as he spent most of his life in poverty. He also found the lack of intellectual support hard to bear. He later wrote: "It was difficult to work alone for many years, under unfavorable conditions without encouragement and support."

From Science Fiction to Science Fact: Development of the Liquid-Fueled Rocket

While at Kaluga, Tsiolkovsky published several science fiction novels that earned him a modest amount of money, but more importantly, they unleashed his imagination about the possibilities of spaceflight. These stories, *An Imaginary Journey to the Moon*, *Is There Life on Other Worlds?*, and *Dreams of the Earth and Sky*, were published between 1893 and 1896. They touched upon futuristic topics such as artificial satellites, space stations and colonies, solar-powered motors, manned asteroids, artificial gravity, gravity-assisted spaceflight, interstellar travel, and the possibility of extraterrestrial life.

These writings revealed an evolution in Tsiolkovsky's ideas about spaceflight. In his early science fiction stories, the spaceship was only able to leave the Earth when gravity mysteriously disappeared. In his 1917 science fiction novel, *Outside the Earth*, the craft employed self-igniting liquid propellants. By 1897, Tsiolkovsky had developed many of the basic formulae needed to describe theoretically the motion of a rocket. These included equations for estimating thrust (the amount of propelling force), flight velocity as a function of fuel consumption, the influence of gravity on a rocket's ascent, and the effects of air resistance.

Although Tsiolkovsky received very little monetary assistance for his research, at the urging of Mendeleyev the Russian Academy of Science granted him 470 rubles in 1899. This allowed him to make further progress, and in 1903 the first of his scientific papers on rockets, "A Rocket into Cosmic Space," was published. In it, the world was introduced to his concept of a spaceship—a vehicle with a teardrop shape that relied upon Newton's law of action-reaction and used liquid oxygen and a hydrocarbon as a fuel source.

The realization that liquid fuels were superior to solid fuels was a major advance in the field of rocketry. Tsiolkovsky wrote, "I do not know of any group of bodies [liquid hydrogen and liquid oxygen] which when combined chemically would yield per unit mass of resultant product such an enormous amount of energy." Liquid fuels also allowed the rocket's speed to be controlled by adjusting the rate at which the liquid was consumed and permitted the engine to be stopped and restarted.

In this 1903 paper Tsiolkovsky presented some of the fundamental formulae for rocket motion. He showed that in space the maximum velocity of a rocket, v_{max}, is directly related to the velocity of the gases expelled from the rocket's combustion chamber. Specifically, he found that:

$$v_{max} = 2.3\ v_{fuel} \log(\ 1 + m_{fuel} / m_{unfueled\ rocket})$$

where v_{fuel} is the velocity of the gas ejected from the rocket, m_{fuel} is the mass of the fuel, and $m_{unfueled\ rocket}$ is the mass of the unfueled rocket and payload.

This relation, called the Tsiolkovsky equation, shows that v_{max} and v_{fuel} are directly proportional, that is, as one value increases so does the other. Therefore, in order to produce high values of v_{max} like those needed to place an object in orbit around the Earth, large values of v_{fuel} are required. Fuels such as liquid hydrogen (H) and oxygen (O) release large amounts of energy and, therefore, produce large exhaust

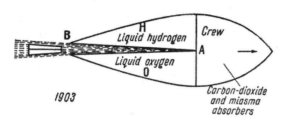

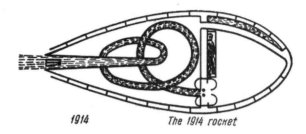

Figure 1-7. (above left) Tsiolkovsky's first 1903 liquid-fueled rocket design. The idea of using liquid fuels was decades ahead of its time.
Figure 1-8. (above right) In 1914 Tsiolkovsky presented this revised design of his 1903 rocket. This updated design also employed liquid hydrogen and oxygen as its fuel.

velocities. In Tsiolkovsky's rocket (Figure 1-7) these liquids are ignited at point A within the rocket and expelled at point B propelling the rocket. The Tsiolkovsky equation is more fully explored in the focus box below.

Tsiolkovsky also demonstrated that in order to reach a high velocity and then land the craft back on Earth, an enormous amount of fuel would be needed. He calculated that if a 100-pound rocket (less than the weight of an average person) was launched, reached the velocity v_{max}, and then landed back on Earth, the amount of fuel needed for the trip would be a whopping 9,900 pounds! Tsiolkovsky was also concerned about the construction of the spacecraft. For example, he suggested that rockets be made of material that had a high melting point so that the vehicle would not be consumed by flames as it sped through the atmosphere.

The above ideas were revolutionary and set forth the foundation of the modern theory of spaceflight. Nevertheless, they received surprisingly little attention from the Russian or international scientific communities because they were not published in the widely read scientific journals of the time.

Tsiolkovsky continued to work on his rocket research, and in 1911 published another paper titled, "Investigation of Universal Space by Means of Reactive Devices." Two additional supplements were published in 1914. This set of futuristic papers discussed the gravitational forces that humans would experience during lift-off, the concept of zero gravity, ways of steering the rocket, the advantages of launching a rocket near the Earth's equator, and the value of multiple rocket stages in reaching space.

The Rise of the Soviet Union

Although Tsiolkovsky's science fiction writings were beginning to gain public recognition following World War I, the academic community was still largely ignorant of his scientific work. Political events in Russia, however, would soon change that situation. The overthrow of Czar Nicholas in 1917 dramatically changed Tsiolkovsky's life. The new Soviet government was anxious to highlight the technical and scientific accomplishments of the common man, and as a result, Tsiolkovsky's efforts were widely publicized. The new communist government elevated him to the status of a national hero, and in 1918, he was elected to the Russian Socialist Academy, a forerunner of the Academy of Sciences, USSR. In 1921, the Soviet Commission for the Improvement of the Lot of Scientists awarded him a lifelong pension in recognition of his efforts. Five years later an expanded version of "Investigation of Universal Space by Reactive Devices" was published in Kaluga at government expense.

The Multistage Rocket and the Jet Plane

After receiving his pension, Tsiolkovsky resigned from his teaching position and devoted himself wholly to solving problems related to spaceflight. He considered futuristic propulsion schemes using light pressure and the radioactive decay of nuclei. By the mid-1920s, at the age of seventy-two, Tsiolkovsky's idea of rocket flight had progressed beyond that of a single-stage rocket to multiple-stage vehicles. He re-

---

Focus Box: The Maximum Velocity a Rocket Can Obtain

$$v_{max} = 2.3 \, v_{fuel} \log(1 + m_{fuel} / m_{unfueled \, rocket})$$

If a spacecraft is to achieve a large enough velocity to orbit the earth, i.e., $v_{max} = 7912$ m/sec, and if the rocket has $m_{fuel} / m_{unfueled \, rocket} = 9$, then according to this equation, one needs a fuel that can produce a $v_{fuel} = 3440$ m/second. It is easier to reach large values of v_{max} if a fuel with a high value of v_{fuel} can be found rather than by altering $m_{fuel} / m_{unfueled \, rocket}$. Since liquid fuels produce large values of $v_{fuel,}$ they are useful as a rocket propellant.

ferred to these vehicles as cosmic trains and introduced them in a 1929 paper, titled "Cosmic Rocket Trains:"

> I visualize the rocket train as a combination of several identical reactive devices. . . . However, only part of this train is carried off into space. The remaining parts, which do not achieve a sufficient velocity, return to Earth.

A similar version of this craft would later be built by Sergei Korolev for use in the Soviet Union's manned space program.

After many years of work in rocket engineering and spaceflight, Tsiolkovsky finally began to win international recognition for his efforts. The German rocket scientist, Hermann Oberth, sent him his congratulations. Although pleased by this acclaim, Tsiolkovsky was angered by the fact that others were often given credit for discoveries that he had made nearly twenty years earlier.

In his later years, Tsiolkovsky concentrated on the theory of jet airplane flight. He published his first paper on this subject, "The New Airplane" in 1929, and in a subsequent paper, "The Reactive Airplane" (1930), he correctly prophesied: "The era of the propeller airplane must be followed by an era of jet or stratosphere airplanes." Two additional papers followed on this subject: "Rocketplane" (1930) and "The Semi-Reactive Stratoplane" (1932).

In 1935, Tsiolkovsky quantified his earlier thoughts about multistage rockets. His paper on this topic, "The Maximum Velocity of a Rocket," described a spacecraft consisting of a linear array of four small rockets. At lift-off these rockets would simultaneously fire their engines and consume their fuel at identical rates (Figure 1-9, row 1). When half of their fuel was exhausted, the remaining fuel would be transferred from the outer craft to a neighboring, inner rocket, and the empty rocket would then be discarded (rows 2 and 3). In this fashion the refueled rocket would possess a full fuel tank to draw upon for its continued ascent. This process would be repeated until only one rocket remained (rows 4 and 5). He believed this final vehicle would be able to achieve a sufficient velocity to enter Earth's orbit or even to escape from the Earth's gravitational pull.

On his seventy-fifth birthday Tsiolkovsky was honored by newspapers and magazines across the Soviet Union. They carried stories about his accomplishments, and two years later, in 1934, a compilation of his papers, titled *Selected Works*, was published. When combined with his earlier works they provided a framework for future advances in astronautical theory.

Principal Contributions

Tsiolkovsky's principal contributions to rocketry were theoretical insights. Because of his humble circumstances, he never had the means to build rocket motors or a spaceship; this work was left to others. He died in Kaluga on September 19, 1935, two weeks after his seventy-eighth birthday. Shortly before his death, he summed up his work as follows:

> The main motive of my life has been to move humanity forward if only slightly. This is exactly why I have been interested in those things that never yielded either bread or strength. But I hope that my labors perhaps . . . may give to society mountains of bread and infinite power.

Tsiolkovsky was one of the principal rocket pioneers. His work was futuristic in its outlook and embodied many of the same ideas that are used in today's space program. His contributions include: (1) the derivation of many of the fundamental equations used in spaceflight, (2) the recognition that liquid fuels were highly desirable rocket propellants, (3) the first theoretical exploration of multistage rockets, (4) discussions about self-sustaining space stations, Earth orbiting solar power stations, space suits, and artificial gravity machines, and (5) futuristic ideas about the possibility of exploring the Moon and Mars, reengineering the environments of

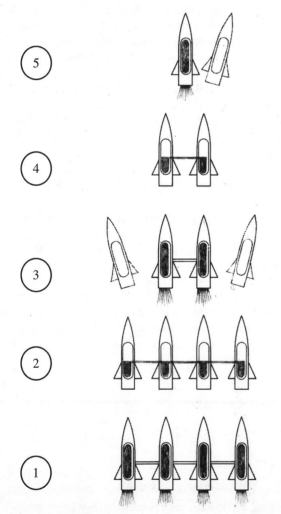

Figure 1-9. Tsiolkovsky's multistage rocket. As the rocket ascends, fuel is transferred to inner rockets, and the outer ones are discarded. Numbers 1–5 illustrate how the combined rockets at 1 ultimately become the single rocket shown at 5.

Focus Box: Tsiolkovsky Time Line

1857	Born September 5, 1857
1865	Jules Verne, *From the Earth to the Moon*
1867	Tsiolkovsky becomes ill and leaves school
1870	Mother dies
1873	Goes to Moscow to continue self-education
	Spaceship propelled by centrifugal force
1876	Completes self-education
1879	Passes Tzarist teaching exam
	Assigned to teach in Borovsk District
	Astronomical Drawings
1881	Alexander II killed by assassin
1883	"Free Space"—discussion of jet propulsion, weightlessness
1885	"I definately decided to devote myself to aeronautics"—Tsiolkovsky
1887	*On the Moon*—fictional journey to the moon
	Lecture at the Moscow Polytechnical Museum
1892	Moves to Kaluga to teach high school
1894	"An Aeroplane or Aviation Flying Machine"
1895	*Dreams of the Earth and Sky*, fiction
	On the Moon, fiction
1903	"Exploring Space with Reaction Devices"
1911	"Investigation of Universal Space by Means of Reactive Devices" published
1914	Supplement "Investigation of . . ." published
1918	Elected to the Socialist Academy
1921	Awarded lifelong pension
1924	"Cosmic Rocket" published
1926	Expanded version "Investigation of . . ."
	The Plan of Space Exploration
1929	"Cosmic Rocket Trains" published. "I am not at all sure of course that my 'Space Rocket Train' will be appreciated and accepted readily at this time. For this is a new conception reaching far beyond the present ability of man to make such things. However, time ripens everything; therefore I am hopeful that some of you will see a space rocket train in action."
1929	"Reactive Engine"
1929	"The New Airplane"
1930	"The Reactive Airplane"
1930	"Rocketplane"
1932	"Semi-Reactive Stratoplane"
1935	"The Maximum Speed of a Rocket"
1935	Dies September 19, 1935

Figure 1-10. A stone obelisk was built in Moscow to honor the Russian space pioneer Konstantin Tsiolkovsky. A statue of Tsiolkovsky is in front.

other planets, populating the asteroid belt, and communicating with extraterrestrial beings.

In all of these areas Tsiolkovsky was decades ahead of his time. His vision soared even beyond that of most science fiction writers of his era. He also served as the inspirational leader of a new generation of Soviet rocket scientists. In 1929 a group of his followers led by Perel'man and Fortikov founded the research institute GIRD (Group for the Study of

Reactive Motion). This organization would later play a central role in the development of the Russian space program.

The Legacy of Tsiolkovsky

On September 24, 1954, the USSR Academy of Sciences established the K.E. Tsiolkovsky Gold Medal. It is awarded every three years to an individual in recognition of outstanding work in the field of interplanetary communications. Monuments were also constructed in his honor in both Moscow and Kaluga.

Despite his pioneering studies on the theory of rocket flight and rocket design, it is difficult to estimate the impact Tsiolkovsky's work had on researchers outside of Russia. Although many of his ideas about rocket propulsion and design were correct and predated those of his colleagues, his results were not widely publicized outside of the Soviet Union. None of his works were translated into a Western language until the 1940s, and by this time, other investigators, such as Hermann Oberth and Robert Goddard, had independently arrived at many of the same conclusions he had discovered. Tsiolkovsky was honored for his work by the Soviet government and its rocket scientists but played only an inspirational role in their rocket program by the end of his life.

These statements, of course, do not detract from Tsiolkovsky's genius. His extraordinary ideas and the concrete results that he achieved, often under very poor circumstances, place him among the great scientists of the early twentieth century. Despite the heavy odds against him, Tsiolkovsky accomplished his childhood goal of doing something big and heroic. He was a founding father of the modern space age.

2   Robert Goddard (1882–1945)

It looked almost magical as it rose, without appreciably greater noise or flame, as if it said, "I've been here long enough; I think I'll be going somewhere else, if you don't mind."

—**Robert Goddard (on the launch of the first liquid-fueled rocket)**

The second of our space pioneers, Robert Hutchings Goddard, was born in Worcester, Massachusetts, on October 5, 1882. Unlike Tsiolkovsky whose parents were impoverished, Goddard was born into a middle-class family. The Goddards were longtime residents of Massachusetts and could trace their roots back to 1666 when William Goddard settled in Watertown, Massachusetts, after leaving Norfolk, England.

Robert's father, Nahum Goddard, married Fannie Hoyt on January 3, 1882. Like the Goddards, the Hoyt family were longtime residents of the area, having lived there for over 200 years. Soon after Robert was born, Nahum moved his family to Roxbury, a small town outside of Boston, and went to work for a knife manufacturer. After a short time he joined resources with fellow employee Simeon Stubbs and together they bought the company.

Even at an early age, Robert displayed a natural curiosity about how things worked. He attempted to make a perpetual motion machine out of wood and string, experimented to see if diamonds could be made out of graphite, and made a balloon out of aluminum. When it was found that the balloon was too heavy to get off the ground he dutifully entered in his diary, "Failior [*sic*] crowns enterprise."

Like Tsiolkovsky, Robert enjoyed reading science fiction stories as a teenager, particularly space adventures. He was fascinated by them and read and reread them several times over. As an adult he recalled,

> In 1898, there appeared daily for several months in the *Boston Post* the story, "Fighters from Mars," or "The War of the Worlds, In and Near Boston." This, as well as the story which followed it, "Edison's Conquest of Mars" by Garrett P. Serviss, gripped my imagination tremendously. Wells's wonderfully true psychology made the thing very vivid, and possible ways and means of accomplishing the physical marvels set forth kept me busy thinking.

Wells' book ignited Robert's imagination at an age when he was beginning to give serious thought to his future occupation. The adventurous side of Robert was enthralled with the idea of traveling to distant worlds. The glorious tales of space travel filled his mind with wonder and made a deep and lasting impression on his thinking.

Goddard's Decision to Study Rockets

In 1898 Fannie was diagnosed with "consumption and complications," the term used then for tuberculosis, and the Goddards moved back to Worcester so she could be near her family. Robert was expected to help with routine chores around the house and at the homes of his relatives. One fall day while working at his Aunt Ward's farm in nearby Auburn, he found himself thinking about his future occupation. His mind drifted to the topic of space exploration, and he wondered what it would be like to build a spacecraft that could travel to Mars,

> On the afternoon of October 9, 1899, I climbed a tall cherry tree at the back of the barn and, armed with a saw and hatchet, started to trim the dead limbs from the tree. It was one of those quiet colorful afternoons

Figure 2-1. Goddard (middle) with his family around 1890. Others in the picture include (left to right) his mother, Fannie, his grandmother, Mary, his great-grandmother, Elvira Ward, and his father, Nahum.

of sheer beauty which we have in October in New England, and, as I looked toward the fields to the east, I imagined how wonderful it would be to make some device which had even the possibility of ascending to Mars, and how it would look on a small scale up from the meadow at my feet. . . . I was a different boy when I descended the ladder. Life now had a purpose for me.

Suddenly the choice of an occupation was clear; he would work to make his dream of space travel a reality. From that moment on, Goddard celebrated that day as his "anniversary day"—the day that he decided to devote the rest of his life to constructing a means of achieving spaceflight.

Goddard's Research Philosophy and Education

Robert began the task of inventing the spaceship while still in high school. He enrolled in classes in mathematics and physics and attempted to apply the principles taught in those fields to spaceflight. As was the case with Tsiolkovsky, he soon realized that the application of Newton's law of action-reaction provided the means of making space travel possible. The spacecraft described by Jules Verne were wonderful to imagine, but the laws of physics and mathematics were the tools that would open the way to space. Robert wrote,

I began to realize that there might be something after all to Newton's Laws. The Third Law made me real-

Figure 2-2. The cherry tree at Goddard's aunt's farm. A ladder Goddard used can be seen at the base of the tree.

ize that if a way to navigate space were to be discovered or invented, it would be the result of a knowledge of physics and mathematics.

During Robert's teenage years at South High School in Worcester, he wrote two papers that are of particular importance in understanding the imaginative and thoughtful sides of his personality. The first paper, titled "The Navigation of Space," revealed his deep interest in space travel and a curious and adventurous spirit. He was not afraid to consider topics that were far ahead of their time.

The second paper, "The Spirit of Inquiry," revealed his thoughtful side. In it he expressed the view that although science was a natural outgrowth of humankind's work, the quest for new knowledge was part of his very being. This philosophy helps to explain, at least partially, Goddard's tenacious attitude toward research and why the investigation and construction of the rocket became such a consuming goal. The rocket was much more than a hobby to him, it became the means by which he would satisfy his quest for adventure and new knowledge.

Niels Riffolt, a longtime friend and assistant, later wrote about Goddard's remarkably positive attitude about the frequent failures of his rocket experiments. During these times Goddard would often tell Riffolt that these efforts were not failures since they provided "valuable negative information." Goddard recognized that once he understood the cause for these failures, new approaches would become apparent. More specifically, through the process of design, testing, redesigning, and retesting the correct solution would ultimately be found. This methodical approach, coupled with his engineering skills and mental toughness, proved crucial to the success of his later rocket research.

After graduating from high school, Goddard enrolled in the Worcester Polytechnic Institution in 1904, and true to his goal of equipping himself to study spaceflight, majored in physics. During this time he experimented with rockets charged with black powder, but the heavy demands of his classwork prevented him from making much progress. In 1908, he graduated with a Bachelor of Science degree and then

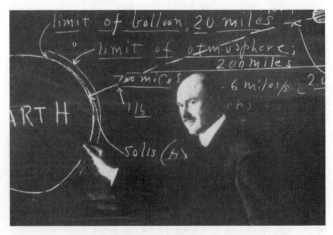

Figure 2-3. Goddard served as a member of the Physics Department at Clark University located in Worcester, Massachusetts.

served as an instructor at the Polytechnic Institution while he pursued advanced degrees at nearby Clark University. He received his master's and doctoral degrees in 1911 and then accepted a post-doctoral fellowship at Princeton University.

Goddard returned home for Easter vacation from Princeton in March 1913 gravely ill. The family doctor was called, and after a short examination left and returned shortly with a specialist from the tuberculosis hospital in Rutland, Massachusetts. Following a second examination, the two doctors concluded that Goddard had tuberculosis in both lungs and had ten days to two weeks to live. This came as a shock to Goddard and he resolved to overcome his illness. With the assistance of his mother, he slowly recovered, but the disease took its toll. By the summer of 1914, his weight had dropped to 136 pounds and he had lost his hair. Nevertheless, by the fall of 1915, the worst was over and he felt well enough to accept a faculty appointment in the Physics Department at Clark University. Here, at last, he was able to continue his experiments with solid-fueled rockets.

Initial Rocket Work—World War I

At Clark University, Goddard began a series of solid fuel rocket experiments using different nozzle designs and combustion chambers. He made two significant discoveries. First, that solid-fueled rockets were very inefficient, converting only about 2 percent of their energy content into useful thrust. By altering the design of the nozzle, he was able to raise this efficiency to nearly 40 percent. Secondly, by using an evacuated metal tube, he found that the velocity of a rocket in space would be 20 percent greater than in the atmosphere.

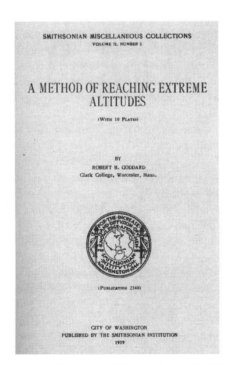

Figure 2-4. The Smithsonian published Goddard's 1916 research proposal, "A Method of Reaching Extreme Altitudes," in 1919. This proposal defined the course that his life would take.

Despite the personal pleasure Goddard received from his work, it was clear that further progress would require more money than his teaching salary allowed. In September 1916 he applied for a research grant from the Smithsonian Institution. Realizing that a request to construct a rocket to explore the solar system was too futuristic and would probably be rejected, he proposed to use the rocket for, "raising recording apparatus to altitudes exceeding the limit for sounding balloons. . . ." He argued that these flights would permit the chemical composition and density of the Earth's upper atmosphere to be measured. Although this goal was far short of his true objective, if the request was granted, it would at least allow him to begin developing a more powerful rocket. This proposal, "A Method of Reaching Extreme Altitudes," was to become one of the most important documents of the space age.

As luck would have it, Goddard's proposal reached the Smithsonian at a particularly opportune time. The institute had recently received a generous bequest from Thomas G. Hodgkins of Long Island. The conditions of the gift stated the Smithsonian could only use half of this money once they had awarded the other half for atmospheric research. This was precisely the type of work that Goddard had suggested. After a favorable technical review of his proposal, Goddard was granted the handsome sum of $5,000.

World War I brought about a temporary redirection of Goddard's efforts when the Army Signal Corps hired him to explore the military uses of the rocket. Relocated to Mount Wilson Observatory near Pasadena, California, he utilized solid-fueled rockets to produce the forerunner to the World War II bazooka. It was interesting but dangerous work. Clarence Hickman, one of his assistants, lost a thumb and parts of three fingers when a charge exploded in his hands. Near the end of the war, Goddard travelled to the army's proving grounds at Aberdeen, Maryland, to demonstrate how his prototypes worked. The demonstration was a success and he thought that the army would continue its support so that these weapons could be refined and manufactured. However, with the surrender of Germany, the army lost interest and the project was terminated. A disappointed Goddard returned to Worcester to resume his teaching duties at Clark University. The military use of the rocket would not be explored again in the United States until World War II.

The "Moon Man"

The post–World War I period was an exciting time. Major scientific discoveries were being made at an unprecedented rate. Einstein had published his theory of relativity, and the wireless telegraph was making its debut. Dr. Arthur Webster, director of the Clark Physical Laboratories, felt that Goddard's rocket research was also worthy of recognition and urged him to publish his results. After some persuasion Goddard agreed and requested that the Smithsonian publish a revised version of his proposal, "A Method of Reaching Extreme Altitudes." The institute agreed, providing Goddard pay the publication costs from his grant. Somewhat surprised by this request, Goddard

Figure 2-5. Goddard's early work led to the WW II "bazooka."

nevertheless agreed, and his proposal was published in 1919. Much to his surprise, at first it seemed to go unnoticed, but this situation changed abruptly. Goddard wrote,

> I thought it odd that no notice of the work was taken by the press, as it seemed that the method was one that they might desire to feature. I had nearly forgotten the press question, when one morning I was startled to learn that the institution itself (or, as I learned afterward, its press representative) had made for the paper the first real suggestion for contact between the planets.

Figure 2-6. Esther was one of Goddard's greatest supporters.

Almost forgotten by Goddard because it had been placed at the end of the manuscript, was a section suggesting that a rocket could be sent to the Moon. Its impact on the lunar surface would be observable from Earth when flash powder placed in the rocket's nosecone ignited. Goddard was shocked to discover that the media completely ignored the bulk of his paper and focused instead on the idea of the Moon rocket. Embarrassing headlines soon appeared in newspapers across the nation including, "Modern Jules Verne Invents Rocket to Reach Moon" *(Boston American),* "Aim to Reach Moon with New Rocket" *(New York Times),* and "Savant Invents Rocket Which Will Hit Moon" *(San Francisco Examiner).*

By this stage in his life, Goddard had developed into a shy, quiet man who shunned publicity and preferred working in seclusion. He was not prepared for the negative tone of these articles. The media even went so far as to label him "the Moon man," and a *New York Times* editorial questioned his knowledge of elementary physics. The whole episode deeply irritated Goddard, and as a result, he became even more secretive about his work. It was useless, he thought, to attempt to explain rocketry to such an ill-informed press, and he concluded that the less they reported about his research, the better off everyone would be.

During this time Goddard met Esther Christine Kisk, a young secretary in the president's office at Clark University. She agreed to type his research papers after school hours and Goddard made several trips to her home to oversee this work and visit. Over time they fell in love and were married on June 21, 1924.

The World's First Liquid-Fueled Rocket

By this time the original $5,000 Smithsonian grant was nearly spent, and with the institute's aid, funds from other sources were sought. These efforts were not successful, but to demonstrate confidence in Goddard and to move his research forward, Clark University awarded him a two-year grant of $3,500 from its own resources.

Goddard's research had focused on the use of solid and powder fuels. By January 1921, however, he was convinced they had serious deficiencies. They produced only intermittent propulsion, left the rocket difficult to control, and often damaged the rocket so badly that its propulsion system failed. A fuel that provided continuous combustion while producing very high expulsion velocities was needed. Faced with this problem, Goddard made calculations, that like those of Tsiolkovsky, showed that a fuel consisting of liquid hydrogen and oxygen would meet his demands. However, liquid hydrogen was virtually unobtainable in the 1920s, so he adopted a mixture of gasoline and liquid oxygen as a cheap, readily available alternative. When mixed, these two liquids formed a very volatile and dangerous substance. As G. Edwards Pendray, one of the founders of the American Rocket Society, wrote,

> Even in the 1930s engineers told us of the frightful things that would happen if we used the stuff. They

said if it got on our clothes and we happened to light a cigarette, we'd go off like a torch. Mixing and igniting liquid oxygen with a flammable substance like gasoline seemed a guarantee of catastrophe. Goddard's first test was a tremendously courageous thing.

Having tried for years to make solid and powder fuels work, Goddard now realized that liquid fuels provided the only practical means of achieving spaceflight. They produced very high exhaust velocities and their rate of consumption could be adjusted. Liquid fuels were difficult to work with, but he viewed this as an engineering concern that could be overcome with perseverance and hard work. About this time, Goddard's longtime friend and colleague Dr. Arthur Webster died, and Goddard was appointed to take his place as head of the Physics Department at Clark.

Goddard knew of the growing international interest in rockets and received letters asking him about his research from both Robert Esnault-Pelterie of France and Hermann Oberth of Germany. Although these European researchers possessed a sound theoretical understanding of how a rocket

should operate, neither had actually built one. Goddard wanted to be the first to do this and felt that if he shared his ideas with others they might use this information to beat him to it. He responded to their inquiries in a courteous, albeit not very forthcoming, manner.

Between 1921 and 1926, Goddard worked with an urgent sense of purpose to construct his rocket. The first prototypes were failures, but they provided him with enough data to refine his design. Finally, on March 16, 1926, his third-generation rocket successfully lifted off from his Aunt Effie's farm in Auburn, Massachusetts, the same place where he had dedicated himself to the study of rocketry twenty-six years earlier. This launch marked a milestone in the history of rocketry and was witnessed by his machinist, Henry Sachs, his wife, Esther, and Percy Roope, an assistant professor of physics at Clark. Like the Wright brother's first airplane flight, this flight did not appear to be very impressive. The rocket reached a height of only 41 feet and then turned and traveled 184 feet in a horizontal direction. The entire flight lasted just 2.5 seconds.

This historic event received little attention in the United States; not a single major newspaper printed a story about it. It did not, however, go unnoticed overseas. But rather than being pleased by this recognition, these reports only increased Goddard's apprehension that others would now be even more anxious to copy his work. He began to fear that the accelerating rocket programs, particularly those in Germany, would soon overtake his effort.

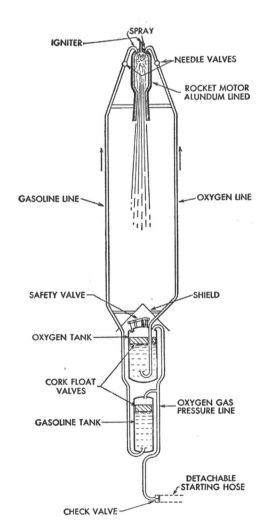

Figure 2-7. In Goddard's rocket design, liquid oxygen and gasoline flowed from the bottom of the craft to the top where it was ignited.

Figure 2-8. The first liquid-fueled rocket was launched from Auburn, Massachusetts, on March 16, 1926.

Figure 2-9. Lindbergh became interested in Goddard's work and arranged for him to receive funds from the Guggenheim Foundation. Left to right, Guggenheim, Goddard, Lindbergh.

Work in New Mexico

Prior to 1926, Goddard had been confident that his research was well ahead of that of other people, but he now wondered how long he would be able to keep ahead of the competition. He described his feelings to Esther, "I'm just a little dog—with a great big bone." This statement masked his fierce determination. Little dog or not, Goddard was determined to retain as much of the field of rocketry for himself as possible. To avoid giving away information that might assist other researchers, he became even more secretive about his work.

On November 23, 1929, an unexpected event occurred that would have a major impact on Goddard's life—a surprise visit from the famous aviator, Colonel Charles A. Lindbergh. Lindbergh had heard of Goddard's research and was fascinated with its potential. He personally arranged for a $50,000 grant to Goddard from the Daniel Guggenheim Fund for the Promotion of Aeronautics. Goddard's good futune continued when a short time later he received a second Carnegie Institution grant earmarked for rocket test facilities. With this support in hand, he took a two-year leave of absence from Clark University to devote himself fully to rocket research.

In order to safely conduct long-distance test flights, Goddard moved to Mescalero Ranch, located about three miles outside of the town of Roswell, New Mexico. Once settled, he undertook one of the most amazing "lone wolf" development programs in the history of technology. On December 30, 1930, just a little less than five years after the launch of his first liquid-fueled rocket, he launched an 11-foot, 35-pound rocket that reached a height of 2,000 feet and attained a speed of 500 mph. This was followed in April 1932, by the launch of the world's first controlled rocket flight. Goddard accomplished this by using a gyroscope to control the rocket's flight path.

After a two-year stay in New Mexico, Goddard's resources were nearly exhausted. The country was in the midst of the Great Depression, and additional funding seemed unlikely. Without the means of maintaining his research effort, he traveled back to Clark University to resume his teaching duties. After only a brief stay, Goddard received a second grant from the Guggenheim Foundation and quickly returned to New Mexico.

Once back at Mescalero Ranch, Goddard attacked his work with renewed vigor. On March 28, 1935, he launched a gyroscopically controlled rocket that rose to an altitude of 4,800 feet, turned horizontally under its guidance system, flew 13,000 feet, and reached a top speed of 550 mph. In May of that same year, one of his rockets reached an altitude of 7,500 feet. Goddard was making significant progress.

Goddard sent a report of his work in New Mexico covering the period from 1930 to 1935 to the Guggenheim and Smithsonian Foundations. The Smithsonian Institution published these reports in 1936, but they attracted little notice. Despite the great advances he had made in the field of rocketry, Goddard began to view his work in New Mexico with mixed emotions. Although intellectually stimulating and technically challenging, his efforts had not produced the type of spectacular results that won public or scientific acclaim. He realized that much more work remained before the rocket could be considered a reliable space vehicle. In a letter dated February 14, 1937, he wrote to his friend C.N. Hickman,

> It is, as you can imagine, a fascinating life. The drawback is that until there has been a great and spectacular height reached, no layman, and not many scientists will concede that you have accomplished anything, and of course there is a vast amount of spade work of much importance that must be done first.

In a sense, Goddard was a victim of his own desire for secrecy. Not only had he shielded his work from his European competitors, but as he had pledged, he did not inform the press of his successes. The reports to the two foundations were not distributed to parties who could promote his work. In this light, it is not surprising that his successes received little recognition.

Goddard and the German V-2 Rocket

Goddard's work was once again interrupted by war, this time by the outbreak of World War II. He became involved with defense work, principally with the rocket-assisted lift-off of heavily loaded airplanes and seaplanes. Solid rockets were attached to the outside of these planes and fired during takeoff to increase their speed. This work was a step backward for Goddard. It interrupted his research on liquid-fueled rockets.

During the war, Goddard learned of Germany's liquid-fueled V-2 program, an effort that dwarfed not only his but the entire American rocket program. After the end of hostilities, he inspected one of these rockets and thought that it was simply a larger version of the rockets he had tested in New Mexico. The injection system, pump assembly, and general layout all looked very familiar. As he continued to view the internal workings of this rocket, his old fears that others might copy his work seemed justified. One of his longtime assistants, Henry Sachs, remembered the scene when Goddard first viewed the V-2,

> I watched him as he stared at the length and girth of the opened rocket, and went over to him. "It looks like ours, Dr. Goddard," I said. "Yes Mr. Sachs," he answered, "it seems so."

Goddard ultimately came to view the V-2 in a different light, as vindication of his claim that rockets, when properly instrumented, could provide new scientific data about the Earth's upper atmosphere and space environment. He also realized that the Germans had not copied his work, but had simply arrived at similar solutions to the same technical problems he had faced.

By the end of the war, Goddard had mellowed and was anxious to share his knowledge with others and to move the field of rocketry forward. He realized that the era of the lone wolf had passed and that the rocket had moved into a new realm that was beyond the ability of a single person to control. However, fate would not be kind to him. Goddard did not live long enough to see one of the captured V-2s launched from the army test facility at White Sands Missile Range in New Mexico. Nor did he participate in the rapid growth of the American rocket program after the war. Following a laryngectomy, he died on August 10, 1945, from throat cancer at the age of sixty-two. He was buried in the family plot at Hope Cemetery in Worcester, Massachusetts, at the beginning of a new era in rocketry that would lead to manned spaceflight.

Goddard's Legacy to Space Travel

Goddard's place in the history of spaceflight is secure. Unlike Tsiolkovsky who could only describe the spacecraft his equations pointed toward through hand drawings, Goddard constructed the first working version of this craft. He deserves credit for inventing the first liquid-fueled rocket and much of the hardware required to control its flight.

Goddard's personality was such that he preferred to work within a small, closely knit group. His shyness made it difficult for him to discuss his work with others. This reluctance was increased by the unfortunate criticism he was subjected

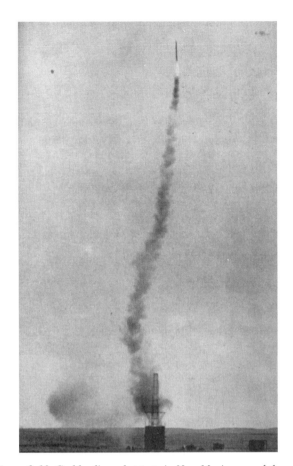

Figure 2-10. Goddard's rocket tests in New Mexico paved the way for the development of more powerful and sophisticated liquid-fueled rockets.

to when his research was still in its infancy. This event and the realization that others were working on the construction of a liquid-fueled rocket caused Goddard not to publicize his results. This decision insured that his work would not receive the timely recognition it deserved. As a consequence, Goddard's impact on the next generation of American rocket scientists was not as great as it could have been.

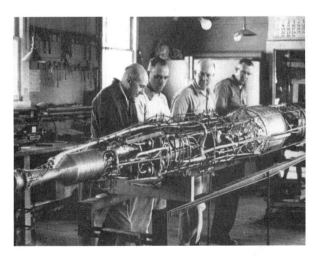

Figure 2-11. The complicated workings of a liquid fueled rocket are illustrated in this picture of Goddard and his colleagues in 1940.

Focus Box: Goddard Time Line

1882	Born Worcester, Massachusetts October 5, 1882	1929	Visit by Colonel Charles Lindbergh
1904	Graduates from South High School	1930	First Guggenheim Foundation award Moves to Roswell, New Mexico
1908	Graduates from Worcester Polytechnic Institute	1932	First gryroscopically controlled rocket flight. Returns to Clark University
1911	Ph.D. from Clark University	1934	Work resumes in Roswell
1913	Diagnosed with tuberculosis	1935	Progress report sent to Smithsonian
1914	Awarded first patents	1936	"Liquid-Propellant Rocket Development" published
1917	First grant from the Smithsonian		
1918	Develops prototype of WW II bazooka	1939–42	Develops high-speed fuel pumps
1919	"A Method of Reaching Extreme Altitudes" published	1941–42	Jet-assisted rocket tests
1924	Marries Ester Kisk	1945	Dies on August 10, 1945

Goddard's social shortcomings do not diminish the technical genius he possessed. Single-handedly he solved the complex problems involved in building a liquid-fueled rocket and was the first person to successfully launch such a device. Fortunately, history has bestowed on him the recognition he did not receive during his life. Barbara DeVoe, in summarizing Goddard's life in *U.S. Senate Document 91-71*, called attention to the stature that Goddard now enjoys in the history of rocketry, "Goddard the man . . . has become Goddard the legend, one of the giants of America's heritage."

3   Hermann Oberth (1894–1989)

The dramatic moment at which a human being first set foot on another heavenly body was a new high point in human history, a high point for which Hermann Oberth laid the foundation and created the prerequisites.

—Wernher von Braun

The third rocket pioneer of the early twentieth century was Hermann Oberth. Like Tsiolkovsky, he advanced the physics of spaceflight, but unlike Goddard, he was unable to overcome the severe technical problems involved in constructing a working liquid-fueled rocket. While the other two pioneers labored in relative isolation, Oberth played a major role in popularizing rocketry. His work led to the formation of groups in Germany that ultimately elevated the rocket from an unreliable and dangerous vehicle into a craft capable of traveling to the Moon.

Oberth's Childhood and Early Education

Hermann Oberth was born on June 25, 1894, in the small Transylvanian town of Hermannstadt. Over the centuries this region had been largely settled by Germans who remained loyal to their homeland even though they were not under Germany's direct rule. Hermann's father, Dr. Julius Oberth, was a skilled surgeon who received his education in Vienna. According to the traditions of the time, it was assumed that Hermann would follow in his father's footsteps and enter the medical profession. Initially, both of his parents hoped this would be the case.

Hermann was an energetic child and well-known to the people of Hermannstadt. When his father accepted the directorship of a hospital in the nearby but larger town of Schaes-

Figure 3-2. Hermann's father and mother, Dr. Julius and Valerie Oberth. His father was a gifted surgeon.

burg, his friends were sorry to lose a valued neighbor but were glad because of the additional recognition it would bring him. As he prepared to leave Hermannstadt to assume his new post, his neighbors, only half in jest, admonished him about Hermann, "Keep that child on a leash! He has too much curiosity." Shortly after leaving Hermannstadt and settling into their new home, Hermann's brother, Adolf, was born.

Hermann had a happy childhood. Around the age of five, his mother gave him a field glass as a present. He thoroughly enjoyed using this wonderful gift, and from the porch of his home, he spent many hours looking down at the town and passing trains. One moonlit evening he found himself watching birds fly across the Moon's illuminated disk. It seemed to him that they were headed for the Moon, and he thought, if birds could fly to the Moon, why couldn't people devise a way of traveling there as well? After pondering this question more fully, he turned his field glass toward the Moon with renewed interest and wondered about its bright and dark areas and composition. One day he asked his father,

"Papa, what's the Moon?"
"A stone ball with mountains," he answered.
"Can we go there? Maybe next summer?"

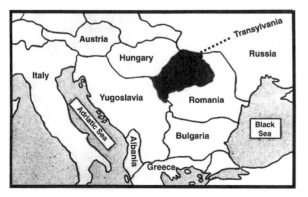

Figure 3-1. In 1894, Transylvania was located in southeastern Europe and bordered Hungary, Russia, Romania, and Yugoslavia. It is now part of Romania.

"Maybe someday somebody will invent a Moonship," he answered, "maybe you."

Without realizing it, both his father and mother had inflamed the desire that was growing in Hermann to be the first person to discover a way of making space travel a reality.

By 1899, Hermann was beginning to draw sketches of vehicles that he thought could take people to the Moon. His interest in spaceflight was solidified in 1906 when his mother gave him a copy of Jules Verne's book, *From the Earth to the Moon*. Verne was a master at making the adventure of space travel seem possible. Hermann devoured the book and then read it seven more times in succession. Like Tsiolkovsky and Goddard, he was mesmerized by science fiction books about space travel and began to ask probing questions about this subject. Was Newton's Third Law correct? Would a bullet-shaped vehicle be superior for space travel compared to the spherical ball described by Verne? Could the human body withstand the stress of the explosive type of launch described by Verne? What kind of fuel would a rocket use? How would this fuel behave in a weightless environment? How would humans react to weightlessness? Hermann was anxious to begin the search for the answers to these and similar questions.

At the age of 17 and while still a schoolboy, Hermann had already reached the conclusion that liquid fuels would serve as excellent rocket propellants. He thought that alcohol and liquid oxygen would be a particularly effective combination. Upon his graduation from school in 1912, and in accord with tradition, he met with his parents to discuss what type of profession he would enter. His mother, perhaps sensing where Hermann's heart lay, suggested that he obtain an advanced degree in mathematics or physics, but his father felt that he should continue the family tradition and pursue a medical degree. His father's position prevailed, and Hermann enrolled at the University of Munich to study medicine.

Much to Hermann's delight, when he arrived at Munich, he found that the curriculum allowed him to attend two lectures outside his field of study. To satisfy his interests in spaceflight, he enrolled in classes in mathematics and astronomy. Hermann was a good student and worked hard at his studies. For relaxation and exercise he walked the streets of Munich and spent time daydreaming about his rockets. Munich was a large, bustling city, a center of commerce and wealth. Other young Germans also flocked to it to pursue their own dreams. One such fellow, Adolf Hitler, aspired to become a famous painter.

The Reality of War

As the school year came to a close, Hermann received word from his parents that the Balkans were becoming a political powder keg and they felt the situation could explode into violence at any moment. Indeed, there were numerous rumors of unrest.

In June 1914, it was reported in the Munich papers that Austrian Archduke, Francis Ferdinand, heir to the thrones of Austria and Hungary and the nephew of Emperor Francis Joseph Ferdinand, would visit Sarajevo in Bosnia. This surprised and troubled Hermann because the archduke was intensely hated by the nearby Serbians, and he realized that such a visit would be very risky. As it would turn out, his concerns were well founded. On June 28th, as the archduke's sedan turned to cross a bridge in Sarajevo, he was shot and killed by Gavrilo Princip, a trained Bosnian assassin.

A month after the assassination of the archduke, Austria-Hungary declared war on Serbia and the flames of World War I were ignited. Since Transylvania was a part of Hungary and Oberth was a patriotic citizen, he joined the Austro-Hungarian Army as a foot soldier. Soon after enlisting in the Army, Oberth was wounded at the eastern front and was judged unfit for further infantry duty. Based on his medical background, he was assigned to a field ambulance unit.

Figure 3-3. Archduke Ferdinand and his wife Sofie (both seated in the vehicle) depart on their fateful last journey from city hall. This scene took place shortly before their assassination by Gavrilo Princip.

Figure 3-4. Dr. Julius Oberth (seated) and his sons Adolf (middle) and Hermann (far right) at the time of WW I. Oberth's brother did not survive the war.

Figure 3-5. Hermann and his wife Mathilde. Happily married, they remained lifelong companions.

Figure 3-6. Hermann Oberth, his wife Mathilde, and their son Julius in 1921.

Oberth was appalled by the carnage he saw in the war and thought that if a rocket could be developed and used, it would hasten the end of the war. In a sort of premonition, he reasoned that if rockets were fired on London, the British would be forced to seek peace since they would have no defense against this weapon. He spent much of his free time working on a rocket design he had begun at Munich, rechecking his calculations. Finally, with the consent of his superiors, these plans were sent to Berlin for evaluation. While awaiting the results of this suggestion at his home in Schaesburg, Oberth met and courted Mathilde Hummell. Mathilde was a talkative, cheerful individual who complemented Hermann well. They were married in July 1918.

Oberth soon found that the military had little interest in rockets. The military reviewer did not believe that a rocket could travel more than a short distance, far from the hundreds of kilometers that would be required to reach Britain. Oberth's suggestion was rejected and he returned to his unit.

Early Rocket Designs

With the surrender of Germany and the end of the war Oberth returned home, but found that life had changed drastically. As a part of the Treaty of Versailles, his beloved Transylvania had been annexed by Romania, and he learned that his younger brother, Adolf, had been killed in the fighting. Oberth found himself in a difficult situation, with a wife and a son, Julius, but no job. His father, sympathetic to his remaining son's situation, offered to support Mathilde and Julius while Hermann finished his education. He even yielded to his

son's desire to change his field of study from medicine to physics and mathematics.

After a short stay at the nearby University of Klausenburg, Oberth returned to Munich to complete his studies. While there he heard a speech by a former Munich art student and corporal in the German Army appealing to the German people to unite behind his leadership. Although they did not speak to each other, Oberth's path had again crossed that of Adolf Hitler. Joseph Goebbels, Hitler's future infamous propaganda minister during World War II, was also a student at the university at this time.

Because of his status as a foreigner, Oberth was denied housing, and his family was not able to join him. In March 1920, largely due to these continuing housing problems, he transferred to Gottingen University. This was actually fortunate because Gottingen was recognized as a first-rate institution in science and mathematics.

By the summer of 1920 Oberth was again hard at work designing a liquid-fueled rocket that he hoped to submit to the university as his doctoral dissertation. Oberth discussed this work with an astronomy professor, but this person saw little overlap with his field and referred Oberth to a professor of geophysics and meteorology. This meeting ended in the same fashion, and he was referred to the noted engineer Ludwig Prandtl. After reviewing this work, Prandtl pointed out some errors and provided Oberth with some technical references to read. He advised Oberth to pursue his work—"There is something to this," he said, "Don't lose your courage." With this encouragement, Oberth threw himself into his rocket design work.

The Two-Stage Model B Rocket

By spring 1921, Oberth had completed the design of a two-stage rocket. The first stage used a mixture of alcohol and oxygen for fuel, and the second stage used liquid hydrogen and oxygen. He referred to this design as his "Model B" rocket.

As in Munich, the laws governing foreigners at Gottingen did not permit him to rent suitable accommodations for his wife and child, so he relocated to the University of Heidelberg, an institute noted for its excellent library and scholarly professors. While at Heidelberg Oberth expanded his dissertation work to include a plan to use his Model B rocket to study the temperature, density, and composition of the Earth's upper atmosphere. Ironically, this proposal included many of the same ideas that Goddard had suggested to the Smithsonian in 1916. Goddard's goal was to construct the rocket, while Oberth presented it as a possible application for his rocket. He lacked the resources to build even a small-scale model of this craft. However, he had no doubts that a full-size version of it could be constructed.

By 1922, Oberth collected his rocket papers into a single document and submitted it to the university for his doctoral degree, but the same problems that arose at Gottingen once again occurred. The astronomy professor did not feel there was sufficient overlap to grant a degree in astronomy. Instead he suggested that Oberth publish it himself. Oberth then approached the physics faculty, but they did not feel his research was worthy of a doctorate in physics. The shock of these rejections caused Oberth to develop a lifelong dislike for the professional scientific establishment and a deep need to prove his worth. With the support of his wife, he accepted the refusal of his dissertation as a challenge. In his autobiography he wrote, "Never mind, I will prove that I am able to become a greater scientist than some of you, even without the title of doctor."

Figure 3-7. Oberth's thesis, The Rocket into Planetary Space, *was printed with the financial help of his wife.*

Oberth never doubted the value of his research and felt that its publication might induce other researchers to explore the capabilities of the rocket. Accordingly, in 1923 he approached the publishing house of R. Oldenbourg of Munich who agreed to publish his dissertation providing that Oberth pay all of the costs. Oberth lacked these funds, but his wife urged him to use her private savings. He accepted her offer and was forever grateful for this kindness. The resulting book was titled *The Rocket into Planetary Space*.

An Unexpected Commercial Success

Oberth divided his book into three parts and based its thesis on four postulates: (1) at the present level of science and technology, it is possible to build machines that can climb higher than the Earth's atmosphere, (2) with further improvements these machines can reach high enough velocities to escape from the Earth, (3) these machines can be built so that men can safely travel in them, and (4) it might be possible to build these machines in a few decades.

Oberth devoted the first section of the book to the mathematical theory of spaceflight. Many of the same equations that Tsiolkovsky had derived can be found here. The second section described Oberth's Model B rocket and contained a summary statement of what he considered to be his two main contributions:

> I consider that in rocketry up to now there are two new essentials: the use of liquid fuel instead of the common explosives, and the two-stage or divided rocket. Both of these I have advanced.

The last section consisted of a futuristic view of mankind's use of space. It discussed travel to the Moon, how rockets parked in Earth's orbit could be employed for weather forecasting, monitoring conditions at sea, observing military build-ups, and how they could be used for global communications. It also discussed how solar energy could be gathered by a large collector in Earth's orbit and then transmitted down to the ground. Finally it addressed the need for space suits and air for the space travelers. All in all, this book was a remarkable document and its publication had a lasting effect on the rest of his career.

While awaiting the publication of this book, Oberth was faced with a pressing matter. How could he support himself and his family? While at school, his father and mother had provided financial backing, but it was now time for him to find a job. The end of World War I brought many changes to his native Transylvania. One such change required teachers to speak Romanian. He quickly learned the language, passed the high-school teaching exam, and was awarded a teaching position in his home town of Schaesburg.

Much to Oberth's surprise and delight, *The Rocket into Planetary Space* was a commercial success. The first printing sold out soon after it appeared and a second printing also sold out quickly. The book had struck a note of nationalist pride in

1894	Born in Hermannstadt on June 25, 1894
1906	Reads science fiction stories by Verne
1912	Graduates from high school
	Completes first rocket design
1913	Travels to Munich to study Medicine
1914	Joins army
1915	Wounded on eastern front
1917	Submits rocket plan to Berlin
1918	Marries Mathilde Hummel
1919	Returns to Munich, enrolls in University of Gottingen
1922	Dissertation rejected at University of Heidelberg
1923	*The Rocket into Interplanetary Space* published
1924	Development of liquid-fuel rocket
1938	Works at the Technical Institute of Vienna
1940	Works at the Technical Institute of Dresden
1941	Assigned to work at Peenemunde
1943	Reassigned to anti-aircraft missile program
1945	Arrested at the end of WW II. Returns to family
1948	Moves to Switzerland
1950	Moves to Italy
1955	Moves to Huntsville, Alabama, to work with von Braun
1958	Returns to Romania

Germany where people were frustrated and humiliated by the war reparations their country was forced to pay under the Treaty of Versailles. However, the popular appeal of his book did not carry over into the scientific community where it was described in such unflattering terms as ridiculous, absurd, and nonsensical.

By this time, Oberth was familiar with Goddard's rocket research in the United States and sent him a letter requesting a copy of *A Method of Reaching Extreme Altitudes*. Goddard reluctantly complied but almost immediately regretted this decision. He considered the field of rocketry as his private domain and was reluctant to share the knowledge he had worked so hard to uncover with anyone. Goddard did not encourage further correspondence. As previously noted, Oberth also wrote to Tsiolkovsky praising him for his efforts.

After publication of *The Rocket into Planetary Space*, Oberth was approached by Max Valier, a World War I aviator turned popular lecturer and writer. Valier asked permission to rewrite this book in a less technical fashion so that it could be sold to a wide audience. With little to lose from this arrangement, Oberth agreed and then turned his attention to his high-school teaching duties.

The book that Valier produced, *The Advance into Space*, was published in 1924 and was so successful that it created a core of amateur rocket enthusiasts in Germany. In 1927, with the encouragement of Valier, these people formed themselves into an organization called the Rocket Society, or *VfR*. Many members of this group would later have important roles in Germany's World War II rocket program.

The Difficulty of Making Hardware

The publication of *The Advance into Space* caught the attention of Fritz Lang, a noted German silent-film maker who was working for the film conglomerate, the UFA. He hired Oberth to serve as technical adviser to his upcoming movie,

Figure 3-8. The moon rocket as seen in the UFA film, Woman in the Moon *(1929).*

Woman in the Moon. Lang also asked Oberth to construct a real rocket that could be launched to promote the movie's premier. Flattered by the offer and confident that this was possible, Oberth readily agreed. He soon realized that it would be impossible to finish this project before the movie's opening. Oberth was a theoretician, not an engineer, and the unanticipated technical problems that arose were overwhelming. Despite his careful theoretical efforts and the years of thinking and planning that had gone into his design, Oberth realized that he had vastly underestimated the difficulty of constructing a working rocket.

He was embarrassed by this failure, but was spared financial retaliation by the film studio when the movie proved to be a huge success. Overjoyed by the large crowds that lined up to see *Woman in the Moon*, the UFA was in a forgiving mood and allowed Oberth to keep his incomplete rocket and unused building materials.

Figure 3-9. Oberth (fifth from the left) with members of the VfR. Wernher von Braun is to the far left.

With this equipment in hand and the growing support provided by the membership of the *VfR*, Oberth continued to try to build a working version of his rocket. Unfortunately, the onset of the Great Depression of the 1930s and the start of World War II prevented much progress from being made. Additional new advances during the war would make his design and hardware obsolete. He would never return to this task.

In 1938, he was invited to participate in a rocket research program at the Technical University of Vienna, and at the outbreak of World War II, he worked in the German wartime rocket research effort at Peenemunde. Once again his status as a foreigner came back to haunt him and he never occupied an important position in that program.

Principal Contributions

Oberth's ideas were ahead of their time, and for most of his life he did not receive the recognition his work deserved. Like the other rocket pioneers he realized that liquid propellants were essential for rocket flight. His two-stage rocket design was much more advanced than that formulated by Tsiolkovsky, and although it was never built, it foreshadowed an era when such vehicles would be commonplace. Most importantly, Oberth inspired a young generation of German scientists who would play an important role in the construction of the rocket. By the end of World War II the resources of entire countries would be required to make the next step toward manned spaceflight.

Introduction to Section II
Beyond the Space Pioneers

If we [the United States and Soviet Union] really cooperated on man-in-space, neither country would have a program because the necessary large support in money and manpower was only because of the competitive element and for political reasons.

—Hugh Dryden, NASA Deputy Administrator, September 7, 1960

Tsiolkovsky, Goddard, and Oberth were all inspired by imaginative stories about space travel by writers such as H.G. Wells and Jules Verne. From their childhood these space pioneer saw themselves as explorers and they longed to experience the thrill of journeying to strange new worlds. As they grew older, this interest developed into a lifelong passion to build a craft capable of interplanetary flight.

In addition to satisfying their inner thirst for adventure, the pioneers also desired to prove their own self-worth by solving a problem that had frustrated previous generations. However, while fictional space stories were interesting to read, they provided little useful information about how actually to construct a spacecraft. This fact did not trouble the pioneers since they knew progress in the field of rocketry would ultimately be based on a firm knowledge of the laws of physics and engineering.

One of the pioneers' greatest contributions was the realization that liquid rather than solid fuels would be required to power the rocket. This was not an easy change to accept because it involved a major departure from past thinking. Solid fuels had several advantages and had been used in rockets for centuries. They were noncorrosive, reliable, readily available, inexpensive, and easy to handle. The rocket pioneers had the foresight to realize that many of those benefits would fade with the passage of time since they were based on historical considerations rather than on the intrinsic properties of the materials themselves.

Despite this position, it was still not obvious liquid fuels provided significant benefits over solid propellants. Liquid fuels provided more thrust per pound than solids, allowed the burning rate to be controlled, and permitted the rocket engine to be turned off and restarted, but they also had numerous disadvantages. They were difficult to use, corrosive, comparatively expensive, required elaborate transfer apparatus, and needed a relatively sophisticated ignition mechanism. So then, what was the key advantage of liquid fuels over solid fuels?

The work of the pioneers provided indisputable proof that the rocket propellant had to produce a high exhaust velocity if a craft was to escape the Earth's gravity. All three men were able to show that liquid fuels could produce these needed velocities. After establishing this fact, the pioneers devoted much of their effort to the engineering problems liquid-fueled rockets posed. These problems were formidable and their solutions were so complex that they soon led to the development of entire new research fields.

The Field of Rocketry Evolves

By the mid-1930s the era of the "lone wolf" research effort had largely passed. The escalating costs and complexity of rocketry required larger organized efforts. To meet this need, rocket clubs arose in the United States (the American Rocket Society), Great Britain (the British Interplanetary Society), Germany (the *VfR*), and the Soviet Union (GIRD).

In addition to providing the necessary funds, these clubs allowed individuals with a variety of skills to work together productively. Some members worked on engines, others on the rocket's structure, some experimented with different kinds of fuels, and others worked on the rocket's guidance and control systems. This pooling of talent yielded fruit as competitions between and within clubs led to new advances.

In order to popularize this work, demonstrations were held before large public audiences. Rockets were attached to automobiles, boats, planes, and trains to increase their speed. Although few of these experiments produced lasting im-

Figure Intro. 2-1. H.G. Wells (left) and Jules Verne (right) played an important role in inspiring the early rocket pioneers to devote their lives to the theory and construction of the rocket.

Figure Intro. 2-2. Once travelers entered space they would be in a weightless environment. This drawing shows how weightlessness was depicted in Around the Moon by Jules Verne.

Figure Intro. 2-4. Adolf Hitler (left) and Benito Mussolini (right). Hitler hoped that the rocket would turn the tide of the second world war in favor of Germany.

provements, they added to the ranks of the amateur rocket enthusiasts.

The work of these clubs soon moved the rocket beyond the level obtained by the pioneers. But as the rocket grew in power and complexity, the financial burdens of supporting this work also increased. Unfortunately, the need for additional resources ran headlong into the reality of the Great Depression. Millions of people found themselves out of work and in desperate circumstances. Unnerved by these conditions, little money was available for hobbies such as rocketry

The Great Depression hastened the transition of the rocket into its next phase. In this era the huge costs of experimentation were assumed by national governments. In this new setting the inspirational writings of Wells and Verne became irrelevant. National leaders were interested in the rocket because of its military value, not as a vehicle for space travel. This attitude was particularly prevalent in Germany. At the end of World War I, Germany was forced to sign the Treaty of Versailles. Under the terms of this treaty the German military was not allowed to possess any artillery. Recognizing this deficiency gave future adversaries an unacceptable advantage, the rocket was explored as an alternative weapon.

The severe economic conditions caused by the Depression and the humiliating terms of the peace treaty caused large numbers of Germans to unite behind leaders who promised better times. When rising discontent caused the collapse of the German government, the military was secretly instructed to begin rearming.

This action gave the Germans a lead in the development of the rocket, but they did not reap any long term rewards for this work. After the end of the war, much of the machinery used in their rocket program and most of their prominent scientists were sent to the United States or the Soviet Union. These two former allies then engaged in a fierce competition for control of the world. In this contest, the rocket became an important tool for measuring the relative strength of these two countries. Both the United States and the Soviet Union now viewed progress in the field of rocketry as a matter of national survival. Faced with the prospect of nuclear annihilation, they spent massive sums of money on their rocket programs.

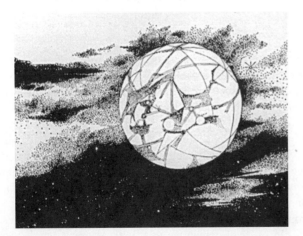

Figure Intro. 2-3. The planet Mars as depicted in H.G. Wells War of the Worlds.

4 World-Wide Interest in Rockets

. . . it has often proved true that the dream of yesterday is the hope of today and the reality of tomorrow.

—Robert Goddard

During the 1920s and 1930s interest in rocketry flourished. People were fascinated by this device and experimental programs were started in Germany and Russia. In Great Britain, theoretical studies were undertaken by the British Interplanetary Society, while in the United States, the American Rocket Society attempted to build liquid-fueled rockets. Popular writers repackaged the ideas of Tsiolkovsky, Goddard, Oberth into books, magazine articles, and movies and sold them around the world.

Although the public's response to rocketry was generally favorable, the academic establishment viewed this new field with disdain. Rocket enthusiasts were ridiculed and often portrayed as misguided fools. In this environment it was impossible for scholars to advocate the use of the rocket as a means of space travel without jeopardizing their career. This hostile attitude would last until the launch of the first Soviet Earth-orbiting satellite, Sputnik.

American Rocket Work

The most active American rocket program in the pre–World War II era was headed by Robert Goddard. As previously mentioned, Goddard was a private person who shunned publicity. He did not encourage exchanges with other researchers in the field and viewed rocketry as his private research area. This attitude extended not only to Europeans such as Oberth, but to American scientists as well. By the mid-1920s his program was the most advanced in the world, but few researchers possessed much information about his discoveries.

Other American researchers joined together and formed the American Interplanetary Society in 1930. Although Goddard was invited to join, he declined. Perhaps he foresaw a danger, born out of his earlier experience with the press, in associating his name with words like "interplanetary." If so, his fears were justified. Soon after the group's formation they found themselves under attack and defending their name. As Goddard had painfully learned, the American public was not prepared for this type of futuristic organization and members of the society were publicly scorned. Unwilling to fight a seemingly futile battle with the press, the society gave in to criticism and changed its name to the American Rocket Society.

Figure 4-1. The three pioneers of the space age: (left to right) Tsiolkovsky, Goddard, and Oberth.

Figure 4-2. The American rocket pioneer, Robert Goddard, was the first person to launch a liquid-fueled rocket.

Figure 4-3. Members of the American Rocket Society prepare a static rocket test in 1934.

Throughout the 1930s, this group conducted experiments with liquid-fuel rockets. The main result of their efforts was to show that cooling mechanisms relying on water could not prevent the rocket from being extensively damaged when firing. This result cleared the way for the development of alternative cooling techniques.

University research groups also began actively to explore rocketry in the 1930s. One such group, the Guggenheim Aeronautical Laboratory at the California Institute of Technology, started a small-rocket program and developed engines using both liquid and solid fuels.

In 1938 General Henry Arnold, commander of the Army Air Corps, asked the National Academy of Sciences for help in developing a method to aid heavily loaded airplanes during takeoff. The Academy suggested he contact the Guggenheim laboratory. Talks were productive and a separate unit, funded by the Air Corps, was created within the laboratory to focus on this task. Established on July 1, 1939 this group was named the Jet Propulsion Research Project. Later it was renamed the Jet Propulsion Laboratory and became one of NASA's premier installations for the construction of unmanned spacecraft.

British Rocket Efforts

By the middle of the 1930s small private rocket groups also were active in Britain. The British Explosives Act of 1875 prohibited test firings of rockets, so their work consisted primarily of theoretical investigations. This restriction was frustrating, but the Explosives Act was strictly enforced. Britain would not play a major role in future rocket construction.

Despite this restriction, the British were not idle spectators. Forced to abandon the idea of building and testing a rocket, they formed the British Interplanetary Society to increase discussion of the physics of spaceflight. The society provided a written forum, the *British Interplanetary Society Journal*, in which astronautical scientists and engineers reported their results. This journal first appeared in 1934 and soon became recognized as the most distinguished publication devoted to spaceflight in the world.

Soon after the British society formed, it encountered public pressure to choose a more modest name. In particular, the word "interplanetary" became a focal point of public ridicule. Some of the more conservative members advocated dropping this word, but the group's president strongly objected. He rallied the membership in a defiant speech attacking past public criticisms and asking the club to remain faithful to its goal.

> Are we to pander to public opinion . . . an opinion which held to ridicule the votaries of heavier than air flight, and which refused to credit the marvel of wireless telegraphy to such good effect that the inventor died heartbroken, deserted even by his friends, who also deemed him mad? . . . It seems to me that a change in name, regardless of the reason for it, would be universally misconstrued as an admission of doubt, as a confession that the interplanetary idea only belongs to the realm of extravagant fiction.

His argument was persuasive. The word "interplanetary" was not deleted and to this day remains part of the society's name.

Figure 4-4. The French rocket pioneer Esnault-Pelterie.

The French Rocket Program

Prior to World War II, France had its own rocket pioneer, Robert Esnault-Pelterie. He was a member of the French Academy of Sciences and a respected airplane inventor. Esnault-Pelterie played an important role in the development of the French rocket program from 1912 to the outbreak of World War II and helped establish an annual prize to reward significant contributions to the field of rocketry. Its first recipient, Hermann Oberth, was given 10,000 francs, twice the stated size of the award, because of his numerous contributions.

The Italian Rocket Program

Prior to World War II, Italy also possessed an active rocket program, but unlike the situation in France, it lacked a notable central figure to champion its cause. Their program continued to proceed in a faltering manner. The Italian General Staff began an experimental solid-fueled rocket program in 1927 but canceled it just two years later. They then supported a liquid-fueled program, but despite some promising results, canceled it in 1930. Two years later, the Italian Air Ministry again funded this program but discontinued its support in 1935. Italy would not play a significant role in the development of the rocket.

Early Soviet Rocket Programs

During the 1920s and 1930s amateur rocket groups were active in the Soviet Union. In 1924 rocket enthusiasts formed an organization called the Study of Interplanetary Flight (OIMS). In the early 1930s two other groups made substantial progress

Figure 4-5. The liquid-fueled Soviet Aviavnito rocket was developed in the mid-1930s.

Figure 4-6. Rockets were used in several different ways in Germany during the 1930s. Here solid rockets were attached to a glider to increase its speed.

investigating liquid-fueled rocket engines. One of these, the Group for Studying Reaction Propulsion (GIRD), was located in Moscow and was under the leadership of Friedrikh Tsander (1887–1933). The other, the Gas Dynamics Laboratory (GDL), was headed by Valentin Glushko (1908–1989). Unlike the GIRD, it received government funding and was located in Leningrad. Both Tsander and Glushko would play an important role in the development of the Soviet space program.

As a result of funding from the Soviet government, steady progress was made in the development of sounding rockets (rockets used to study the Earth's upper atmosphere), solid-propellant military rockets, and rocket engines for airplanes. The liquid-fueled Aviavnito rocket completed in 1936 was 10 feet long, weighed 213 pounds, and could reach an altitude of 3.5 miles (see Figure 4-5).

Soviet solid-fueled military rockets made their debut in World War II. These weapons played an important, though not a decisive, role in the war serving as a form of artillery. Plans were also underway to construct a guided missile, but that project was canceled when Germany invaded Russia shortly after the start of World War II.

German Interest in Rockets

Five years after Goddard's launch of the liquid-fueled rocket, its first foreign counterpart was successfully flown. This German rocket, the Huckel-Winkler 1, was 2 feet long, 1 foot across, and used liquid methane and oxygen. It lifted off from Gross Kuhner, a region located near Dessau, Germany, on March 14, 1931, and reached an altitude of nearly 1,000 feet. This successful launch gave a misleading impression of the state of German rocket technology. For example, its successor, the Huckel-Winkler 2, caught fire immediately after lift-off, reaching an altitude of only 10 feet.

The construction of these liquid-fueled rockets reflected the high level of interest on the part of the general population in Germany in this device. Rocket research proceeded at a

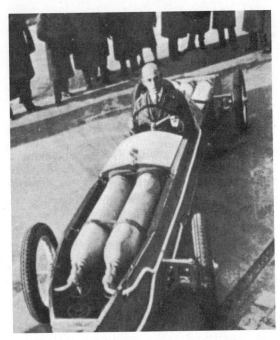

Figure 4-7. Max Valier in a car powered by solid rockets.

much faster pace in Germany than in any other country and this work was conducted by private groups without government support. By far the most important of these organizations was the Rocket Society, or *VfR*. Germany also had active promoters who increased the public's awareness of the capabilities of the rocket. Among these individuals were Willy Ley, a widely respected popular science writer, and Max Valier, a former World War I aviator.

The fascination that the Germans had with rockets seemed to have no boundary. In 1928, Valier collaborated with automaker Fritz von Opel and sounding rocket maker

Figure 4-8. Gerhard Zucker tried to deliver mail over long distances using a rocket. The unsuccessful attempt shown here was made on July 31, 1934.

Figure 4-9. The Nazi party came to power in Germany in the late 1930s. The increasing interest of the German military ushered in a new era in rocketry.

Friedrich Sander to create the world's first rocket glider and automobile (see Figures 4-6 and 4-7). By May 23, 1928, this rocket car reached a speed of nearly 125 miles per hour. Rockets were attached to railway cars, airplanes, seaplanes, and were even used in a scheme to deliver mail. Gerhard Zucker believed that the rocket could be used to quickly transport mail between Europe and Britain. This particular idea met with disaster when his rocket exploded during a test on July 31, 1934 (Figure 4-8).

Fortunately for the Allied forces during World War II, work on a rocket airplane generated little interest in Germany. This project was viewed as too far ahead of its time and was terminated before any significant progress had been made. Few rocket projects lasted more than a few years, and none produced a practical device. This work was not wasted, however, because it produced valuable insights into the construction of the rocket and added to the number of Germans who saw the real potential of the rocket as a means of space travel. Many Germans possessed a great deal of engineering and technical expertise, and they joined the *VfR* in large numbers. The engineering genius of this group was Klaus Riedel, an exceptionally talented, self-taught engineer, who had no formal technical education.

The financial pressures created by the great depression of the 1930s stopped the experimental work of the *VfR* and its dues-paying membership rapidly declined. In desperation, the remaining core members of the society sought financial support from the German Army. Germany was prohibited by the Treaty of Versailles from having any long-range artillery, and the army saw the rocket as a means of circumventing this restriction. As a result, Wernher von Braun, a member of the *VfR*, was hired by the army to transform the rocket from a technological toy into a potent weapon.

As Adolf Hitler rose to power, the international ties of the *VfR*, which up to that time had been very good, were severed. The German rocket program was now in the hands of the military, and they wanted to keep their activities secret.

5   Wernher von Braun (1912–1977)

It is man's nature that he always wants to explore—to move on, to develop, and to advance. If we stop doing that we will no longer be human.

—Wernher von Braun (address to the National Space Institute in July 1975)

Wernher von Braun was born on March 23, 1912, in Wirsitz, Germany. His parents, Baron Magnus and Baroness Emmy von Braun, were both well known and respected in that region of the country. The Baron served in a number of government positions, first as a councillor of the province of Posen, and then at the national level during World War I.

In 1931, he served as the Secretary of Agriculture in President von Hindenburg's administration and remained in that position until Adolf Hitler came to power two years later in 1933. Wernher benefited by having a father who believed in duty, personal responsibility, and honesty and a mother who loved the arts and music.

Von Braun's Early Childhood

During his early childhood, von Braun displayed a flair for linguistics, a skill that was encouraged by his mother who was fluent in six different languages. He was also an inquisitive child and full of energy. His father recalled Wernher's early life,

What I remember most vividly from the years of his childhood is the absolute futility of all my attempts to apply a bit of parental guidance to him. His growth rate was exorbitant, and I often thought that I should channel his outbursts of activity toward the more civil goals that were the accepted standards of society. Determination, fatherly strictness, diplomacy—nothing worked.

Besides his early zeal for life, there was little to suggest that von Braun would become one of the world's premier rocket engineers. Initially he was a poor math student and showed so little interest in either physics or engineering that his father decided to place him in another school. The school headmaster agreed with this decision but felt strongly that Wernher had the potential to make a significant contribution in some area of study. He told the Baron, "The boy will be a credit to you. I know his kind—there is genius in him and he will go far." Neither were there any hints of Wernher's future

Figure 5-1. The birthplace of Wernher von Braun in Wirsitz, Germany. His father was an influential member of the community.

Figure 5-2. Dr. Heise's classroom at the Franzoesisches Gymnasium, Berlin, 1922. Wernher is seated in front of Dr. Heise and to his left.

profession in his family background. His ancestors included knights, jurists, governors, and even generals, but no technically oriented people.

Von Braun's passion for space travel evidently was nurtured by the Baroness, who was a self-taught amateur astronomer. She provided him with a continual stream of encouragement. In Ernst Stuhlinger's biography of von Braun, her pivotal role in molding Wernher's early development is described:

> To those who knew Wernher's mother, Baroness Emmy von Braun, it was obvious that she must have been the source of a good portion of his innate gift for science and technology. However, her influence was by no means limited to the remarkable arrangement of his genes. Throughout Wernher's formative years, she was an inexhaustible well spring of encouragement and stimulation, of straightforward knowledge, and of wise counsel.

Perhaps the spark that ignited Wernher's fascination with space exploration was provided when the Baroness presented him with the gift of a telescope upon his confirmation into the Lutheran Church. With this new tool, he explored the heavens and began dreaming about the day when people would be able to travel to the Moon, planets, and distant stars. In this respect his childhood paralleled that of Hermann Oberth, whose mother also gave him an optical instrument to view the heavens.

The Baroness's recollections of Wernher's childhood were different from those of her husband. She recalled,

> It was exciting to have Wernher as a child. He was like a dry sponge, soaking up every bit of knowledge as eagerly as he could. There was no end to the questions he asked; he really could wear you out quickly.

At the age of sixteen, Wernher's interest in space was reinforced while attending the Hermann Lietz boarding school. He convinced the school administration to purchase a 5-inch refracting telescope and, along with his classmates, built an

observatory, to house it. His math skills increased rapidly during this time, he even taught mathematics to the senior class in the absence of one of the institute's instructors.

After completing his high school education, Wernher enrolled in the Charlottenburg Institution of Technology and worked at the Borsig Engineering Works (a Berlin machine factory) as an apprentice in its locomotive shop. While working towards his advanced degree, he maintained his interest in rockets and spaceflight and contacted Willy Ley, a popular science writer living in Berlin. Through Ley's connections, he was introduced to Hermann Oberth and enthusiastically agreed to assist him with his rocket experiments.

In the late 1920s and early 30s, the rocket mania discussed in the prior chapter swept over Germany, and von Braun was caught up in the excitement. He became an active member in the newly formed German amateur rocket society, the *VfR*, in which both Oberth and Ley were leading members.

The Military Recruit von Braun

As the experiments of the *VfR* grew in size and complexity, it became obvious that a more remote test site was needed. Rudolph Nebel, one of Oberth's assistants, obtained permission from the military in 1930 to use a 300-acre abandoned ammunition dump on the outskirts of Berlin as a test site for the club.

While helping the *VfR* members with their experiments, Von Braun pursued his own education at the Charlottenburg Institute and received a bachelor's degree in mechanical engineering in 1932. It was about this time that the German military began to take an interest in the work of the *VfR*. Two army ordnance officers, Colonel Karl Becker and Captain Walter Dornberger, were assigned the task of exploring whether the rocket could serve as a long-range weapon. Dornberger would later play a major role in von Braun's wartime activities.

Becker, Dornberger, and a third army officer, Wolfram von Horstig, met with the members of the *VfR* to see if they possessed the skills the army needed. Following this initial session, the army offered the *VfR* a modest contract of 1,000 marks to continue their research, but this funding was contingent on the successful firing of their prototype rocket, the *Mirak II*, at the nearby army proving grounds at Kummersdorf. The society accepted this offer and conducted the required test.

The *Mirak II* blasted off from Kummersdorf, reached an altitude of 200 feet and then crashed back to Earth. Although this flight was a fair demonstration of the technological state of rocketry in Germany, the officers attending the test were unimpressed. They were looking for a long-range weapon, and the short flight of the *Mirak II* was clearly inadequate to meet that need. The army probably would have cancelled the *VfR's* contract altogether except for a follow-up meeting between von Braun and Colonel Becker.

Becker realized that the rocket was still in its infancy and took a more sympathetic view of the *VfR's* demonstration flight. He was also very impressed with von Braun's obvious knowledge about rockets and tried to persuade him to work

Figure 5-3. Rudulf Nebel (left) and eighteen-year-old von Braun (right) carrying rockets at a private testing ground near Berlin in 1930.

Figure 5-4. The VfR Repulsor 2 rocket reached a height of 2,000 feet on May 23, 1931.

Figure 5-5. Captain and later Major-General Dornberger would serve as the military overseer of von Braun's work during WW II.

for the army. Von Braun had only intended to argue for the army's continued support of the *VfR* and was surprised by the unexpected direction the meeting took. An arrangement was reached by November 1932 that allowed him to continue his education in rocketry at the University of Berlin while working as a civilian employee of the army. As part of this deal, von Braun was allowed to use the army proving grounds at Kummersdorf to test his ideas on rocket propulsion.

Other members of the *VfR* had mixed feelings about working with the military, but after much discussion, it was clear that they had little choice. The Great Depression of the 1930s had decimated the group's membership, and its funds were nearly exhausted. As von Braun described the situation, "It became obvious that the funds and facilities of the army would be the only practical approach to space travel."

Von Braun viewed his collaboration with the army not as an effort to create a destructive weapon, but as a means of financing what he believed would be a protracted research effort to build a vehicle that was capable of spaceflight. His plan to use the military to fulfill his own purposes would turn out to be incredibly naive.

Initially, the place he was given to work at Kummersdorf was neither elaborate nor sophisticated. Von Braun had the assistance of one mechanic and a workshop that was nothing more than a small concrete pit with a sliding roof. These facilities reflected the modest level of interest that the military had in rockets at this time. Nevertheless, even with this pitiful level of support von Braun made substantial progress on rocket propulsion, completing his doctoral degree in only two years.

The university required that all of its Ph.D. dissertations be published, but the military did not want von Braun's research to become public. A compromise was reached by giving Wernher's thesis the vague title, "About Combustion

Tests." The military felt confident that such an uninformative title would insure that very few people would take note of it.

The A-Series Rockets

Finally able to work full time at Kummersdorf, von Braun began to construct a developmental sequence of rockets called the "A-series." The first such rocket, the A-1, weighed 300 pounds, was 4 2/3 feet tall, and 12 inches in diameter. When fired, it developed 600 pounds of thrust. Although this rocket demonstrated technical progress, it was not impressive enough to maintain the support of the army. Von Braun was told that continued funding was contingent on the successful test flight of the A-1's more powerful successor, the A-2.

Von Braun worked hard on the A-2 and felt confident that it would perform satisfactorily. He was pleased and relieved when two of these powerful rockets successfully completed their demonstration flights. Finally impressed with the military potential of the rocket, the army dramatically increased its support, and the research group headed by von Braun grew rapidly. The third-generation, 20-foot long, 1,500-pound A-3 was soon constructed and launched. Von Braun was elated by these successes. The A-1, A-2, and A-3 all provided important new data that allowed more powerful and reliable rockets to be built.

By the spring of 1937, the German rocket program had grown to about eighty people, and von Braun realized that a more remote site was needed to conduct long-range flight tests. After months of searching, he selected a peninsula off the island of Usedom located in an isolated area on the Baltic seacoast called Peenemunde. In April 1937, the German rocket program was moved to this site. Work on the A series of rock-

Figure 5-6. A German rocket launch site at Pennenmunde. The letter A shows a V-2 rocket resting on its side. The letter B shows the location of a mobile test stand.

Figure 5-7. The launch of the V-2 during World War II against civilian targets in Britain opened a new and ugly aspect of warfare.

ets continued at a brisk pace, and the first successful launch of the 12-ton, 45-foot long, liquid oxygen/alcohol-fueled A-4 occurred on October 3, 1942. This rocket accelerated to over twice the speed of sound and, just sixty seconds after launch, reached an altitude of 54 miles. Upon witnessing this event, von Braun's military supervisor, now General Dornberger, made a prophetic statement: "Today the spaceship was born."

The successful launch of the A-4 caused a great deal of excitement among the scientists at Peenemunde. Like Dornberger, von Braun was able to see beyond the military implications of this flight, and he spoke of his long-held dream of using the rocket to explore space. These discussions, even though informal, later came back to haunt him when the Gestapo (the German military police) claimed that he was not adequately focusing the German effort on the rocket's military use. Progress on the A-4 was uneven, but each launch provided additional data that allowed the rocket to be further refined and improved.

The V-2 Rocket

Although the military continued to fund the development of the A-4, it was not a high priority program, and Hitler considered terminating the entire effort. On July 7, 1943, von Braun and now Major-General Dornberger met with Hitler and pleaded at length to be allowed to continue their work. Their arguments were so convincing that Hitler agreed not only to continue this program but gave it a high priority. He was convinced by the end of the meeting that the rocket could alter the course of the war and ordered its mass production. For propaganda purposes, the A-4 was renamed the Vengeance 2 , or V-2. A production target of 2,000 V-2s per month was set, but wartime difficulties limited the actual figure to about 600.

The first tactical V-2 was launched on September 8, 1944, from the outskirts of the Hague, Netherlands, toward England where it impacted with nearly a ton of high explosives near London at Chiswick-on-Thames. On that same day, Paris was bombarded with V-2s. Dorette Kersten, von Braun's secretary, recalled his reaction upon learning of this news over public radio:

Von Braun was completely devastated. In fact, never before or afterwards have I seen him so sad, so thoroughly disturbed. "This should never have happened," he said. "I always hoped the war would be over before they launched an A-4 against a live target. We built our rocket to pave the way to other worlds, not to raise havoc on this Earth. Should this really be the fruit of our work?"

Von Braun faced the possibility of death many times during the war. On one occasion, he and Dornberger were trying to discover the nature of technical problems encountered during the descent of a V-2 and took a position in the center of the V-2 impact area to observe the rocket as it sped toward the Earth. The first missile fired for their observation fell within 300 feet of their position, and the explosion hurled von Braun through the air and into a ditch where he landed shaken but uninjured. Although it is not clear what if anything was learned about the V-2's flight, he quickly realized that this observation method was a bad idea.

Von Braun also escaped the saturation bombing of Peenemunde by 600 aircraft of the British Royal Air Force in August 1943. This raid took the lives of 735 men, women, and children, among whom were Dr. Walter Thiel, a close colleague of von Braun, and Thiel's entire family. A third narrow escape from death occurred in February 1944, when he was arrested by the Gestapo. Himmler, the feared head of this organization, felt that the Gestapo could more efficiently run the V-2 program than the army and attempted to gain control of it. Von Braun's resistance caused Himmler to question his loyalty and orders were issued for his arrest and imprisonment at Stettin. He avoided execution only because Dornberger personally intervened and convinced Hitler that von Braun's work was essential to the war effort.

Von Braun's Surrender

By March 1945, it was clear to von Braun that Germany was going to lose the war, but he was determined that the progress made in rocketry would not be lost. A secret meeting was convened of key Peenemunde scientists and he asked if they should surrender to the Russians or the Americans. The decision was made in favor of the Americans, but this hardly solved matters. How could he move 10,000 personnel and 2,000 tons of materials into the path of the invading American armies?

Near the end of the war, von Braun received many conflicting orders. One order directed him to move his research group to Bavaria and set up emergency production of the V-2s. Von Braun realized that this location would conveniently place his group in the path of the advancing American forces. He decided to act. To fool the authorities during this move, he devised a fake operation named "Project for Special Dispositions." With self-printed armbands, phony vehicle stickers and papers, the group cautiously moved south, traveling only at night to avoid Allied air attacks.

Figure 5-8. Even when not carrying explosives, the impact of an A-4 created a crater that was 50 feet deep and 130 feet across.

One night while en route, von Braun's driver fell asleep, causing the car to crash over an embankment and breaking von Braun's arm in two places. Upon arrival in Bavaria, the rocket engineers found themselves virtual prisoners of Himmler's men. As the American Army approached, however, these guards disappeared, and von Braun sent his brother to contact the Americans. After traveling only a short distance he ran into an advance infantry patrol and surrendered the rocket group.

America Captures the German Rocket Scientists

The Americans were well aware of the German rocket program and desperately wanted to capture its key scientists.

Focus Box: Specifications of the German V-2 missile

The V-2	
Maximum velocity	3,466 mph
Impact velocity	1,800 mph
Height of trajectory	60.3 miles
Unfueled weight	10,300 lbs
Alcohol	8,304 lbs
Liquid oxygen	10,800 lbs
Fueled weight	28,380 lbs
Warhead	2,230 lbs
Length	46.1 ft
Diameter	5.4 ft

Figure 5-9. The British bombing raid of August 17–18, 1943 did much damage to the German complex at Peenemunde.

A special program headed by Colonel Holger Toftoy, Chief of Ordnance Technical Intelligence in Europe, named "Operation Paperclip," was established solely for this purpose. Toftoy was informed of the surrender of the German rocket group and hurried to meet them. Officially he was allowed to select 100 German scientists and relocate them in the United States, but after meeting with von Braun, he increased this number to 127. By September 20, 1945, this group of displaced Germans found themselves in Boston, Massachusetts. Their arrival in the United States marked an important new beginning for the American rocket program.

Von Braun in America

Shortly after arriving in Boston, the German scientists were sent to Fort Bliss, Texas, to prepare test firings of the

Figure 5-10. Von Braun broke his arm while fleeing toward the American forces.

captured V-2 rockets at the nearby White Sands Missile Test Range in New Mexico. One of the most important products of their efforts was the Bumper, a two-stage rocket that used a V-2 as its lower stage and a solid-fueled U.S. Wac Corporal missile as a second stage. This program gave America a number of "firsts" in rocketry, namely: (1) the first extremely high-altitude atmospheric research, (2) the first man-made object sent beyond the Earth's atmosphere and into space, (3) the first animal launched in a rocket with live recovery, (4) the first high-altitude television transmission, and (5) the first ground control of a ballistic rocket by radio command.

Focus Box: Von Braun (front, seventh from the right) and the German scientists in Huntsville, Alabama

On March 1, 1947, von Braun traveled back to Europe to marry his eighteen-year-old cousin, Maria Louise von Quistorp, in Landshut, Bavaria. For security reasons, he was accompanied by a bodyguard of the U.S. Secret Service. Together with his wife and parents, he then returned to work at White Sands. On December 9, 1948, the von Braun's first child, Iris Careen, was born.

Toftoy, now the military overseer of von Braun's group, repeatedly asked that the missile development facilities at Fort Bliss be expanded to accommodate more sophisticated tests, but these requests were turned down. Permission was then asked of Vice Chief of Staff General Matthew Ridgeway to relocate the group to Redstone Arsenal in Huntsville, Alabama. This request was approved, and in the spring of 1950, the group moved to Huntsville where von Braun was appointed Technical Director of the army's Guided Missile Development Group. In 1956 this unit was renamed the Army Ballistic Missile Agency (ABMA). The second von Braun child, Margarit Cecil, was born in Huntsville on May 8, 1952.

The first technical success at Huntsville occurred on August 1, 1953, with the launch of the Redstone rocket. The joy over this success, however, was overshadowed by another event. Less than three weeks later, on August 20, 1953, the Soviet Union exploded their first hydrogen bomb. American intelligence services reported that the Soviets were also engaged in a massive effort to develop a ballistic missile. Government policy makers realized that the United States and Soviet Union were now entering an era of technological competition—the winner of which might have the capability to militarily dominate the world.

In response to this threat, the American military branches each established their own rocket programs. The army focused on short- and intermediate-range missiles, the navy concentrated on missiles that could be fired from ships at sea, and the air force worked on the development of the cruise missile and the Atlas intercontinental ballistic missile (ICBM).

The Jupiter Rocket

When a mutual need arose, the military branches conducted joint research efforts. The army and navy were engaged in one such effort, the construction of an intermediate-range missile that could be fired from either land or sea. The army assigned its role in this project to von Braun's Huntsville group. The main product of this joint army/navy effort was the Jupiter series of rockets. The first rocket produced under this program, the Jupiter A, relied heavily on the technology developed for the Redstone.

Work was also pursued on a three-stage rocket, the Jupiter C. This rocket employed an enlarged Redstone for the first stage, a cluster of eleven solid-fueled rockets for the second stage, and three solid-fueled rockets as a third stage. Three Jupiter C test flights were conducted between September 1956 and August 1957 with impressive results. The first test rocket set a new flight record, reaching an altitude of 682

miles and landing 3,400 miles down range from its launch site at Cape Canaveral. The sheer weight and size of the Jupiter, however, made it an impractical missile for the navy. In December 1956 the Navy officially terminated their involvement in this program and focused on the development of a lighter missile, the Polaris.

The 1957 International Geophysical Year

In 1950 scientists from around the world proposed a campaign to gather data about the Earth. To give countries time to prepare, the International Geophysical Year (IGY) was set to begin several years later in July 1957. The Eisenhower administration saw this activity as a way of showcasing American technology and decided to participate. Accordingly, on July 29, 1955, the White House announced that as part of America's contribution to the International Geophysical Year, it would launch a satellite to collect data about the Earth and its near space environment. Not to be outdone by its political rival, on the following day, the Soviet Academy of Sciences announced its intention to place a satellite into Earth's orbit.

America's First Step toward an Earth-Orbiting Satellite

At the direction of President Eisenhower, Donald Quarles, Assistant Secretary of Defense for Research and Development, appointed a committee headed by Dr. Homer Steward, a physicist at the California Institute of Technology, to review satellite proposals from the military branches. Two major satellite proposals were received, one from the army and the other from the navy. The army effort was headed by von Braun's Huntsville

Figure 5-11. Von Braun's marriage to Maria Louise von Quistorp in Landshut.

Figure 5-12. Jupiter C rocket developed by von Braun's group at Redstone Arsenal in Huntsville, Alabama.

group and proposed to use a Redstone rocket with upper stages consisting of Loki rockets to place a 5-pound scientific payload into orbit. This plan, called Project Orbiter, competed against a navy proposal called Project Vanguard, headed by Milton Rosen of the Naval Research Laboratory.

The Rosen group proposed to use a modified version of the navy's Viking rocket. It called for a three-stage rocket. The first stage would use liquid oxygen and kerosene, the second stage nitric acid and dimethyl hydrazine, and the third stage would employ a solid propellant. When completed this multistage rocket would stand 72 feet tall, have a diameter of 45 inches, and weigh 22,600 pounds.

On September 9, 1955, after evaluating the two proposals the committee voted 7 to 2 in favor of Project Vanguard. They realized that the construction of the modified Viking would delay the launch date compared to Project Orbiter, but this was not viewed negatively. The 1957–58 IGY was still several years away and ample time remained for the Viking to be tested. The Steward Committee's decision to procede with Project Vanguard came as a shock to von Braun. He later wrote,

We could hardly believe it. After all, we wanted to launch an American satellite, not an army or a Huntsville satellite. We knew how close the race with the Soviets was and how difficult it would prove for the Vanguard people to make good on their promise, with all their brand new, untried components that had to be developed, including a new rocket motor.

Eisenhower, however, was pleased with this choice. The Viking rocket was to be developed by private industry, whereas the Redstone was the brain child of the military. In Eisenhower's mind this difference was important because it made it less likely that this project would be viewed as a military effort. The selection of a project that was based on a civilian-designed rocket therefore had a great appeal to him. Project Vanguard was officially authorized to begin work on September 9, 1955.

Von Braun tried hard to convince the government to continue work on Project Orbiter as a backup to Project Vanguard. He knew that if a fourth stage were added to the Jupiter C, an orbital launch was within America's grasp. The Department of Defense was concerned by von Braun's continued requests to launch a satellite. When the first Jupiter C was readied for launch on September 20, 1956, the Redstone Arsenal base commander, General Medaris, was instructed to personally inspect the rocket's nose cone to insure that it did not contain an additional stage that might "inadvertently" ignite and place an American payload into orbit.

As work on Vanguard proceeded, von Braun and his Huntsville associates became increasingly nervous that the Soviet Union would beat America into space. Von Braun lobbied his superiors for permission to launch, but his pleas were turned down. He later recalled his frustration, "In various languages our fingers were slapped and we were told to mind our own business, that Vanguard was going to take care of the satellite problem."

Victor Stuhlinger, a close associate of von Braun, also asked General Medaris to once again seek permission for a launch. After listening to Stuhlinger's fears that the Soviet Union might upstage the American effort, Medaris replied:

Now look, don't get tense. You know how complicated it is to build and launch a satellite. Those people will never be able to do it! Through all my various intelligence channels, I have not received the slightest indication of an impending launch. As soon as I hear something, I will act. When we learn something of their activities, we will still have plenty of time to move. Go back to your laboratory, and relax!

6 Sputnik and the American Response

While it is true that the Vanguard group does not expect to make its first satellite attempt before August 1957, whereas a satellite attempt could be made by the Army Ballistic Missile Agency as early as January 1957, little would be gained by making such an early satellite attempt.

—E. Murphree, Department of Defense Special Assistant for Guided Missiles

On October 4, 1957, an international meeting was being held at the National Academy of Sciences Building in Washington, DC, to discuss rocket and satellite plans for the International Geophysical Year (IGY) which had started on July 1. The aim of this global research effort was to acquire more knowledge about the Earth's physical properties and environment. In support of this program both the United States and Soviet Union had pledged to launch an Earth-orbiting satellite and scientists were anxious to learn what, if any, progress had been made toward that goal. However, neither country appeared to have any news to report.

Immediately after the meeting, the Soviets hosted a party for the participants. Several American scientists attended, including Richard Porter of General Electric, William Pickering, head of the Jet Propulsion Laboratory (JPL), Lloyd Berkner, chairman of the National Academy of Science's Space Science Board, James van Allen of Iowa State University, and Herb Friedmann of the Naval Research Laboratory.

Shortly after arriving, Walter Sullivan, a science correspondent for the *New York Times*, received a telephone call from the *Times* night desk which he took in the embassy's hall. He was told that Reuters was reporting that the Soviet Union had launched an Earth-orbiting satellite called *Sputnik 1*. Sullivan was shocked by this news and hurried back to the party. As he reentered the room and looked around, it became apparent no one else knew of the launch, not even the Soviets. He approached Porter, Pickering, and Berkner, and dropped the news like a "hand grenade." After quickly recovering from his surprise, Berkner promptly clapped his hands to gain the attention of the crowd and offered a toast to his hosts for placing the world's first satellite into orbit.

The Soviet delegates seemed as surprised as anyone else by this announcement, and after regaining their composure, they politely accepted the congratulations of the assembled scientists. For the rest of the night, they smiled "like they had swallowed a thousand canaries."

Soviet Events Leading to Sputnik

The fact that the Soviet delegates were unaware of the launch of *Sputnik 1* was not all that surprising. Sergei Ko-

rolev, the genius behind the Russian manned-space program, had only received Khrushchev's approval for this attempt less than a year earlier, in January 1956. At that time Khrushchev was visiting the Soviet intercontinental ballistic missile (ICBM) center at Kaliningrad. Upon completion of his tour, he asked if there were any other issues to be discussed. Korolev broached the idea of using one of the new ICBMs to orbit a satellite and reminded Khrushchev that the Americans had already made this commitment. Khrushchev was attracted to large, high-profile projects, and the thought of gain-

Figure 6-1. The two major superpowers were about to engage in a race for space dominance. The United States was led by President Dwight Eisenhower, and the Soviet Union was led by Nikita Khrushchev.

Figure 6-2. Sputnik 1 *was the first object placed into earth's orbit. It was launched October 4, 1957, and remained in orbit 98 days. The diameter of the satellite was 22.8 inches and it weighed 183 pounds.*

ing the world's attention while simultaneously giving America "a punch in the nose" appealed to him. He gave his permission to proceed provided it did not interfere with the ICBM work.

The rocket that Korolev planned to use to orbit the Soviet satellite was the R-7. But initial tests of this rocket had not been encouraging. Its first flight on May 15, 1957, failed just 50 seconds after lift-off. A second R-7 exploded less than a month later on June 9, and a third failure occurred on July 12. Undeterred, Korolev continued his efforts, and the next flight on August 3 was a success. This was followed by a second successful launch on September 7.

Buoyed by these successes, Korolev mentioned in a Moscow speech commemorating the birth of Tsiolkovsky, that Russia would soon attempt to place a satellite in orbit. The audience sat passively as this announcement was made, apparently not grasping what was being said or dismissing it as propaganda. After this speech, Korolev hurried back to the Russian launch complex at Tyuratam.

On October 4, 1957, an R-7 rocket with the *PS* (Preliminary Satellite) placed in its nose-cone was readied for lift-off. However, each time the moment to launch approached a mechanical problem arose causing a delay. As sunset approached Korolev was growing increasingly frustrated, but resolved to make one last attempt. As he watched from his underground bunker, a lone Soviet bugler walked to the pad

and in clear tones, blew a prelaunch warning. Moments later the R-7 came to life in a blaze of fire, and its light illuminated the surrounding terrain. Workers at the site were elated as the R-7 rose high into the night sky: they embraced, danced, sang, and shouted, "She's off! Our baby is off!" The scene was one of unrestrained joy.

There was only one way to be certain that the *PS* had reached orbit. They must wait and see if its signal could be detected once it had circled the Earth. Korolev and the other site personnel gathered at the rocket assembly hanger and anxiously waited as technicians tuned radio receivers to the 20 MHz frequency at which the *PS* transmitted. As the time for the satellite's reappearance arrived, the signal was received and he instructed that it be broadcast throughout the complex on loud speaker so that all could hear it. There was no doubt in anyone's mind, the *PS* was in orbit! Workers carried Korolev on their shoulders to an improvised platform where he addressed the assembled crowd:

> Today we have witnessed the realization of a dream nurtured by some of the finest men who ever lived, including our outstanding scientist Konstantin Eduardovich Tsiolkovsky. Tsiolkovsky brilliantly foretold that mankind would not forever remain on the Earth. The Sputnik is the first confirmation of his prophecy. The conquering of space has begun. We can be proud that it was begun in our country.

Despite the joy at Tyuratam, the political leadership of Russia did not fully appreciate what had been accomplished. When Korolev telephoned Khrushchev late at night to inform him of their success, Khrushchev listened to the news, congratulated him and his engineers, and then went back to sleep. After it was confirmed to be in orbit, the *PS* was renamed *Sputnik 1*, which in English means *Companion 1*.

In hindsight, the lift-off of *Sputnik 1* on October 4, 1957, should not have come as a great surprise. This day marked the 40th anniversary of the Russian Revolution, and it was understandable for the Soviets to call attention to this fact by attempting to create some incident of international significance.

The American Response to Sputnik

In contrast to the joy felt by the Soviets, American public reaction was swift and negative. Americans had been led to believe that their technology was the most advanced in the world and they were not pleased to see this view challenged. In the wake of World War II, the United States basked in the glory of its new consumer society: television sets, man-made fibers, and an automobile in every driveway. In contrast, the Soviets had experienced catastrophic losses during the war. Over 21,000,000 military personnel and civilians had been killed and 70,000 towns and villages had been destroyed. It was inconceivable to the average U.S. citizen that a war-ravaged country could upstage the world's most technologically advanced nation. Little solace was found in the fact that an American program to place an object in orbit was underway. In the harsh Cold War environment of the late 1950s, coming in second was the same as coming in last, and this type of finish had potentially serious consequences.

Americans from all walks of life expressed their shock and anger at this Soviet accomplishment. It was hard enough to accept the fact they had been humbled by the Soviets, but the constant radio signal broadcast from *Sputnik 1* could be heard by anyone using a simple radio receiver. Its signal was a steady, humiliating reminder that America had been beaten.

Congressional leaders in the United States were particularly angered by the success of Sputnik. U.S. Senator Henry Jackson of the state of Washington described the launch of *Sputnik 1* as a "devastating blow" to the United States and called upon President Eisenhower to proclaim "a week of shame and danger." Senate majority leader Lyndon Johnson expressed a new and troubling doubt about America's leadership in the world, a view that was shared by many others when he said,

> That sky had always been so friendly, and it had brought us beautiful stars and moonlight and comfort; all at once it seemed to have some question marks all over it. . . . I guess for the first time I started to realize that this country of mine might perhaps not be ahead in everything.

Johnson saw *Sputnik 1* as more than a slap in the face of the United States. He believed it presented a real security threat. To yield such an unassailable observation post to an avowed enemy was unthinkable. He was ready to accept this Soviet challenge and openly expressed this intention to force an American response, "Maybe it was all right with others in the government, but he for one didn't care to go to bed by the light of a Communist Moon."

While many Americans agreed with Johnson's sentiment, others held a more restrained view: "They have their Germans, and we have our Germans, and our Germans are behind their Germans. That is all there is to it." The American scientific community was also shaken by the launch of Sputnik but was confident America still enjoyed a sizable technological lead over the Soviet Union. Recalling *Sputnik 1*, Pickering said,

> Until Sputnik flew, neither the Russians nor the Americans understood the importance of it. And then all of a sudden Sputnik was up there, and it scared the daylights out of the Americans to realize that this Russian thing was going over, happily beeping away; anybody could hear it. And there wasn't a thing they could do about it. Those peasants over there had put this thing up while we were fooling around. . . . And so it was natural, I think, that the Americans would respond by saying, "Okay, you wait, we'll do better!"

Wernher von Braun, whose Huntsville team had been repeatedly frustrated in their efforts to place an American satellite in orbit, was not surprised by the Soviet triumph. In his opinion, there had been ample warnings an attempt was imminent. The U.S. government had simply chosen not to take them seriously. At a presentation in Washington, DC, on October 29, 1957, a few weeks after the successful launch of *Sputnik 1*, von Braun expressed sadness for the fact that the United States had been beaten into space and attempted to rally support for the American program:

> October 4, 1957, the day when Sputnik appeared in the skies will be remembered on the planet as the day on which the age of spaceflight was ushered in. For the United States, the failure to be the first in orbit is a national tragedy that has damaged the American prestige

Focus Box: As *Sputnik 1* orbited the earth it transmitted a radio signal at 20 MHz. This signal was strong enough that it could be heard using a modest radio receiver. The figures below show the pulse pattern of the radio signal, and Moscow school children listening to this signal.

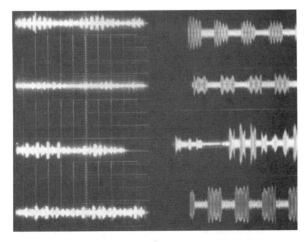 

Moonglow

Figure 6-3. Cartoonists contrasted Eisenhower's passion for golf versus the Soviet focus on space. Here Khrushchev is seen in golf attire as Sputnik 1 *flies overhead (Simon and Schuster, 1958).*

Figure 6-4. Cartoonists portrayed an America caught napping by the launch of Sputnik.

around the globe. . . . We failed to recognize the tremendous psychological impact of an omnipresent artificial Moon, visible to anyone with a simple radio receiver. . . . The real tragedy of Sputnik's victory is that this present situation was clearly foreseeable two years ago, when the separate U.S. satellite program was established. . . . We have lost a battle, and we may lose a few more, but we have not yet lost the war.

Adding insult to injury, the American press portrayed *Sputnik 1* as the "Pearl Harbor" of the space race. By equating its launch to the relatively recent devastating surprise attack on America, the implications of being caught napping again were put into an unsettling context. Most news reports also gave the impression that Sputnik had somehow given the Soviets a military advantage that had to be countered.

In the days following the Sputnik launch, American anxiety grew as news organizations continued to make unflattering comparisons between the American and Soviet space programs. A wave of mortification, frustration, and deep-felt anger swept over the nation. As in the aftermath of Pearl Harbor public opinion quickly converged on the conclusion that prompt action was needed. An irresistible demand arose for the government to act and to act quickly!

President Eisenhower's Response to Sputnik

Prior to *Sputnik 1*, there was little political pressure to launch an American satellite. While the Army Ballistic Mis-

sile Agency had the technology to place an object in orbit as early as January 1957, there was no sense of urgency to do so. Even after the success of *Sputnik 1*, President Eisenhower did not see the need of accelerating the American space effort. He was genuinely baffled by the reaction of his countrymen and expressed his puzzlement: "I can't understand why the American people have got so worked up over this thing."

Unlike Senator Johnson, Eisenhower did not see Sputnik as a security threat and refused to view it as a potential weapon. As a former military commander, he realized that in this context Sputnik was inconsequential: "One small ball in the air does not raise my apprehensions, not one iota." He thought of the Soviet feat as little more than a publicity stunt. Despite these feelings, Eisenhower was soon compelled by a "wave of near-hysteria" to respond. He told his Cabinet, "We can still destroy Russia. We know it. We've got to convince the world we're doing well."

In reaction to the public outcry, the Eisenhower administration announced the United States would launch its own satellite by December. The American press seized on this proclamation and promoted the upcoming launch as America's answer to the Soviet challenge.

A Second Soviet Satellite: *Sputnik 2*

While the United States was busy preparing for the launch of its first satellite, the Soviet Union placed a second satellite, *Sputnik 2*, into orbit on November 3, 1957. This craft was considerably more advanced than *Sputnik 1* and possessed scientific instruments to measure solar radiation and the intensity of cosmic rays. What really caught the public's attention, however, was the fact that it carried a passenger, a dog named Laika (see Figure 6-5). Included in *Sputnik 2's* payload was life-support equipment, and instruments to monitor the animal's pulse, blood pressure, respiration, and movement. The satellite weighed an astonishing 1,120 pounds. American rocket scientists were impressed by this lift capability and began to admit that they might have underestimated the Soviets' level of technology. This admission came as no surprise since it was now obvious to everyone that their rocket program was quite advanced. To American political leaders, the fact that Laika was flown was proof that the Soviets intended to place a human into orbit.

America Responds to the Soviet Launches

By late 1957, the first three Viking test flights to validate various components of the rocket had been successfully completed. The entire rocket had not been flown when the orbiting of *Sputnik 1* prompted President Eisenhower to move the planned launch date forward to December. In reaction to this announcement, work on Project Vanguard was accelerated.

The first multistage Viking blasted off from Cape Canaveral on October 23, 1957, but only its first stage was fueled, the upper two stages were empty. This successful flight lifted a 4,000-pound payload to a height of 109 miles and propelled it 335 miles down range. It was a promising start. Under normal circumstances, additional Viking flights would have taken place to verify the operation of the entire rocket, but the time for testing had run out.

On December 3, 1957, at the invitation of the Eisenhower administration, the world press came to Cape Canaveral to witness the lift-off of America's first satellite, *Vanguard*. American political leaders wanted to contrast the open manner in which this flight took place to the secret manner in which the Sputniks were launched. Unfortunately, another difference between these launches would also soon be apparent. As the press watched, the Viking rocket rose just three feet off its pad and then collapsed back to Earth in

Figure 6-6. Sputnik 2 *could be seen orbiting the earth with the unaided eye. This photograph showing the path of* Sputnik 2 *was taken by Paul Donaldson on November 7, 1957, from Arlington, Massachusetts.*

an inferno of flames. To add to the embarrassment of this scene, the 3-pound, grapefruit-sized *Vanguard* broke free from the top of the exploding rocket and lay pitifully beeping by the wreckage.

Press reports of this fiasco were merciless, referring to the destroyed American mission as the "Kaputnik," "Stayputnik," "Puffnik," "Dudnik," and "Flopnik." The editors of *Nation* wrote, "We managed so successfully to focus the eyes of the world on the effort that when it exploded, the whole world was watching." To say that the launch was a huge embarrassment was an understatement.

A comparison between the American and Soviet programs was indeed revealing—*Sputnik 1* orbited a payload of 183 pounds, and just about a month later *Sputnik 2* orbited 1,120 pounds and carried a living dog as well as scientific instruments. *Vanguard* weighed a puny 3 pounds, and its Viking rocket barely left the ground. Political pressure immediately mounted for a successful launch to prove to the world that America was still competitive in the space race.

Explorer 1

After the Soviet launch of *Sputnik 1*, Army Secretary Wilber Brucker once again approached the Eisenhower administration and offered to have the Huntsville group place a satellite in orbit. This time he found the government much

Figure 6-5. Sputnik 2 *carried a live dog named Laika. There were no plans to recover this animal, it died in space.*

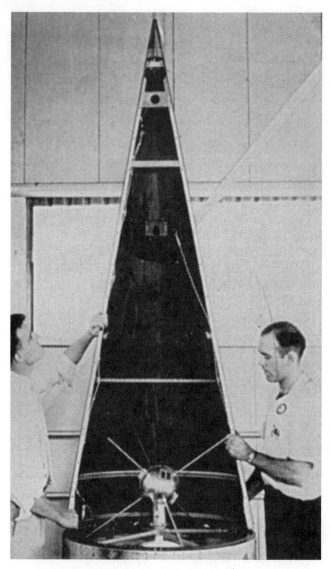

Figure 6-7. Vanguard *was a small satellite having a diameter of only 6 inches. It weighed 3 pounds and was encased in a cone-like structure that was attached to the top of the Viking rocket.*

Figure 6-8. America's first attempt to place an object into orbit ended in a catastrophe when the Viking rocket blew up shortly after liftoff.

more receptive and approval of the program was given on November 8, 1957, just five days after the launch of *Sputnik 2*.

Since much of the hardware needed for von Braun's Juno rocket had already been built, and many of the key personnel were stationed in Huntsville, work proceeded rapidly. The Jet Propulsion Laboratory was given responsibility for the construction of the satellite, tracking and ground support functions, and for supplying the solid-fueled Sergeant rockets that would propel the Juno's three upper stages. This cooperative arrangement was successful, and on January 31, 1958, one month ahead of schedule, *Explorer 1* was placed into orbit atop Juno 1.

Explorer 1 was a fairly modest satellite, measuring 80 x 6 inches and weighing about 31 pounds, 11 pounds of which were devoted to a scientific package. The satellite was placed into a very elliptical orbit, with an apogee of 1,580 miles and

a perigee of 233 miles. It would remain in orbit around the Earth for 12.3 years.

After *Explorer 1* was successfully blasted into orbit, the three principle architects of the mission, von Braun, van Allen, and Pickering, were ushered into the Great Hall of the National Academy of Sciences building in Washington, DC, for a brief meeting with the press. Here these scientists released the built-up frustration and tension associated with America's belated entry into space by jubilantly holding a replicate of *Explorer 1* above their heads.

The 31-pound weight of *Explorer 1* was considerably less than either of the Soviet Sputniks, but it carried a scientific package of three instruments. One instrument was designed to measure the temperature of the rocket and its environment, the second recorded the impact of micrometeorites, and the third measured the intensity of charged particles in the satellite's environment. The last of these experiments, developed by James van Allen of Iowa State University, produced the first major discovery of the space age.

The van Allen Radiation Belts

Based upon data obtained from rockets sent high into the Earth's upper atmosphere (sounding rockets), van Allen expected that the cosmic ray flux would simply increase with altitude and then level off once outside the Earth's atmosphere in a rather predictable manner. He was therefore quite surprised when his instrument on *Explorer 1* revealed the presence of an intense region of radiation that circled the Earth. This result was reported at a joint symposium of the American Physical Society and the National Academy of Sciences on May 1, 1958. In his address, van Allen said:

> Above an altitude of about 1,000 kilometers we have encountered a very great increase in radiation intensity which is vastly beyond what could be due to cosmic rays alone. . . . In fact, it is of the order of, or greater than, 1,000 times the intensity of cosmic rays as extrapolated to these greater altitudes.

Although the information provided by *Explorer 1* allowed the radiation belts to be discovered, this flight was not sufficient to characterize them. Fortunately, additional flights had already been planned. *Explorer 2* was launched on March 5, 1958, but failed to reach orbit when its fourth stage did not ignite.

Three weeks later *Explorer 3* was launched and confirmed the existence of these belts and collected additional measurements. Since the laws of physics state that charged particles flow along magnetic field lines, van Allen was able to use data from *Explorers 1* and *3* to map the size, shape, and strength of the Earth's magnetic field, a feat that had never before been possible.

Vindication but the Troubling Consequences of van Allen's Discovery

With the launch of *Explorers 1* and *3* America made two major discoveries, the presence of intense radiation belts surrounding the Earth, later called the van Allen radiation belts, and the first mapping of the Earth's magnetic field. William Pickering had vowed that America would soon make its mark in space, and thanks to the Explorer program and its German-American scientists, he was proven correct.

The discovery of high levels of radiation outside of the Earth's atmosphere provided useful information about our planet. However, their discovery raised serious questions about the future of manned spaceflight itself. How extensive were these regions, and could a human survive in this environment? If the high level of radiation found by van Allen was representative of that everywhere in space, the dream of

Figure 6-9. The Jupiter C, developed by von Braun's group (the Army Ballistic Missile Agency) in Huntsville, Alabama, placed America's first satellite, Explorer 1, into orbit on January 31, 1958.

Figure 6-10. Left to right, Pickering, van Allen, and von Braun celebrate the successful launch of the satellite by holding a replica of Explorer 1 above their heads at a Washington press briefing.

Tsiolkovsky, Oberth, and Goddard of placing a man in space would be impossible.

Rocketry: A New Cold War Measuring Stick

The Soviet Union gained a great deal of prestige as a result of the launch of *Sputnik 1*. In a rapid turnabout, many countries now viewed the Soviet Union as a technological

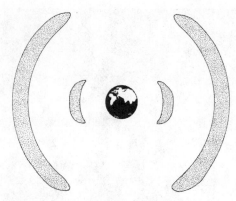

Figure 6-11. Explorer 1 *transmitted data back to earth that revealed the presence of high levels of radiation.*

equal of the United States. Their stature was further enhanced by the launch of *Sputnik 2* and the subsequent *Vanguard* fiasco. People everywhere began to view the Soviet Union with renewed respect. American political leaders were concerned about this situation because of its potential impact on the formation of political alliances in the Cold War.

There is little doubt that America had the capability to launch the world's first satellite several months before *Sputnik 1*, but the Soviet Union probably could have launched a satellite sooner as well. In the mid-1950s, neither government considered this to be a particularly important goal. Khrushchev's permission to Korolev was given in an almost offhanded manner and Eisenhower certainly did not view the launch of a satellite as a priority. Neither leader foresaw the impact that an Earth-orbiting satellite would have on people around the world, and both were astonished at the world's reaction. Nonaligned countries were quick to adopt rocketry as a way of judging the technological superiority of the two superpowers. This fact was recognized by a young senator from Massachusetts, John F. Kennedy. In a 1961 speech he spoke of the importance of being preeminent in space,

It seems to me, and this is most dangerous for all of us, that we are in danger of losing the respect of the people of the world. . . . The people of the world respect achievement. For most of the 20th century they admired American science and American education, which was second to none. But now they are not at all certain about which way the future lies. The first vehicle in outer space was called Sputnik, not Vanguard.

7 The Space Race Begins in Earnest

I believe this nation should commit itself to achieving the goal, before this decade is out, of landing a man on the Moon and returning him safely to Earth.

—President John F. Kennedy, May 25, 1961

The last chapter discussed how the launch of *Sputniks 1* and *2* dramatically affected American political policy. In Eisenhower's view these flights posed no military threat to the United States and were little more than publicity stunts; he was inclined to dismiss them. However, the reaction in Congress and across the country was so intense he soon found himself under enormous pressure to respond. Since the launch of Laika aboard *Sputnik 2* suggested the Soviet Union's goal was to place a man into orbit around the Earth, Eisenhower felt compelled to match that feat by launching an American into space. In order to accomplish this he decided to consolidate America's fragmented space effort into a single agency. The question was not whether such a reorganization would take place, but rather where the various units would be collected.

Three agencies were considered: the Department of Defense, because of its Huntsville/navy programs, the Atomic Energy Commission, because of its work on nuclear propulsion, and the National Advisory Committee for Aeronautics (NACA), because of its long history of aviation research. Among these three organizations, NACA appeared to be the logical choice. Eisenhower did not want to extend the Cold War into space, so assigning the nation's space program to the Department of Defense did not make sense. Such a move might be interpreted by the Soviet Union as a military challenge and they might respond by developing space-based weapons. The Atomic Energy Commission was a relatively new agency and did not possess much expertise with either aircraft or rocket technology. If this task were assigned to them, a major change in their direction would be required. In contrast, the assignment to NACA seemed like a natural fit. This agency had been designing aircraft for several decades and during the 1950s it had also worked with missiles. In addition, NACA had already developed a solid, long-term strategy for aeronautic and space research.

After considering these alternatives, the Eisenhower Administration formally requested that Congress assign America's space effort to NACA. The bill enacting this change (PL 85-568) passed through Congress with little debate, and on July 29, 1958, Eisenhower signed it into law. NACA was officially renamed the National Aeronautics and Space Admin-

Figure 7-1. President Eisenhower, a retired Army general, fought to place America's space program under a civilian agency.

istration (NASA) and began operations on October 1, 1958, the start of the federal government's next fiscal year.

Dr. T. Keith Glennan, President of Case Institute of Technology and a former member of the Atomic Energy Commission, was selected to head the new agency, and Dr. Hugh Dryden, former Director of NACA, was appointed Deputy Chief Administrator. NACA's facilities and 8,000 employees as well as several existing military programs were transferred to NASA. These included the Langley Memorial Aeronautical Laboratory, the Ames Aeronautical Laboratory, the Lewis Flight Propulsion Laboratory, the Wallops Island Pilotless Aircraft Station, the Muroc Flight Test Unit, Project Vanguard, the Army ABMA program in Huntsville, and Air Force programs on lunar probes and rocket propulsion.

Glennan was anxious for NASA to commence operations as soon as possible. Just one week after the agency came into existence, he authorized the start of Project Mercury, America's first manned spaceflight program. In February

Figure 7-2. Eisenhower (center) presenting commissions to NASA newly appointed Deputy Director Hugh Dryden (left) and Chief Administrator T. Keith Glennan (right).

1960, NASA submitted its first long-range plan to Congress. In keeping with its NACA heritage, the plan called for continued aeronautical research, but also for the development of huge rockets that could place massive objects into orbit. These "heavy-lift" rockets would be used to launch Earth-orbiting environmental and weather satellites, lunar and planetary probes, and manned Earth and Moon orbiting missions. NASA's plan included an annual budget of 1 to 1.5 billion dollars for each of the next ten years. America's new space agency was off to an ambitious start.

The X-15 Program

When NACA officially became NASA, it was already conducting a research program involving spaceflight and high-speed aircraft. The heart of this effort was the *X-15*. This remarkable aircraft was capable of a top speed of 4,534 miles per hour, or 6.72 times the speed of sound (Mach 6.72).

Unlike past efforts that tested aircraft designed and built by others, the *X-15* was a NACA project from its beginning. In this regard, it represented a transitional program, changing NACA from a testing organization into an agency with a research, development, and management capability.

Figure 7-3. The X-15 *was capable of flying at very high speeds in the thin upper atmosphere of the earth.*

Several of the procedures and equipment developed in the X-15 program would be carried over to manned spaceflight. These included the criteria a craft had to meet before it would be considered safe enough to be flown by a man (to be "man-rated"); a system of pilot training using simulators; the high-altitude pressure flight suit; and tracking stations to monitor long-distance flights. The X-15 program also collected useful data on high-speed aerodynamics and special alloys of titanium and Inconel-X that were capable of withstanding temperatures as high as 1300° F. The *X-15* was a transitional craft, it had wings to fly in the atmosphere, but employed reaction controls to maneuver in space. It was also used to test various atmospheric reentry procedures and conduct-limited Earth science experiments.

The New Space Program

Because of its new role in space research, NASA had to greatly expand its scope of operations. New test facilities and powerful new rockets had to be built. The meaning of long-distance flight also changed from missions covering a few thousand miles to orbital flights. In order to maintain contact with these craft, NASA had to place tracking sites around the world equipped with antennae, telemetry equipment, computers, radio and land communication devices. In addition, a more sophisticated contracting office needed to be established to ensure these projects ran smoothly and that the collected data were properly analyzed.

The construction of heavy-lift rockets occupied much of NASA's early years. These new rockets included (1) the Scout, a low-cost, solid-propelled rocket designed to put small payloads into orbit; (2) the Centaur, a multistage, liquid-fueled rocket used for planetary and lunar missions; (3) the Saturn, a heavy-lift rocket; and (4) the Nova, a development rocket intended for future manned lunar flights. Rockets available for immediate use included the Redstone, Thor, and Atlas. While NASA pursued work on these launch vehicles, the Atomic Energy Commission continued its research into a nuclear-powered upper-stage rocket, the Nerva.

The Military Pioneer Program

A few weeks before the Eisenhower administration submitted its bill to Congress to create NASA, the Department of Defense created the Advanced Research Projects Agency (ARPA). The goal of this program was to provide seed money for innovative projects to enhance the effectiveness of the armed services. Soon after being established, ARPA funded five rocket launches to explore the Earth–Moon environment.

Although the military need for data about the Earth space environment was, at best, questionable, ARPA argued that these missions would improve the technology used in America's ICBMs. In reality, ARPA feared that the formation of a civilian agency, like NASA, might preclude future military space research. The military viewed this situation with concern. If the Soviets were to develop space-based weapons,

Figure 7-4. The "Thor" series of boost vehicles often proved to be unreliable. Used by the air force, all four of their attempted Pioneer launches failed.

Figure 7-5. The Juno 2 was used as a replacement for the Thor series boosters in the later Pioneer launches. Note the characteristic ICBM shape that American rockets of the period possessed.

America and its allies would be defenseless. These flights were an effort to place the military and NASA on a more equal footing.

ARPA called their five-mission program Project Pioneer and awarded three lunar launches to the air force and two to the army. Space Technology Laboratories (later to become TRW) constructed the satellite payload for the air force. Its 30 × 29 inch payload contained 86 pounds of scientific instruments, a television camera, a transmitter and antenna, and a small retro-rocket to slow the satellite's speed so that it could enter lunar orbit. JPL built a more modest probe for the army. It consisted of a slow scanning television camera, a magnetic tape recorder, and a transmitter to send images back to Earth.

The first Pioneer flight, an air force launch, took off from Cape Canaveral on August 17, 1958, using a Thor-Able rocket. It was an inauspicious start. A little over a minute after leaving the pad, the rocket blew up when a turbo-pump in its motor failed. Perhaps appropriately, this launch was designated *Pioneer 0*.

Pioneer 1, the second Air Force attempt, was launched from Cape Canaveral about two months later on October 11, 1958. After the Thor-Able rocket successfully jettisoned its first and second stages, its third stage cut off sooner than expected and the fourth stage failed to ignite because the craft's interior temperature was too cold for its mercury batteries to work properly. As a result, the probe failed to reach escape

velocity. After climbing to an altitude of over 70,000 miles and returning some useful data about the van Allen radiation belts, *Pioneer 1* reentered the Earth's atmosphere and disintegrated. The entire mission lasted forty-three hours.

Pioneer 2 was the last air force ARPA launch. It lifted off on November 8, 1958, less than three months after the launch of *Pioneer 0*. This probe also failed to reach the Earth's escape velocity because of a third-stage ignition problem. It reached an altitude of only 963 miles, and then like its predecessor, reentered and disintegrated in the Earth's atmosphere. This flight ended the air force portion of the Pioneer program.

Pioneer 3 was the first of the army's two ARPA-funded Moon missions. Launched using a army Juno rocket on December 6, 1958, about eleven months after the Vanguard disaster, its mission was to explore the space environment of the Earth and Moon. The mission failed when the first stage of its carrier rocket prematurely shut down. Although the first-stage burn time was only four seconds shorter than planned, the probe did not achieve escape velocity. After reaching a lofty height of nearly 64,000 miles, the probe reentered and disintegrated in the Earth's atmosphere. The total duration of the flight was 38.6 hours. This flight was not, however, a total failure. It provided some useful data about the inner van Allen radiation belt surrounding the Earth and discovered a second belt located about 10,000 miles above the Earth's surface.

Figure 7-6. Luna I was the first man-made craft to achieve a solar orbit. The satellite is pictured inside the top cone.

Pioneer 4, the last of the two army ARPA experiments, was launched on March 3, 1959, by *Juno 2*. Like *Pioneer 3*, its payload also weighed 13 pounds and had dimensions of 20 × 9 inches. Its mission objectives were (1) to achieve a proper Earth–Moon trajectory, (2) obtain radiation data using two Geiger-tubes, (3) provide ground crews with experience tracking space vehicles, and (4) acquire the Moon in its radiation sensing scanner.

This was the most successful of the Pioneer flights. Forty-one hours and thirteen minutes after launch, it passed within 37,300 miles of the Moon and then went into a solar orbit. It discovered that the amount of radiation beyond the outer van Allen belt fell below the value considered dangerous to a human. Ground control stations tracked the probe for 82 hours and 4 minutes to a distance of 407,000 miles. On the negative side, the craft did not pass close enough to the Moon for its light scanning sensor to activate.

Despite the secondary successes just described, on the whole the Pioneer program provided very little data about the near-Moon environment. Rather than the triumph the military had hoped for, these launches fell far short of expectations. The Eisenhower administration continued to decrease the military's role in space research and soon transferred many of its remaining rocket programs to NASA. The military would never again challenge NASA's role as the country's premier space agency.

Soviet Space Activities

While the American Pioneer program performed in a lack-luster fashion, the Soviet effort continued to make significant progress. On May 15, 1958 *Sputnik 3* was launched with a payload weighing more than a ton! Instrumentation on this satellite measured the solar and cosmic radiation field, the Earth's magnetic field, upper atmospheric ion density and composition, and the near-Earth micrometeorite intensity.

In addition to this instrument package, the Soviets placed a recorder on *Sputnik 3* to store data until it could be downloaded when the craft passed over the Soviet Union. This approach differed from the scheme being developed by the United States which relied on a worldwide system of ground stations. *Sputnik 3* would continue to orbit the Earth for two more years.

About two months prior to the launch of *Pioneer 4,* the Soviet Union sent *Luna 1* toward the Moon on January 2, 1959. When this craft was about 3,600 miles above the Moon, a cloud of sodium gas was released that could be detected by telescopes on Earth. *Luna 1* then passed by the Moon and went into solar orbit. The Soviets launched a second lunar probe, *Luna 2,* on September 12, 1959. This craft measured the radiation levels in the van Allen belts, proceeded to the Moon, released a second cloud of sodium gas, and then impacted on the Moon. It was the first man-made object to hit the Moon.

The Soviets continued to make notable advances in their space program. To mark the second anniversary of *Sputnik 1,* *Luna 3* was launched on October 4, 1959. This probe successfully reached the Moon, took pictures of its back side, and then transmitted them back to the Earth. Claiming the honor of the discoverer, the Soviets named the newly revealed lunar features. The public was soon barraged by news reports about the Sea of Moscow, the Gulf of Cosmonauts, the Tsiolkovsky crater, and the Sovietsky Mountain Range.

American political leaders believed this latest Soviet accomplishment portended a serious threat to the nation. Senator Lyndon Johnson, who was still tormented by the embarrassment of *Sputnik 1*, called yet again for an increased American space effort.

State of the American Space Program

A major goal of NASA's initial work was to improve the reliability of its boosters. Despite Senator Johnson's efforts to dramatically increase funding in this area, these rockets frequently failed to perform satisfactorily. By 1959 the United States had attempted 37 satellite launches, but less than a third reached orbit. Even when successful, the weight of the satellites placed in orbit were unimpressive when compared to their Soviet counterparts.

NASA engineers were faced with an enormous amount of work. The stresses placed on the boosters during flight had been underestimated and stringent manufacturing constraints would also have to be developed, adopted, and imposed before the rate of successful launches could be expected to increase. Getting into space was proving to be much more difficult than supporters of the space program had anticipated.

8 The Early Stages of Project Mercury

We set sail on this new sea because there is new knowledge to be gained and new rights to be won, and they must be won and used for the progress of all people.

— President John F. Kennedy, September 12, 1962

In September 1958 an agreement was reached between NASA and ARPA to form a committee, the Joint Manned Satellite Panel, whose purpose was to formulate a strategy to move America's manned space program forward. The steering committee consisted of eight people, six from NACA/NASA and two from ARPA. The plurality of nonmilitary people reflected Eisenhower's philosophy that the exploration of space should be restricted to peaceful purposes.

NASA had just been created and did not have any experience in the field of manned spaceflight. In contrast, the armed services had devoted considerable time to this endeavor. "Man in Space Soonest" was a four-part air force plan whose goal was "to achieve an early capability to land a man on the Moon and return him safely to Earth." The estimated cost of this project was $1.5 billion and it was expected to be completed by 1965. The navy proposed to place a man in orbit around the Earth as part of its Manned Earth Reconnaissance program. It envisioned using a two-stage rocket and a novel spacecraft that changed into a delta wing glider once in orbit. The army had the least ambitious plan. Its effort, called Project Adam, consisted of a suborbital flight using a modified Redstone rocket. The simplicity of the proposed project was also its strength. The air force and navy plans required considerable research and development efforts whereas the army proposal offered a quick way of reaching space. Despite this appeal, it had a powerful opponent, namely Dr. Hugh Dryden (1898–1965), Director of NACA and later NASA Deputy Administrator. In testimony before the House Space Committee he spoke disparagingly about Project Adam, stating that "tossing a man up in the air and letting him come back . . . [had] about the same technical value as the circus stunt of shooting the young lady from the gun. . . ."

In August 1958 responsibility for placing a man in orbit was officially assigned to NASA. This decision prompted Deputy Secretary of Defense Donald Quarles to terminate the military's involvement in this effort and transfer all such funding to NASA. Nevertheless, the Joint Manned Satellite Panel played an important role in directing NASA's future efforts. As a result of the panel's deliberations, a three-stage plan, "Objectives and Basic Plan for Manned Satellite Project" was developed. The first stage involved testing space hardware on the ground to the point where it consistently performed satisfactorily. The next step required testing of the entire craft during an unmanned flight. The final phase of the plan called for a manned mission. In the panel's view, this scheme offered the quickest and safest route to space. It would serve as the standard operating procedure for all of NASA's subsequent manned space programs.

The Beginning of Project Mercury

In November 1958, NASA established a formal program to place a man in space that would ultimately be called Project Mercury. Named after the wing-footed messenger of the gods in Greek mythology, its primary goal was to place a man in orbit around the Earth. Responsibility for this effort was given to Robert Gilruth of NASA's Langley Aeronautical Laboratory, who in turn, created a working committee called the Space Task Group. The first major task of this group was to select McDonnell Aircraft Corporation as the prime contractor of the single-person Mercury capsule and General Dynamics as the builder of the Atlas rocket, the booster that would place America's first person in space.

Figure 8-1. The Atlas rocket being assembled at the Convair Astronautics Plant in Sorrento, California.

Initial Steps toward Spaceflight

In early 1959 NASA faced three major safety issues related to manned spaceflight. Identified in a NASA report to Congress titled "Ten Years in Space, 1959–1969," they involved (1) the extent and intensity of the high-energy radiation discovered by van Allen as part of the International Geophysical Year, (2) the effects of weightlessness on the human body, and (3) the level of destructive heat that a spacecraft would generate during atmospheric reentry. Answers were needed in each of these areas. Was the radiation level outside the Earth's atmosphere prohibitively high? Could the human body function in a weightless environment? How would an astronaut return to Earth?

NASA made rapid progress in addressing each of these concerns and by the middle of 1959 had arrived at some partial answers. To avoid the problem posed by the presence of high energy radiation, manned craft would be placed in orbits that were lower than 120 miles, far below the height at which the radiation reached harmful levels. The second area was addressed by implementing a rigorous test program to determine whether pilots could withstand the stresses associated with rocket flight. Before humans would be committed to a flight, animals would be flown to determine the physiological effects of weightlessness on living objects. Finally, NASA officials were well aware of the need to protect the craft from frictional heating because of its work in the X-15 program. An extensive effort was implemented to incorporate this knowledge in the design of the Mercury capsule.

The Mercury Capsule

Because of its heritage as a former aircraft development agency, NASA possessed a great deal of knowledge about high-speed flight. In order to reach high speeds, these aircraft had sleek, aerodynamic shapes. Although these designs were well suited for atmospheric flight, the Mercury capsule would fly in a different environment—the vacuum of space. In this environment an aerodynamic shape had no special value. Therefore NASA experimented with a number of unconventionally shaped vehicles. Harvey Allen of Ames Laboratory and Max Faget of Langley were the first to propose the "blunt body" design that eventually was adopted for the capsule. Its unusual shape invited criticism, Gilruth recalled:

> Some people felt that such a means for flying man in space was only a stunt. The blunt body in particular was under fire since it was such a radical departure from the airplane. It was called by its opponents, "the man in the can," and the pilot was termed only a medical specimen.

The unusual bell-like shape of the Mercury capsule turned out to be practical because the craft was not actually intended to fly in the atmosphere. After completing its mission, the astronaut would fire a set of small rockets, called retrorockets, attached to the bottom of the capsule. These rockets would slow

Figure 8-2. A McDonnell test pilot demonstrates how to climb out of a Mercury capsule.

the craft's speed, causing it to descend back to the Earth. As it passed through the atmosphere, parachutes, stored in the top portion of the craft, would be released to slow the capsule's velocity to a level that was safe for a manned landing.

The capsule was designed to land in water or on the ground. Under normal conditions it would land in the ocean and had to be watertight. However, if an emergency arose while in orbit, the capsule might be forced to land on the ground. In that case, it had to be sturdy enough to withstand the impact.

Outwardly, the Mercury capsule was surprisingly tiny (see Figure 8-2), just 11 feet tall with a maximum diameter of 6 feet. Its small size was necessary because of the modest lift capability of the American rockets. At that time, it was simply not possible to place a more substantial object into orbit. Considering the amount of space taken up by instruments, it was hard to believe that a person could actually fit inside it. It was a far cry from the spaceships that the rocket pioneers or science fiction writers had envisioned.

Astronaut Selection

Once the decision was made to place a person in orbit, pilots were needed to fly the Mercury capsule. In deference to previous explorers, these people would be called astronauts for they were to "sail into a new, unchartered ocean." The Langley Space Task Group established a number of criteria for the selection of these individuals, but before the plan was finalized, President Eisenhower vetoed the entire selection process. Much of the technology used in the Mercury program had a military origin, and he was concerned that opening the astronaut corps to civilians posed a serious security risk to the nation. To avoid this possibility, he directed that the astronauts be chosen from among military test pilots. The armed services gladly cooperated with this effort and a check of personnel records found over 100 suitable candidates.

The Space Task Group readily accepted Eisenhower's decision to restrict the astronaut selection to test pilots, but not for security reasons. The group felt strongly that the success of a space mission might well depend on the intercession of the astronaut, and military test pilots had been trained to handle in-flight emergencies. In *This New Ocean: A History of Project Mercury,* a similar rational is given, "The astronauts were first and foremost test pilots, men accustomed to flying alone in the newest, most advanced, and most powerful vehicles this civilization had produced. They were talented specialists . . . who had survived the natural selection process in their profession." Their prior flight experience would also allow them to make valuable suggestions on how the capsule could be improved. These skills would be especially valuable in a program where the performance of the flight hardware was unknown.

A search of 508 military service records identified 110 pilots who passed the initial NASA screening process. By the middle of February 1959 a group of these pilots were invited to Washington for an interview. NASA was pleased to find that the majority of them were interested in joining the Mercury program. As one astronaut candidate remarked, "How could anyone turn down a chance to be a part of something like this?" Based on this strong response, the decision was made not to interview any further candidates. George Low explained the reason for this:

> During the briefing and interviews it became apparent that the final number of pilots should be smaller than the twelve originally planned. The high rate of interest in the project indicates that few, if any, of the men will drop out during the training program. It would, therefore, not be fair to the men to carry

Figure 8-3. The first American astronauts. From left to right: Carpenter, Cooper, Glenn, Grissom, Schirra, Shepard, and Slayton.

along some who would not be able to participate in the flight program. Consequently, a recommendation has been made to name only six finalists.

Through a series of written tests, additional interviews, medical exams, and physical endurance tests, the initial group of sixty-nine pilots was cut to eighteen. The NASA selection committee found it extremely difficult to narrow this field further and eventually recommended that seven be selected. On April 9, 1959, NASA Director T. Keith Glennan introduced these men to the nation: John H. Glenn, Jr.; Donald K. Slayton; M. Scott Carpenter, Jr; Virgil I. Grissom; Walter M. Schirra, Jr.; Alan B. Shepard; and L. Gordon Cooper, Jr.

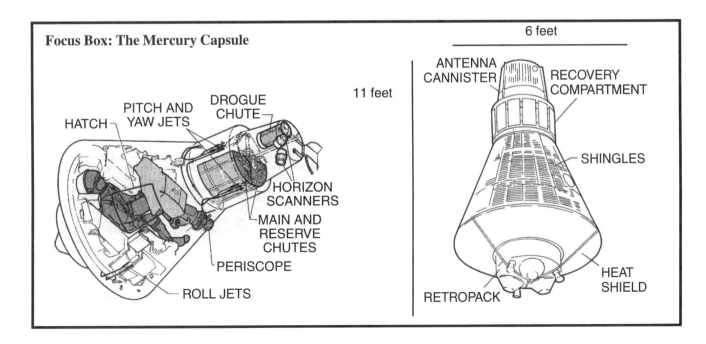

Focus Box: The Mercury Capsule

HATCH
PITCH AND YAW JETS
DROGUE CHUTE
HORIZON SCANNERS
MAIN AND RESERVE CHUTES
PERISCOPE
ROLL JETS

11 feet

6 feet

ANTENNA CANNISTER
RECOVERY COMPARTMENT
SHINGLES
HEAT SHIELD
RETROPACK

Figure 8-4. NASA's Big Joe *flight tested the effectiveness of the Mercury capsule's heat shield.*

Figure 8-5. At the end of a flight, the heat shield had a scarred, burnt appearance.

There was much criticism outside of NASA that none of these astronauts were scientists. In a talk before the Society of Experimental Test Pilots on October 9, Donald Slayton defended this choice:

> As in any scientific endeavor the individual who can collect maximum valid data in minimum time under adverse circumstances is highly desirable. The one group of men highly trained and experienced in operating, observing, and analyzing airborne vehicles is the body of test pilots. . . .

The logic of this argument, especially as it dealt with the largely unknown domain of spaceflight, was persuasive. The seven men selected were among the most experienced and dauntless pilots in America. If a problem arose during flight, it was hard to imagine who else would be more capable of dealing with it. The astronauts were lauded by the press for their courage and immediately given the status of national heroes by the American people.

Astronaut Safety

The astronauts had agreed to work with notoriously unreliable rockets and to fly in a completely new craft. There was no guarantee that the whole system would work, and if it failed, chances were high that the pilot would be killed. The courage of the astronauts was never questioned, but they were not reckless. Having flown experimental aircraft, they were aware that even minute matters, if overlooked, could produce tragic results. To help reassure the astronauts, McDonnell Aircraft issued a report called "Project Mercury Indoctrination." It was required reading. The section on safety specifically reminded them of their former days as test pilots; it stated:

> The problem of attaining a high degree of reliability for project Mercury has received more attention than has any other previous missile or aircraft system. Reliability has been a primary design parameter since the inception of the project.

These were more than just empty words. McDonnell Aircraft had built more than sixty separate redundancies into the capsule which would allow the pilot to take alternative actions in the event of a malfunction. This was rarely the case in experimental military aircraft. In addition, systems would be extremely ground tested before a manned flight would be attempted.

Big Joe: The Capsule Heat Shield

One item beyond the control of the astronaut was the intense heat that the capsule would generate during reentry into the atmosphere. As the capsule descended, its Earth-facing side would be rapidly heated to over 3000° F. This high temperature would destroy a normal craft. The solution that NASA adopted was to coat the bottom of the capsule with a special material, that when heated, broke into tiny flakes. The passing air would then carry these heated flakes away from the capsule, keeping the temperature inside the craft within tolerable bounds. The process, where flaked material carries heat away from the craft, is called ablation. In theory, the use of an ablation heat shield solved the reentry problem, but to

Figure 8-6. The launch escape tower is seen on top of this rocket prior to a test.

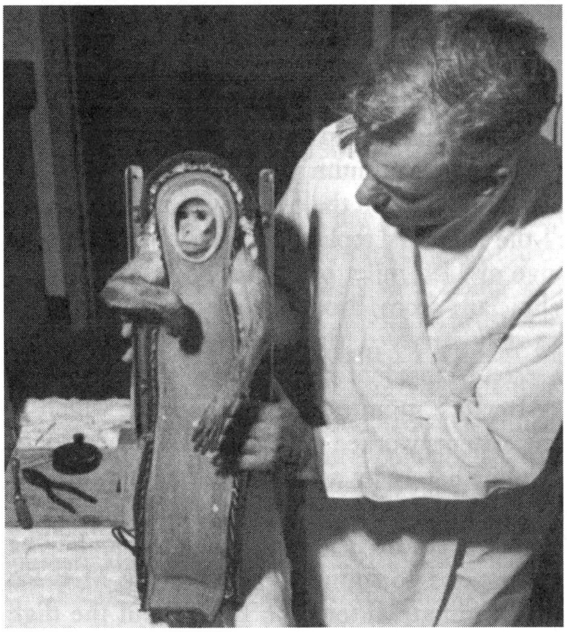

Figure 8-7. SAM, the Rhesus monkey used in Little Joe-2, *is shown in his flight suit prior to the test of the Mercury escape system.*

confirm this, test flights were needed. This was the objective of the *Big Joe* program.

As part of this effort, over 100 thermocouples were attached to the capsule's interior to measure its temperature during a flight. Fifty other instruments, including microphones, pressure gauges, and accelerometers, were also installed to provide additional flight data. After some minor delays, *Big Joe* was ready for launch, and on September 9, 1959, it roared off from Cape Canaveral in what appeared to be a perfect flight. Two minutes later, telemetry from the rocket revealed that the now exhausted booster engines had failed to separate as planned. This meant that the craft's velocity would be slowed by 3,000 feet per second and it would fall short of its intended impact site by 500 miles. After the capsule separated from the Atlas and began its descent, ground controllers were also uncertain if the shield was properly oriented. Navy ships rushed to the new impact site to retrieve the capsule before it sank. Seven hours later, it was safely recovered by the destroyer *Strong*.

Information from the on-board recorders allowed the rocket's flight to be reconstructed. The capsule separated from the Atlas rocket 135 seconds later than expected, at an altitude of 345,000 feet and a speed of 15,000 miles per hour. The capsule's shape and center of gravity were such that the proper reentry orientation was obtained. Engineers were delighted to discover that the heat shield reached a temperature of 3,500° F, very close to the temperature they had hoped for.

Despite this high temperature, an examination of the on-board sensors revealed that the capsule's interior temperature had remained quite livable. In a post-flight ceremony, a letter signed by fifty-three members of the Big Joe team was retrieved from the capsule and handed to Gilruth. It read, "This

note comes to you after being transported into space during the successful flight of 'Big Joe' capsule, the first full-scale flight operation associated with Project Mercury. The people who have worked on this project hereby send you greetings and congratulations."

The Big Joe test proved to be a great start to Project Mercury. The heat shield was found to be so effective that a second launch was canceled and the thickness and weight of the heat shield was reduced by almost half. This flight effectively ended NASA's concerns about the ability of the ablative shield to protect the astronaut during a normal reentry.

Little Joe Tests: The Astronaut Escape System

NASA engineers soon realized that the least reliable part of the space vehicle was not the capsule itself, but the Atlas launch rocket. The air force judged the performance of the Big Joe test as a failure. The success rate of the Atlas was estimated to be less than 80 percent. This relatively low standard of acceptability reflected the rocket's military heritage. In the age of "overkill," it was not of great concern that a few rockets would fail, since many were sent to destroy a single target. This approach, although militarily acceptable, was clearly inappropriate for manned flight. The rocket's reliability had to be improved and safety measures had to be developed that would allow the astronaut to escape injury in the event of a malfunction.

The Space Task Group conducted a study of over 60 Redstone and about 30 Atlas launches to identify the most likely manner in which these rockets would fail. Relatively

| Table 8-1. Mercury Suborbital Test Flights |||||
Mission	Launched	Height (miles)	Goal
LJ-1	Aug. 21, 1959	0.4	heatshield test
Big Joe	Sept. 9, 1959	95	heatshield test
LJ-6	Oct. 4, 1959	37	booster test
LJ-1A	Nov. 4, 1959	9	heatshield test
LJ-2	Dec. 4, 1959	53	Sam, medical
LJ-1B	Jan. 21, 1960	9	Miss Sam/abort escape plan

few causes were found that led to a situation where the astronaut was endangered. Based on this data, automatic abort-sensing systems were installed in both the Redstone and Atlas rockets.

If an abort condition was detected, a rocket attached to the top of the capsule would be fired to pull the capsule away from the rocket. A series of suborbital flights, called "Little Joes," were conducted at Wallops Island to test this escape system. After the hardware had been tested in unmanned dummy capsules, live rhesus monkeys were to be flown to see how they responded to weightlessness.

The first monkey to be flown as part of this test program was named SAM, after its home at the Air Force *S*chool of *A*viation *M*edicine in San Antonio, Texas. SAM was launched aboard *Little Joe-2* on December 4, 1959 with astronauts Alan Shepard and Virgil Grissom on hand at Wallops to witness the flight. The craft reached an elevation of 280,000 feet and SAM experienced weightlessness for about 3 minutes. The escape system worked as planned and SAM was retrieved unharmed. On January 21, 1960, a second successful test of the escape system was performed using a female monkey who had been named Miss SAM.

Soviet Space Missions

Soviet scientists were also hard at work on their own space program. On May 15, 1960, a massive 10,000 pound capsule called *Sputnik 4* was placed into orbit around the Earth. After completing a series of tests and orbiting the Earth sixty-eight times it attempted to reenter the atmosphere, but there was a problem with its infrared sensor. *Sputnik 4* would remain in orbit until 1965.

Three months after the launch of *Sputnik 4*, the Soviets launched *Sputnik 5*. This craft carried two live dogs, named Strelka and Belka. After orbiting the Earth eighteen times, the capsule returned to Earth and the dogs were recovered unharmed. This marked a new space milestone, the first time that living specimens had been recovered from orbit. In the race to place a man in orbit, the Soviets were still far ahead of the United States.

Criticism of Project Mercury

The fact that the Soviet program continued to outstrip its American counterpart was not lost on the news media. As far back as October 1959, *Newsweek* published an article titled "How to Lose the Space Race." The article claimed that the approach being employed by the United States was fatally flawed and noted that the way to insure failure was to "start late, downgrade Russian feats, fragment authority, pinch pennies, think small, shirk decisions."

This assessment was not completely off the mark. America obviously trailed the Soviets and appeared to be falling further behind. What it had achieved seemed modest in comparison to the impressive Soviet successes. The tone of articles about Project Mercury reflected a growing public apprehension that the United States was losing the space race. James Barr in *Missiles and Rockets*, a weekly defense industry trade journal, severely criticized Project Mercury in its August 15, 1960 issue:

NASA's Mercury manned-satellite program appears to be plummeting the United States toward a humiliating disaster in the East-West space race. . . . The program today is more than one year behind its original schedule and is expected to slip to two. Therefore, it no longer offers any realistic hope of beating the Russians in launching the first man into orbit around the Earth—much less serve as an early stepping stone for reaching the Moon. Mercury originally had the supposed advantage of being cheap, an attribute that made it particularly attractive to the Administration. However, Mercury has proven to be a trip down a dead-end road that U.S. taxpayers are finding themselves paving with gold. . . .

Progress on Project Mercury continued to be uneven. On July 29, 1960, the Mercury-Atlas flight, *MA-1*, failed one minute after its launch due to a catastrophic failure within the Atlas rocket. A few months later, on November 21, 1960, a second major failure occurred when the Redstone booster of Mercury test flight *MR-1* automatically shut down. As project engineers watched through periscopes in their nearby concrete bunker, the rocket ignited, wobbled slightly, rose less than 4 inches, and then settled back onto its fins. The engines stopped, and within seconds, the main and reserve parachutes shot into the air and landed around the base of the rocket. Langley's Space Task Group immediately sent a report of this failure to NASA Headquarters.

Apparently, sufficient thrust had developed to lift the booster at least 3.32 inches, thereby activating all the systems. The booster settled back down on the pad, damaging the tail fins, and perhaps the structure as well (some wrinkles are visible in the shell). The reason for this shutdown is unknown. . . . At the time of this writing, the booster destruct system is still

Figure 8-8. As NASA launch personnel watched, MR-1 failed to lift off and the capsule's parachutes were deployed. Here the drogue chute is seen settling to earth.

armed, and cannot be disarmed until the battery depletion during the morning of November 22. Capsule pyrotechnics are also armed. The problem is further complicated by the fact that the main parachute is still hanging from the capsule; thus the booster could be blown over in a high-wind condition.... The extent of damage to the capsule has not yet been assessed. Assuming a minimum of damage, it is planned to use the same capsule, together with the *MR-3* booster, for the *MR-1* firing. It will probably take a month before this launching can take place.

This latest failure led to a new wave of criticism of NASA and caused a number of people to write off America's chances of beating the Soviets into space. *Time* magazine reported: "Project Mercury's latest failure, ... just about evaporated the last faint wisp of hope that the United States might put a man into space before Russia does." A despondent NASA official later remarked about this flight, "All we did was launch the escape tower."

A double standard seemed to exist in the space race. Failures in the American program were harshly criticized,

Table 8-2. Additional Mercury Test Flights

Mission	Launched	Height (miles)	Goal
LJ-5	Nov. 8, 1960	10.1	escape mechanism
MR-1	Nov. 21, 1960	0	qualify systems
MR-1A	Dec. 19, 1960	131	abort system
MR-2	Jan. 31, 1961	157	Ham/abort system
MA-2	Feb. 21, 1961	114	hardware interface
LJ-5A	Mar. 18, 1961	7.7	escape system
MR-BD	Mar. 24, 1961	114	Redstone test
MA-3	Apr. 25, 1961	4.5	orbital attempt

whereas Soviet setbacks received scant coverage. For instance, when on December 1, 1960, the Soviets launched two more dogs into orbit aboard *Sputnik 6* and failed to recover them, little notice was taken by the press.

Finally, on December 19, 1960, the American public received good news when *MR-1A* was successfully launched in an unmanned suborbital flight. Post-flight inspection of the capsule showed that it had survived the mission in good shape. Despite some minor technical problems (the Redstone rocket produced slightly more acceleration than was expected), this test demonstrated that the United States was finally making real progress.

A Mid-Program Evaluation

Three and a quarter years after the launch of *Sputnik 1*, the United States and Soviet Union had launched a combined total of forty-two rockets. Of these, thirty-eight were intended to be Earth-orbiting satellites, three were solar-orbiting satellites, and one was a lunar probe. America launched thirty-three of these craft, while nine were Soviet.

Despite this 3 to 1 superiority, American rockets had carried only 34,240 pounds of payload compared to 87,000 pounds for the Soviets. This substantial weight difference demonstrated that the Soviet rockets could lift considerably heavier objects than their American counterparts. The Soviets had also crashed a probe into the Moon, photographed the Moon's back side, and recovered two dogs from the Earth's orbit. To the dismay of the American political leadership, the lead the Soviet Union enjoyed at the time of *Sputnik 1* was still in place, and seemed to be growing larger.

MR-2: Could a Man Function on Space?

Project Mercury was rapidly approaching the point where it was necessary to fly additional live animals on suborbital flights to test the capsule's life-support systems and gather more data on the physiological effects of weightlessness. The goal of the suborbital flight, *MR-2*, was to carry a live chimpanzee to obtain more data in these two areas. This

Figure 8-9. The chimpanzee, Ham, was flown aboard MR-2 *to test man's ability to operate instruments in the weightless of environment.*

animal was named HAM, in honor of its home at the *Hollo-man Aerospace Medical* Center, located near Alamogordo, New Mexico. HAM was trained to pull various levers in response to illuminated white and blue lights. By measuring the chimpanzee's response to these signals during flight, researchers hoped to gain information on the medical effects of severe gravitational (g) forces and weightlessness.

On January 31, 1961, *MR-2* blasted off from Cape Canaveral, reaching a height of 157 miles and achieving a speed of 5,860 mph. HAM experienced g forces of 14.7 and weightlessness for 6.6 minutes. During this time, he successfully moved the test levers as instructed by the colored flashing lights. The capsule's life-support system operated as planned. The entire flight lasted 16.5 minutes and landed in the sea, 60 miles from the nearest recovery ship, the destroyer *Ellison*. When recovery helicopters arrived at the capsule's landing site, they saw that it had tilted on its side and was taking in water. After having survived his suborbital flight and performing all the tasks asked of him, HAM was in danger of drowning! Rescue crews frantically worked to save both HAM and the capsule. It was a close call, but the operation was successful.

HAM became quite famous as a result of the successful *MR-2* flight. However, when he was later shown his spacecraft, it was "visually apparent that he had no further interest in cooperating with the spaceflight program." HAM was glad to retire from the space program and return to a normal life on Earth.

The Langley Space Task Group was quite pleased with the postflight analysis of *MR-2* and felt that after some small improvements, the Mercury capsule would be reliable enough to carry an astronaut. McDonnell Aircraft had devoted over 2.6 million man-hours of engineering time, 380,000 man-hours of tooling, and 1.5 million hours of production time to the Mercury capsule. Based upon its performance, they also felt it was ready.

However, Von Braun's Huntsville group was not satisfied with the performance of the Redstone rocket. They had completed three separate studies of its reliability, and in their opinion, the probability that a Redstone-Mercury combination could be successfully launched was in the range of 88 percent. In the event of a failure, they estimated that the crew had a 98 percent chance of surviving. Although these values suggested that a manned mission was likely to succeed, they judged the risk to be too high. Von Braun requested and was granted one additional unmanned Redstone flight to solve the remaining problems. This delay, although understandable, was to have a major and adverse effect on America's effort to catch up with the Soviet Union.

MA-2: A Final Heatshield Test Flight

While von Braun's group worked on the Redstone rocket, NASA conducted an additional test of the Atlas rocket and the capsule's ablation shield. Although this rocket would not be needed until later in Project Mercury, it was necessary to verify that the heat shield could withstand the *maximum* heat that would be created by the reentry of a capsule.

The mission to gather this data, *MA-2*, was launched on February 21, 1961, and reached a maximum velocity of around 12,000 mph. All of its stated objectives were met and all systems performed satisfactorily. At the post-flight news briefing, Gilruth stated that a human could have survived this flight.

Down to the Wire

The United States was unaware that the Soviet Union was preparing another series of heavy Sputnik launches in preparation for their own manned flight. On March 9, 1961 *Sputnik 9* was launched carrying a payload of 10,300 pounds and a live dog named Chernushka. After one orbit, the capsule automatically descended to Earth and Chernushka was recovered in good health. It was clear that the Soviet's were very close to achieving their goal of placing a human in orbit.

The Redstone test sought by von Braun (*MR-BD*) occurred on March 24, 1961 and was a success. The modifications made to the rocket worked and it was declared "manrated." NASA was ready to proceed with a manned suborbital flight. On the very next day, March 25, 1961, the Soviets launched and recovered yet another dog, named Zvezdochka, aboard *Sputnik 10*. A manned Soviet orbital flight seemed imminent.

On April 25, 1961, NASA launched *MA-3*. The goal of this mission was to place a "mechanical astronaut" into orbit. Forty seconds after lift-off the rocket failed to achieve its proper trajectory and the on-board abort sensors fired the escape rockets. The capsule was pulled free of the rocket, coasted to a height of 24,000 feet, and then landed in the ocean about 2,000 yards from the launch pad.

9   Yuri Gagarin: The First Man in Space

A ballistic flight will cost a lot of money and will be risky. The results will be very meager. Science needs a flight around the Earth. Not a small step forward but a stride forward that is decisive and daring.

—Sergei Korolev, Soviet Chief Designer

On the morning of April 12, 1961, a gray-green bus sped toward a launch pad at the Soviet space complex at Tyuratam. Inside were Yuri Gagarin and Gherman Titov and several other cosmonauts who had come to wish Gagarin a safe mission. All realized they were participating in a historic occasion—Gagarin would shortly attempt to become the first human to orbit the Earth. In the event of a sudden illness or accident, Titov stood ready to take his place.

As they neared the launch pad, the occupants became more anxious. They sang, talked, and joked in an attempt to help Gagarin relax. This seemed hardly necessary because he seemed to be perfectly at ease. When the silvery Soviet R-7 rocket was finally in sight, Gagarin looked at it through the bus window and exclaimed, "Beautiful!"

When the bus stopped alongside the launch pad, Titov rose and tapped his helmet against Gagarin's in a farewell gesture. He then returned to his seat feeling a little like an unwanted guest at a party. He later recalled,

We were both fighter pilots, so we understood each other. He was commanding the flight, and I was his backup, just in case. But we knew "just in case" wasn't going to happen. What could happen at this late stage? Was he going to catch the flu between the bus and the launch gantry? Break his leg? It was all nonsense. We shouldn't have gone out to the launch pad together. Only one of us should have gone.

Walking awkwardly in his spacesuit, Gagarin left the bus and was greeted by waiting space officials, dignitaries, and technicians. General Nikolai Kamanin, head of the Soviet space program commission, recalled being somewhat annoyed at the cosmonauts' behavior,

When Gagarin left the bus, everyone let themselves release their emotions and started to hug and kiss one another. Instead of wishing him a nice journey,

Figure 9-1. Gagarin (front) and Titov (background) on their way to the launch of Vostok 1. Gagarin was confident and relaxed prior to his flight.

Figure 9-2. Yuri Gagarin (standing) was formally informed that he had been selected to be the first Soviet in space just days before his launch.

Figure 9-3. Gagarin's capsule, the Vostok, consisted of a spherical object with a retrorocket package attached.

Figure 9-4. Gagarin greets Chief Designer Korolev just prior to the liftoff of Vostok 1.

some of them were shedding tears and saying good-bye as if forever. We had to apply force to pull the cosmonauts out of their embraces. Although others may have had doubts about the outcome of the flight, Gagarin exuded confidence and was anxious to get the flight underway.

Gagarin's Selection

Less than a week had passed since the cosmonauts stood before the Soviet State Commission for Spaceflight to learn which one of them would fly this mission. Gagarin and Titov were thought to have the best chance of being selected. Earlier, the cosmonauts had been asked by Sergei Korolev, Chief Designer, "In your opinion, who should be sent into flight first and why?" When he examined their written assessments, 60 percent had recommended Gagarin. Among the reasons cited were he "never loses heart," is "bold and steadfast," "modest and simple," and "decisive."

The final cosmonaut selection was largely based on the recommendation of Korolev and Kamanin. When later asked to explain how he had reached his decision, Kamanin replied,

A real man is characterized by at least five traits: courage, sober judgement, a strong will, diligence, and dedication to his objectives. All of these traits are abundantly represented in Gagarin's character. But in addition to character, medical and psychological fitness must be taken into account. . . . In all

these respects both Gagarin and Titov were fine specimens.

He went on to add,

We had to give thought not only to the first flight but to the second as well. Titov was a more likely choice for the second flight. With regard strictly to their merits as human beings the two were evenly matched. . . . Gagarin is from a kolkhoz (collective farm) family. Titov comes from a hard-working Siberian family. . . . Both are first-rate pilots. . . .

Kamanin also emphasized an important characteristic that Gagarin possessed:

Gagarin has a great capacity for analyzing every-thing that interests him. He is a born investigator. And for the first flight it was very important that the cosmonaut be able to see a lot in a short time and an-alyze what he had seen.

Korolev was also influenced by many of these same factors but foremost in his mind was that the first Soviet to fly in space had to possess the qualities of "patriotism, courage, modest, iron will, knowledge, and love of people." Although disappointed at not being chosen, Titov accepted Gagarin's selection without bitterness. When later asked his opinion about this matter, he modestly replied, "Yuri was trained to perfection. He was the worthiest one among us. The choice was absolutely right."

The Launch of Vostok 1

At the launch site, Gagarin was first greeted by General Kamanin. As he proceeded toward the awaiting rocket, Gagarin briefly met with Korolev, who had just returned from personally inspecting the capsule. Korolev handed him a duplicate of the medallion carried to the Moon by *Luna 2* and wished him a successful flight. These farewells completed, Gagarin climbed to the base of the rocket, waved, and addressed the assembled crowd,

> Dear friends, you who are close to me, and you who I do not know, fellow Russians, and people of all countries and continents. In a few minutes a powerful space vehicle will carry me into the distant realm of space. What can I tell you in these last minutes before the launch? My whole life now appears to me as one beautiful moment. All that I previously lived through and did, was lived through and done for the sake of this moment. You can understand that it is difficult for me to analyze my feelings right now, when the critical moment is close at hand: that moment for which we have long—and passionately—been preparing ourselves. I wonder whether it is worthwhile to tell you of the feelings I experienced when I was offered the chance to make this flight? Joy? No, it was not merely joy. Pride? No, it was not merely pride. I felt great happiness. To be the first man in space—to meet nature face to face in an unprecedented encounter—could one dream of anything greater? But immediately after that moment I thought of the tremendous responsibility I bore: to be the first to pave the way into space for mankind. . . . Just tell me if there is any more complex task than the one that has fallen to my lot! This responsibility is not toward one person, not toward a few dozen, not toward a group. It is a responsibility toward all mankind—toward its present and future. . . . Am I happy as I set off on this spaceflight? Of course I'm happy. After all, taking part in new discoveries has brought the greatest happiness to people of all times and all ages.

Noticing the time, he concluded,

> There are not many minutes left, now, before liftoff. Dear friends, I say to you, "Until we meet again"—as people always say to others when leaving on a long trip. How I would like to embrace all of you—both those I know and those I have never met; both those who are far away and those who are near.

With these parting words Gagarin turned, proceeded up the gantry elevator, and entered the single-person Vostok capsule. After a small adjustment to the capsule hatch was made and other routine prelaunch checks were completed, Korolev spoke briefly with Gagarin to ask how he was feeling, "I feel fine," he replied. "How are you feeling?" Gagarin was calm, confident, and even able to joke. He was ready.

In the nearby underground bunker controller Leonid Voskresensky peered through a periscope at the R-7 and called out the commands, "Bleed-off . . . Hold . . . Ignition . . . Switch to launch . . . Liftoff. . . ." As *Vostok 1* blasted off on its historic flight, Gagarin reported, "We're off!" He later recalled this moment,

> A faint noise started. . . . When the engines entered their main, primary stage, the noise intensified, but it wasn't so sharp that it deafened me or interfered with work. . . . I was prepared for much more noise. Then the rocket smoothly, lightly rose from its place. I didn't even notice when it started. Then it felt like there was a slight shiver in the structure of the rocket. The vibrations were high frequency, low amplitude. . . .

From a nearby observation platform Titov watched as the R-7 began its journey, "I could hear the high pitched whine of the fuel pumps pushing fuel into the combustion chambers, like a very loud whistle . . . I saw the base of the rocket belching fire when the engines ignited, and there were stones and pebbles flying through the air because of the blast. . . . It hammers your ribcage, shaking the breath out of you. You can feel the solid concrete bunker shaking with the noise. . . ."

As the R-7 sped upward, Gagarin remained calm, mentally noting the changing sensations he was experiencing,

> The g-load was increasing steadily, but it's completely manageable, as in normal airplanes. About 5 gs. At that g-load, I reported and communicated with the ground the whole time. It was somewhat difficult to talk, since all the muscles of my face were drawn. There was some strain. The g-load continued to rise, then reached its peak and began to steadily decrease. Then I felt a sharp drop in g-load. I felt as if something had separated from the rocket. I felt something like a knock. Then the noise dropped sharply. The state of weightlessness began to emerge, although the g-load was about 1 at that time. Then the g-load came back and began to increase. I began to be

Figure 9-5. Yuri Gagarin waves farewell to the assembled personnel just prior to lift-off.

Figure 9-6. Sergei Korolev as he talked to Gagarin moments before lift-off.

pressed to the seat, and the noise level was substantially lower. At 150 seconds, the nose fairing separated. The process was very crisp.

The three-stage R-7 performed perfectly and eight minutes after lift-off, Gagarin entered into a 109 × 196 mile orbit with an revolution period of 89 minutes. The capsule's controls had been locked in place because doctors were afraid weightlessness might cause the pilot to become disoriented. Only in the event of an emergency would the cosmonaut be told the special code needed to take control of the craft. Mark Gallai, a former pilot himself and then head of cosmonaut flight training, later wrote that this procedure was not popular with the cosmonauts.

All of the test pilots believed that these concerns were stupid. . . . None of us pilots liked this system and we made a great noise about it. It was our feeling that the possibility of a pilot going crazy was much smaller than the possibility of failure in the radio communication.

Before the flight, Gallai violated mission rules by giving the code to Gagarin. Later it was learned that this was a poorly held secret. At least three other people, rocket engineer Oleg Ivanovsky, Korolev, and Kamanin, had all separately given the code to Gagarin.

Impressions from Space

While in orbit, Gagarin observed the Earth and sky through viewports and monitored the capsule's instruments. His memories of the flight were vivid and they had an enormous impact on him:

Of all the nights I had seen in my lifetime none was remotely comparable to night in space. I have never forgotten it. What was so different about it? It was

most unusual to see night fall so swiftly, and dawn follow it so quickly—blindingly bright and full-bodied. The sky was blacker than it ever appears from Earth, with the real, slate-blackness of space. At times I had the impression that everything around me was happening in a kind of gigantic planetarium where one can turn night into day and day into night. After the darkness, the horizon began to brighten. But it was not the same kind of thing I had seen in the forest or on the open plain, when the sun rises slowly and even majestically, still hidden from view as the first predawn rays reach the Earth, and the woods, meadows, and streams are suffused with light. In space everything happened differently—the sun abruptly shot up over the horizon, and its fierce light stabbed into my eyes. The cosmic dawn is instantaneous; and in space the hues of beginning day, both soft and brilliant, thrill you with a kind of magic.

While in orbit Gagarin recorded his observations using a pencil and notepad and was in frequent contact with Soviet controllers. Later, after the pencil and notepad floated under his seat, he recorded his spoken observations using on-board instruments and transmitted them to Earth. He was constantly asked questions from the ground, such as, "What does the hydrosphere look like?" and somewhat surprisingly, "Is the Earth's roundness apparent?" to which he replied, "Yes, of course."

Prior to his flight, medical doctors had expressed concerns about the effect of weightlessness on the human body. Would the absence of gravity affect the body's circulatory system or the cosmonaut's sense of sight, hearing, and touch? Could a person eat and drink without the aid of gravity? As Gagarin circled the Earth, these concerns quickly faded since he was not experiencing any adverse symptoms or disorientation. His heartbeat and breathing were normal, and he could eat and drink without any difficulty.

Gagarin's Landing

As *Vostok 1* neared the coast of Africa, Gagarin reported to a ground controller, "The flight is proceeding normally. I am having no trouble with weightlessness." A short time later, the capsule's retrorockets fired on schedule, but then a potentially serious situation arose:

I felt the braking rocket kick in . . . as soon as [it] shut off there was a sharp jolt, and the craft began to rotate around its axis at a very high velocity . . . about 30 degrees per second, at least. I was an entire "corps de ballet:" head, then feet, head, then feet, rotating rapidly . . . I waited for separation. There wasn't any. I knew that, according to plan, that was to occur 10–12 seconds after the braking rocket switched on . . . I decided that something was wrong. . . . I estimated that all the same I would land normally . . . the Soviet Union was 8,000 km long, which meant I would

Figure 9-7. The historic launch of Vostok 1 carrying Yuri Gagarin into orbit around the Earth.

Figure 9-8. Vostok 1 landed charred but in one piece. It was extensively cleaned prior to its display.

land somewhere in the far east. . . . I reasoned that it was not an emergency situation. I transmitted the all-normal signal. . . . The separation occurred at 10 hours and 35 minutes—and not at 10 hours and 25 minutes, as I had expected.

Although the connections between the descent and rear equipment module had properly separated, a dense umbilical cable of electrical wires still connected them. When the combined craft reentered the atmosphere, this configuration caused it to tumble end-over-end. As it encountered thicker layers of air, the craft then began to oscillate right and left and its exterior temperature began to rise. Gagarin now heard some unsettling cracking noises.

I felt the oscillations of the craft and the burning of the coating. I don't know where the crackling was coming from; either the structure was cracking, or the thermal coating was expanding as it was heated—but it was audibly crackling . . . I felt the temperature was high. . . .

The intense heat generated by reentry eventually disintegrated the cable, but as the two modules separated, the descent module was given such a violent spinning motion that Gagarin almost lost consciousness. He later recalled, "I felt as if the g-load was 10g. There was a moment for about two–three seconds when the indicators on the instruments began to become fuzzy. Everything seemed to go gray."

Gagarin was not suppose to land in his capsule but to eject and parachute to the ground. At a height of around 23,000 feet the hatch was automatically blown off. This momentarily startled him.

I'm awaiting the ejection . . . at an altitude of 7,000 meters hatch No. 1 was shot off. . . . I'm sitting there thinking, that wasn't me that was ejected, was it? Then I calmly turned my head upward, and at that moment, the firing occurred, and I was ejected. It happened quickly, and went without a hitch . . . I flew out in the seat. Then a cannon fired, and the stabilizing chute deployed.

Gagarin's landing was witnessed by Anna Takhtarova and her six-year-old daughter Rita as they were planting potatoes in a collective farm near the village of Smelovka. In the book Our Gagarin, Soviet space journalist Yaroslav Golovanov provided the following account of this landing based on interviews with Gagarin.

Stepping onto firm ground again, I caught sight of a women and a little girl standing near a dappled calf and looking curiously at me. I was still in my bright-orange spacesuit, and they were a bit frightened by its strangeness. "I'm a friend, comrades! A friend!" I shouted, taking off my helmet and feeling a slight shiver of nervousness. "Can it be that you have come from outer space?" the woman asked. "As a matter of fact, I have!" I replied.

Later Golovanov admitted the woman and child actually had a much different concern, "when they saw Gagarin's orange protective suit, the women became frightened, because

As cosmonaut Yuri Gagarin marched along a red carpet laid for him at the Vnukovo Airport located outside Moscow, thousands of Russians chanted, "Ga-gar-in! Ga-gar-in!" Just two days had passed since his orbital flight. Average Russian citizens took enormous pride in his accomplishment—he was truly a national hero.

there was all this business about Powers only a year before. They said, 'Where are you going? Where are you off to?' They thought maybe he was a spy."

Aftermath of the Flight

Gagarin's flight was historic and the Russian people were elated. He was flown to a dacha in the Zhiguli Hills on the Volga for medical tests and a brief recovery period. On April 14, he was flown to Moscow for a celebration at the Vnukovo Airport and then at Red Square. At the airport he was greeted by his wife, mother, and government officials. Finally, the impact of the flight began to take its toll, and he grew nervous as he marched along the ceremonial red carpet that had been placed down for him. He recalled: "Never—not even in the spacecraft—had I been so nervous as I was at that moment. The path seemed to stretch on for ever and ever. But while I was walking along it, I managed to keep my presence of mind. I knew that all eyes were on me."

Gagarin then proceeded by automobile to Red Square where millions of Russians were waiting to give him a hero's welcome. Flowers and banners were everywhere. Live cov-erage outside of Russia was provided by the European Broadcasting Union. There were tributes to Tsiolkovsky, the Soviet scientists, and mocking comments about the American space effort. Much to the consternation of American political leaders, it was clear to the world that America had been beaten again.

Khrushchev was extremely pleased by the worldwide acclaim that Gagarin's spaceflight brought to the Soviet Union, and he took great personal pride in it. Unlike Gagarin, however, who saw his flight as a triumph for all mankind. Khrushchev viewed Gagarin's flight as a purely Soviet accomplishment that held almost inestimable political benefits in the Cold War with America. He moved immediately to make this political point clear. At a Moscow event he attributed the success of Gagarin's flight to the Soviet system of government:

> In forty-three years of Soviet government, formerly illiterate Russia . . . has traveled a magnificent road. . . . This victory is another triumph of Lenin's idea. . . . This exploit marks a new upsurge of our nation in its onward movement toward communism.

Introduction to Section III
America Plans to Place a Man on the Moon

Do we have a chance of beating the Soviets by putting a laboratory in space, or by a trip around the Moon or by a rocket to land on the Moon, or by a rocket to go to the Moon and back with a man? Is there any other space program which promises dramatic results in which we could win?

—President John F. Kennedy (April 20, 1961, memo to Vice President Johnson)

President Eisenhower was intrigued with the idea of space exploration, but not entirely captivated by it. Philosophically, he was opposed to expanding the size of the government and had agreed to the creation of NASA largely due to public pressure and because of the fundamentally different manner in which scientific and military research are conducted. In his memoirs Eisenhower wrote, "Information acquired by purely scientific exploration could and should be made available to all the world. But military research would naturally demand secrecy." The launch of Sputnik terrified the American public and created a national panic. Eisenhower assented to the formation of NASA in response to public demand—not as a deepfelt desire to create a civilian space agency.

Originally, Eisenhower favored a modest civilian effort whose budget would be less than half a billion dollars. The goal of this organization would be to conduct "scientific studies," not propaganda exercises that he characterized as "stunts." Unlike many politicians of his time, Eisenhower was not convinced that American prestige was at stake. General Andrew Goodpaster, his longtime aide recalled, "The president's approach was if we're doing the right thing in about the right way we'll let the prestige work itself out." This view was supported by T. Keith Glennan, chief NASA administrator, who also did not favor "spectacular" missions whose main purpose was to match the Soviets. Glennan wrote, ". . . we must avoid the undertaking of particular shots, the purpose of which would be propagandistic rather than directed toward solid accomplishments in understanding the environment with which we are dealing."

Eisenhower insisted that space missions had to have "intrinsic merit" beyond the space race competition with the Soviet Union and this philosophy caused him to favor unmanned missions. His preference for unmanned missions was reinforced by an ad hoc presidential panel on manned spaceflight. When responding to the issue of "whether the presence of a man adds to the variety or quality of the observations which can be made from unmanned vehicles—in short whether there is a scientific justification to include man in space vehicles," it answered "No." The panel concluded that "man's senses can be satisfactorily duplicated at remote locations by the use of available instrumentation. . . . It seems,

Figure Intro. 3-1. President Eisenhower supported a modest civilian space program focused on scientific exploration. He opposed the establishment of a program designed solely to compete with the Soviet Union.

therefore, to us at the present time that man-in-space cannot be justified on purely scientific grounds."

The committee argued that the rationale for manned flight was not scientific but emotional, "it may be argued that much of the motivation and drive for the scientific exploration of space is derived from the dream of man's getting into space himself." This latter argument clashed with Eisenhower's belief that the government should support space exploration only if it contributed to the nation's defense or scientific knowledge. In this context, his opposition to Project Mercury is understandable. David Calladan and Fred Greenstein write in *Spaceflight and the Myth of Presidential Leadership*, "Project Mercury represented everything Eisenhower claimed he wanted to avoid in space policy. It was hugely expensive, driven almost entirely by

Figure Intro. 3-2. The launch of Sputnik 1 *was a defining moment in the American space program. It led to the creation of NASA and to the space race between the United States and the Soviet Union.*

competition with the Soviets, and lacked a compelling scientific rationale."

After John F. Kennedy was elected president, concern deepened within NASA that substantial cuts were imminent. Initially, Kennedy was not an avid supporter of a greatly expanded space program. Kennedy's ambivalence about the space program was reinforced when his science advisor,

Figure Intro. 3-3. T. Keith Glennan was NASA's first chief administrator. Appointed by Eisenhower, he shared the president's opinion that the space program should be based on its scientific merit rather than political considerations.

Jerome Wiesner, issued a report critical of NASA's progress and future prospects. Wiesner later expressed the opinion that, "If Kennedy could have opted out of a big space program without hurting the country . . . he would have."

Roger Launius, author of *NASA: A History of the U.S. Civil Space Program*, hypothesized that Kennedy's attitude about the American space program was dictated more by cold-war politics than by a quest for knowledge. He writes that "[Kennedy] was not a visionary enraptured with the romantic image of the last American frontier in space and consumed by the adventure of exploring the unknown. . . . The Soviet Union's nonmilitary accomplishments in space . . . forced Kennedy to respond and to serve notice that the United States was every bit as capable in the space arena as the Soviets."

On April 12, 1961, cosmonaut Yuri Gagarin was launched from the Soviet space center at Tyuratam and became the first person to orbit the Earth. The shock that this flight produced among the American public was similar to that of *Sputnik 1*. Coincidently, Gagarin's flight also occurred less than a week before another political disaster, the failure of the American-supported invasion of Cuba—the Bay of Pigs. In Kennedy's mind, these two events, combined with the Soviet's highly successful Sputnik and Luna missions, required an immediate response. The situation, as Kennedy saw it, was summarized by Hanson Baldwin in an article in *The New York Times,* "Even though the United States is still the strongest military power and leads in many aspects of the space race, the world—impressed by the spectacular Soviet firsts—believes we lag militarily and technologically." For Kennedy, the thought of the United States coming in a distant second to the Soviets was unacceptable.

The Decision to Land a Man on the Moon

President Kennedy responded to these Soviet triumphs by searching for ways in which America could achieve a notable space first. His thinking on this matter was greatly influenced by the cold-war competition between the United States and the Soviet Union and the effect these successes had on world opinion. In May 1961, NASA Administrator Webb and Secretary of Defense McNamara sent a letter to Johnson arguing that a vigorous space program was a cold-war necessity,

> This nation needs to make a positive decision to pursue space projects aimed at enhancing national prestige. . . . The non-military, non-commercial, non-scientific but "civilian" projects such as lunar and planetary exploration are, in this sense, part of the battle along the fluid front of the cold war. . . . The orbiting of machines is not the same as the orbiting or landing of man. It is man, not merely machines, in space, that captures the imagination of the world. . . . If we fail to accept this challenge, it may be interpreted as a lack of national vigor and capacity to respond.

Since the Soviets had already placed a man in orbit around the Earth, NASA was asked to propose a plan that would redirect the world's attention to the United States. James Webb (1907–1992), the newly appointed NASA administrator, took advantage of this politically charged atmosphere by arguing for an accelerated effort to launch communication, meteorological, and planetary satellites, as well as a manned Moon expedition. These goals were put into a formal proposal and delivered to Vice President Johnson who was a strong advocate of a vigorous space program.

Johnson endorsed Webb's plan, forwarded it to President Kennedy, and then took a leading role in building a political consensus to push the plan ahead. Johnson was a savvy politician and clever debater; when asked by congressmen to justify the enormous costs of Webb's proposal, he responded, "Now would you rather have us be a second-rate nation or should we spend a little money?"

On May 25, 1961, President Kennedy took advantage of astronaut Alan Shepard's successful suborbital flight (discussed in Chapter 10) by declaring in a speech before a joint session of Congress titled "Urgent National Needs":

Now is the time to take longer strides—time for a great new American enterprise—time for this nation to take a clearly leading role in space achievement, which in many ways may hold the key to our future on Earth. I believe that this nation should commit itself to achieving the goal, before this decade is out, of landing a man on the Moon and returning him safely to Earth. No single space project in this period will be more impressive to mankind, or more important for the long-range exploration of space; and none will be so difficult or expensive to accomplish.

The idea of landing an American on the Moon seemed to inspire the country, and it received substantial financial support from both political parties. Backed by a strong national consensus, funding requests from NASA for a vastly expanded space program easily passed in the House of Representatives and the Senate. President Kennedy eagerly signed these appropriation bills into law.

However, there were still some people who had severe reservations about this commitment. Eisenhower was quoted as saying he "couldn't care less whether a man ever reached the Moon." He saw little value in this idea when he was in office and nothing had happened to change his mind. He wrote to a friend that the decision to implement a crash program to land a man on the Moon was "almost hysterical" and "a bit immature." He was troubled by the fact that this announcement came just after the failed Bay of Pigs and that the costs of a Moon landing would adversely affect other space efforts. In response to a letter from astronaut Frank Borman, Eisenhower wrote,

What I have criticized about the current space program is the concept under which it was drastically revised and expanded just after the Bay of Pigs fiasco. [It] immediately took one single project or experiment out of a thoughtfully planned and continuing program involving communication, meteorology, reconnaissance, and future military and scientific benefits and gave the highest priority—unfortunately in my opinion—to a race, in other words, a stunt.

T. Keith Glennan also saw Kennedy's pledge as a mistake. In a meeting with Eisenhower he said, "this is a very bad move . . . we are entering into a competition which will be exceedingly costly and which will take up an increasingly large share of that small portion of the nation's budget which might be called controllable." In a July 1961 letter to James Webb, Glennan questioned the lasting political benefits from a manned Moon program as well as the wisdom of competing against the Soviets in this area, "No Jim, I cannot bring myself to believe that we will gain 'prestige' by a shot we may make six to eight years from now. I don't think we should play the game according to the rules laid down by our adversary."

The Dye Is Cast—To the Moon

Dr. Webb and other high-level NASA administrators were surprised by the sudden change in Washington's attitude. The tasks they were being asked to pursue were far beyond the normal projects that the nation had previously supported. The resources needed for a Moon landing were on

Figure Intro. 3-4. Cosmonaut Yuri Gagarin became a worldwide hero after making mankind's first orbit around the Earth.

Figure Intro. 3-5. NASA Chief Administrator James E. Webb (left) and Associate Administrator Hugh L. Dryden (right). Webb led the American space program from 1961 to 1968.

Figure Intro. 3-6. In a speech before Congress, Kennedy committed the United States to a manned lunar landing before the end of the 1960s.

the same level as those required to build the Panama Canal or the atomic bomb. Webb wondered whether the political leadership fully appreciated the magnitude of the effort they had requested.

By advocating many of its new projects, NASA itself had taken a great leap of faith, greater perhaps than the American political leaders realized. When Webb proposed a lunar landing, NASA did not have the necessary operational procedures, launch facilities, or space hardware to carry out this mission. NASA scientists did not even know if the Moon had a solid surface on which to land! Would the landing craft simply disappear into a cloud of lunar dust and sink out of sight? The lessons taught by Vanguard and Pioneer, whose premature launches ended in disaster, weighed heavily on the minds of NASA officials.

In addition to these technological problems, Webb worried that once the sense of crises created by the Soviet successes passed, Congress would fail to allocate the funds it had promised. This fear was reinforced when his request for a long-term financial commitment from both the president and Congress was turned down. Despite these problems, Webb realized there could be no turning back. NASA had agreed to undertake an ambitious program whose outcome was far from certain.

Steps to the Moon

In the public's mind, the cornerstone of Webb's plan was the goal of placing a man on the Moon and returning him safely back to Earth. NASA had begun work on a Moon program soon after the agency came into existence in 1958 and their studies suggested that with patience and perseverance the problems associated with a landing could be overcome.

Logistically, an important operational question had yet to be answered: Should a single craft fly to the Moon, or should a vehicle first be launched into Earth's orbit and then a subcraft proceed to the Moon? The idea of first placing a vehicle into orbit around the Earth, then proceeding to the Moon in a subcraft, and descending to the lunar surface in a third vehicle seemed "risky" and "far-out." However, by July 1962 it became NASA's chosen plan. The program to place a man on the Moon was named Project Apollo.

Prior to attempting a lunar landing, NASA had to first determine if a man could survive in a weightless environment for an extended period of time. The equipment and the procedures for the rendezvous and docking of spacecraft also had to be tested and perfected. NASA initiated two programs to acquire this information. The first program, Project Mercury, would test space hardware and man's reaction to weightlessness. It would be followed by Project Gemini, whose main goals were to perfect rendezvous and docking procedures and determine if a person could work in space.

Before astronauts could journey to the Moon, unmanned probes had to be built and launched to collect additional data. NASA needed to know more about the Moon's space environment, possible landing sites, and the nature of the Moon's surface. If the United States was to complete all of these tasks before the end of the decade, an enormous amount of work lay ahead.

The Evolution of Kennedy's Vision of Manned Spaceflight

With the passage of time, historian Michael Beschloss has questioned Kennedy's motives in announcing that America would place a man on the Moon. Did Kennedy believe that a lunar landing held "the key to our future on Earth" in a scientific sense or in terms of the outcome of the cold war competition with the Soviet Union? Beschloss believes the latter to be true and views the President's decision not as a visionary pronouncement in "the New Frontier spirit of discovery," but "ultimately a political decision made in terms of cold-war strategy." He offered the following assessment of the President's declaration,

> It is a measure of Kennedy's aversion to long-term planning and his tendency to be rattled by momentary crises that one may conclude that in the absence of the Gagarin triumph and the Bay of Pigs fiasco in April 1961, he might never have gone to the length of asking Congress to spend $20 billion on a crash Moon program. Kennedy's desire for a quick, theatrical reversal of his administration's flagging position, especially just before a summit with Khrushchev, is a more potent explanation of his Apollo decision than any other.

Even if this was initially the case, by 1963 Kennedy was having second thoughts about whether the mounting costs of the space program were justified solely by cold-war considerations. When he asked Vice President Johnson about this matter, Johnson's response in stark cold-war terms, ". . . our space program has an overriding urgency that cannot be calculated solely in terms of industrial, scientific or military development. The future of society is at stake." Nevertheless, Kennedy remained unconvinced and his thoughts about the space program began to undergo a radial change.

On August 5, 1963, cold-war tensions between the United States and Soviet Union had declined and both countries agreed to sign a treaty banning nuclear weapon tests. Trying to build on this peaceful foundation, Kennedy met with NASA Administrator Webb to discuss turning the Apollo program into a joint American/Soviet effort. Two days later after this meetings, Kennedy publicly invited Soviet participation in an address before the United Nations:

> Why, therefore, should man's first flight to the Moon be a matter of national competition? Why should the U.S. and Soviet Union, in preparing for such expeditions, become involved in immense duplications of research, construction, and expenditure? Surely we should explore whether the scientists and astronauts of the two countries—indeed of all the world—cannot work together in the conquest of space, sending some day in this decade to the Moon not the representatives of a single nation, but the representatives of all our countries.

In accordance with these statements, on November 12 Kennedy asked Webb to develop a set of projects, "with a view of their possible discussion with the Soviet Union as a direct outcome of my September 20 proposal for broader cooperation . . . in outer space, including cooperation in lunar landing programs." Kennedy clearly was intent on pushing this initiative, not just as a cost-savings measure, but as a way of establishing a long-term working relationship with the Soviets. We will never know what outcome this enlightened view of space exploration might have produced because Kennedy was assassinated just ten days later on November 22. The Johnson administration tried to keep this idea alive and on January 21, 1964 suggested a number of joint small-scale cooperative space projects to the Soviets. These suggestions were rebuffed. With the death of Kennedy, the opportunity for this type of program had passed.

10   Project Mercury's Suborbital Flights

This country should be realistic and recognize that other nations, regardless of their appreciation of our idealistic values, will tend to align themselves with the country which they believe will be the world leader—the winner in the long run. Dramatic accomplishments in space are being increasingly identified as a major indicator of world leadership.

—Vice-President Lyndon Johnson (April 28, 1961, memo to President Kennedy)

Gagarin's Earth-orbiting flight had an immediate and positive effect on the world's perception of the Soviet Union: it was viewed with increased respect and admiration. Gagarin's flight provided tangible evidence that the Soviet Union's technical expertise was equal to or perhaps even greater than that of the United States. The possible impact that this viewpoint might have on global alliances became a great concern to American political leaders. So it was no coincidence that on April 12, 1961, one day after Gagarin's historic flight, NASA officials, including Director James Webb, were called before a special hearing of the House of Representatives Space Committee. Angered by this latest disaster, they were anxious to scold NASA for its failure to gain a lead over the Soviets in the space race. Representative Fulton (R-PA) was particularly aggravated and had harsh words for NASA officials,

> I am tired of being second to the Soviet Union. I want to be first. . . . Do you gentlemen realize you have a responsibility of the way the capitalistic system looks to the world—on its efficiency and on its progress in scientific development? . . . Don't you think the Soviet Union having the first man in space gives them a tremendous advantage at the bargaining table . . . [that] this has placed immense strength . . . in Soviet hands?

Deputy NASA Administrator Dryden responded by stating that work on Project Mercury was proceeding on an "urgent basis," and that the American space program was "moving as rapidly as we know how to move." This response failed to placate congressmen who wanted to see visible results.

The American public was already chagrined by the Soviet's launch of the world's first satellite, *Sputnik 1*, and now they were further annoyed by Gagarin's successful flight. Their frustration at the government's inability to catch up to the Soviets was not helped when a newsman telephoned "Shorty" Powers, spokesperson for Project Mercury at Cape Canaveral and asked him to comment about Gagarin's flight. Awakened at 3 a.m., Powers replied, "We are asleep down here." This statement, taken out of context, was broadcast across the country as his assessment of the nation's space program and further inflamed the public's anger.

The American Response to Gagarin's Flight

On the evening of April 14, President Kennedy met with key space advisors to discuss possible ways in which America could respond to this latest Soviet feat. He asked his assembled space experts, "Is there any place we can catch them? What can we do? Can we go around the Moon before them? Can we put a man on the Moon before them? . . . Can we leapfrog?"

Kennedy was told by Deputy NASA Administrator Hugh Dryden that an all-out space effort, like the atomic bomb Manhattan Project of World War II, would close the gap with the

Figure 10-1. In response to Soviet space successes, President John F. Kennedy initiated the most ambitious space effort ever undertaken.

Soviets but such a program would cost nearly $40 billion. This was an enormous amount of money and was far beyond what NASA's current budget would allow. Kennedy paused and said, "The cost, that's what gets me. . . . When we know more, I can decide if it's worth it or not." As the meeting progressed and one option after another was discussed and discarded, Kennedy became exasperated and said, "If someone can just tell me how to catch up. Let's find somebody—anybody, I don't care if it's the janitor over there, if he knows how." His advisors finally grasped the depth of the President's feelings on this matter when he stated, "There's nothing more important."

Although Kennedy had initially questioned the value of the space program, he was now firmly convinced of its political importance. One of the president's most influential advisers, Theodore Sorensen, sensed that by the end of the meeting the president had already decided to commit the nation to a lunar expedition. Sorensen recalled, "[Kennedy] was more convinced than any of his advisers that a second-rate, second-place space effort was inconsistent with this country's security, with its role as a world leader, and with the New Frontier (the theme of his presidential campaign) spirit of discovery." A similar assessment of the President's thinking was reached by historian John Logsdon. Logsdon concluded that Kennedy saw that "if he wanted to enter the duel for prestige with the Soviets, he would have to do it with Russian's own weapon, space achievement." About a week after this historic meeting, the CIA-backed invasion of Cuba, referred to as the Bay of Pigs, failed, and Kennedy's desire to act increased.

On April 20, Kennedy asked Vice President Lyndon Johnson to construct a report on the current status of the American space program. He wanted to determine if the United States was making a "maximum effort" in the space race. He inquired, "Do we have a chance of beating the Soviets by putting a laboratory in space, or by a trip to the Moon, or by a rocket to land on the Moon, or by a rocket to go to the Moon and back with a man? Is there any other space program which promises dramatic results in which we could win?"

Johnson canvassed the scientific and political leadership for information and on April 21 presented his reply. Johnson concluded that the Soviets were ahead of the United States in the development of large rocket engines, and although we had greater resources to devote to space than the Soviets, we simply had chosen not to do so. The data he had collected showed that neither the United States nor the Soviet Union currently had the capability of placing a man on the Moon. However, with a strong effort, America's space experts felt we could probably beat the Soviets to the Moon and possibly land a man there as early as 1966 or 1967. He concluded his report by stating if the United States were to challenge the Soviets seriously, a more energetic program was required. As to the present time, he wrote, "We are neither making maximum effort nor achieving results necessary if this country is to reach a position of leadership." Von Braun contributed to this report but he also expressed some reservations, "I do not believe," he wrote," that we can win this race unless we take at least some measures which thus far have been considered acceptable only in times of a national emergency."

The Soviet success also demanded an expression of public congratulations, and four astronauts sent statements. The message from John Glenn is representative,

The Russian accomplishment was a great one. It was apparently very successful, and I am looking forward to seeing more detailed information. I am, naturally, disappointed that we did not make the first flight to open this next era. The important goals of Project Mercury, however, remain the same—ours is the peaceful exploration of space. These first flights, whether Russian or American, will go a long way in determining the direction of future endeavors. I hope the Russians have the same objectives and that we can proceed with mutual dissemination of information so that these goals, which all mankind shares, can be gained rapidly, safely, and on a progressive scientific basis.

NASA's First Post-Gagarin Flight

Under political pressure to accelerate its work and produce higher profile results, NASA changed the goal of the upcoming unmanned Mercury *MA-3* mission from that of a suborbital flight to a full-scale one-orbit mission.

MA-3 was launched from Cape Canaveral on April 25, 1961, about two weeks after Gagarin's historic flight, but the results were disappointing. Shortly after lift-off the rocket's guidance control system malfunctioned and the entire vehicle was destroyed by the range safety officer. Trying to put the

Figure 10-2. Vice President Johnson urged Kennedy to support an ambitious space program.

best face possible on this failure, NASA reported that during the abort the Mercury launch escape system had worked perfectly. Given that this was not the objective of the flight, the press was not fooled and correctly deemed this success to be of little consequence.

On a more positive note, *Little Joe 5-B* was successfully launched on April 28 and demonstrated that the Mercury capsule escape system would work under the most extreme conditions expected to arise in an aborted Mercury orbital flight.

America's First Suborbital Manned Flight

Despite this recent record of mixed successes, NASA felt it was now ready to move ahead with its first manned suborbital flight. As director of Project Mercury, Robert Gilruth was responsible for deciding which astronaut would fly this mission. In January 1961, he gathered the astronauts at Langley and announced that this mission would be manned by Alan Shepard. Although this decision did not come as a surprise to many of the astronauts, it did to Shepard. Upon hearing his name, Shepard immediately felt a mixture of feelings ranging from excitement to concern about his companions. In recounting this meeting he said,

> I honestly never felt that I would be the first man to ride the Mercury capsule. I knew I had done well on the tests and the simulator rides. And I thought there was a good chance that I would get one of the early flights. But I had conducted my own private poll, and frankly I figured that one of the others would probably go first. . . . I was excited and happy of course, but it was not a moment to crow. Each of the other fellows had very much wanted to be first himself. And now, after almost two years of hard work and training, that chance was gone. I thanked Mr. Gilruth for his confidence. Then the others, with grins on their faces covering up what must have been their own great disappointment, came over and congratulated me.

Shepard earnestly threw himself into the mission's training program, completing forty simulated flights. If a problem occurred, he was determined it would not be caused by an error on his part. As an experienced test pilot he was well aware that a single mistake could cost him his life. He also realized the best way to insure a successful flight was to focus on his own responsibilities. He later recalled,

> Once in a while the thought of making an unsuccessful flight would get to me too much, and I could feel it in my stomach. I was not going around shaking, but there were some butterflies. There was so much work to do, however, that I never took time out specifically to worry.

Shepard named his Mercury capsule, *Freedom 7*. Freedom to symbolically represent America and "7" to pay tribute

"Fill 'Er Up——I'm In A Race"

Figure 10-3. Gagarin's successful flight spurred the United States to dramatically increase its space efforts.

to the original 7 Mercury astronauts. Each capsule in Project Mercury would be given a different first name, but it was always followed by the number "7."

The launch of *Freedom 7* was planned for May 2, 1961, but poor weather forced its cancellation. Rescheduled for May 4, weather problems again forced a delay until the next day. A little after 1 a.m. on Friday, May 5, Shepard was awakened for his third attempt and began his preflight preparations. Like Gagarin he would remember every detail of his mission,

> The medical exam and the dressing went according to schedule. There were butterflies in my stomach again, but I did not feel that I was coming apart or that things were getting ahead of me. The adrenaline was surely pumping, but my blood pressure and pulse rate were not unusually high. A little after four we left the hanger and got started for the pad. Gus (Grissom) and Bill Douglas were with me. Dee O'Hara, our nurse, went as far as the hall with us, and I said, "Well, Dee, here I go."

The ascent up the gantry elevator to the awaiting capsule was uneventful. Shepard lowered himself into the couch, and after receiving some encouraging words by the support crew, he was sealed inside. Once settled, he fortified himself for the flight: "O.K., buster, you volunteered for this thing. Now its up to you to do it."

After several delays due to weather conditions and last-minute adjustments to the vehicle support equipment, *Freedom 7* lifted off at 9:34 a.m. EST. Shepard's pulse rate shot up from 80 to 126 beats per minute. Indeed the pulse rate of the entire nation shot up as millions viewed the launch on television sets. Astronaut Schirra circled overhead in his F-106 chase plane to monitor the rocket's ascent.

There had been some concern among the astronauts and engineers as to how much vibration the pilot would feel during ascent. Shepard was relieved to find that the initial ascent was very smooth, but this soon changed.

> One minute after lift-off the ride did get a little rough. This was where the booster and the capsule passed from sonic to supersonic speed and then immediately went slicing through a zone of maximum dynamic pressure as the forces of speed and air density combined at their peak. The spacecraft started vibrating here. Although my vision was blurred for a few seconds, I had no trouble seeing the instrument panel. I decided not to report this sensation just then. We had known that something like this was going to happen,

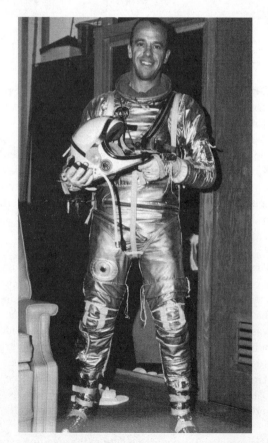

Figure 10-4. Alan Shepard would become the first American to fly a suborbital mission in the Mercury Program.

and if I had sent down a garbled message that it was worse than we had expected and that I was really getting buffeted, I think I might have put everybody on the ground into a state of shock. . . . So I waited until the vibration stopped and let the Control Center know indirectly by reporting to Deke [a fellow astronaut] that it was "a lot smoother now, a lot smoother."

In addition to checking out the launch hardware, Shepard's mission had two other objectives: to determine the effect of weightlessness on an astronaut's ability to complete tasks in space and to test the capsule's control system. Data about weightlessness was obtained soon after the capsule separated from the booster:

> All through this period, the capsule and I remained weightless. And though we had a lot of free advice on how this would feel—some of it rather dire—the sensation was just about what I expected it would be: pleasant and relaxing. It had absolutely no effect on my movements and efficiency.

Shepard took control of the craft and tested the rocket control system—no problems. Later he considered this result to be the most encouraging aspect of the entire mission. He then prepared the craft for reentry by placing it in the proper orientation and switched the controls back to automatic pilot. Taking one last look through the capsule's periscope he reflected,

> The view was spectacular. The sky was very dark blue; the clouds were a brilliant white. Between me and the clouds was something murky and hazy which I knew to be the refraction of the various layers of the atmosphere through which I was passing.

As *Freedom 7* plunged to Earth, Shepard endured forces of 11 g's. This was less than he had been subjected to during training, and he was able to talk to ground communicators throughout his descent. From exchanges with Slayton, he knew that his capsule was on course for its intended impact site. The capsule's parachutes opened on schedule, and the craft landed close to the awaiting recovery ships.

According to Shepard, the ocean landing was abrupt, but not particularly hard, comparable to what he had experienced when his Navy jet had been catapulted from the deck of an aircraft carrier. Shortly after landing, Shepard was retrieved by a recovery helicopter and brought to the deck of the aircraft carrier the USS *Lake Champlain*. While on his way, he had a chance to review his mission,

> . . . I felt relieved and happy. I knew I had done a pretty good job. The Mercury flight systems had worked out even better than we had thought they would. And we had put on a good demonstration of our capability right out in the open where the whole world could watch us taking our chances.

As he neared the carrier, he could see sailors waving their caps and cheering his safe arrival. As a Navy man, he was moved by this greeting and felt a "real lump in my throat." Before starting his medical tests, one final obligation remained to be fulfilled—to pay his respects to the vehicle that had carried him safely through his mission, "I went back to the capsule, which had been lowered gently onto a pile of mattresses on the carrier deck . . . I wanted to take one more look at *Freedom 7*. I was pretty proud of the job that *it* had done too."

The entire flight of *Freedom 7* lasted a little over 15 minutes, during which time Shepard experienced 5 minutes of weightlessness and briefly piloted the capsule. It reached a maximum speed of 5,036 miles per hour, achieved an apogee of 116 miles, and landed in the ocean 302 miles down range from the launch site. A postflight inspection found that the capsule was essentially undamaged.

Shepard's successful flight provided an immediate boost to NASA's morale. It also raised the spirit of the American people who had been annoyed by a barrage of Soviet space achievements. Congratulatory messages streamed in from people as diverse as the King of Morocco to a group of Peruvian scientists. The secret versus the public manner in which the Soviet and American launches had taken place increased worldwide respect for the U.S. program.

President Kennedy took advantage of the favorable climate created by the success of *Freedom 7* to make a dramatic proclamation. In a joint session of Congress, he committed the United States to place a man on the Moon and return him safely to the Earth by the end of the decade. He directed NASA to change the goal of Project Apollo from a circumlunar mission to a manned lunar landing. The American public readily embraced this challenge and on August 7 the Congress agreed to the expanded mission. NASA's budget was increased to $1.7 billion. The success of *Freedom 7*, the boldness of Kennedy's vision, and the desire to regain its prior dominance over the Soviet Union, inspired America to accept a lunar expedition as a national goal.

A Near Disaster: America's Second Suborbital Flight

A second American suborbital flight was planned to build upon Shepard's successful mission. Virgil Grissom, the astronaut selected for this flight, named his capsule the *Liberty Bell 7*—to symbolically represent freedom and the bell-like shape of the capsule. Lift-off occurred on July 21, 1961. Grissom described the ascent:

The lift-off was very smooth. I felt the boosters start to vibrate and I could hear the engines start. Seconds later, the elapsed-time clock started on the instrument panel. I punched the Time Zero Override to make sure that everything was synchronized . . . and reported over the radio that the clock had started. I could feel a low vibration at about T plus 50 sec-

onds, but it lasted only about 20 seconds. There was nothing violent about it. It was nice and easy just as Al had predicted.

Although slightly shaken by the lift-off, Grissom quickly gained confidence and was impressed with the sights he saw out his capsule window as he sped through the lower atmosphere toward space. He was particularly struck by the changing color of the sky,

Shortly after lift-off I went through a layer of cirrus clouds and broke out into the sun. The sky became blue, then a deeper blue, then—quite suddenly and abruptly—it turned black. Al had described it as dark blue. It seemed jet black to me. There was a narrow transition band between the blue and the black—a sort of fuzzy gray area. But it was very thin, and the change from blue to black was extremely vivid.

Once the capsule separated from the Redstone rocket, Grissom assumed manual control of the craft, "I did have some trouble with the attitude controls. They seemed sticky and sluggish to me, and the capsule did not always respond as well as I thought it should."

These problems were annoying, but they did not jeopardize the mission. As Grisson noted, "It just wasn't perfect." Further work, he felt, would easily correct this problem. After

Figure 10-5. On May 5, 1961, America entered the space race with the launch of Freedom 7, *the first manned flight of Project Mercury.*

Figure 10-6. Shepard was hoisted aboard a helicopter for the ride to the awaiting aircraft carrier USS Lake Champlain.

completing as many maneuvers as his time and fuel allowed and contacting ground control, he oriented the capsule for reentry. The reentry proceeded as planned, and the capsule landed within two miles of its intended impact site,

I hit the water with a good bump. This came at T plus 15 minutes and 37 seconds. The capsule nosed over in the water, and the window went clear under. Almost immediately, I could hear a disconcerting gurgling noise. But I made a quick check and could see no sign of water leaking in. . . . I decided to see if it would right itself. And sure enough, it came around in about 20 or 30 seconds.

After recording instrument settings, Grissom radioed the recovery helicopter to come overhead and pick up the craft. He was astonished when the capsule's hatch suddenly blew open and water poured in. He reacted quickly,

I made just two moves, both of them instinctive. I tossed off my helmet and then grabbed the right edge of the instrument panel and hoisted myself right through the hatch. I have never moved faster in my life. The next thing I knew I was floating high in my suit with the water up to my armpits.

Grissom quickly swam a short distance away and watched as the helicopter attempted to grab the capsule. He

swam towards the *Liberty Bell 7* to help, but soon realized the helicopter crew had snagged it. Grissom watched the rescue attempt unfold, "The top of the capsule went clear under water then. But the chopper pulled up and away and the capsule started rising gracefully out of the water. Now I thought, 'These boys have really saved us after all.'"

The recovery helicopter valiantly tried to lift the capsule, but the *Liberty Bell* had taken on too much water. As the struggle continued, the helicopter's engine became dangerously overheated and the craft was pulled so low that its three wheels were touching the water's surface. Realizing that the capsule was about to pull the helicopter into the water, the crew released the *Liberty Bell* and it promptly sank in 15,000 feet of water.

While this drama was being played out, Grissom realized that he was in trouble. His space suit had been taking on water and he was sinking! Waves leaped over him and it was difficult just to keep his head above the water. He anxiously sought help from a second recovery helicopter but to little avail,

I suppose the crew did not realize how much trouble I was in. I was panting hard, and every time a wave lapped over me I took a big swallow of water. I tried to rouse them by waving my arms. But they just seemed to wave back at me. I wasn't scared now. I was angry.

Despite what seemed like a long time to Grissom, only 4 minutes passed before he was plucked from the water by a third recovery helicopter. He was now safe, but the *Liberty Bell 7* was seemingly lost forever.

We had worked so hard and had overcome so much to get *Liberty Bell* launched that it just seemed tragic that another glitch had robbed us of the capsule and its instruments at the very last minute. It was especially hard for me, as a professional pilot. In all of my years of flying—including combat in Korea—

Figure 10-7. Surrounded by other astronauts and NASA Director Webb, President Kennedy presents Shepard with the NASA Distinguished Service Medal.

Figure 10-8. Virgil Grissom preparing for his Mercury flight. The goal of each astronaut was to "fly" a perfect mission. The Liberty Bell *was the only craft he ever lost.*

Figure 10-9. A rescue helicopter tries to retrieve the Liberty Bell 7 *from the ocean. The recovery carrier, the USS* Randolph, *can be seen on the left horizon.*

this was the first time that my aircraft and I had not come back together. In my entire career as a pilot, *Liberty Bell* was the first thing I had ever lost.

Despite weeks of intense postflight tests in which Grissom repeated his actions aboard the *Liberty Bell*, the cause of the hatch blowout was never discovered. Although no one wanted the cause of this failure to be resolved more than Grissom, he was philosophical about the inconclusive investigation. As an experienced test pilot, he knew that sometimes unique problems arise whose cause is almost untraceable. "It was just one of those things."

Need for Another Suborbital Flight?

Although Grissom's mission nearly cost him his life, the flight hardware had performed satisfactorily. Coupled with the successful flight of Shepard, NASA had proven that an astronaut could successfully function in a weightless environment. The original Mercury plan called for two more suborbital flights but it was not clear what new data they would produce.

While NASA was examining the possibility of moving on to an orbital flight, the Soviets launched another cosmonaut, Gherman Titov, into orbit. The launch of *Vostok 2* occurred on August 6, 1961. Unlike Gagarin's one-orbit flight, this mission circled the globe 17.5 times, and it was reported that Titov had exercised manual control of the craft during the flight. Despite this attention-grabbing flight, NASA chose to take a cautious approach toward subsequent missions. In fact, two days after Titov returned to Earth, the Langley Space Task Group announced that America's first orbital flight might be delayed until January 1962.

11 Project Mercury's Manned Orbital Flights

There is nothing superhuman . . . about being an astronaut. There is nothing spooky or supernatural about flying in space.

—John Glenn, Jr. in *We Seven*

Von Braun's Redstone rocket had been used to propel Shepard's and Grissom's suborbital missions. It had an impressive lift capability, but to place a Mercury capsule into orbit, its more powerful successor, the Atlas rocket, would be needed. Between July 29, 1960 and September 13, 1961, four unmanned tests of this rocket were conducted and, as seen in Table 11-1, they provided mixed results. However, the impressive success of *MA-4* was so encouraging that several NASA officials believed the next launch should be a manned orbital mission. Postflight analysis of the *MA-4* capsule showed it to be in such good condition that Gilruth, director of Project Mercury, himself offered the opinion that a man would have survived the flight.

As the amount of time between the orbital flight of Gagarin and a comparable American mission grew longer, NASA Administrator Webb grew increasingly anxious to place an American astronaut into orbit. Some Project Mercury managers objected to such a plan and urged that a live chimpanzee be placed into orbit first, following the model that led to Shepard's successful suborbital flight. This idea met with political resistance. There were rumors that the president's advisors feared another animal flight would invite unflattering comparisons between the U.S. and Soviet programs and subject the public to another round of ridicule from the Soviets. Advocates of the more cautious approached won. Paul Haney, a public affairs officer at NASA headquarters, announced, "The men in charge of Project Mercury have insisted on orbiting the chimpanzee as a necessary preliminary checkout of the entire Mercury program before risking a human astronaut." Deputy Administrator Hugh Dryden further explained NASA's decision,

Figure 11-1. MA-4 flight monitors (left to right) Ralph Gendielle, Donald Arabian, and Walter Kapryan. Because the Soviet lead in space seemed to be getting larger, there was considerable pressure for this mission to succeed.

"You like to have a man go with everything just as near perfect as possible. This business is risky. You can't avoid this, but you can take all the precautions you know about."

MA-5 would be America's first attempt to place a live animal in orbit, a chimpanzee named Enos ("man" in Greek or Hebrew). Enos was selected because of his calm personality. Captain Jerry Fineg, chief veterinarian for this mission, described Enos as "quite a cool guy and not the performing type at all." Preparations went smoothly, and five hours prior to lift-off, Enos was strapped into his special couch in the capsule. Thirty minutes before launch, the countdown was stopped and the capsule opened to correct a switch that had been improperly turned off. Ground controllers chuckled that Enos had flipped the switch after talking to HAM and deciding that the flight was not such a good idea after all.

MA-5 blasted-off into a thin layer of cirrus clouds on November 29, 1961. As in HAM's suborbital flight, Enos pulled levers in response to flashing colored lights. His mental capabilities were tested using patterns of circles, triangles, and squares which he had trained with on Earth. Enos did not seem disoriented by his 181 minutes of weightlessness. While in orbit the capsule's attitude control system developed a problem, and the flight was terminated after completing two of the three planned orbits. Officials later judged this condition would

Table 11-1. MA Flight Data

Mission	Launched	Duration	Results
MA-1	July 29, 1960	3 min	Partial success
MA-2	Feb. 21, 1961	18 min	Successful
MA-3	April 25, 1961	7 min	Aborted
MA-4	Sept. 13, 1961	1 h 49 min	Successful

"We're a Little Behind the Russians and A Little Ahead of the Americans"

Figure 11-2. John Fischetti of the Newspaper Enterprise Association drew attention to the fact that America had only orbited a chimpanzee while the Soviets had placed two men in orbit.

not have endangered an astronaut. He would have simply switched off the automatic system and taken manual control of the capsule. After completing a flight of 3 hours and 21 minutes Enos and the capsule were recovered by the destroyer *Storms*. A postflight examination of the capsule showed that it had not been seriously damaged by reentry. Enos was rewarded for his participation in this flight by being retired to his home at Holloman AFB in New Mexico. NASA had proven that America was now prepared for its first manned orbital flight.

The Selection of America's First Man to Orbit the Earth

At a news conference following the successful flight of *MA-5*, Gilruth announced that John Glenn would be the first American to attempt an orbital flight. Scott Carpenter would serve as the mission's backup pilot in the event that Glenn became sick or was otherwise unable to fly.

Competition for Mercury flights was friendly, but intense. All of the astronauts were accomplished pilots, all had made a major career commitment to the space program, and all had trained hard. Glenn was both relieved and pleased with his selection. The suborbital flights of Shepard and Grissom had paved the way, and now he was being given the opportunity to complete what they had started. Glenn recalled his thoughts at that time,

> I had hoped all along that I would be the first man in the world to go into space. I felt I was qualified, and

I think I must have tried a thousand times since I joined the program to imagine what it would be like to sit on top of the booster, ready to go. Some disappointments were in store for me, however. First, the Russians worked faster as a nation than we did and beat us to it. Then two of my colleagues were picked to go ahead of me on the Redstone ballistic flights. . . . I guess I am a fairly dogged competitor, and getting left behind twice in a row was a little like always being a bridesmaid but never a bride

It was now Glenn's turn to lead America into space.

America's First Man in Orbit

NASA felt completely confident in its ability to place a man in orbit. *MA-6*, America's first manned orbital mission, was planned for early 1962. It was to be a three-orbit mission. This was modest compared to Soviet cosmonaut Titov's 17.5-orbit flight, but its duration was in keeping with the cautious approach NASA was determined to follow.

Significant differences existed between Glenn's flight and the earlier suborbital missions. One of the major differences was in the booster. This would be the first time an Atlas rocket was used in a manned flight. It was nearly five times more powerful than a Redstone and three times faster. A second difference was the duration of the mission. Unlike the prior ballistic flights that lasted less than 20 minutes, this flight would last over 4 hours and Glenn would fly the Mercury capsule for a significant portion of this time. Thirdly, Glenn would make a number of scientific observations. During the flight he would search for comets, note the frequency of meteor flashes in the Earth's atmosphere, and describe the shape and intensity of the atmospheric emission caused when charged particles entered the upper atmosphere (the aurora).

NASA invited the world media to cover the launch of the capsule, which Glenn had named *Friendship 7*. After several delays, conditions were favorable, and on February 20, 1962, the countdown commenced. About 20 minutes prior to lift-off, live television coverage began with an estimated audience of 100 million people. An additional 50,000 had gathered on the beaches near Cape Canaveral.

Friendship 7 roared off its launch pad at 9:47 a.m. As it rose from the pad, Glenn analyzed the lift-off,

> I had always thought from watching Atlas launches that it would seem slow and a little sluggish, like an elevator rising. I was wrong; it was not like that at all. It was a solid and exhilarating surge of up and away.

No problems occurred during lift-off and the capsule quickly reached the orbital speed of 17,544 mph. Initially, all of its systems functioned properly, but soon a problem developed with the automatic thruster controls. Glenn switched them off and took manual control. As his craft sped around the Earth, Glenn watched his first sunrise in space and was

Figure 11-3. John Glenn poses with Thomas O'Malley of General Dynamics and Paul Donnelly of STG.

Figure 11-4. On February 20, 1962, Friendship 7 *lifted off carrying John Glenn, the first American to orbit the earth.*

surprised to find his vehicle surrounded by thousands of small, luminous particles. He was puzzled by them but they were clearly visible each sunrise and sunset. Recurring thruster problems required more of his time than had been anticipated so many of the activities scheduled for the second and third orbits were cancelled.

As Glenn continued his piloting and observational duties, ground controllers received a signal from an on-board sensor that indicated the heat shield and compressed landing bag were not securely joined to the capsule. This was very serious. If the shield was lost, the capsule would burn up as it tried to return to Earth. Ground controllers knew the heat shield was only being held by the straps of the retropackage, but did not communicate this information to Glenn. They wanted to determine what options remained and felt there was nothing Glenn could do to remedy the situation anyway.

Following discussions with the capsule's manufacturer, the operations team concluded the retrorocket straps were adequate to hold the heat shield in place long enough during reentry for air pressure to take over this job. They sent a stream of messages to Glenn not to jettison the retropackage prior to reentry. Based upon these instructions and other ground station inquiries, Glenn deduced the serious nature of the problem he faced,

It was clear to me now that the people down on the ground were really concerned or they would not be asking such leading questions. I was fairly certain, however, that everything was in good shape. It occurred to me that if the heat shield were really loose, I would almost certainly have been able to hear it

shaking behind me or banging against the edge of the capsule. . . . Still there was room for concern. This was the only thing that stood between me and disaster. . . .

As the capsule began its descent, neither Glenn nor mission control were totally certain he would survive. Shortly after reentry, Glenn heard small objects brushing against the capsule and thought this noise came from the retropackage breakup. Moments later he saw a strap from the retropackage swing into view and become consumed by fire,

A loose strap burned off at this point and dropped away. Just at that moment I could see big flaming chunks go flying by the window. . . . I thought the heat shield was tearing apart. This was a bad moment. But I knew that if the worst was really happening it would all be over shortly and there was nothing I could do about it. . . .

Despite seeing a "real fireball" outside his window, Glenn realized that the moderate temperature inside the capsule meant that the ablation shield must be attached and working. A life-threatening situation had been averted, but he

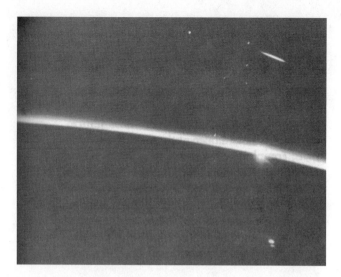

Figure 11-5. Glenn's photograph of a sunset from his Mercury capsule.

Figure 11-6. Glenn receives a hero's welcome upon his return to Earth.

now faced another problem. Shortly after passing the region of peak g, the capsule had begun to oscillate to such an extent that it became uncontrollable. Having just survived one problem, Glenn now wondered if this motion would interfere with the deployment of the parachutes. He was about to manually eject the first chute, when it automatically deployed and the oscillations began to diminish.

At a height of 17,000 feet, the main chute successfully opened and *Friendship 7* settled gently into the Atlantic Ocean about 40 miles from its projected impact area. The capsule was recovered and hoisted aboard the destroyer *Noa*. After warning the crew to step back, Glenn blew open the capsule's hatch and exited. His first words on leaving the Mercury craft, "It was hot in there."

Physicians on the scene described Glenn as hot, sweating profusely, and fatigued, but otherwise in good condition. During the flight he had lost a little over 5 pounds. President Kennedy was informed of this successful landing and talked to the media from the Rose Garden of the White House,

I know that I express the great happiness and thanksgiving of all of us that Colonel Glenn has completed his trip, and I know that this is particularly felt by Mrs. Glenn and his two children. I also want to say a word for all of those who participated with Colonel Glenn at Canaveral. They faced many disappointments and delays—the burdens upon them were great—but they kept their heads and they made a judgment, and I think their judgment has been vindicated. We have a long way to go in this space race. But this is the new ocean, and I believe the United States must sail on it and be in a position second to none.

The nation was elated by Glenn's successful flight, Lloyd Swenson wrote in *This New Ocean: A History of Project Mercury*,

Honors and celebrations consumed several days. Glenn, his family, Vice-President Johnson, and the Mercury entourage passed in review on February 26 before an estimated 250,000 people lining rainy streets in Washington, after which the astronaut gave a 20 minute informal report to a joint session of Congress. New York City proclaimed March 1 "John Glenn Day," and Mayor Robert Wagner presented medals to Glenn and Gilruth. The next day there was an informal reception in honor of the orbiting American at United Nations Headquarters. Glenn then journeyed to his home town, New Concord (population 2,300), Ohio, where about 75,000 greeted him on March 3.

After a thorough postflight analysis, the capsule was exhibited in seventeen countries. Millions of people viewed it close-up and marveled at the astronaut's courage. After this "fourth orbit of *Friendship 7*," the capsule was sent to the Smithsonian Institution in Washington, DC, and displayed next to Lindbergh's *Spirit of St. Louis*.

MA-7: A Potential Disaster

Donald (Deke) Slayton was scheduled to be the next Mercury astronaut to orbit the Earth. On March 15, 1962, NASA announced that an irregularity in his heartbeat had been detected, so Scott Carpenter would take his place.

Slayton's condition had been detected in August 1959 during a centrifuge ride to test his flight endurance. Now that NASA faced an imminent launch, Slayton's medical condition became an issue. Medical experts disagreed as to whether or not his condition posed any threat, but NASA could not afford to jeopardize the mission's success. The final decision on who would pilot *MA-7* rested with Administrator

Focus Box: A Child's Perspective on Glenn's Mission

After returning from space, Glenn was flown to Cape Canaveral where he relaxed and vacationed with his family. President Kennedy had come down from the capital to congratulate him on his successful flight. During this time Glenn also worked on a speech to be delivered to a joint session of Congress. Glenn would soon depart for Washington, DC, along with the president. The following story, recounted in *John Glenn: A Memoir*, illustrates Glenn's sense of humor. Nearly 40 years after his historic flight, this incident was still fresh in his mind.

I worked on my speech, and on Monday morning we flew to the airport at West Palm Beach to meet the president for the trip back to Washington on *Air Force One*. Jackie was going to be staying in Palm Beach with Caroline and their infant son, John, Jr. Jackie wanted Caroline to meet us, so they came out to the airport with the president to see us off.

We were on the plane, and the president boarded, and behind him came Jackie with little Caroline, holding her by the hand, and Jackie said, "Caroline, this is the astronaut who went around the Earth in the spaceship. This is Colonel Glenn."

Caroline looked at me, and then all around the plane. Finally she turned back to me, her face disappointed, and said in a quavering voice, "But where's the monkey?"

Webb himself. He decided that although Slayton's condition might not cause a problem, there was no way to be absolutely certain. Caution dictated if another astronaut was available, that person should pilot *MA-7*.

Scott Carpenter was the logical person to serve as Slayton's replacement. He had trained as Glenn's backup pilot, so his selection allowed the flight to proceed on schedule. His mission was intended to be similar to Glenn's flight. He would orbit the Earth three times and collect scientific data. However, some additional experiments were planned. While in orbit, a multicolored, tethered balloon would be released from the capsule and observed to determine if an astronaut's perception of color was affected by weightlessness. The balloon would also be used to measure aerodynamic drag as the capsule orbited the Earth. An additional task was to obtain blue and red filtered photographs of the boundary line between the lower atmosphere and the Earth's surface, the Earth–atmosphere limb. These pictures would be used for two purposes: (1) to precisely determine the location of the Earth–atmosphere boundary, and (2) to determine the best wavelength for meteorological satellite photography.

Carpenter would also observe how a liquid behaved in a weightless environment, and measure the transmission of light through the atmosphere by observing flares launched from the ground. *MA-7* had a full schedule of events considering it would only last three orbits. Carpenter named his craft *Aurora 7* to symbolically represent a light for all mankind.

After several weather-related delays, *Aurora 7* was launched on May 24, 1962, as an estimated 40 million people watched on television. Carpenter later provided a firsthand account of his journey into space,

The launch was just a snap. It was the shortest 5 minutes I had ever experienced. We rose with very little vibration inside the capsule. In fact, the ride up was much gentler than I had anticipated. The engines made

a big racket, but there was no violent trembling of the whole structure. . . . There had been no problem when I went through the area of "maximum g" where the aerodynamic forces piling up against the capsule had reached their peak. There was a barely perceptible vibration, but no more than I had experienced on lift-off.

One of his first activities in orbit was to test the capsule's control systems. He found the manual system performed well, but the automatic system needed further refinements. Although much of the first orbit was devoted to these tests, he found time to make some visual observations of the Earth and was deeply impressed by what he saw,

The sight was overwhelming. There were cloud formations that any painter could be proud of—little rosettes or clustered circles of fair-weather cumulus down below. I could also see the sea down below and the black sky above me. I could look off for perhaps a thousand miles in any direction, and everywhere I looked the window and the periscope were constantly filled with beauty.

As the capsule passed from daylight into darkness, Carpenter attempted to observe flares launched from the Great Victoria Desert in Australia. Unfortunately, dense clouds covered this region and in the course of maneuvering the craft to observe these lights, a great deal of fuel was consumed. This would later have a serious impact on the mission.

Carpenter reported that the capsule lights were so bright that he was not able to see many more stars than when on Earth, and in his opinion, stellar positions would not be useful for gaining heading data for navigational purposes. A problem arose when the temperature in his space suit increased, causing him to perspire profusely. With the help of ground controllers, the needed adjustments were made.

Figure 11-7. The liftoff of Aurora 7 on May 24, 1962, carrying astronaut Scott Carpenter.

Figure 11-8. Scott Carpenter and his wife in a parade given in honor of his successful mission. This sort of celebration was typical of that given to the returning astronauts after a successfull flight.

During the second orbit, Carpenter performed numerous capsule maneuvers, one of which was to observe and photograph airglow better. By this time, the capsule's fuel supply was seriously depleted, and he was instructed to conserve his remaining propellant. The tethered balloon experiment yielded disappointing results. The multicolored balloon did not totally inflate, so only two colors were observable, orange and a dull aluminum. Of these two, Carpenter noted that orange was much easier to see. The drag measurements proved difficult to obtain, and the balloon refused to jettison. In Carpenter's words, "All in all, the balloon was sort of a mess."

By this time, ground controllers noted that Carpenter's body temperature had risen to 102 degrees. He was again forced to devote valuable time adjusting the temperature of his space suit. To conserve fuel for reentry, the capsule was now allowed to drift while in orbit, and Carpenter had few activities to occupy his time. This unanticipated downtime later proved useful to NASA for planning the duration of astronaut rest and sleep periods for future, long-duration flights.

Near the end of the flight, the fuel supply aboard Aurora 7 was so low that there was barely enough left to keep the capsule in a proper orientation during reentry. If the craft entered the atmosphere at the wrong angle, the chances of surviving reentry were zero. Carpenter was somewhat philosophical about this situation and remembered thinking,

This has been the greatest day of your life. You have nobody to blame for being in this spot but yourself. If you do everything correctly from now on you may make it. If you do not, you just won't.

At the proper time, the capsule entered the atmosphere at very nearly the exact angle, but other problems began to arise. As was the case during Glenn's descent, the capsule began to undergo increasingly large oscillations, finally reaching nearly 270 degrees. Carpenter decided to deploy the initial parachute in an attempt to damp them out. This procedure was successful.

As the craft continued to descend and passed 10,000 feet, the main chute failed to deploy. Within seconds, Carpenter hit the manual deploy switch and was relieved to see the chute come out as a glorious orange and white canopy.

Because of the need to conserve fuel at the end of the flight, it was impossible to precisely align the capsule for reentry, so it landed 210 miles from its intended impact site. Upon landing, Carpenter exited the capsule, deployed a raft and waited for the recovery craft. He took time to review the flight,

I couldn't believe that it had all happened. It had been a tremendous experience, and though I could not ever really share it with anyone, I looked forward to telling others about it as much as I could. I had made some mistakes and some things had gone wrong. But I hoped that other men could learn from my experiences. I felt that the flight was a success, and I was proud of that.

Three hours after landing, Carpenter was picked up by a recovery helicopter. After returning to the United States he received numerous congratulatory messages and was given a hero's welcome in his hometown of Boulder, Colorado. Later he was cheered by a crowd of 300,000 people in Denver.

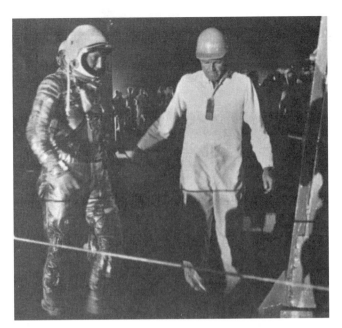

Figure 11-9. Schirra is assisted by astronaut Cooper as he enters the elevator that will carry him to the top of the launch gantry.

Figure 11-10. Sigma 7 being recovered by the USS Kearsarge. *After a tense launch, Schirra was glad to be home.*

MA-8: An Intermediate Duration Flight

The flights of Glenn and Carpenter essentially fulfilled the objectives of Project Mercury, so there was little need for another short duration mission. However, it was felt that a one-day flight could provide useful data, so the scope of Project Mercury was expanded to include this new goal. An intermediate flight of six orbits was planned to help bridge the gap between the three-orbit flights of Glenn and Carpenter and a full-day mission.

On June 27, 1962, NASA announced that the next Mercury flight, *MA-8*, would be six orbits long, with a target launch date in September. Walter Schirra would pilot the mission, and L. Gordon Cooper would serve as his backup. *MA-8* would primarily be an engineering mission, although a few science experiments were included. A second attempt would be made to observe flares launched from Australia, and a 140 million candlepower xenon light in Durban, South Africa, would be lit. Photographs of interesting terrestrial features and cloud formations would be obtained as opportunities arose.

As Schirra trained for *MA-8*, the Soviets launched the 5-ton *Vostok 3* on August 11, 1962, with Major Adrian Nikolayev on board. On the following day, *Vostok 4* carried Lt. Col. Pavel Popovich into a nearly identical orbit with an apogee of 156 miles and a perigee of 113 miles. After 64 orbits, or four days in space, Nikolayev returned safely to Earth. Popovich landed within 3 miles of him just 15 minutes later after completing 48 orbits. This double flight demonstrated significant progress in achieving a rendezvous between the spacecraft, a task that would be needed in a lunar mission.

NASA discovered that it was not a simple matter to increase the number of orbits from three to six, and in fact, it presented many difficulties. Additional electrical power was needed to power the capsule's instruments, more oxygen was required for the astronaut, and a larger amount of hydrogen peroxide was needed to purify the air. Since each of these items added weight to the capsule, some modifications to the interior structure had to be made. Despite these and other technical changes, the outside of the capsule appeared to be nearly identical to *Aurora 7*. Schirra named his craft *Sigma 7*. Sigma is the mathematical symbol for summation.

Sigma 7 was launched on October 3, 1962. Seconds into the flight a tense moment occurred when the rocket unexpectedly rolled to within 20 percent of an abort condition. Fortunately the roll stopped and the mission was able to proceed. Once in orbit, Schirra tried to observe flares launched from Woomera, but as in the other missions, the area was covered by clouds. Later in his flight, he attempted to observe the xenon light in Durban, but again cloudy conditions were present. On a more positive note, the capsule's automatic flight control system worked beautifully. Nine hours and thirteen minutes after lift-off *Sigma 7* landed within 4.5 miles of its intended impact area. Postflight inspection of the capsule found very few problems to correct, and with the exception of the problem at launch, the flight was textbook perfect.

MA-9: The Last Mercury Flight

The ultimate objective of the expanded Mercury program, a one-day manned flight, was now within NASA's grasp. This mission, designated *MA-9*, would be piloted by L. Gordon Cooper. Although this one-day, 22-orbit mission was still far short of the 64-orbit record held by cosmonaut Nikolayev, it was long enough to provide NASA with valuable new data. Cooper named his 3,026-pound capsule *Faith 7*.

Table 11-2. The Manned Mercury Flights

Mission	Astronaut	Capsule	Launched	Type	Duration
MR-3	Alan Shepard, Jr.	*Freedom 7*	May 5, 1961	suborbital	15 min
MR-4	Virgil Grissom	*Liberty Bell 7*	Aug. 21, 1961	suborbital	16 min
MA-6	John Glenn, Jr.	*Friendship 7*	Feb. 20, 1962	3 orbits	4 h 55 min
MA-7	Scott Carpenter	*Aurora 7*	May 24, 1962	3 orbits	4 h 56 min
MA-8	Walter Schirra, Jr.	*Sigma 7*	Oct. 3, 1962	6 orbits	9 h 13 min
MA-9	L. Gordon Cooper, Jr.	*Faith 7*	May 15, 1963	22 orbits	34 h 20 min

While preparing for this flight, NASA was also accelerating work on its two other manned programs, Gemini and Apollo.

Faith 7 experienced a one-day launch delay and then roared from its pad on May 15, 1963. The lift-off went perfectly, and there was little difference between the prelaunch target orbit and that actually achieved. The highest priority experiments of *MA-9* were medical, and Cooper took care that these were completed in a timely fashion.

A second tethered balloon experiment was attempted, but failed to work as planned. During his 16th orbit Cooper observed and photographed the zodiacal light and night air glow. On orbits 17 and 18, he obtained infrared weather photographs. Late in the mission some capsule problems arose, so Cooper decided to make a manual rather than the planned automatic reentry. Thirty-four hours and twenty minutes after lift-off, Cooper splashed down in the Pacific Ocean.

Like previous astronauts, Cooper was given a hero's welcome with parades held in Honolulu, Washington, New York City, and Houston, and like Glenn, he addressed a joint session of Congress. The flight of *Faith 7* was so successful that there was little need for additional Mercury flights. On June 12, 1963, Administrator Webb announced that Project Mercury had met all of its goals and officially ended the program.

The First Woman in Space

A few days after Webb's announcement, on June 14, 1963, the Soviet Union launched *Vostok 5* carrying cosmonaut Valery Bykovsky, and two days later *Vostok 6* was launched carrying cosmonaut Valentina Tereshkova, the first woman in space. After completing 81 and 48 orbits, respectively, they returned to Earth within minutes of each other. Not only was Tereshkova the first woman in space, her flight was longer than all of the Mercury flights combined. Obviously, the Soviets still enjoyed a sizeable lead in the space race.

The mission of *Vostok 5* and *6* essentially duplicated the earlier dual cosmonaut flight and further demonstrated the ability of the Soviets to simultaneously launch and support multiple flights. After landing, Tereshkova attended the International Congress of Women held in Moscow to make the political point that men and women enjoyed equal status in the Soviet Union. Despite this claim, another nineteen years would pass before the launch of the next Soviet woman. It would be even longer before an American woman was placed in space.

Conclusion

From its inception to its successful conclusion, Project Mercury lasted fifty-five months. During this time, almost 2 million Americans dedicated their time and skills to the effort. The project cost a little over $400 million and six astronauts had flown in suborbital or orbital missions (see Table 11-2).

Among its main findings were proof that a human could perform complex tasks in space. "Zero g," or weightlessness, did not limit the astronaut's ability to pilot the craft or make scientific observations. After returning from space, the astronaut required a few days to completely readjust to Earth's gravity, but exposure to weightlessness did not appear to have any long-term health effects.

In terms of the program's logistical operation, launch preparations were found to be much more time consuming than originally anticipated. Automated systems would be required to shorten flight preparation time in the future.

An additional lesson revealed by Project Mercury was the need for a sophisticated ground support system that required state-of-the-art communication devices and computers. The most important lesson learned from this program, however, was that man had an invaluable role in space. Mercury Flight Director Christopher Kraft summed up this finding:

> Man is the deciding element. . . . As long as man is able to alter the decision of the machine, we will have a spacecraft that can perform under any known conditions and that can probe into the unknown for new knowledge.

Unmanned Lunar Reconnaissance Missions
Part 1: Project Ranger

... the culture of the lab [JPL] still remained that of a center for battlefield missiles, where failure in test flight was acceptable because other rounds would be available for future launches. As a result, Ranger became the exercise that taught JPL how to build successful spacecraft.

—T.A. Heppenheimer in *Countdown: A History of Space Flight*

As Project Mercury's astronauts were taking America's first tentative steps into space, other programs were underway to send unmanned spacecraft to explore the Moon. Initially the goal of these missions was purely scientific—to learn more about the Moon's physical properties and space environment. NASA realized before a manned landing could take place, many questions had to be answered. For instance, was the Moon's surface covered by a thick layer of dust or was it firm enough to support a landing craft? If a dust layer was present, would a craft sink into it and become buried? If a firm surface existed, was it covered by boulder-size rocks and numerous small craters that would cause a craft to tip over as it attempted to land?

Prior to 1960, much was known about the main properties of the Moon. Its surface consisted of rough mountainous regions called highlands and relatively flat regions called maria. The highlands consisted of dark material which reflected about 18 percent of the sunlight striking it. The maria reflected about 5 percent of the incident sunlight and, as a result, were even darker. The highlands were heavily pitted with craters, whereas the maria were relatively free of these objects. Radio studies had found that only 7 percent of the Moon's reflected light was polarized, a value that was much smaller than would be produced by a solid surface. These measurements suggested that the density of the lunar surface was less than half that of solid rock. Ground-based infrared thermal measurements also supported the view that the Moon's surface was covered by a layer of dust whose thickness varied from place to place. Around young craters such as Tycho, the dust layer was thought to be about one-third of an inch deep. In other regions the dust was thought to be deeper, but depths of more than a few centimeters seemed unlikely anywhere on the Moon.

Although these ground-based radio and infrared measurements strongly suggested that the lunar dust layer was not deep enough to endanger a landing craft, additional direct evidence was needed. Perhaps other explanations would be found to explain these results. Before a manned mission could be undertaken, any questions about the nature of the surface had to be answered. NASA could not risk being wrong on this matter.

Figure 12-1. The Moon as seen from Earth. The prominent crater with the light-colored material coming from it is Tycho.

Figure 12-2. The crater Tycho is an excellent example of a young lunar feature.

NASA's Unmanned Lunar Probes

Initially, NASA intended to explore the Moon using two probes. The first probe was to obtain data about conditions between the Earth and the Moon and acquire high-resolution pictures of the Moon's surface. A small seismic instrument package would also be hard-landed to measure lunar activity. This craft was named Ranger.

The second probe was envisioned as having two stages. The first stage would orbit the Moon and obtain a high-resolution photographic map of a substantial region of the Moon. These images would be used to identify future Apollo landing sites. The second stage would contain a lunar lander that would then descend to the Moon and deploy instruments to measure day/night temperatures, radiation levels, surface rigidity, and soil composition. This two-stage craft was called the Surveyor Lunar Orbiter.

NASA believed that the information collected in these two missions would be sufficient to construct a vehicle that could land a man on the Moon. By early 1960, plans to build the two probes were underway. As work progressed it became obvious that the weight of the Surveyor Lunar Orbiter would exceed the lift capability of the current generation of American rockets. Therefore, its mission was divided into two parts, each of which would employ a single-stage probe. One probe was assigned the task of orbiting the Moon and photographing its surface. This satellite was appropriately called the Lunar Orbiter. The goals of the second craft were to test the hardware needed to make a soft landing and to collect the surface data mentioned above. Reflecting these objectives, this craft was named the Surveyor.

The task of turning these conceptual plans into space hardware was delegated to two of NASA's field centers, the Jet Propulsion Laboratory (JPL) and the Langley Research Center. JPL would be responsible for the Ranger and Surveyor probes, and Langley would work on the Lunar Orbiter.

Project Ranger

Late in 1959 NASA headquarters asked William Pickering, the director of JPL, to develop a three-year program to explore the solar system from space. Project Ranger would serve as the lunar component of that plan and was originally envisioned by NASA headquarters to consist of five missions. The objectives of these flights were to produce a high-resolution photographic map of the Moon, measure the particle density and field strengths between the Earth and Moon, and land a seismic capsule on the Moon's surface.

The Ranger missions were divided into two segments. The first two flights were engineering missions whose goals were to test the launch vehicles, the probe, and ground control procedures. They would also collect scientific data about the Earth–Moon space environment using eight different detectors placed aboard the spacecraft. They were not intended to gather data about the Moon itself. The goals of the remaining three Ranger craft were to obtain close-up images of the Moon, measure the quantity of radioactive minerals in its soil using a gamma-ray spectrometer, and hard-land a seismometer to measure lunar activity.

The Ranger spacecraft was a six-sided probe that used two large solar panels to collect energy for the probe when in space. Extending outward from the body of the craft was a

Figure 12-3. NASA Chief Administrator Keith Glennan (center), Deputy Administrator Hugh Dryden (left), and Associate Administrator Richard Horner (right) directed NASA activities at the time of Project Ranger.

Figure 12-4. JPL Director William Pickering in 1954.

Figure 12-5. JPL Lunar Program Director Clifford Cummings (left) and JPL Ranger Project Manager James Burke (right).

towerlike structure with instruments mounted on it to measure the space environment. This gave the craft a distinctive "T" shape (see Figure 12-6).

The launch vehicle to be employed in this effort was the Atlas-Agena rocket. The Atlas would be used at lift-off and once its fuel was expended, it would be jettisoned. The upper Agena stage would then ignite and place the probe into an orbit around the Earth. The Agena would then be fired a second time to propel the Ranger into a high orbit or to the Moon. Prior to the launch of the first Ranger, NASA added four additional flights to the Ranger program. These probes would provide further data on the topology and hardness of the Moon's surface.

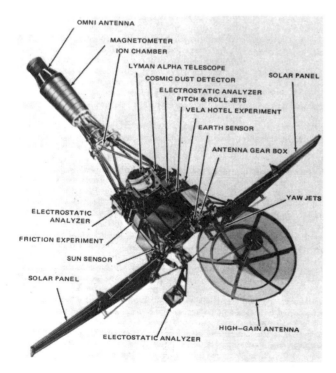

Figure 12-6. Components of the engineering Rangers: Rangers 1 and 2.

The *Ranger 1* and *2* Engineering Flights

Ranger 1 was ready for launch on July 18, 1961, but its lift-off was postponed when numerous difficulties were encountered. It finally blasted-off on August 23, about three months after President Kennedy's address to Congress committing America to a manned lunar mission and approximately one month after Virgil Grissom's suborbital Mercury flight.

Initially, the launch looked good, but telemetry received at NASA Goldstone's radio station in California revealed that a valve in the second stage of the Atlas-Agena launch vehicle had prematurely closed. As a result of this failure, *Ranger 1* was placed into a much lower orbit than planned. About one week after lift-off, it reentered the Earth's atmosphere and was destroyed over the Gulf of Mexico.

The second engineering flight, *Ranger 2*, was launched on November 18, 1961, but another problem with the second stage of the Atlas-Agena rocket caused the probe to be placed into a very low orbit. This time the craft reentered the Earth's atmosphere just 6 hours after leaving its launch pad. Little useful data was obtained by either of these two flights. The Ranger program was off to a poor start.

Rangers 3 through *5*

Ranger 3 was launched on January 26, 1962, more than two years after the Soviet probe, *Luna 2*, had impacted the Moon, and about a month before John Glenn's Mercury flight. Unlike its predecessors, the goals of *Ranger 3* were to photograph the Moon and hard-land a seismometer on its surface. It was the first American craft whose purpose was to impact the Moon.

This time the Atlas rocket's radio guidance system failed and the Ranger went into a higher orbit than planned. An error in the Agena's flight program then caused a further deviation from the planned trajectory. The probe missed the Moon by over 23,000 miles. On the brighter side, some aspects of this mission were successful. The craft successfully

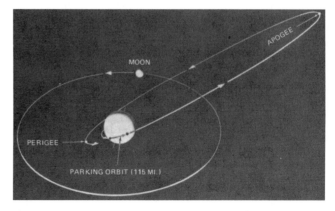

Figure 12-7. The flight path of Rangers 1 and 2 would carry them well beyond the orbit of the moon. Their goal was to measure the properties of space in this region.

Figure 12-8. *NASA's Goldstone radio antenna received telemetry transmitted from the Ranger craft. This data was then sent to JPL.*

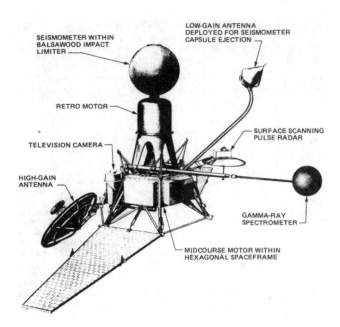

SEISMOMETER WITHIN BALSAWOOD IMPACT LIMITER

LOW-GAIN ANTENNA DEPLOYED FOR SEISMOMETER CAPSULE EJECTION

RETRO MOTOR

SURFACE SCANNING PULSE RADAR

TELEVISION CAMERA

HIGH-GAIN ANTENNA

GAMMA-RAY SPECTROMETER

MIDCOURSE MOTOR WITHIN HEXAGONAL SPACEFRAME

Figure 12-9. Rangers 3 *through* 5 *contained a seismic package (seen at the top of the craft) that was to be hard-landed on the Moon.*

made its programmed mid-course correction, reorienting itself using the sun and the Earth. The on-board gamma-ray spectrometer also operated for 50 hours and sent back data about the environment between the Earth and the Moon. As the craft sped by the Moon, an attempt was made to image the lunar surface using the craft's camera system, but a failure in its central computer caused this activity to be unsuccessful.

The partial success of *Ranger 3* raised hopes at JPL that the next mission would produce real results. As *Ranger 4* was prepared for launch, project personnel were upbeat and public interest was high. The April 8 edition of the *New York Sunday News* stated the upcoming flight, "will blaze a trail for the first American—and, hopefully, the first Earthling—to follow to the Moon."

Ranger 4 lifted-off on April 23, 1962. This time the Atlas and Agena rockets worked well, and the Ranger was placed into a trajectory that guaranteed it would hit the Moon. Despite this promising start, it soon became apparent that the mission was in serious trouble. Telemetry from the spacecraft indicated the master clock had failed. This meant that the craft would not conduct its pre-timed operations, and even worse, it could not accept any commands from Earth to alter this situation. *Ranger 4* was essentially useless.

After crashing into the Moon, NASA Administrator Webb gave the mission the best possible spin by stating *Ranger 4* was "an outstanding American achievement." This positive assessment was echoed in newspapers across the country despite the fact that no information about the Moon's environment or surface was obtained. At least it had hit the Moon.

The last of the original Ranger missions, *Ranger 5*, was launched on October 18, 1962. Its payload weighed 727 pounds and contained an instrument package that was to transmit data back to the Earth about the lunar surface. The Atlas-Agena

rocket performed well, and after reaching orbit, the Agena fired and sent the Ranger on its way to the Moon. Approximately 38 minutes after leaving the Earth's orbit, electrical power supplied by the probe's solar panels abruptly stopped. The spacecraft automatically switched to its on-board battery, but it only had sufficient power for a few hours of operation. Halfway through the Ranger's mid-course maneuver the battery died. The craft's radio transmitter then ceased to operate and the ship began to tumble. As a result of the incomplete mid-course maneuver, *Ranger 5* missed the Moon by 450 miles.

NASA Boards of Inquiry

On October 22, JPL Director Pickering ordered an internal board of inquiry. Homer Newell, director of the NASA Office of Space Science convened a non-JPL NASA review panel, headed by Albert Kelly, head of the Electronics and Control Division of NASA's Office of Advanced Research and Technology. After a three-week investigation, the Kelly board issued a report that harshly criticized JPL. The board charged that the Ranger craft was far too complex for its intended mission and was overdesigned to such an extent that the failure of noncritical components jeopardized the craft's primary mission. They concluded that the Ranger problems were so serious "the present hardware comprising *Rangers 6* through 9 . . . is unlikely to perform successfully." The Kelly board found that rather than relying on extensive ground testing, the attitude at JPL was one of "shoot and hope." There appeared to be a complacent acceptance that some component of the craft would fail and that this fault would be fixed in a subsequent flight.

The board recommended that more exhaustive test procedures be adopted to identify hardware problems prior to a flight. They also recommended that the sterilization of the Ranger, intended to prevent Earth microbes from being transported to the Moon, be discontinued. Although the aim of preventing the contamination of the Moon by Earth organisms was noble, the board felt this process compromised the operation of the probe's electronic components.

The board also addressed the poor performance of the Atlas-Agena rocket by recommending that responsibility for this vehicle, which up to that time was jointly shared between the Air Force and NASA, be given to just one of these organizations. They recommended that the goals of Project Ranger be more clearly stated and all activities not directly related to those objectives be eliminated. Finally, the Kelly board recommended that the project management be strengthened and streamlined. To its credit, the internal JPL review reached many of these same conclusions. It agreed that unless "specific and forceful action" was taken, future Rangers were likely to fail.

JPL administrators felt some of the criticisms of the Kelly board were unfair. When the mission was first undertaken, Ranger was recognized by NASA as a high-risk endeavor. Headquarters, it seemed, had forgotten that fact. JPL also explained that some tests could not be done on the ground since the required facilities did not exist. Nevertheless, the recommendations of the Kelly board were accepted. *Rangers 6* through *9* would have a single goal: to obtain high-resolution pictures of the lunar surface. In order to implement these and other suggested changes, further Ranger launches were postponed until 1964.

The Refocused Ranger Program: *Ranger 6*

Due to the failure of the first five Ranger flights, public pressure for a successful Ranger mission was intense. NASA headquarters wanted assurances from JPL that the next flight would succeed. JPL tried to comply by imple-

Figure 12-10. JPL engineers at work testing a Ranger prior to its launch.

menting the recommendations of the Kelly board. Pickering personally insured the full cooperation of JPL personnel by announcing that Ranger had the highest priority of any project at the laboratory.

Ranger 6 arrived at Cape Canaveral on December 23, 1963. Project scientists then contacted the Manned Spaceflight Center in Houston to discuss which region of the Moon should be photographed to provide the most useful data to the Apollo Program. Much to their surprise they found their NASA colleagues at the Office of Manned Spaceflight in Houston "frankly didn't care where we put *Ranger 6*." JPL engineers were told the Apollo landing craft hardware had progressed beyond the point where it could be modified, and the best the Ranger photographs could do was to confirm the appropriateness of that design. Despite this rather disturbing attitude, Ranger scientists selected a target area they believed would provide useful information to Apollo. As the launch

Focus Box: *Ranger 3's* Path to the Moon

The manner in which a Ranger would be launched depended on the time of day. The basic plan called for the Ranger to be placed into Earth's orbit by its Atlas/Agena rocket. Once in orbit, the Agena would fire a second time and send the Ranger on its way to the Moon. The Ranger would then perform a midcourse maneuver to adjust its trajectory so that it would arrive at the proper place on the Moon.

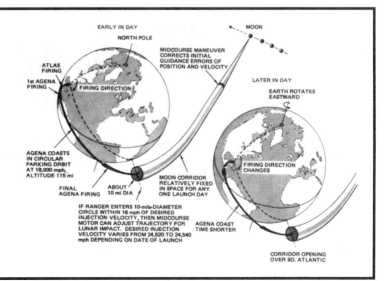

Figure 12-11. In the aftermath of the failed Ranger missions, Robert Parks (left) replaced Clifford Cummings as JPL lunar and planetary director and Harris Schurmeier replaced James Burke as JPL Ranger project manager.

date approached, news crews arrived at both JPL and NASA headquarters to cover the upcoming lift-off.

Ranger 6 was launched on the morning of January 30, 1964. Its ascent proceeded smoothly, and the Atlas rocket performed perfectly. Two minutes into its flight, the Atlas detached itself from the speeding craft and fell back to the Earth. During this separation, the Ranger's TV system unexpectedly turned on and then, one minute later, turned off. Although troubled by this unexpected event, ground controllers were still optimistic. The Agena rocket ignited, placing the craft into orbit. This was followed by a second Agena burn that raised the craft's velocity above the Earth's escape velocity. *Ranger 6* separated from the Agena, deployed its solar panels, oriented itself, and sped toward the Moon. News about the flight continued to be good and the craft's mid-course correction was so precise that no further maneuvers were required. *Ranger 6* was on target for the Moon's Sea of Tranquility!

Nineteen minutes before impact, the craft responded to a signal from Earth to begin warming-up the first of its two cameras. A short time later a similar signal was sent to the

Figure 12-12. The audience at JPL was stunned after hearing the news of the failure of Ranger 6 *and its impact on the Moon.*

Ranger's second camera system. Newsmen crowded into the conference rooms at JPL and NASA headquarters to listen to reports on the Ranger's final moments as it raced toward the Moon. Then the unthinkable happened. As *Ranger 6* descended toward the Moon, no images were returned to Earth. The craft's transmitting system had failed.

Newton Cunningham, NASA Ranger program chief, watched this drama unfold as he sat at his console in the control center at JPL. Stunned by the irrefutable evidence of yet another failure, he pushed his chair back and murmured, "I don't believe what's happened." Silence settled over perplexed audiences in both Pasadena and Washington, DC, as it was announced that *Ranger 6* had impacted the Moon. The sixth consecutive Ranger mission had failed. It was difficult to take much solace from the fact that all of the systems, except for the TV system, had performed perfectly. JPL Director Pickering recalled the mood at JPL:

> It was a beautiful shot until the last fifteen minutes. That really was, in a sense, the low point, and in a sense a high point, at JPL. It was the low point in our realization that we failed. But there was also a realization that we know how to fix this and we're going to pull ourselves together and do it.

Pickering convened two panels of inquiry to investigate this latest failure. They concluded that the high voltage power supply aboard the craft had been damaged when the Ranger TV telemetry channel had unexpectedly come on soon after launch, but were unable to determine why this occurred. NASA convened its own review board which determined the camera's failure was probably due to multiple causes. This report, coupled with the prior Ranger failures, led to a Congressional investigation. After completing their hearing, the committee recommended that NASA strengthen its oversight of JPL activities. Morale at the laboratory plummeted.

Ranger 7

As the launch date for *Ranger 7* approached, JPL realized another failure would almost certainly end the Ranger program and possibly lead to its own dismantling. The pressure at JPL "was unbelievable" and project personnel felt that their careers were on the line. There could be no more mistakes.

The target area of *Ranger 7* was chosen to be the northern rim of the Sea of Clouds. This area is transversed by material ejected from the large crater Copernicus and appeared suitable for an Apollo landing. Again, the Office of Manned Spaceflight was asked for input but no response was received.

The launch countdown for *Ranger 7* began at 6:47 a.m. EDT on July 27, 1964. News agencies from around the world gathered in Washington and Pasadena. The launch was postponed until the next day. This time, prelaunch checkouts went well and the Atlas roared off into the Florida skies at 12:50 p.m. EDT. More than one year had passed since the last manned Mercury flight.

Figure 12-13. Ranger 7 *begins its journey to the Moon.*

Figure 12-14. *The success of* Ranger 7 *brought tremendous relief and joy to JPL. In the above picture, the audience at JPL roared its approval upon hearing the news of the successful* Ranger 7 *flight.*

As with *Ranger 6*, the launch went as planned and by 1:20 p.m., the Agena propelled *Ranger 7* out of the Earth's orbit and toward the Moon. As it neared the Moon, ground controllers instructed the probe's cameras to turn on. Soon, excellent lunar pictures began to arrive at NASA's Goldstone receiver. JPL spokesperson George Nichols excitedly reported the Ranger's progress to the assembled media:

> Twelve minutes before impact, excellent video signals continue. . . . Ten minutes . . . no interruption of excellent video signals. All cameras appear to be functioning. . . . all recorders at Goldstone are "go". . . . Seven minutes . . . all cameras continue to send excellent signals. . . . Five minutes from impact, video still continues excellent. Everything is "go," as it has been since launch. . . . Three minutes . . . no interruption, no trouble . . . Two minutes . . . all systems operating. Preliminary analysis shows pictures being received at Goldstone. . . . One minute to impact. . . . Excellent. . . . Excellent. . . . Signals to end . . . IMPACT!

The crowd reaction at NASA headquarters and JPL was immediate. Some employees shouted with joy while others threw papers into the air. Controllers shook hands and slapped each other on the back. For other employees the emotion was so great they just cried. As Pickering, Newell, and Project Manager Schurmeier entered the auditorium during this celebration, they were greeted by a standing ovation. A newsman asked Pickering, "How do you view the labora-

tory's future after the success of *Ranger 7*?" To the laughter of nearby JPL employees, he replied, "I think it's improved."

In all, 4,308 high-quality images of the Moon were obtained by *Ranger 7* from distances ranging from 1,500 miles to 100 feet. Objects as small as six inches in diameter were visible in the closest images. Later that day, a news conference was held by the Ranger science team. Team member Gerald Kuiper began the briefing by saying, "This is a great day for science, and this is a great day for the United States." He then spoke about the quality of the Ranger photographs:

Figure 12-15. *The stress that JPL employees felt was intense. In the above picture one employee could not contain her feelings after this success.*

We have made progress in resolution of lunar detail not by a factor of ten, as was hoped would be possible with this flight, nor by a factor of 100, which would have been already very remarkable, but by a factor of 1,000.

Project scientist Eugene Shoemaker, of the U.S. Geological Survey, then went on to explain how these images provided evidence that the lunar surface was firm enough to support a landing craft. This result was a great relief to NASA and lessened apprehension over the adopted design of the Apollo landing craft.

The success of *Ranger 7* had favorable repercussions in Congress and within NASA itself. On August 5, the agency's long delayed $5.3 billion 1965 appropriation bill was passed by the Senate, and all efforts to cut it were defeated. Senator J. William Fulbright who had planned to submit an amendment to trim the NASA appropriation by 10 percent, suddenly decided not to. When asked to explain his change of mind, he tersely replied, "The climate had changed." The Apollo manned program office in Houston also changed its attitude about Ranger from one of indifference to one of extreme interest.

Soviet Progress

Nearly five years had passed since the Soviet *Luna 3* had returned the first pictures of the Moon's far side. Although the Soviet lead in the space race appeared firm at that time, their subsequent progress had been uneven. In April 1963, *Luna 4*, a craft that was most likely intended to soft-land, missed the Moon by 5,100 miles. After this failure, two years would pass before another Soviet lunar launch was attempted.

Rangers 8 and *9*

Ranger 7 had demonstrated that the numerous problems associated with the Ranger flights had finally been overcome, so hope was high that *Rangers 8* and *9* would also be successful. *Ranger 8* was launched on February 17, 1965, and per-

Figure 12-16. The target of Ranger 9 *was the crater Alphonsus. The central peak of this crater was suggested to be volcanically active. The white circle shows the impact area of* Ranger 9.

formed superbly, transmitting 7,137 images covering more than a square mile of the Sea of Tranquility.

About a month later, on March 21, the final Ranger, *Ranger 9*, was launched toward the crater Alphonsus. This crater was of particular scientific interest because the Russian astronomer N.A. Kozyrev had reported seeing glowing gases coming from its central peak. If true, this region might reveal unique information about the Moon's interior conditions. *Ranger 9* successfully reached the Moon and returned 5,814 images of this region. A close examination of them, however, failed to find any evidence for the volcanic activity reported by Kozyrev.

Ranger Results

The Ranger program lasted nearly four years and cost approximately $260 million. Individual mission results are

Table 12-1. Ranger Flights

Mission	Launch Date	Outcome	Crash Site	Coordinates
Ranger 1	Aug. 23, 1961	Agena fails to reignite		
Ranger 2	Nov. 18, 1961	Agena fails to reignite		
Ranger 3	Jan. 26, 1962	Agena fires too long		
Ranger 4	April 23, 1962	Timer fails on probe		
Ranger 5	Oct. 8, 1962	Probe's solar panel fails		
Ranger 6	Jan. 30, 1964	Impacts Moon, no pictures	Sea of Tranquility	9° 24′ N 21° 30′ E
Ranger 7	July 28, 1964	Impacts Moon, pictures	Sea of Clouds	10° 38′ S 20° 36′ W
Ranger 8	Feb. 17, 1965	Impacts Moon, pictures	Sea of Tranquility	2° 43′ N 24° 38′ E
Ranger 9	March 21, 1965	Impacts Moon, pictures	Crater Alphonsus	12° 58′ S 2° 22′ W

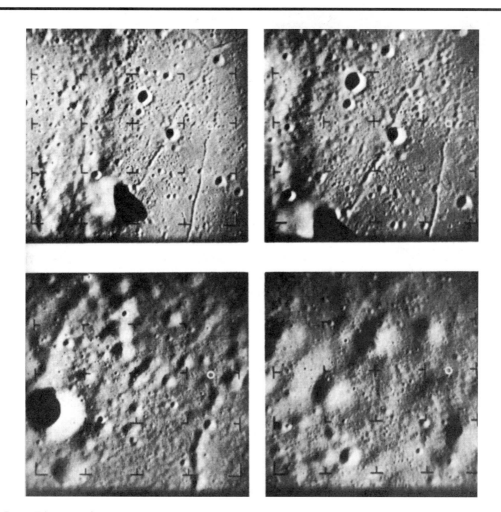

Focus Box: Pictures from a Ranger Impact

As a Ranger descended to the Moon's surface, its cameras automatically took and then transmitted pictures back to NASA's Goldstone receiver. This series of four pictures shows the view that one of the Rangers had as it streaked toward its impact point (shown by a circle in the middle right) on the Moon. The boxes show the field as the craft approached the surface. The + symbols were placed on the film by the camera system and were used for scale measurements. The crater near the middle right is seen in the first three images.

summarized in Table 12-1. Ranger provided over 17,000 photographs of the Moon's surface and revealed craters as small as half a foot in size. This resolution was over 1,000 times better than obtained by telescopes on Earth. Based on an examination of *Ranger 7* photographs and older Earth-based observations, Eugene Shoemaker developed the following model of the Moon's surface:

A layer of shattered and pulverized rock covers more than 95 percent of the mare. It is of variable thickness and rests with irregular contact on the underlying substance of the mare. . . . The fragments in this layer or blanket of shattered rock have been derived by ejection from craters. . . . The upper surface of the debris layer is pockmarked by craters. . . . The uppermost few millimeters of the debris layer is conceived as a fragile open network of loosely stacked, very fine grains. . . . The bearing strength and shear strength of the material increase rapidly with depth, and at depths of a few tens of centimeters are probably similar to the bearing and shear strength of moderately consolidated dry alluvium on Earth.

Dr. Gerald Kuiper studied the high-resolution photographs obtained by *Ranger 9* and agreed with Shoemaker's

Figure 12-17. Project Ranger research scientists Eugene Shoemaker (left) and Gerard Kuiper (right).

view. He saw rocks on the Moon's surface that had apparently been blasted free by a falling object. Based on the size and likely composition of these rocks, he calculated that the lunar surface could support a mass of about 1 ton per square foot, more than enough to prevent a spacecraft from sinking into the lunar surface. He also noted that a large number of well-defined, small craters were visible on the Ranger photographs. If a thick lunar dust layer existed, these craters would have collapsed in on themselves and would have become nearly invisible. Clearly, the Ranger photographs, coupled with ground-based observations, provided strong evidence that the Moon's surface would support a landing craft.

The Ranger missions also demonstrated that NASA now possessed rockets powerful enough to send a craft to the Moon. The probe's flight could be remotely controlled, its instruments instructed to transmit data back to Earth, and it could be impacted at a predetermined location. These were impressive feats, but NASA officials were painfully aware that the Soviets had completed very similar tasks over five years earlier with *Luna 2*. It was premature to celebrate since an enormous amount of work lay ahead.

13 Unmanned Lunar Reconnaissance Missions
Part 2: Projects Lunar Orbiter and Surveyor

The results of Project Ranger were not entirely clear in 1965. Although the scientists studying Ranger images pretty much agreed that the three regions targeted by *Rangers 7, 8,* and *9* were smooth enough for a lunar landing and that the surface was fairly solid, they still argued about the origins of specific features and how to interpret many of the details seen in the photographs.

—**William F. Mellberg in *Moon Missions***

Project Ranger had helped to develop navigation and spacecraft guidance systems, but the real objective of those flights was relatively simple: to acquire pictures as the probes crashed into the Moon. For a manned expedition, NASA would have to place a spacecraft into lunar orbit, land, and then lift-off.

In addition to developing flight expertise, NASA needed more information about the Moon. A global, high-resolution map was required to identify Apollo landing sites, and there was still no direct evidence the lunar surface could support a massive landing craft. Two missions were authorized to gather this data. The job of performing a photographic reconnaissance and orbital flight would be assigned to a mission called Lunar Orbiter and a lunar landing would be assigned to Surveyor.

The Soviet Lunar Missions

While the United States methodically worked on these craft, the Soviet Union was also experiencing mixed success with its lunar program. Less than two months after the successful launch of *Ranger 9*, the Soviets resumed their efforts to soft-land an instrument package on the Moon. *Luna 5* was launched from the Soviet space complex at Tyuratam on May 9, 1965, but crashed into the Moon when its retrorockets failed to fire. *Luna 6* was launched on June 8, 1965, but once its mid-course correction had been completed, its engine failed to turn off. This malfunction caused the probe to

miss the Moon by 97,000 miles. *Luna 7* and *8* both crashed into the Moon when their retrorockets failed to fire for the proper duration. Finally, on January 31, 1966, *Luna 9* soft-landed a probe on the Moon. This craft transmitted three panoramic views of its landing site, taken at different times during the lunar day, that showed objects as small as a few inches in size.

Luna 9 provided other valuable data about the Moon's surface. As scientists had suspected, the landing site was covered by a thin layer of dust that was very dark and porous. It was also evident that the surface was firm enough to support the 220-pound probe. NASA had once again been upstaged by the Soviets.

Establishing a Plan of Action

The main goal of Lunar Orbiter was to photograph potential Apollo landing sites. Researchers at NASA's Manned Spaceflight Center had found that the Ranger photographs did not provide enough information to allow a site to be selected. These pictures were useful in ruling out certain locations, but they did not help to differentiate between seemingly good sites. NASA planned to address this problem by acquiring high-resolution images from five Lunar Orbiter missions. Each of these vehicles would be launched by an Atlas/Agena-D rocket, the same rocket that had been used in Project Ranger. Once the probe was in orbit around the Earth, the Agena-D would be fired to propel it toward the Moon. When near the Moon, the probe's retrorocket would be used to place the craft into a highly elliptical orbit which would then be gradually altered until the craft dropped to within 24 miles of the Moon's surface (see Figure 13-1).

In order to coordinate requests from the scientific community and Apollo managers to place instruments on the Lunar Orbiter and Surveyor spacecraft, Homer Newell, NASA's director of the Office of Space Sciences, appointed the Surveyor/Orbiter Utilization Committee in early 1965. Because of the urgent needs of the manned program, the committee recommended that the initial Lunar Orbiters give priority to photographing candidate landing sites rather than addressing scientific questions.

Table 13-1. Soviet Luna Flights

Mission	Launched Date	Result
Luna 5	May 9, 1965	Crashes on Moon
Luna 6	June 8, 1965	Misses the Moon
Luna 7	Oct. 4, 1965	Crashes on Moon
Luna 8	Dec. 3, 1965	Crashes on Moon
Luna 9	Jan. 31, 1966	Landing

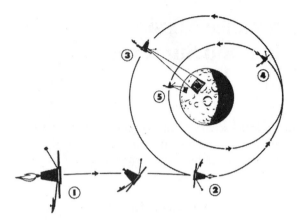

Figure 13-1. The Lunar Orbiter initially went into a lunar orbit that was reduced in size by firing the craft's retrorockets. The numbers indicate important events during the craft's trip to the Moon.

The Lunar Orbiter Spacecraft

NASA had originally planned to assign Lunar Orbiter to JPL. However, that facility was already fully committed to Ranger, Surveyor, and other projects. Instead, Langley was asked to head this project. After a competitive bid process, Boeing Aerospace was awarded a subcontract for $76 million to build the five Lunar Orbiters, and Eastman Kodak was contracted to build their photographic systems.

The Lunar Orbiter was a modest craft compared to the massive Soviet Luna probes. *Luna 9* weighted 3,500 pounds compared to the Lunar Orbiter's weight of 840 pounds. This substantial difference reflected the vastly larger lift capability of the Soviet rockets.

Figure 13-2. Homer Newell, NASA director of the Office of Space Sciences.

The Lunar Orbiter had a cone-like shape with four square solar panels attached to its base, giving it the appearance of a four-leaf clover (see Figure 13-3). Communication between the craft and Earth was maintained using low- and high-gain antennae. The most vital piece of equipment on the spacecraft was a self-contained photographic system that could obtain both wide and narrow field images (see the Focus Box on page 99). In addition to this equipment, the craft also carried instruments to measure the radiation level in lunar orbit, the density of micrometeorites, and anomalies within the Moon's gravitational field.

The Lunar Orbiter Missions

Lunar Orbiter 1 was launched on August 10, 1966. Four days and twenty-three hours later, it began its photographic mission. All systems performed well, and 207 pictures, mainly directed toward the Moon's front equatorial region, were taken. These photographs included nine possible Apollo landing sites. When NASA scientists viewed these images they were concerned because they showed the maria, which looked flat from Earth, were actually littered by numerous small craters. The new images also cast doubt on some of the interpretations made from the *Ranger 7* photographs. The Ranger images suggested that the bright ray regions extending from young craters were heavily pitted by craters, but in the Lunar Orbit images many of these regions appeared fairly smooth. It was important to resolve this matter because if the lunar surface was rough, exceptional care would have to be taken in selecting Apollo landing sites.

Early in the Orbiter's mission, NASA considered maneuvering the craft in order to obtain an oblique (glancing) angle picture of the Moon's surface. Boeing Project Manager Robert Helberg opposed this idea, fearing control of the craft might be lost and the primary objective of the mission jeopardized. NASA officials determined the risk was small and the craft was instructed to obtain these images. The resulting photographs provided a three-dimensional quality that the normal Lunar Orbiter photographs lacked. The oblique views revealed such interesting details about the Moon's topology that more of them would be acquired on subsequent flights. *Lunar Orbiter 1* was the first American spacecraft to orbit the Moon. After completing its photographic mission, it continued in lunar orbit transmitting data on the Moon's gravitational field, radiation levels, and density of meteoroids.

Lunar Orbiter 2, launched on November 6, 1966, also targeted potential Apollo landing sites. In all, thirty sites were imaged at medium and high resolution. These photographs revealed that some of the lightly colored rayed regions extending from the centers of several large craters were suitable landing sites. Oblique angle photographs were also taken of a few selected areas to allow cartographers to construct contour maps that would later aid astronauts in identifying landmarks as they descended toward the Moon. The spectacular image of the crater Copernicus seen in the Focus Box on page 101 is an excellent example of an oblique angle photograph. Even

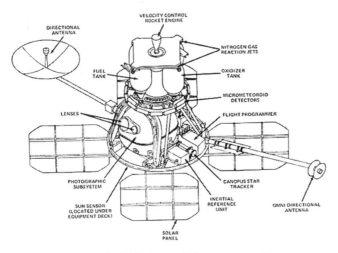

Figure 13-3. The Lunar Orbiter spacecraft.

NASA was impressed by this picture. Oran Nicks, deputy associate administrator, Office of Space Sciences and Applications, remarked: "On first seeing this oblique view of the crater Copernicus, I was awed by the sudden realization that this prominent lunar feature I have often viewed by telescope was a landscape of real mountains and valleys, obviously

fashioned by tremendous forces of nature. It is no wonder that some writers immediately classified it as the picture of the year!"

NASA planners were so happy with the results of the first two Lunar Orbiter flights they felt only one additional mission would be needed to select the first few Apollo landing sites. On January 5, 1967, the Surveyor/Orbiter Utilization Committee approved the mission plan of *Lunar Orbiter 3,* and the craft was readied for launch.

Lunar Orbiter 3 lifted-off on February 5, 1967, and was placed into a more highly inclined lunar orbit than had been the case for the other two missions. *Lunar Orbiter 2* was still in orbit around the Moon at this time and this allowed ground controllers to demonstrate an important new capability, the simultaneous tracking of multiple craft in lunar orbit. *Lunar Orbiter 3* obtained stereoscopic views of the Moon's surface by reimaging the same lunar region from different orbital perspectives. Two hundred and twelve images were obtained, but due to a failure in the on-board camera's film advance motor, only 182 were transmitted to Earth.

Despite the partial failure of *Lunar Orbiter 3*, sufficient photographs existed to identify promising Apollo landing sites. The Surveyor/Orbiter Utilization Committee decided to devote a larger number of the photographs of the two remaining missions to scientific exploration. At the top of this list

---

Focus Box: The Lunar Orbiter Photographic System

The most vital instrument on the Lunar Orbiter was its photographic system. It consisted of three main components: a dual-lens camera, a film developer, and a scanner/transmission system that allowed images to be electronically produced and returned to earth.

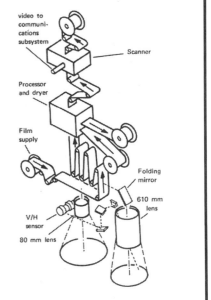

The Lunar Orbiter carried nearly 100 yards of film allowing hundreds of photographs to be obtained. The camera contained two lenses, an 80mm focal length lens with a field of view of 44 × 38 degrees and a 610mm focal length lens that imaged a 20 × 5 degree region. Images were simultaneously obtained with both of these cameras. Frame numbers, a nine-level gray scale, exposure times, a resolution chart, and a calibration grid were automatically placed on the film to aid in image reconstruction. The orbiter's camera system could obtain single photographs or sequences of photographs. To compensate for the craft's motion, a velocity/height sensor in the camera directed the film to move in such a way as to compensate for the craft's motion. Once a dual frame of near and wide field images was obtained, the film was then conveyed to an onboard developer.

After the photograph had been developed, it was conveyed to the photographic readout unit. This system consisted of a linescan tube, a photomultiplier, and supporting optics and electronics. The linescan tube produced a 6.5-micron diameter spot that moved horizontally along the film. After completing a segment, the spot moved vertically and the process was repeated until the entire dual-frame image was scanned.

As the light passed through the film, its intensity fluctuated depending upon the content of the picture. Dark areas transmitted a small amount of light, whereas the lighter colored areas transmitted much more light. This varying light level was detected by a photomultiplier and converted into an electronic signal that was amplified and transmitted to earth. Once received, the signal was enhanced and the picture was electronically reconstructed.

Figure 13-4. The two cameras aboard the Lunar Orbiter fit into a volume about the size of a large watermelon. The cameras are located near the center of this picture of a Lunar Orbiter.

was the completion of a high-resolution map of the entire Moon's surface. On May 3, 1967, the Surveyor/Lunar Orbiter Utilization Committee implemented this expanded goal when they directed *Lunar Orbiter 4* to

> . . . perform a broad systematic photographic survey of lunar surface features in order to increase the scientific knowledge of their nature, origin, and processes, and to serve as a basis for selecting sites for more detailed study by subsequent orbital and landing missions.

In order to meet this directive, *Lunar Orbiter 4* was to be placed into a polar orbit. This would allow it to acquire images covering a minimum of 80 percent of the Moon's front side.

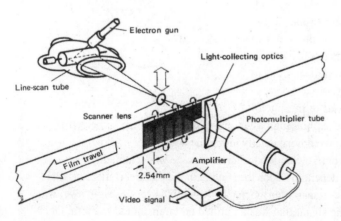

Figure 13-5. The scanner system used light from an electron gun to convert film images into a digital image.

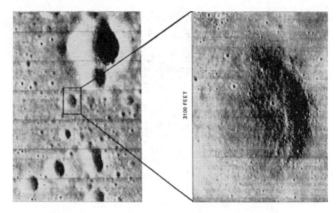

Figure 13-6. The improved resolution provided by the Lunar Orbiter photographs was important in determining if candidate Apollo landing sites were smooth.

Lunar Orbiter 4 blasted-off on May 4, 1967. During its flight, 99 percent of the Moon's near side was photographed at a resolution ten times better than could be obtained from Earth, and coverage of the back side of the Moon was increased to nearly 80 percent. These images revealed previously unknown geological features in the Moon's polar and limb regions.

The last of the orbiters, *Lunar Orbiter 5,* was assigned two sets of objectives. The first included: (1) the acquisition of additional photographs of Apollo landing sites using near-vertical, stereoscopic, and oblique angle images, (2) the completion of the Moon's far side mapping, (3) photographs of selected Surveyor landing sites and future Apollo landing sites, and (4) high-resolution photographs of lunar features that were of special scientific interest. The second set of objectives was similar to those of prior missions. They included (1) measuring the Moon's gravitational field, (2) acquiring data about the radiation field and micrometeorite density in lunar orbit, and (3) providing tracking data to improve the computer program used to plan lunar orbital missions. Obviously, *Lunar Orbiter 5* would be the most complex of the lunar missions.

Lunar Orbiter 5 successfully lifted-off from Cape Kennedy on August 1, 1967, and a few days later entered a highly elliptical, polar orbit. As it orbited the Moon, it obtained a total of 212 photographs of five Apollo landing sites, thirty-six locations of scientific interest, twenty-three regions of the Moon's far side that had not been photographed, and an image of the nearly fully illuminated Earth. After its mission was complete, this orbiter, like its predecessors, was instructed to crash into the Moon.

Lunar Orbiter Results

The Lunar Orbiter program took less than one year to complete but provided information that was vital to the success of the Apollo program. By March 1967, NASA was able to decrease the number of candidate Apollo landing sites to eight. As additional Lunar Orbiter data became available, this

Table 13-2. Lunar Orbiter Flights

Mission	Launch date	No. of Pictures	Notes
Lunar Orbiter 1	Aug. 10, 1966	211	First U.S. probe to orbit the Moon
Lunar Orbiter 2	Nov. 6, 1966	184	Examines 13 candidate Apollo sites
Lunar Orbiter 3	Feb. 4, 1967	182	Completes Apollo site search
Lunar Orbiter 4	May 4, 1967	163	Images Moon's south pole
Lunar Orbiter 5	Aug. 1, 1967	213	Completes mapping of Moon

number was further reduced to five. Statistics about each mission are provided in Table 13-2.

Lunar Orbiter also contributed to the scientific knowledge of the Moon. Its near side was completely mapped with 10–100 times better resolution than had been obtained from Earth, and for the first time ever, a high-resolution map of the Moon's far side was obtained. Other instruments on the Lunar Orbiters measured the radiation level and micrometeorite density in the lunar environment, as well as anomalies in the Moon's gravitational field. As we will see, this latter data would lead to new theories about the composition and formation of the lunar maria. In addition to these results, NASA successfully tested the hardware and procedures required to place a craft into lunar orbit.

The Surveyor Program

As work progressed on Lunar Orbiter, the Surveyor program, designed to soft-land a craft on the Moon, was also underway. Its goals were to (1) test the space hardware needed to make a lunar descent, (2) obtain detailed photographs of the surface of the Moon, and (3) collect data about the Moon's surface composition and strength. The latter objective would provide a definitive answer to the nagging question about whether the Moon's surface could support a *substantial* manned vehicle. Although the 220-pound *Luna 9* had landed on the Moon, the Apollo landing module would weigh nearly 15,000 pounds. Could the Moons's surface support this weight? Primary responsibility for the Surveyor program was assigned to JPL with Hughes Aircraft serving as the spacecraft's principle contractor.

The Surveyor Craft

The Surveyor craft had a squat, triangular appearance (see Figure 13-7). Attached to its main body were two storage bays and a long, mast-like structure extended from its center. Located at the top of this mast was a high gain antenna array

---

Focus Box: The Crater Copernicus

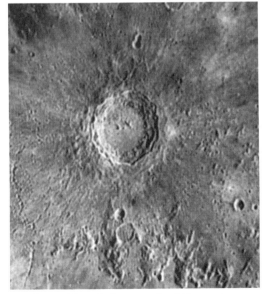

Ground-based photograph of Copernicus from Lick Observatory.

Oblique view of Copernicus from *Lunar Orbiter 2*.

Figure 13-7. The Surveyor spacecraft (1 - antenna array, 2 - solar panel, 3 - camera, 4 - instrument box).

and large, moveable 0.7 × 1 meter solar panels. The Surveyor sat on three symmetrically placed landing legs. At the base of the craft was the main descent rocket that produced up to 10,000 pounds of thrust and three smaller rockets, called vernier rockets, that each produced 103 pounds of thrust.

Once on the lunar surface, Surveyor would photograph its landing site and transmit these images back to Earth. Its camera was capable of obtaining both narrow and wide-angle views and was pointed toward a mirror that could be rotated

Figure 13-8. Footpad of Surveyor 1, *firmly planted on the surface of the moon.*

and tilted. By looking at this mirror, the camera could view different areas of the local lunar surface without having to be moved. At launch, the total weight of the craft was about 2,200 pounds, but after consuming much of its fuel during flight, it weighed only 622 pounds on the Moon.

The original design of Surveyor allowed for a 400-pound payload, but problems with the Atlas-Centaur launch rocket forced engineers to reduce this weight to 44 pounds. In later missions the payload weight was increased to 251 pounds.

The Surveyor Flight Plan

The seven Surveyors were launched from Kennedy Space Center (formerly Cape Canaveral) by an Atlas-Centaur rocket. Once separated from its launch vehicle, the Surveyor oriented itself using on-board sensors that determined the location of the sun and selected bright stars. During its 60-hour-plus flight to the Moon, vernier rockets were employed to make small corrections to the flight trajectory.

When 240 miles from the Moon, the Surveyor's main retrorocket fired to decrease the craft's velocity to around 240 miles per hour. An on-board radar system then sensed the craft's distance from the lunar surface and its vernier rockets reduced the landing speed to 7 miles per hour. Once safely on the lunar surface, ground controllers sent messages to turn on its television camera and scientific package.

Early Problems with JPL

As with Project Ranger, relations between JPL's Surveyor team and NASA headquarters were anything but harmonious. NASA Administrator Homer Newell was astonished to find that many of the lessons learned as a result of Project Ranger appeared to have been forgotten. He recalled, "After all that had happened following the *Ranger 6* failure . . . one would have thought that JPL had the message." He was shocked to learn that "JPL considered Surveyor a low key project which could be kept on the back burner, with the contractors left pretty much to their own devices." NASA headquarters moved quickly to correct this misunderstanding and informed JPL that Surveyor was "one of the highest priority projects in the space science program."

The Surveyor Flights

Surveyor 1 was primarily an engineering mission. Its payload consisted of a camera system and over 100 sensors to monitor the probe's internal systems. No scientific instruments were carried on this flight.

Surveyor 1 was launched on May 30, 1966, and after a nearly flawless journey, soft-landed inside a 60-mile diameter crater in the mare, Ocean of Storms. This craft continued to operate until July 14, 1966, and during its six-week lifetime sent 11,150 images back to the Earth. More importantly, *Surveyor 1* put to rest any remaining doubts as to whether the Moon's surface could support a substantial craft.

Surveyor 2 lifted-off on September 20, 1966. One of its vernier rockets failed to fire during a mid-course maneuver, putting the craft into an uncontrollable spin. The probe crashed near the crater Copernicus. A few months later, on December 1966, the Soviets soft-landed their second craft, *Luna 13*, on the Moon's Ocean of Storms. Like *Luna 9* it transmitted panoramic views of its landing site, but also deployed an extendable 1.5-m-long mechanical arm to measure the soil density and radioactivity.

Surveyor 3 was launched on April 17, 1967, and successfully landed two days later in the Ocean of Storms. This craft was equipped with a 5-inch-long scoop mounted on a 5-foot-long retractable arm designed to measure the mechanical properties of the lunar soil. In all, seven bearing tests were completed, four trenches were dug, and thirteen impact experiments were performed. The soil was found to behave like fine-grained terrestrial soil that has a small amount of cohesion. The top soil was somewhat crusty to a depth of 2 inches. Its density was similar to that found on the surface of the Earth and the density increased with depth. In addition to studying the properties of the soil, *Surveyor 3* returned 6,315 images of its landing site. Contact was lost with this craft after its first lunar night.

Surveyor 4 lifted-off on July 14, 1967, but crash-landed when contact was lost 2.5 minutes before it touched down on the lunar surface.

Surveyor 5 was sent on its way to the Moon on September 8, 1967, and landed in the Sea of Tranquility. It returned 19,000 pictures. This craft also carried an instrument that shot a stream of alpha particles into the lunar soil and measured the number of particles scattered back toward the craft. The number of detected particles revealed information about the chemical composition of the lunar soil. The soil was found to be composed of basalt type material, confirming the theory that the maria were large lava flows.

Surveyor 6 was launched on November 6, 1967, and landed close to its intended impact site in the Sinus Medii (Central Bay). It returned 30,000 images and tested the chemical composition of the material at its landing site. After completing these tasks, it was instructed to reignite its vernier rockets. This burn was successfully carried out and the craft rose about 11 feet from the lunar surface. This was an important experiment because it demonstrated that it was possible to reignite the rocket engine after landing on the Moon. In order for the Apollo astronauts to leave the Moon, their vehicle would have to perform a similar firing.

Surveyor 7, the last of these craft, lifted-off toward the Moon on January 6, 1968. Unlike the previous Surveyors that landed in the lunar maria, this craft was sent to the lunar highlands. After setting down near the crater Tycho, it transmitted over 21,000 images back to Earth. On board instruments found that the highland soil was rich in calcium and aluminum, but contained fewer "iron-group" elements (atoms between titanium and copper) than the maria. Since minerals in the former group have a lighter color, the highlands appear brighter from the Earth than the maria. Contact was lost with *Surveyor 7* on its second lunar day.

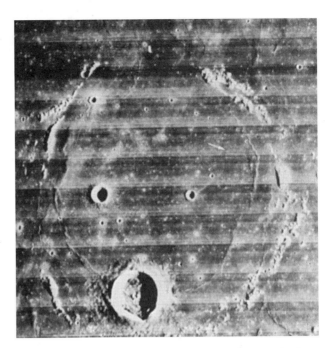

Figure 13-9. Surveyor 1 *landed in a lunar maria region, the Ocean of Storms.*

Surveyor Results

The primary goals of the Surveyors were to: (1) obtain detailed pictures of the lunar surface, (2) determine whether the lunar surface could support a manned craft, (3) provide information about the composition of the lunar soil, and (4) test the ascent rocket on the craft to verify it still worked after a landing. All of these objectives were met. Of the seven Surveyors launched from Earth, five soft-landed and two crashed into the Moon. Four of the five landing sites were located near the Moon's equator. The maria locations were chosen

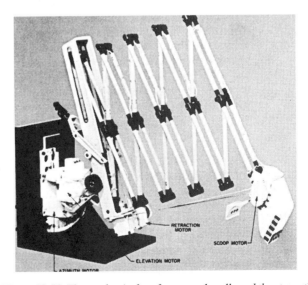

Figure 13-10. The mechanical surface-sampler allowed the strength of the lunar soil to be determined.

Table 13-3. Surveyor Flights

Mission	Launch Date	Destination	Coordinates		Photographs	U.S. Firsts
Surveyor 1	May 30, 1966	Ocean of Storms	2° 27′ S	43° 18′ W	10,150	Soft lunar landing
Surveyor 2	Sept. 20, 1966	Crashed	5° 30′ N	12° 00 N		
Surveyor 3	April 17, 1967	Ocean of Storms	20° 56′ N	23° 20 N	6,316	Bearing test
Surveyor 4	July 14, 1967	Crashed	0° 26′ N	1° 30 W		
Surveyor 5	Sept. 8, 1967	Sea of Tranquility	1° 30 N	23° 11′ E	18,000	Chemical soil test
Surveyor 6	Nov. 7, 1967	Central Bay	0° 28′ N	1° 29′ W	30,000	Lift-off from Moon
Surveyor 7	Jan. 7, 1968	Crater Tycho	40° 53′ N	11° 26′ W	21,000	Highland landing

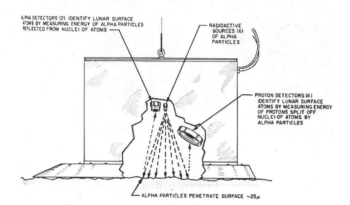

Figure 13-11. The Surveyor Alpha particle experiment.

Figure 13-12. This set of two images show the landing site of Surveyor 7 *from orbit (top) and at ground level (bottom).*

because they were candidate Apollo sites. The fifth site was in the lunar highlands, allowing scientists to form a more global view of the Moon's surface properties. All of the Surveyor landing sites, with the exception of that of *Surveyor 1*, were selected based on photographs returned by Lunar Orbiter. *Surveyor 1* was launched prior to the Lunar Orbiter missions.

Unlike the Earth's soil, which is made of a variety of colors, the Moon's surface was found to be an uninviting mixture of light and dark grays. Rock fragments littered the surface and numerous small craters were present. The mechanical properties of the lunar soil resembled those of terrestrial soil. Although the material in the maria regions was somewhat more brittle than in the highlands, the soil properties in these regions were otherwise quite similar. To a depth of several millimeters the top soil was less dense, lighter in color, and more compressible than the underlying layers.

The lunar soil was found to be similar to that of the Earth, consisting largely of oxygen, silicon, and aluminum atoms and resembling the composition of terrestrial basalts (lavalike rocks). This result was somewhat surprising. It differed dramatically from what was expected if the Moon had simply condensed out of early solar system material and then segregated into a crust, mantle, and core. A lunar origin for most of the meteorites found on the Earth was ruled out. Since its formation, the Moon's surface had apparently been substantially changed by meteorite bombardments and extensive lava flows. The Surveyor results supported the view that the Moon

was either once part of the Earth, or that the Earth and Moon formed together as a double planet and then underwent similar geological changes.

The *Surveyors* also fulfilled three other important objectives: (1) they verified the hardware and control procedures required to make a soft-landing on the Moon, (2) they demonstrated NASA could land a vehicle at a predetermined location on the Moon, and (3) they confirmed that a craft's rockets could be reignited after landing. This last finding insured the astronauts would be able to leave the Moon once their mission had been completed.

14 The Mercury 13 and the Selection of Women Astronauts

The men go off and fight the wars and fly the airplanes and come back and help design and build and test them. The fact that women are not in this field is a fact of our social order. It may be undesirable. It obviously is, but we are only looking [for] people with certain qualifications.

—John Glenn (July 1962, testimony before the Special Subcommittee on the Selection of Astronauts)

In the summer of 1962 the Soviet Union enjoyed a sizable lead over the United States in the space race. They had been the first to place a satellite into orbit (*Sputnik 1*), the first to orbit a live specimen (*Sputnik 2*), the first to recover live specimens from orbit (*Sputnik 5*), the first to place a man in space (Gagarin), the first to complete a manned one-day orbital mission (Titov), and the first to rendezvous two manned spacecraft (*Vostok 3* and *4*).

American political leaders were growing uneasy about this mounting list of accomplishments compared to the agonizingly slow progress being made by NASA. In the cold-war environment of the late 1950s and early 1960s Americans were concerned that these Soviet triumphs posed a threat to their security. As early as January 1958, presidential candidate Lyndon Johnson ominously warned people, "The Roman Empire controlled the world because it could build roads. Later when men moved to sea, the British Empire was dominant because it had ships. Now the communists have established a foothold in space."

After the successful flights of *Vostok 3* and *4*, the Soviets planned to test their new generation spaceship, the Soyuz. Unfortunately, the construction of this craft was far behind schedule. To take advantage of the time this made available, Soviet Chief Designer Sergei Korolev decided to launch another dual mission. The first craft, *Vostok 5*, was to be flown by cosmonaut Valeri Bykovsky. However, it was the second craft that would make this mission special. At the direction of Soviet Premier Nikita Khrushchev, it would be flown by a woman, Valentina Tereshkova. Not only would this flight allow Soviet doctors to compare the medical problems that confronted men and women cosmonauts, it would provide an immense propaganda coup over the West.

The absence of female astronauts in the American space program made the capture of this space first seem easy. However, unknown to Khrushchev, a group of American women pilots, later to be called the Mercury 13, had secretly passed the same initial battery of medical exams given to the Mercury 7. The greatest obstacle preventing these women from qualifying as astronauts was a newly created requirement established by NASA itself. If it could somehow be satisfied, the first woman in space might well be an American.

Figure 14-1. Soviet cosmonaut Valentina Tereshkova.

The Beginning of the Mercury 13

In 1959 America recruited its first group of astronauts. NASA required that all candidates had to: (1) be less than 40 years old, (2) be less than 5 feet 11 inches in height and in excellent physical condition, (3) possess a bachelor's degree or equivalent, (4) be a graduate of a test pilot school, (5) have 1,500 hours of flight time, and (6) be a qualified jet pilot. President Eisenhower further insisted these candidates be members of the armed services because sensitive military hardware was to be employed in the space program. Each candidate began the process of qualifying for the astronaut corps by passing a series of thirty medical tests given at the Lovelace Clinic in Albuquerque, New Mexico.

In 1959 Dr. Randolph Lovelace, founder of the clinic, attended an Aviation Conference in Miami, Florida. During the course of the meeting he met Air Force Brigadier General Donald Flickinger. Flickinger believed the Soviets were

training women cosmonauts and asked Lovelace what advantages or disadvantages women might have for space flight compared to men. Neither man knew much about the capabilities of women pilots so they talked with Jerrie Cobb, a pilot for the commercial aircraft company called Aero Commander. Impressed with Cobb's flight credentials, they asked if she would come to Albuquerque and take the same tests given to the male astronaut candidates. Cobb agreed, and in February 1960 she traveled to the clinic and passed these tests.

Lovelace and military officials then decided to test a larger group of women. Cobb helped construct a list of candidates from Federal Aviation Administration (FAA) records and information provided by the Ninety-Nines, an organization of women pilots. Twenty-four women were invited to participate in this program. Travel and living expenses were paid by the famed pilot Jacqueline Cochran. Between February 1961 and the summer of the same year, twelve of these women passed the medical exams. Their names were "K" Cagle, Jan Dietrich, Marion Dietrich, Wally Funk, Jane Hart, Jean Hixson, Gene Jessen, Irene Leverton, Sarah Ratley, "B" Steadman, Jerri Truhill, and Rhea Woltman. With the inclusion of Cobb, this group came to be called the Mercury 13.

While the women were undergoing their exams at the Lovelace clinic, Cobb passed two additional phases of astronaut testing, one given through the cooperation of the Veteran's Administration and the other by the U.S. Navy. She was then appointed to be a NASA consultant by Administrator James Webb. Two other women passed the second phase of exams consisting of psychological, psychiatric, and isolation studies. These tests required the use of special government equipment such as underwater tanks, centrifuges, and pressure chambers. However, before the remaining women could be tested, the program was brought to an abrupt halt. When later asked why, Jane Hart, one of the Mercury 13, replied, "I have no idea . . . That is one of the mysteries of the year."

Post-Mercury Astronaut Qualifications

In order to serve as an astronaut in the post-Mercury period, a candidate had to meet six criteria. This person had to (1) be a U.S. citizen, (2) be less than 35 years old, (3) be in excellent physical health, and less than six feet tall, (4) possess a bachelor's degree in a physical or biological science or an engineering field, (5) be a graduate from one of the military jet test pilot schools, and (6) have the recommendation of their present organization. A major difference between these criteria and those used in Project Mercury was the requirement that a candidate had to be a graduate of one of the military's jet test pilot schools. In the early 1960s women were not allowed to fly military jets and no civilians were permitted to attend these schools. Given this "catch-22" situation, an alternative means of satisfying NASA's requirements had to be found.

Cobb and others believed this requirement was unnecessarily restrictive. They managed to gather enough support in Congress to have a hearing scheduled on the matter of astronaut qualifications. Could a compelling case be made to train women astronauts? If so, there was still a chance America could place the first women in space. The primary question remained, would NASA be willing to accept this new challenge?

Congressional Hearings

Beginning on July 17, 1962 hearings were held in Washington, DC, by the Special Subcommittee on the Selection of Astronauts. The committee was chaired by Congressman Victor Anfuso, a supporter of the inclusion of women in the astronaut corps. He opened the proceedings with these remarks,

Ladies and gentlemen, we meet this morning to consider the very important problem of determining to the satisfaction of the committee what are the basic qualifications required for the selection and training

Figure 14-2. Jerrie Cobb served as an eloquent spokesperson for the Mercury 13.

of astronauts. There is no question that our manned space flight program must make use of every available resource that can contribute to its success. As we look into the future, we can see greater and greater demands for special talents placed upon the people from whom future space travelers will be drawn. We are particularly concerned by the fact that the talents required should not be prejudged or prequalified by the fact that these talents happen to be possessed by men or women. Rather we are deeply concerned that all human resources be utilized.

He then announced the committee would first hear testimony from three prominent women pilots: Jerrie Cobb, Jane Hart, and Jacqueline Cochran.

Jerrie Cobb began her testimony by giving a brief history of the Mercury 13's attempt to enter the astronaut program. She then made three points: (1) the women pilots were not interested in becoming astronauts as a part of the battle of the sexes, but were seeking, "a place in our Nation's space future without discrimination . . . to participate . . . in the making of history . . . as women have in the past"; (2) there were practical reasons for including women as astronauts such as women weigh less and consume less food and oxygen than men, lessening "the grave obstacle in the cost and capability factors of manned space vehicles"; and (3) the inclusion of women in the space program would allow the

United States to claim a space milestone, the first nation to place a woman in space.

The subcommittee then heard from Jane Hart, another of the Mercury 13. Like Cobb she emphasized that women could provide real benefits to the space program. She said, "I am not arguing that women be admitted to space merely so that they won't feel discriminated against. I am arguing that they be admitted because they have a very real contribution to make." Addressing the historical resistance women had met when trying to enter new fields, she added, "It seems to me a basic error in American thought that the only time women are allowed to make a full contribution to the nation is when there is a manpower shortage." She admitted women's attempts to become astronauts would be greeted by some with "condescending little smiles and mildly humorous winks." Indeed, when the Navy had asked for permission to let Cobb fly one of their aircraft to help determine differences between men and women pilots, the return wire noted, "If you don't know the difference already, we refuse to put money into the project."

Hart then pleaded that the Lovelace Clinic be allowed to continue its research into the medical questions associated with women astronauts. In her opinion, "even the extreme view that women will have no place in outer space for many years does not justify the cancellation of a research program that had already begun and that would doubtlessly supply information useful right now as well as in the future."

Although Hart had no desire to be the "Susan B. Anthony of the space age" and believed the role of wife and mother could be tremendously rewarding, she added, "I don't think, either, that it is unwomanly to be intelligent, to be courageous, to be energetic, to be anxious to contribute to human knowledge."

The Committee's Questioning of Cobb and Hart

Following the opening statements by Cobb and Hart, the committee asked questions in three primary areas: (1) was it important for an astronaut to be an engineer? (2) was it important for an astronaut to be a jet test pilot? and (3) should a separate program be established to train women astronauts? In response to an inquiry about the first question Cobb stressed the role of pilot over that of engineer,

> I think it is true, that the astronaut has more duties to perform in space than just to operate the spacecraft. It is to observe and perform other duties, but the primary function is still that of flying the aircraft. That is why it is easier to take a pilot and teach him the other jobs which need to be done in space than to take an engineer or a geologist, or some other scientist and teach them to be a pilot. I don't think the qualification of an engineering degree and jet test pilot experience should just be knocked out, but that NASA should realize there is an equivalent experience which we can offer. . . .

Figure 14-3. Jacqueline Cochran (right) and future Mercury 13 member Jean Hixson (left) holding a model of an F-86 in 1955.

The committee's questions then quickly moved to the next area, must an astronaut be a jet test pilot? When asked to explain the women's objection to this requirement Cobb answered,

> Some of us have worked as test pilots but it is impossible for a woman in this country to be a jet test pilot because there are no women pilots in the military services and the test pilot schools are operated solely by the military services. There are no other test pilot schools except those of the Navy and the Air Force, and since there are no women pilots in the services they do not have the opportunity to go to these schools to learn to be a jet test pilot. . . .

Since it was impossible for women to become a jet test pilot, Cobb argued they should be allowed to use an "equivalent experience" to meet this requirement. In defense of this position she stated,

> I suggest there is an "equivalent experience" in flying that may be more important in piloting a spacecraft. Pilots with thousands of hours of flying time would not have lived so long without coping with emergencies calling for microsecond reactions. What counts is flawless judgment, fast reaction, and the ability to transmit that to the proper control of the aircraft. We would not have flown all these years, accumulating thousands and thousands of

hours in all types of aircraft without accumulating this experience. This experience is the same as acquired in jet test piloting. I think you might acquire it faster as a jet test pilot, but it is by no means the only way to acquire it. Some have 8,000 to 10,000 hours—have flown a million miles in all types of airplanes—this is the hard way to acquire that experience, but it is the same experience.

Cobb believed a training program for women could be completed "within a few months on an all-out basis" and would not delay NASA's efforts to reach the Moon. When asked by a committee member, if she felt this program should be conducted at the risk of delaying the Moon landing or other space activities, she emphatically answered, "No sir; I do not. . . ."

The Surprising Testimony of Jacqueline Cochran

The next person the committee heard from was Jacqueline Cochran. Cochran was one of the most skilled women pilots in the world, holding over 100 international speed, altitude, and distance records in conventional and jet aircraft. During World War II she was responsible for the recruitment, training, and operation of the WASPS, a group of up to 1,000 women who flew noncombat air missions. Since Cochran had provided the women with financial assistance during their Lovelace tests, the committee assumed she supported their position and were therefore taken aback by her opening remarks.

Cochran stated (1) she did not believe there had been any intentional or actual discrimination against women in the astronaut program to date, (2) she believed that in the selection of astronauts, "it was natural and proper to sift them from the group of male pilots who had already proven by aircraft testing and high precision flying that they were experienced, competent, and qualified to meet possible emergencies in a new environment," (3) that because NASA required so few astronauts in the immediate future and since a large body of male pilots already existed, the training of female astronauts could not be justified on the basis of current needs, and (4) although she believed women would ultimately be proven to be as qualified as men for spaceflight, there was no proof of this. Until such evidence could be produced, women should not be trained within the astronaut program.

To provide more data about a woman's suitability for space flight, Cochran advocated that a large group of women should be given tests, "short of and apart from any astronaut trainee program." When asked by the committee to expand upon on her statement that it was proper to select astronauts solely from among the male pilots she added, ". . . I don't have any doubt about women. I am thinking with the great rush that is necessary now to maybe catch up, from all I have been told by the newspapers, that we do not want to slow down our program, and you are going to have to, of necessity, waste a great deal of money when you take a large group of women in because you lose them through marriage." In regard to her proposed test program she added, "I think it must be a very large group, to determine many things, before we jump into something where we do not know what we are doing."

Cochran's comments struck many on the subcommittee as surprisingly negative. Chairman Anfuso tried to elicit more supportive responses by asking several pointed questions, but Cochran's answers remained stubbornly elusive:

ANFUSO: Miss Cochran, you do believe that women belong in the space program?
COCHRAN: I certainly think the research should be done. Then I can tell you afterward. . . . If they prove successful under the test, then I could answer that question. . . .
ANFUSO: Two of the principal requirements are a test pilot experience and an engineering background. What is your comment on that?
COCHRAN: It seems logical to me. I think if you have a group of people with more knowledge you are going to take the best you can. I don't think it is necessarily mandatory—I don't know, we have not tested that out.
ANFUSO: Do you think women should be trained as test pilots?
COCHRAN: If we are economically sound enough—and I don't think we are—to spend that type of money, and take the chance that about the time we are ready to use that person, she starts a family, then, I am all for it, but I am against waste, because I don't think we can afford it.

It was clear from her testimony that Cochran did not agree with Cobb and Hart that women astronauts should be trained and incorporated into NASA's program as soon as possible. In Cochran's view the time of the female astronaut was still far in the future. She also attacked Cobb's argument that America should be the first country to place a woman in space: ". . . I would rather see us program intelligently and with assurance, and with surety, than to rush into something because we want to get there first, whether the Moon or a satellite. I would like to see us do it properly and successfully rather than to make a mess of it." She then added, "nothing should interfere . . . with the present research program that is being conducted." Cochran was certainly not ready to recommend that the Mercury 13 be trained as astronauts.

The NASA Testimony

Given the mixed messages the committee had received from the women pilots, NASA's presentation on the next day of the hearings was crucial in determining what role, if any, women would play in the American space program in the foreseeable future. Unless NASA was willing to accept Cobb's argument about "equivalent experience," there was little chance women would be trained as astronauts. NASA was represented by George Low, director of Spacecraft and Flight Missions, and by astronauts John Glenn and Scott Carpenter. During the course of the day the committee focused on three areas: (1) the requirement for an astronaut to be a jet test pilot, (2) the notion of equivalent experience, and (3) the feasibility of establishing a parallel program to train female astronauts.

Low began the session by defending NASA's requirement that an astronaut possess jet test pilot experience. He stated, "Careful examination and evaluation of the tasks that an astronaut must perform, and the emergency situations with which he must be prepared to cope, have led to the conclusion that of all existing occupations, the testing of jet aircraft most nearly approximates the piloting of spacecraft." He testified that these individuals were trained to make rapid decisions and to take prompt action under stressful conditions. He noted that,

In many ways, manned spacecraft can be considered as a next generation of very high performance jet aircraft. Their velocity and altitude are very great. A spacecraft has life-support systems, control systems, landing systems, power and fuel systems, and many other similarities with high-performance jet aircraft. Thus, there is a logical reason for selecting jet test pilots. . . . In order to limit the selection to those applicants that have demonstrated their capabilities, the further qualification that the applicants be experienced jet test pilots was established.

Low's opening statements were then supported by both Glenn and Carpenter. Glenn told the committee,

The astronaut's function is actually then to take over full control, to analyze, assess, and report the various things that he encounters, or new situations in

Figure 14-4. NASA Administrator George Low testified in defense of the new astronaut requirements.

Figure 14-5. John Glenn also served as a NASA spokesperson. Widely respected for completing America's first orbit flight, his testimony carried considerable weight.

Figure 14-6. The testimony of astronaut Malcolm Scott Carpenter agreed in every important detail with that presented by Low and Glenn.

which he finds himself. In doing this he must perform these functions under periods of high stress, both mentally and physically, and observe many complex functions under these stresses . . . the test pilot program is built around people who continually demonstrate the emotional, physical, and mental stability to do this. To bring all this back to what type of person you want, we felt that the person who can best perform all of these functions is still represented most nearly by the test pilot background.

When asked for his opinion, Carpenter agreed with these statements. Low was then asked to comment on Cobb's suggestion that long hours in conventional aircraft could be substituted for jet test flight experience. Low replied that he disagreed with this statement, "The type of emergency situation that test pilots get into daily in their own flying experience is not matched by the piloting community as a whole. It is true that other pilots also get into stressful situations, but not as often or as frequently . . . as the experimental test pilots do." Carpenter agreed with Low and used an example to refute the notion of equivalent experience. He stated that a person couldn't qualify for a backstroke swimming race by demonstrating he could swim twice the required distance using the crawl. Glenn also strongly argued against the use of equivalent experience, suggesting its acceptance might jeopardize the safety of the astronaut crew,

To say that a person can [fly] around in light planes or transports for—I don't care how many thousands of hours, you name—and run into the same type of emergencies that he is asked to cope with in just a normal six-month or one-year tour in test flying is

not being realistic. If I am on a space mission and I have someone with me, I want the highest qualified person I can over there, whether it is a woman—without regard to color, race, creed or whatever . . .

Glenn's comment also partially addressed the unspoken charge that the criterion of jet pilot test experience had been purposely inserted to keep women out of the astronaut corps. Chairman Anfuso then asked Low if these qualifications applied to men as well as women and was assured they did. Anfuso realized this was an extremely important point, so he paused and again addressed the NASA representatives,

ANFUSO: Gentlemen, I believe it necessary to set this testimony straight, for the record; as far as it has gone. According to your new regulations, there is absolutely no discrimination against women?
LOW: That is correct.
ANFUSO: If women can meet these conditions you have set out, then they will qualify?
LOW: Yes, sir.
ANFUSO: And you have set up these regulations, I gather, first of all, to achieve success in this program because you want the best?
LOW: Yes, sir.
ANFUSO: . . . whether they be women or men is immaterial, but you want the best?
LOW: Yes, sir.

Low admitted under further questioning that the present set of astronaut requirements made it impossible for a woman to qualify as an astronaut, but was unapologetic about this fact. Even when the subcommittee reminded him that this would continue to be true as long as these requirements were in force, Low simply agreed.

The committee was obviously concerned the Mercury 13 were not being treated fairly, particularly in regard to the cancellation of their medical test program. Low responded that to his knowledge this project was under the direction of the Lovelace Clinic and the U.S. Navy. NASA, he said, was not involved and had not been consulted by these organizations. When NASA officials were finally asked about this program, they responded that there was "no requirement" for women astronauts.

Glenn then tried to give the committee a more accurate perspective of the Lovelace tests,

I think . . . the write-ups on this have been a little misleading . . . the tests mainly are run to see if there is anything wrong with a person physically. It isn't that it qualifies anybody for anything. It just shows that they are a good healthy person. These things are such a minimum thing, I think it has been overemphasized . . . as if once they had passed the physical examination they were automatically astronauts.

The committee realized that the current system would not allow women to qualify as astronauts. Therefore, the next set of questions concentrated on alternative routes. Glenn was asked if it was impossible for NASA to expand its present program? Glenn replied, "Nothing is impossible." Seizing on this statement, a committee member then replied, "All right, then, it is possible to expand it in certain directions, and why not in the direction of women?" Perhaps annoyed by this rhetorical trap, Glenn responded, ". . . I am not anti anybody; I am just pro space. We have not seen the idea of women in space put forward with the idea that they are better qualified, which is what we are looking for. The only thing we have seen thus far is women coming in space just by the very fact they are women."

Realizing this type of linguistic trickery would ultimately prove unproductive, Chairman Anfuso sought other roles for a women astronaut that might not require jet test pilot experience. Since both the Gemini and Apollo spacecraft would use multiperson crews, he asked Low if NASA had established a set of qualifications for crewmembers. Low replied, "The term 'astronaut,' for Project Gemini, at least, involves and concerns all crewmembers . . . the work involved in one of these space missions is such that each and every crewmember must be trained to do every function in the spacecraft." Anfuso realized that not only would the spacecraft pilot have to be a former jet test pilot, all crewmen would have to possess this skill.

Anfuso then asked Low if NASA was opposed to offering a parallel astronaut training program for women assuming it did not interfere with their current programs. Low responded that this type of program would not be feasible for several years,

. . . we are certainly not opposed to anything like that, in the future. . . . On the other hand . . . we have at this time a large pool of men who have gone through a preselection process by virtue of having become experienced test pilots. . . . We don't foresee in the near future, talking about the next five or ten years now—the need . . . for more than forty or fifty space pilots in the NASA program. . . . We, therefore, have no plans . . . to start a major training program for space pilots, be they men or women. . . .

He then went on to add, "One more point . . . the equipment available for training pilots for our flights, the centrifuges, the vacuum chambers, all of this equipment is very much loaded up at the present time . . . we would be interfering with the current program." This last remark had a profound impact on the committee. The last thing they wanted to do was to create a situation that would delay the American effort to reach the Moon. Later in his testimony Low drove home this crucial objection. When asked, "If you got a directive today that women astronauts are to be trained . . . would your program in any way be impeded by this directive?" Low responded, "Very much so."

The First Woman in Space

On June 14, 1963, one month after the last flight of the Mercury program and about a year after the conclusion of the Congressional hearings, cosmonaut Bykovsky lifted-off from the spaceport near the Soviet town of Tyuratam in Kazakhstan. Two days later, Valentina Tereshkova followed him into orbit. In the Soviet Union the announcement that a Soviet woman was in space was received with tremendous joy. Not since the flight of Gagarin had the level of public excitement been so high. Pictures of a smiling Tereshkova were transmitted from her Vostok capsule and distributed around the world.

Tereshkova's flight came as a complete surprise to the American public who saw it as yet another Soviet triumph. Ironically, Tereshkova did not even know how to pilot an ordinary plane when she started her training. To Soviet officials this was not considered to be an important matter.

Impact of the Anfuso Hearings on the Selection of Female Astronauts

A number of factors contributed to the failure of Cobb and her supporters to convince the congressional committee that the Mercury 13 or any other women should be allowed to participate in the astronaut qualifying process. First, the female witnesses did not hold a common position. In contrast to the testimony of the women, the statements by Low and the two NASA astronauts were in harmony. Chairman Anfuso's initial concern that talents should not be "prejudged or prequalified" on the basis of gender was repeatedly addressed by Low who told the committee that none of the astronaut requirements were adopted with the aim of discriminating against women. Their only purpose was to identify the most qualified people.

Secondly, Cobb was forced to argue against a seemingly logical requirement, that astronauts possess jet test pilot experience, without being able to provide supporting evidence. Unfortunately her position was at odds with NASA's actual flight experience. Cobb's attempt to substitute "equivalent experience" was effectively challenged by people who had flown space missions. Glenn's statement that this substitution was "unrealistic" and if adopted would pose a threat to the safety of the astronaut crew, was particularly damaging. Cobb was faced with an unsolvable dilemma. She had to prove women could perform adequately in space before women were allowed to fly in space.

Thirdly, the insinuation that the Mercury 13 were well on their way to becoming astronauts when their program was cancelled was severely undercut by Glenn's testimony. Passing the Lovelace tests did not mean one had almost completed the astronaut qualification process. Since NASA never requested these tests, the agency also could not be accused of a breach of faith.

Fourth, Cobb's argument that women astronauts were needed so America could place the first women in space generated surprisingly little support. Cochran belittled this idea and Glenn used it to suggest the primary reason to place a woman in space was to achieve a propaganda victory. Low's statement that a parallel female qualifying program would interfere with NASA's ongoing efforts was also a serious obstacle. In the end, the committee was unwilling to accept the responsibility of advancing a new project that might delay America's Moon program.

Another Possible Outcome?

The outcome of the committee's hearings might have been different if other witnesses had been called to testify. Cochran was a professional aviator and a businesswoman. Her statements were influenced by the practical aspects of training women pilots and the costs associated with this endeavor. Low's primary focus was to overtake the Soviets in the race to the Moon. He feared the training of women astronauts might hinder these efforts. Glenn and Carpenter were former jet test pilots. They were concerned with making NASA's space hardware as reliable as possible. None of these witnesses were interested in the political benefits that would have resulted from placing the first woman into space.

It would not have been difficult to find powerful people who thought this objective was a worthy endeavor. In January 1961, Jerome Wiesner, Special Assistant for Science and Technology, pointed out the political value of successful space firsts. He wrote to President Kennedy, ". . . during the next few years the prestige of the United States will in part be determined by the leadership we demonstrate in space activities." Weisner then added that recent American accomplishments had "not been impressive enough" when compared to Soviet achievements.

The connection between space firsts and national prestige was even more strongly promoted by Kennedy's secretary of Defense, Robert McNamara. On July 21, 1961, he wrote Vice President Johnson, "Major achievements in space contribute to national prestige. This is true even though the scientific, commercial or military value of the undertaking may, by ordinary standards, be marginal or economically unjustified. What the Soviets do and what they are likely to do are therefore matters of great importance. . . ." Vice President Johnson was in total agreement with this assessment; he wrote Kennedy, "In the eyes of the world, first in space means first, period; second in space is second in everything."

The Aftermath

In hindsight, even if more favorable, politically inclined witnesses had been called to testify before the Afuso committee, it is doubtful whether the year between the end of the hearings and the Soviet launch of Tereshkova would have been sufficient for America to capture this space first. A crash program was contrary to NASA's methodical approach toward manned space flight.

The 1962 Congressional hearings on the qualifications of astronauts were a significant, but often overlooked event in the history of the American space program. As a consequence of these hearings, America relinquished any hope of placing the first woman in space and a rationale was established that essentially prevented a woman from serving as a pilot/astronaut for the next thirty-seven years. It was not until 1978 that NASA first advertised for women astronauts. The first American female astronaut to journey into space, Sally Ride, flew aboard the space shuttle Challenger (*STS-7*) on June 18, 1983 as a mission specialist. It was not until July 1999 that the first female astronaut, Eileen Collins, a jet fighter pilot and graduate of the U.S. Air Force Test Pilot School, flew an American spacecraft (*STS-93*).

15   Project Gemini: Part I

As only Project Mercury could have made Project Gemini possible, so without Gemini we could not now be aiming for the Moon with Project Apollo.

—Virgil Grissom, astronaut

Project Mercury had demonstrated that the United States now possessed the technology to place a single man into orbit around the Earth for a daylong flight. This was a promising start, but far more work remained to be done before a crew could be sent to the Moon. Although the Mercury flights had been completed without the loss of a single astronaut, they had not been error free. Shepard's flight had thruster problems, Grissom's capsule hatch had unexpectedly blown open, Glenn's control system had malfunctioned, Schirra's Atlas had come very close to an abort condition, and Cooper had to perform a manual landing. After *MA-8* Schirra returned to Cape Canaveral to gather his belongings and was reminded of the dangers spaceflight still held,

> I was ready to leave but there was an Atlas launch, and I stayed to watch. I went to Hanger S and up an iron-rung ladder to the roof, where there is a viewing platform. It's about four miles from the pad. We listened to the count on a radio speaker, three, two, one, lift-off. At about 100 feet off the ground the missile was destroyed. There were pieces of the Atlas all over the horizon. I was thinking, oh my God!

President Kennedy had eloquently committed the nation to an aggressive space program by stating that it was America's destiny to set sail on this "new ocean." The destination the President had selected, however, was a long way off, and America's efforts had so far only managed to get the ship just beyond the harbor entrance. NASA was now ready to take the next step. The program to move the United States closer to its goal of landing a man on the Moon was called Project Gemini, named for the twin sons of Jupiter, Castor and Pollux.

Gemini had four primary objectives. The first goal was to construct a larger capsule. The capsule used in Project Mercury was only large enough to accommodate a single astronaut, but Gemini activities required a two-person crew. Spacewalks, or Extra-Vehicular Activities (EVAs), would be a prominent part of Gemini. While one astronaut was outside the capsule, a second astronaut would control the craft. These EVAs were important in determining whether a man could work in space. A two-person crew was also needed for multiday missions so that one astronaut could monitor the capsule's control systems while the other slept. Finally, a two-person crew served as a bridge between the single-manned Mercury capsule and the planned three-person crew of the Apollo craft.

The second Gemini objective was to learn more about how the human body reacted to prolonged periods of weightlessness. Soviet cosmonaut Gherman Titov had completed a 17-orbit, 25-hour mission aboard *Vostok 2* in August 1961 but suffered from nausea. It was vital to the Apollo program to learn if even longer periods of weightlessness caused an astronaut to become incapacitated.

The third Gemini objective was to rendezvous and dock with another spacecraft. In theory this procedure did not present any special difficulties, but in practice, unexpected problems might arise. As part of the Apollo program, rendezvous and dockings would be required, so these skills had to be learned and perfected. The last Gemini objective, that of developing precise landing procedures, would allow the astronauts to be quickly recovered from the ocean with a minimum number of supporting ships and aircraft.

The U.S. Congress was also interested in Project Gemini for national security reasons, it developed the skills required to intercept and inspect suspicious Soviet satellites. During

Figure 15-1. The official emblem of the Gemini program. The design denotes the fact that two astronauts would fly the Gemini missions.

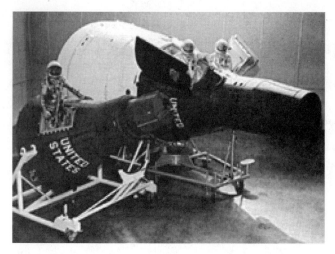

Figure 15-2. The single-person Mercury and the two-person Gemini capsules.

Project Gemini, new rockets would also be built that could be used to launch more sophisticated satellites to monitor Soviet military movements.

To accomplish these four objectives, NASA could not simply upgrade the hardware built for Project Mercury. The Mercury capsule was too small to accommodate a two-person crew, and there was insufficient room to store the food, water, and air needed to sustain the crew during multiday flights. A new craft was also desirable for other reasons. When a Mercury capsule had a prelaunch problem, a substantial amount of its interior fixtures had to be removed in order to fix it. This was a time-consuming task that could damage the removed equipment and make the situation even worse. In a redesigned capsule, important spacecraft instruments could be built into the outside wall of the craft for easy access by technicians. The above changes increased the weight of the Gemini capsule by a large amount, so a more powerful rocket was needed to place it into orbit. The Air Force's Titan II was therefore selected as the launch vehicle for Project Gemini.

The Gemini Spacecraft

The Gemini spacecraft consisted of three sections: a reentry module, a retrograde module, and an equipment module. The astronauts occupied the reentry module which provided about 50 percent more room than a Mercury capsule. During a mission the astronauts would sit beside each other. In between them was an instrument panel that controlled a series of small maneuvering rockets. To the front of the astronauts was a second control panel, and above it were individual windows. A long, cylindrical neck extending from the front of the capsule contained the descent parachute, the radar used for rendezvous and docking, and two independent sets of eight reentry thrusters. Above each astronaut was a hatch to enter and exit the capsule. Immediately behind the crew seats was a reentry heat shield that was essentially a larger version of the one used on the Mercury capsule. Behind the heat shield was the retrograde section of the craft. It contained four solid-fueled retrorockets employed for reentry and eight thrusters that allowed the craft to be maneuvered in space. Just before reentry, this part of the craft was jettisoned.

The last section, the equipment module, had an abbreviated cone-like shape whose maximum 10-foot diameter matched that of the Titan rocket. It carried instrumentation, propellant tanks, and batteries that supplied the entire craft with electrical power. Four pairs of maneuvering rockets were located around the rear of this module, and its end was covered by a thin gold foil to shield the fuel canisters from direct sunlight (see the Focus Box below).

The Titan II Rocket

The Titan II rocket was a modified two-stage Air Force ICBM. It had a diameter of 10 feet, and when the Gemini capsule was attached, its total height was 109 feet. Both rocket stages used the same liquid fuel: unsymmetrical dimethyl hydrazine and nitrogen tetroxide. When these exotic substances are combined, they spontaneously ignite, negating the need

---

Focus Box: The Gemini Capsule

Main parts of the Gemini capsule:
a) the reentry module
b) the retrograde module
c) the equipment module

Launch weight	8,000 pounds
Length	18 feet
Maximum diameter	10 feet
Reentry module length	11 feet
Reentry module diameter	7.5 feet
Reentry module air	pure oxygen

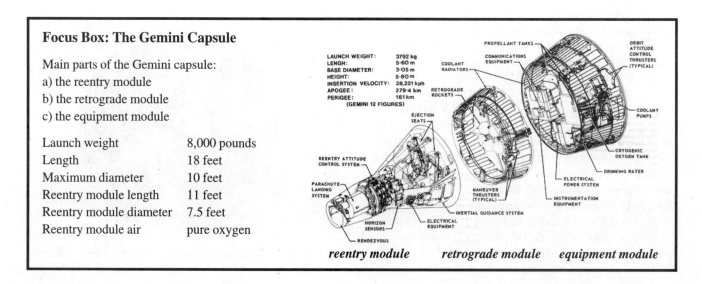

LAUNCH WEIGHT: 3792 kg
LENGH: 5·60 m
BASE DIAMETER: 3·05 m
HEIGHT: 5·80 m
INSERTION VELOCITY: 28,221 kph
APOGEE: 279·4 km
PERIGEE: 161 km
(GEMINI 12 FIGURES)

reentry module   *retrograde module*   *equipment module*

for a complicated ignition system. The first stage of the Titan employed two rocket engines with a combined thrust of 430,000 pounds. The upper stage was powered by a single engine that produced 100,000 pounds of thrust. The nozzles attached to these engines could be tilted to direct the rocket's motion when in flight.

The Agena Rendezvous and Docking Unit

One of the primary missions of Project Gemini was to perfect the procedures needed to rendezvous and dock with another spacecraft. An Agena-D rocket with an attached docking mechanism (see Figure 15-4) was to be used for this purpose. The Agena was 5 feet in diameter and 23 feet long and normally served as an upper stage for the Air Force's Thor and Atlas rockets. It was considered to be a reliable rocket and was employed to propel the Ranger crafts to the Moon and the Mariner probes to Mars and Venus.

Once the Gemini capsule entered the Agena-D docking port, hydraulic latches pulled the two craft together and held them firmly in place. When joined, the Gemini astronauts could use the powerful Agena engine to propel the combined spacecraft to greater speeds and higher orbits.

Only three of the original seven astronauts, Cooper, Grissom, and Schirra, would fly during Project Gemini. Slayton, Shepard, and Carpenter were grounded for medical reasons and Glenn would leave the astronaut corps. Other astronauts who would participate in this program were Edwin Aldrin, Jr., Neil Armstrong, Frank Borman, Eugene Cernan, Michael Collins, Charles Conrad, Richard Gordon, Jr., James Lovell, Jr., James McDivitt, David Scott, Thomas Stafford, Edward White II, and John Young.

By the mid-1960s the Gemini hardware was ready for flight tests and the astronaut training program was well underway. Excitement within the nation was building and the momentum generated by Project Mercury seemed unstoppable. Astronaut Schirra wrote about the air of confidence that existed.

> We had come of age, suddenly. In the aftermath of Mercury's stunning success and with our sights set on a lunar landing, we were confident and full of enthusiasm. As Gemini got going, we brought together the ablest people who had learned the ropes in Mercury, astronauts and engineers who had matured together, and added a superb class of newcomers. It was a championship team.

The Flights of *Gemini 1* and *2*

The goal of the first Gemini flight, *Gemini 1*, was to test the structural integrity of the newly designed capsule and Titan rocket. This unmanned orbital flight blasted-off from Cape Kennedy on April 8, 1964. The Titan was an awesome booster. After just two and a half minutes, it had consumed 130 tons of propellants, and was at a height of 38 miles and 55

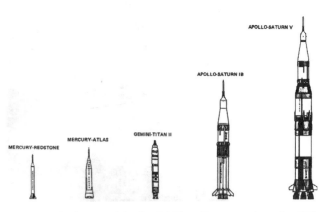

Figure 15-3. The size of the Gemini Titan II compared to other U.S. rockets. With its capsule attached, the Titan II was 109 feet tall.

miles down range. The first stage was then jettisoned and the second stage burned an additional 30 tons of fuel, lifting the craft to a height of 96 miles and a down-range distance of 600 miles. The Gemini capsule and second stage of the Titan were now in orbit traveling at 4.9 miles per second.

Gemini 1 performed flawlessly and the booster/capsule combination was declared to be sound. Flight Director Walter Williams reviewed the mission objectives with the press and proclaimed, "There's no question these objectives were met." After completing nearly 65 orbits, the combined spacecraft reentered the Earth's atmosphere 3.5 days after its launch and was consumed in flames over the South Atlantic.

NASA's enthusiasm over this successful flight was, however, tempered by some harsh facts. It had occurred nearly two and a half years behind schedule. Officials were also aware that the entire project was significantly overbudget. Many people criticized NASA for being too conservative and for moving too slowly. Others disliked the emphasis being placed on a "race" with the Soviet Union. In the April 9, 1964, edition of *The New York Times*, Governor Nelson Rockefeller said,

Figure 15-4. The Agena-D was used to practice rendezvous and docking maneuvers during the Gemini project. The docking port is on the far side of the Agena in this picture and its engine is on the near side.

Figure 15-5. The launch of Gemini 1 *on April 8, 1964. This unmanned flight tested the integrity of the rocket and capsule combination.*

"Our exploration of space should be based on our own purposes, not on a mythical race with an opponent whom sometimes we race and other times we invite them to participate." These latter voices, however, were in the minority; most Americans were anxious to win the Moon race. The press repeatedly questioned NASA officials on how close the United States was to launching a manned lunar mission.

NASA was unswayed by the public's desire to rush forward and continued with its methodical approach to spaceflight. The second Gemini mission, *Gemini 2*, was intended to be an unmanned suborbital flight whose goals were to refine prelaunch checkout and launch procedures, demonstrate that the capsule could withstand the intense heat of reentry, and confirm that all major spacecraft subsystems functioned properly. As the Titan awaited the arrival of its delayed Gemini capsule, a severe thunderstorm pounded the Cape and lightning struck the launch pad. Although no damage to the rocket was visible, NASA decided to conduct a thorough test of the craft's electrical system. Before these tests could be completed, Hurricane Cleo approached the Cape, and the Titan's second stage was moved to a nearby building to prevent possible damage. There was not enough time to remove the first stage, so it remained in place until the storm passed. On September 11 an even stronger storm, Hurricane Dora, approached, and the entire rocket was placed in storage. A

further delay occurred when a third storm, Hurricane Ethel, passed through the launch area.

Finally, on December 9, 1964, *Gemini 2* was reassembled and readied for launch. As the countdown reached zero, the Titan's first-stage engines ignited, but one second later automatically shut down when a loss of hydraulic pressure in the rocket's primary control system was detected. The mission seemed to be cursed.

On January 19, 1965, *Gemini 2* finally blasted-off from the Cape and, during its 18-minute flight, travelled 2,058 miles down range and reached an altitude of 96 miles. All systems worked properly, and the capsule was recovered from the South Atlantic by the Navy carrier the USS *Lake Champlain*. A postflight examination revealed that the capsule was undamaged by the heat of reentry. Despite the frustrating delays, the mission's results were reassuring. Gemini astronaut Virgil Grissom expressed the relief of many when he stated, "there was no question about it, Gemini was moving again."

Gemini 3: The First Manned Gemini Flight

Virgil Grissom noted: "no two Gemini flights were alike, and each built upon the know-how gained by its predecessors, so each flight was a bit more sophisticated." NASA was pleased by the flights of *Gemini 1* and *2* and was now ready to move on to a manned mission.

The crew of *Gemini 3* consisted of former Mercury astronaut Virgil Grissom and a rookie astronaut, John Young. This was Grissom's first orbital flight and Young's first spaceflight. On his prior Mercury flight, Grissom's capsule, the *Liberty Bell*, had been lost when, shortly after landing, its hatch unexpectedly blew open and ocean water rushed in. He therefore named his new capsule the *Molly Brown* after the "unsinkable" heroine of the hit Broadway show. NASA officials thought this name lacked dignity, but when informed that his second choice was the *Titanic*, gave their approval.

Gemini 3 had several objectives: to (1) simulate rendezvous maneuvers, (2) land within a predetermined area, (3) test the worldwide ground tracking network, (4) refine prelaunch and launch systems, and (5) evaluate recovery procedures.

Grissom and Young entered their capsule in the early morning hours of March 23, 1965, and blasted-off into orbit at 9:24 a.m. Their three-orbit mission would be relatively short, lasting a little less than three hours. They nevertheless were able to simulate a rendezvous by performing several orbit changing maneuvers. Their mission completed, they splashed down in the Atlantic ocean and were recovered by the aircraft carrier the USS *Intrepid*. One disappointing result of the flight was that the *Molly Brown* missed its intended landing site by 52 miles because it did not provide as much aerodynamic lift as expected. Nevertheless, NASA was happy with this mission and felt that the hardware had proven itself to be reliable. This flight also provided the United States with another space first: the first time a capsule had been maneuvered while in orbit using substantial thrusters.

Figure 15-6. The Gemini boosters being assembled.

Gemini 4: America's First Spacewalk

Gemini 3 concluded the brief developmental phase of Project Gemini, and the focus now changed to gathering medical data about the human body's reaction to spaceflight, developing the skills needed for rendezvous and docking, and completing a precise landing. Up to this time, the longest American spaceflight was only 34 hours long and even a minimal Moon mission would last at least eight days. More data about the body's reaction to prolonged weightlessness was needed.

Gemini 4 was launched on June 3, 1965, and carried astronauts James McDivitt and Edward White. It was the first American mission to be shown live in Western Europe. *Gemini 4* had four objectives: to (1) evaluate the astronauts' response to prolonged weightlessness, (2) complete an Extra Vehicular Activity (EVA), or spacewalk, (3) test spacecraft control systems, and (4) rendezvous with the second stage of the Titan. In addition to these primary goals, a number of other medical, engineering, and scientific experiments were scheduled. Millions of people around the world watched the lift-off of *Gemini 4* on television, and over 1,000 media representatives were present at the Cape to report on its progress.

Although Soviet cosmonaut Aleksey Leonov had performed the first spacewalk nearly three months earlier during the *Voshkod 2* mission on March 18, 1965, the EVA planned by White was not intended as a response to this Soviet feat. A spacewalk had been part of Gemini planning as far back as 1962. There is little doubt, however, that Leonov's spacewalk caused NASA to accelerate its time line for an EVA.

Soon after reaching orbit, McDivitt attempted a visible rendezvous with the second stage of the Titan booster. He used the capsule's thrusters to move directly toward his target, but found that instead of getting closer, the booster moved further away. McDivitt was puzzled by this strange behavior, and after consultations with controllers, it was decided to conserve the craft's remaining fuel. Prior to lift-off he had been told by Mission Director Christopher Kraft that the primary objective of *Gemini 4* was a successful EVA, not a rendezvous. Since the spacewalk might require some maneuvering fuel, an adequate supply had to be held in reserve. Andre Meyer, of the Gemini Program Office, later explained the reason for the failed rendezvous by noting that, prior to this flight, the astronaut training teams "just didn't understand or reason out the orbital mechanics involved." It was clear that a more thoughtful training program had to be put in place as soon as possible.

The activity that captured most of the public interest during *Gemini 4* was the spacewalk by astronaut Edward White. This EVA began while the capsule was passing over Carnarvon, Australia. After opening the capsule's hatch, installing a camera to record his movements, and attaching an umbilical cord to hold him to the capsule and provide oxygen, White moved away using a small hand-held gun that expelled compressed oxygen. White described his spacewalk this way,

> What I tried to do was actually fly with the gun, and when I departed the spacecraft there was no push-off whatsoever. Seconds later I knew we had something with the gun because it was providing the impulse for my maneuvers, giving me the opportunity to control myself where I wanted to go out there.

White maneuvered across the top of the capsule and beyond its nose. He did not experience vertigo or disorientation and, in fact, greatly enjoyed the experience. When finally told to return to the capsule, he sighed, "It's the saddest moment of my life."

After reentering the craft, White realized that he was physically drained—sweat streamed down his forehead and fogged his helmet's face plate. Before beginning any new activities, both he and McDivitt took a well-deserved rest. Once

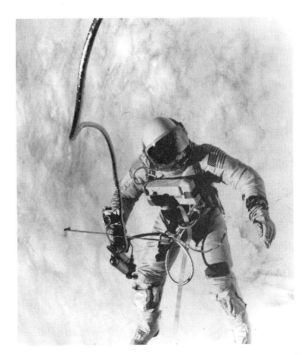

Figure 15-7. Astronaut Edward White makes America's first spacewalk. The maneuvering gun is in his right hand. His EVA took place 4 hours and 18 minutes into the flight of Gemini 4 *and lasted 20 minutes.*

recuperated, measurements of the radiation level within the South Atlantic Anomaly (a localized region of increased radiation) and navigation observations using a hand-held sextant were obtained. They also photographed clouds, storms, and interesting land formations for later analysis back on Earth.

Gemini 4 logged 98 hours in space, nearly four times the length of the longest Mercury flight, and provided America with its first EVA. However, McDivitt and White were not able to significantly improve on the landing precision obtained by *Gemini 3*. They missed their intended impact site by 50 miles. Recovery aircraft monitored their descent, and the astronauts were retrieved within minutes of landing. About an hour after splash down, both astronauts were welcomed aboard the aircraft carrier the USS *Wasp*.

A medical examination found that McDivitt and White showed little adverse effect from their prolonged exposure to weightlessness. Postflight activities included a ticker tape parade in Chicago attended by an estimated one million people. President Johnson then directed both men to attend the Paris Air Show, where they had the pleasure of meeting cosmonaut Yuri Gagarin, the first man to orbit the Earth.

Figure 15-8. Christopher Kraft, Jr., Robert Gilruth, and George Low (left to right) show their concern at Mission Control in Houston when the Gemini 5's *fuel cell problem was detected.*

The Flight of *Gemini 5*: A New Space Record

Gemini 5 was successfully placed in orbit on August 21, 1965, carrying astronauts L. Gordon Cooper, pilot of *MA-9*, and Charles Conrad, Jr. The flight plan called for an eight-day mission, the longest yet attempted and the same time needed to make a round trip to the Moon. Its principal objectives were to (1) determine the astronauts' effectiveness during a long-duration flight, (2) test the reliability of the onboard fuel cells, and (3) test the capsule's guidance and navigation systems. In addition, seventeen secondary experiments were also planned.

Two hours into the mission Cooper was to eject a 76-pound radar evaluation pod (REP) from the nose cone of the Gemini capsule. He was to then back 52 miles away and rendezvous with it. This pod was released, but then a problem arose. The Gemini capsule carried a new fuel cell power system that converted hydrogen and oxygen into electrical power while producing water. Due to a malfunction with a heater, the oxygen pressure in the cell system decreased to a very low level. The crew was forced to switch off all noncritical systems, and as a result, the rendezvous attempt was cancelled. *Gemini 5* drifted in orbit while engineers tested various schemes to restore power. During this time, the astronauts had little to do, so they chatted, ate, looked out their windows, and joked with controllers by singing "over the ocean, over the blue, here's *Gemini 5* singing to you." The next day the heater unexpectedly began to function properly and the oxygen pressure returned to more normal values. NASA officials decided to allow the flight to continue and the astronauts resumed their experimental program.

On the third day of the flight, Cooper completed a "phantom" rendezvous with a predetermined location in space using the capsule's maneuvering system. During this rendezvous five orbital changes were made, demonstrating that the techniques developed for space rendezvous worked. On the fifth day of their flight, six attitude-control thrusters failed and the astronauts were once again forced to let their capsule drift. Despite these problems they still were able to complete sixteen out of seventeen planned experiments, including photographing celestial objects, Earth surface features, zodiacal light, and gegenschien (a diffuse light in the opposite direction of the Sun). They also collected data on cardiovascular conditioning, visual acuity, and horizontal perception.

As the time for their landing approached, the astronauts saw that Hurricane Betsy was rapidly approaching their intended landing site. Based on this information, officials decided to end the mission one orbit early. After preparing the capsule for reentry, Cooper and Conrad landed about 78 miles away from their planned impact site due to a programming error in the capsule's computer. *Gemini 5* completed 120 orbits and the flight lasted 7 days and 22.9 hours. The mission was hailed as a success by the American press and the problems that had been encountered were minimized. Newspaper accounts, reflecting the impatient mood of the country, overstated the importance of this flight and proclaimed, "Man has got what it takes to fly to the Moon and back. Now for the Moon shot."

Postflight medical exams of Cooper and Conrad revealed that their bodies had lost a larger amount of calcium and plasma than the crew of *Gemini 4*. However, after a few days rest, both astronauts felt healthy, and President Johnson asked them to make a six-nation goodwill tour that included a stop at the International Astronautical Federation Congress in Athens. Here they met the crew of *Voshkod 2*, Aleksey Leonov, the first person to walk in space, and Pavel Belyayev.

16 Project Gemini: Part II

I didn't sleep much the night before launch. I had a stabbing awareness, even sharper than I had ever felt in all my Air Force years, that I was in a very dangerous business. I didn't feel fear; I felt agonizing concern for the wife and sons I loved. I didn't want to be a heroic casualty in man's conquest of space and I was not oblivious to the hazards involved.

—Frank Borman prior to the flight of *Gemini 7*

Gemini 6 was scheduled to lift off on October 25, 1965, with astronauts Walter Schirra, Jr., and Thomas Stafford. Its major objective was to rendezvous with an Agena that would be launched prior to their flight. As the astronauts waited in their capsule, the Atlas-Agena target vehicle lifted-off. A few minutes into the flight the Agena separated from the Atlas but then developed a severe wobble. As the aerodynamic forces built, the Agena broke into at least five pieces. Without a rendezvous vehicle to chase, the mission was cancelled.

With the failure of the *Gemini 6* Agena, NASA decided on a bold plan, to use *Gemini 7* as a rendezvous target. This idea appealed to the astronauts. Schirra proclaimed it "a hell of an idea," and Borman, speaking for himself and Lovell, the crew of *Gemini 7,* "wanted" to go. The original mission of *Gemini 7* was to gather data on the long-term medical effects of weightless. In all they were scheduled to complete eight experiments related to space medicine. As Frank Borman ruefully remarked, ". . . in space we were going to be hooked up to more electrodes and other measuring devices than Frankenstein's monster."

Frank Borman and James Lovell lived up to the name of the Gemini program, "the Twins." Both astronauts were 37 years old, both were born in March 1928, both had blond hair, and both had blue eyes. Their flight lifted-off on December 4, 1965, and once in orbit they devoted some time to house cleaning chores, Borman recalled,

> The crew compartment was roughly the size of a Volkswagen Beetle's front seat, which meant we didn't have a hell of a lot of room for a fourteen-day trip involving twenty experiments and a midspace rendezvous. . . . Almost everything in the cabin was tiny, including its two windows. It was simultaneously a workroom, kitchen, bathroom, and bedroom—the only facilities of any size were the narrow but deep and long stowage compartments.

While *Gemini 7* orbited the Earth, *Gemini 6* was being reconditioned for launch. Finally, on December 12 all was ready, and the main engine of its Titan booster roared to life. At that moment, *Gemini 7* was orbiting overhead and Borman

excitedly radioed controllers, "I see ignition!", then "Oh-oh, I see shutdown." Inexplicably, two seconds after ignition, the engines had automatically shut down! In an incredible display of courage, neither astronaut ejected from the capsule despite the extreme danger of the situation. The NASA history book *On the Shoulder of Titans* vividly states the seriousness of this situation,

> If ever there were a time to use the spacecraft ejection seats to get away from a cocked and dangerous rocket, this seemed to be it. . . . Had it [the rocket] climbed only a few centimeters, the engine shutdown would have brought 150 tons of propellants encased in a fragile metal shell crashing back to Earth. There could be no escape from the ensuing holocaust.

Neither astronaut had felt the rocket leave the pad, so they did not eject. If they had been wrong, both would have been killed. With this failed launch, a new question arose, could *Gemini 6* be recycled a second time and launched before *Gemini 7* had to return to Earth? The normal turnaround for the Titan rocket was 96 hours, and although *Gemini 7* could remain in orbit for another 144 hours, this left little room for error. Ground crews worked as quickly as possible, and in the "can do" spirit that typified NASA workers, *Gemini 6* was ready for launch just 72 hours later.

On December 15, 1965, *Gemini 6* successfully reached orbit. Once in space, Schirra dimmed the capsule lights to better see *Gemini 7* against the blackness of space. Five hours after leaving the pad, Schirra saw a bright object in the sky and told Mission Control, "My gosh, there is a real bright star out there. That must be Sirius." The "star" was, in fact, *Gemini 7* illuminated by reflected sunlight, and once told of his mistake, he maneuvered his craft toward it. The final climatic moments of this rendezvous were described in *On the Shoulders of Titans,*

> Following a braking and translation maneuver, VI-A [*Gemini 6*] coasted until the two vehicles were 40 meters apart, with no relative motion between them.

Figure 16-1. The aborted liftoff of Gemini 6.

The world's first manned space rendezvous was now a fact. In Mission Control, the cheering throng of flight controllers waved American flags, while Kraft, Gilruth, and others of the jubilant crowd lit cigars and beamed upon the best of all possible worlds.

During the next three orbits the two craft approached to within 13 inches of each other. Stafford was impressed by the very small velocity adjustments he could make and felt that a future docking would be easy.

After 16 orbits *Gemini 6* returned to Earth, leaving *Gemini 7* to continue its long-duration flight. *Gemini 7* later returned to Earth after 13.8 days in orbit on December 18, 1965. It landed about 7.3 miles from its intended impact area, slightly better than the 8-mile difference achieved by *Gemini 6*, and substantially better than that obtained by prior missions. A postflight medical examination of Borman and Lovell showed that their long exposure to weightlessness had not resulted in a corresponding larger loss of calcium from their body, and remarkably, their total blood volume was unchanged from preflight values. Apparently the mission was long enough for the body

to adjust to its new space environment. This was indeed good news for the upcoming Apollo program.

Gemini 8: America's First Orbital Docking

Gemini 8 would be manned by Neil Armstrong and David Scott. Armstrong was a former Navy pilot and had flown seventy-eight combat missions in Korea. He later joined NACA and helped test several advanced aircraft, including the X-15. Having joined the astronaut corps in 1962, this would be his first spaceflight although he had served as the backup pilot for *Gemini 5*. David Scott was a relative newcomer to NASA. He was a graduate of the Experimental Test Pilot School and Aerospace Research Pilot Schools at Edwards Air Force Base and joined NASA in October 1963. Their mission had two major goals: to rendezvous and dock with an orbiting Agena and to conduct an Extra Vehicular Activity (EVA) using the newly constructed Astronaut Maneuvering Unit (AMU). The AMU contained its own life-support and propulsion systems and was designed to allow the astronaut to operate in space independently of the capsule.

The flight plan called for the target Agena and Gemini capsule to be launched on the same day. The Agena was successfully placed into a 179-mile circular orbit on March 16, 1966. While awaiting their own lift-off, Neil Armstrong and David Scott were told this good news and Armstrong replied, "Beautiful, we will take that one." A short time later the engines of their Titan roared to life, and they were propelled into a 96 × 163 mile orbit. Once in space, the automatic rendezvous program started and the capsule's orbit was adjusted so that the two craft gradually came closer together.

Five hours after launch, the crew of *Gemini 8* could see the Agena about 76 miles away. Scott radioed, "We have some object in sight. It looks like the Agena! We've got a real winner here." Armstrong maneuvered the capsule ahead of the Agena and then rotated the capsule's nose around so that it pointed at the Agena's docking collar. He then closed the remaining distance at the rate of three inches per second.

Figure 16-2. *This is how the crew of* Gemini 6 *saw* Gemini 7 *during America's first successful space rendezvous.*

Figure 16-3. *Astronauts Lovell (left) and Borman (right) are welcomed aboard the USS* Wasp *after their 14-day* Gemini 7 *flight.*

Moments later, he reported to Mission Control, "Flight, we are docked! It's . . . really a smoothie, no noticeable oscillations at all." Mission Control erupted with congratulatory backslaps and handshakes, broad smiles were everywhere. America's first space docking had been completed.

Once docked, Scott tested their maneuverability by firing the Agena engine to rotate the combined craft through a 90-degree angle. This maneuver was successful, but shortly afterward the ship unexpectedly began to roll. This had never happened in simulations, and none of the adjustments made by the crew to stop it appeared to work.

The situation aboard *Gemini 8* continued to deteriorate as their craft rotated faster and faster. Armstrong and Scott thought the problem was with the Agena, so Armstrong radioed Houston, "We're backing off." The astronauts were surprised to find that after this was done, the capsule began to rotate even faster. Scott radioed ground controllers,

> We have a violent left roll here at the present time, and we can't turn the RCSs (Retrorocket Control Systems) off, and we can't fire it, and we certainly have a roll . . . stuck hand control.

By this time, the situation was becoming critical, the capsule was rotating at the rate of one revolution per second. The astronauts were becoming disoriented and Armstrong radioed, "We've got serious problems here . . . We're tumbling

Figure 16-5. The astronaut maneuvering unit (AMU) was intended to allow tasks to be completed outside of the capsule.

end over end." Their vision was beginning to blur. Something had to be done, and quickly.

Armstrong deactivated the capsule's orbital attitude and maneuvering system and began firing the reentry control system. Using these thrusters, he was able to regain control of the capsule, but not before using 75 percent of the reentry system's fuel. He then reactivated the maneuvering thrusters and found that one of them was stuck in its open position. This explained why the capsule had rotated, but the dangerously low level of reentry fuel required that the mission be terminated. *Gemini 8* was instructed to prepare to return to Earth, and command of the discarded Agena was taken over by ground controllers.

Gemini 8 landed within 1.3 miles of its backup site, located 690 miles southeast of Okinawa. The flight had lasted 10 hours and 41 minutes. Although many of its objectives had not been met, its primary goal of docking with the Agena had been successful. A postflight analysis determined that the rotation was caused by a short circuit in the capsule's control system. Overall, NASA saw no need to delay future missions and preparations continued for the launch of *Gemini 9*.

Gemini 9: A Failed Docking Mission

Astronauts Elliot See and Charles Bassett II were scheduled to fly *Gemini 9*. Their objectives were similar to those of *Gemini 8*, to rendezvous and dock with an Agena target vehicle and conduct an EVA using the AMU.

Figure 16-4. The launch of the Gemini 8 Agena target vehicle. This vehicle was used to practice rendezvous techniques.

Figure 16-6. The failure of the lower shield of the Agena to jettison made a docking with Gemini 9 *impossible.*

The astronauts trained for their rendezvous mission at a McDonnell facility in St. Louis. On February 28, 1966, See and Bassett, and backup crewmen Thomas Stafford and Eugene Cernan, left Ellington Air Force Base for St. Louis in two dual-seat T-38 jets. The weather in St. Louis was poor. Light snow and rain were falling, and the area was covered by a low-lying fog. As the jets descended out of the murky skies, the astronauts realized that they were too far down the runway to land and decided to circle the airport and try again. See attempted to fly below the cloud layer and crashed into the airport building housing the *Gemini 9* capsule, instantly killing himself and Bassett. Stafford and Cernan landed safely on their next attempt. A review board, headed by astronaut Alan Shepard, Jr., was convened to investigate this tragedy, and Stafford and Cernan were reassigned as the prime crew to *Gemini 9*.

Tom Stafford was a veteran of *Gemini 6* so this would be his second Gemini flight. Cernan was a native of Chicago and had been a Navy pilot since 1956. At the age of 32, he was the youngest American yet assigned to a spaceflight.

The *Gemini 9* target Agena was launched on May 17, 1966, and initially its flight appeared normal. A few minutes later, however, its Atlas rocket malfunctioned, and the craft crashed into the Atlantic Ocean 119 miles down range. With the loss of the Agena, NASA prepared to use a backup target vehicle called the ATDA (Augmented Target Docking Adapter). This craft was successfully placed into a 179-mile-high orbit on June 1. Telemetry signals, however, indicated that its protective covering had not been properly jettisoned. If this was the case, officials feared that a docking might not be possible. Hoping that the telemetry signals were caused by some other malfunction, *Gemini 9* was readied for launch.

Just two minutes before lift-off a failure in the ground control equipment occurred scrubbing the mission. A dejected Stafford and Cernan had no alternative but to leave the capsule and return to their quarters. Much later Stafford re-

called his string of bad luck, beginning with *Gemini 6*, "Frank (Borman) and Jim (Lovell) may have more flight time, but nobody had more pad time in Gemini than I did." This was the fourth time in five attempts that his booster had failed to leave the pad.

After the Titan rocket was recycled, Stafford and Cernan lifted off two days later at 8:40 a.m. on June 3. Three hours and twenty minutes later the astronauts finally sighted the ATDA at a distance of 56 miles. As they approached, Stafford reported seeing its flashing acquisition lights and thought that the shroud had, after all, been properly discarded. However, as the distance to the ATDA decreased, this hope was dashed. Stafford radioed to ground controllers: "Look at that moose. . . . The shroud is half open on that thing! . . . It looks like an angry alligator out here rotating around!" (see Figure 16-6). The presence of the shroud made it impossible to dock since it covered the port. After flying around the ATDA Stafford suggested that the nose of his capsule be used to nudge the covering off, but officials considered this to be too risky. Instead he was instructed to test two new rendezvous methods, both of which worked well.

Once these maneuvers were complete, the crew rested for a day and then moved onto their next objective, an EVA by Cernan. After both astronauts put on their pressure suits, Cernan left the capsule but found that it was much more difficult to perform tasks in space than during simulations on Earth. His body heat began to overwhelm the spacesuit's environmental control system and by the time 80 percent of his assigned tasks were complete, his faceplate was badly fogged. He considered terminating the spacewalk, and after consulting with Stafford and Mission Control, reentered the capsule. Once back inside, Stafford reported that he could not see Cernan's face through his fogged face plate even though their helmets were almost touching. All-in-all the spacewalk posed more questions to NASA than it answered.

After a rest period, Stafford and Cernan completed the six experiments assigned to their mission and then prepared for reentry. *Gemini 9* landed less than a half a mile from its intended impact site east of Bermuda after a flight of 72.3 hours. When they opened the hatch of their capsule they were close enough to wave to the crew of the recovery ship, the USS *Wasp*.

Gemini 9 achieved some notable successes during its flight, three different rendezvous techniques had been demonstrated to work, and their inspection of the damaged ATDA proved that the Gemini capsule could make delicate maneuvers. On the other hand, the Agena had failed to reach orbit in two of its last three attempts. The ATDA docking (highlighted by the widely publicized picture of the "angry alligator") was an embarrassment, and a fogged faceplate prevented Cernan from testing the AMU. Public reaction to these shortcomings was surprisingly negative and showed that NASA was becoming a victim of its own success. Anything less than a perfect mission was now being viewed by the public as a failure.

Figure 16-7. An Agena target vehicle is prepared for launch to support the flight of Gemini 10.

Figure 16-8. India and Ceylon as seen from Gemini 11 from a height of 540 nautical miles. The Arabian Sea is to the left and the Bay of Bengal is to the right.

Gemini 10: A Successful Docking

Astronauts John Young and Michael Collins were selected to pilot *Gemini 10*. Their main objectives were to rendezvous and dock with their Agena target vehicle, perform a second rendezvous with the still orbiting Agena used in *Gemini 8*, test the Agena propulsion system, and carry out EVA activities. This would be Young's second Gemini flight and the first for Michael Collins. Collins was a former test pilot at Edwards Air Force Base and had served as the backup pilot for *Gemini 7*.

The *Gemini 10* Agena target vehicle lifted off at 3:39 p.m. on July 18, 1966, and successfully reached orbit. About 100 minutes later the astronauts were launched and entered a 156-mile-high orbit. Young and Collins then began rendezvous maneuvers and 5 hours and 52 minutes later docked with the Agena. The Agena's 16,000-pound thrust rocket was then used to increase their orbit to 458 × 176 miles. The ignition of the Agena gave the astronauts quite a jolt, causing Young to remark, "It may be only 1g, but it's the biggest 1g we ever saw!" They then fired their Agena engine a second time to place them into an orbit that would bring them closer to the *Gemini 8* Agena.

After these maneuvers, Collins conducted his first EVA. As the capsule entered night, he opened the hatch, stood up,

and took twenty-two ultraviolet pictures of the southern Milky Way. At the beginning of daylight, he took additional pictures of a color patch so scientists could determine whether film accurately recorded colors in space. Collins was not able to complete this activity because, for some unknown reason, both his and Young's eyes were becoming irritated. Since this posed a potentially serious problem to the rest of the flight, it was decided to terminate this activity. Ground controllers later suggested that their eye irritation might have been caused by fans in their space suit. This seems to have been correct, for once one of the fans was turned off, the eye irritation problem stopped.

The next task of *Gemini 10* was to undock from their Agena and closely approach *Agena 8*. Young moved his craft to within 10 feet of it, and after a brief time Collins emerged to conduct an EVA. He first retrieved a package from the capsule's exterior wall and then propelled himself toward the Agena. Once there, he lost his grip and used his hand-held air gun to return to the capsule. After one more try, he reached the Agena and retrieved an experimental package from its exterior but lost his camera while trying to maneuver.

Having fulfilled their mission objectives, Young and Collins began the process of preparing to return to Earth. Reentry started on their 43rd revolution. They landed in the western Atlantic 70.7 hours after lift-off on July 21, 1966, just a little over 3 miles from their intended impact site and were recovered by the USS *Guadalcanal*.

Gemini 11: Rendezvous and Docking

Gemini 11 was launched on September 12, 1966, with astronauts Charles Conrad and Richard Gordon aboard, both of whom had served as backups for *Gemini 8*. They worked

Figure 16-9. Astronaut Richard Gordon, Jr., returns to the Gemini capsule after completing an EVA during Gemini 11.

well as a team and were confident of their abilities. In *A Man on the Moon* Chaiken described this crew,

> Even before launch, they had earned a reputation as the cockiest, not to mention the most fun-loving, team of astronauts ever to fly; Conrad, short, balding, and wisecracking, and Gordon, not much taller, but more formidable, with the rugged face and build of a boxer. Gordon's friends knew him as a man of strong emotions.

Their mission was intended to extend the orbital height that man had reached and to rendezvous with the Agena during their *first* orbit. Nine scientific and three technology experiments were also planned. Two problems arose prior to launch, but they were quickly fixed, and on September 12, 1966, the Agena target vehicle lifted off. Conrad and Gordon anxiously awaited their launch. There was no room for error, a successful rendezvous with the Agena during their first orbit necessitated a launch within 2 seconds of their planned lift-off time. In an impressive display of the technical state of the Gemini program, they blasted-off just ½ second into this launch window.

Soon after reaching orbit, Conrad and Gordon began their rendezvous maneuvers and the distance between the two craft quickly decreased. After only 85 minutes, they drew to within 49 feet of the Agena and then docked with it. They then repeatedly docked, and once satisfied with this exercise, they fired the Agena engine to change their orbital plane. After this successful firing, both astronauts rested in preparation for an EVA by Gordon.

As in the previous mission, the first EVA consisted of standing up in the hatch of the capsule. Gordon then propelled himself to the Agena and attempted to connect a 98-foot tether line to it. As he was sitting on the Agena working at this job, the scene prompted Conrad to shout from the capsule "Ride 'em cowboy." On the ground Gordon had been able to complete this task in about 30 seconds, but in space it was much more difficult. He worked hard to keep from floating off the

Agena and began to perspire heavily. He radioed Conrad, "I'm pooped Pete . . . I still can't see. The sweat won't evaporate." Gordon succeeded in tying the two craft together but was completely exhausted. As a result, his EVA was terminated after only 33 minutes rather than its planned 107 minutes.

On the next day, Conrad and Gordon were to use the Agena to lift the combined craft to a substantially higher altitude. The Agena was fired on schedule and the two astronauts were rapidly propelled into space. As they looked back toward the receding Earth, Conrad remarked,

> The world's round! . . . I can see all the way from the end, around the top . . . it really is blue, that water really stands out and everything looks blue . . . the curvature of the Earth stands out a lot, there's no loss of color and details are extremely good.

As they sped away from the Earth, they took over 300 pictures of weather formations and ground features for scientists on Earth. After reaching an altitude of 823 miles and completing two orbits they fired their Agena to decrease their altitude to 182 miles.

On the 29th orbit of their mission, Gordon completed a second stand-up EVA, the main objective of which was to obtain ultraviolet photographs of a number of star fields. The next goal of *Gemini 11* was to conduct an experiment to test whether the tether line that Gordon had attached earlier could be used to make joint flying, or station keeping, easier. This experiment was only partially successful. Conrad and Gordon found that they could only keep the tether line tight by rotating the two craft. Without this rotation, the line bowed.

The tether experiment was the last major activity planned for *Gemini 11*, but since the capsule still had a substantial amount of unused fuel, an additional station-keeping activity was undertaken. After releasing the connecting cord, Conrad placed their capsule in an identical orbit with the Agena but trailing it by several miles. Once this task was accomplished they altered their orbit to close the distance between the two vehicles. This alternative station-keeping method worked well and was very fuel efficient. After completing this maneuver, they prepared their craft for reentry.

The final task of *Gemini 11* was to test a newly developed computer-controlled reentry procedure. As Conrad and Gordon began their descent into the atmosphere, the computer took control of the capsule, and as the astronauts looked on, it sent instructions to the maneuvering rockets. This system worked so well that Conrad and Gordon landed within three miles of their recovery ship, the USS *Guam*.

Gemini 12: The Final Flight of Project Gemini

Gemini 12 was the final flight planned for Project Gemini. It was manned by astronauts James Lovell, crewman of *Gemini 7* and backup command pilot of *Gemini 9,* and Edwin

Aldrin. Aldrin was a veteran pilot and held a doctorate in astronautics from M.I.T. Interestingly his thesis dealt with manned orbital rendezvous. During this mission he would have a firsthand opportunity to test his ideas. Their mission objectives were to finalize rendezvous, docking, and EVA procedures.

As Lovell and Aldrin walked to their capsule prior to lift-off, workers in the area saw small signs attached to the astronaut's backs that read, respectively, "The" and "End," in recognition of the fact that no further Gemini flights were planned. Project personnel took pride in what they had accomplished and were sad to see this project end. Support teams that had worked closely together would soon be broken up and friends would be transferred to other NASA projects. There were mixed feelings of sadness and satisfaction at this lift-off.

The launch of the *Gemini 12* and its Agena target vehicle occurred on November 11, 1967. However, shortly after achieving orbit, the capsule's rendezvous radar system failed. Aldrin then used an onboard sextant and charts he had made to complete their rendezvous. Using this backup system, they rendezvoused and docked with the Agena vehicle just 3 hours and 45 minutes after lift-off. They then practiced docking and undocking maneuvers.

As in *Gemini 11*, the flight plan called for Lovell and Aldrin to use the Agena's main engine to lift the combined craft into a higher orbit. However, sensors alerted ground controllers to a potential problem with the Agena's engine. Rather than risking the safety of the astronauts, it was decided to cancel this exercise. In its place, the astronauts were asked to view a solar eclipse that would occur during their time in orbit. This activity was part of the original flight plan but had been deleted because of time constraints. The cancelled high-altitude boost allowed it to be reinstated. After photographing this event while over the South Atlantic, Aldrin proceeded with the first of three EVAs. During this stand-up EVA he photographed star fields using an astronomical ultraviolet camera and then retrieved a micrometeorite collection package from the exterior of the capsule. In all, this EVA lasted 2 hours and 20 minutes.

The next day, Aldrin performed a second EVA. This would be the only one of his EVAs where he left the craft and it was the last opportunity NASA would have to iron out the problems encountered during the prior spacewalks. During this EVA he tied a cord between the capsule and the Agena to prepare for another station-keeping test and also performed several other tasks: torquing bolts, cutting metal, and assembling and disassembling electrical conductors as a test of man's ability to work in space. When he floated near Lovell's window he playfully paused to wipe it clean. Amused by this gesture, Lovell radioed him, "Hey, would you change the oil, too?" After these activities were complete, Aldrin returned to the capsule and rested. His only complaint was that his feet had gotten slightly cold. On the fourth day of the mission, he performed a third, stand-up EVA, to obtain additional ultraviolet pictures of a number of constellations.

Between the second and third EVA, Lovell and Aldrin again tested the tethered flight concept of station keeping.

Figure 16-10. Astronauts Lovell and Aldrin placed small signs on their backs that read "The" and "End" to commemorate the last of the Gemini flights.

They aligned the two craft perpendicular to the Earth, and after some trouble tightening the cord, they obtained a stable configuration without the need to rotate the craft. This exercise lasted about 4 hours. An additional problem arose during the flight when one of the capsule's fuel cells nearly failed, but the astronauts were able to keep it operating throughout the mission. On their 59th orbit, *Gemini 12*, like *Gemini 11* before them, began a computer-controlled reentry. Although the ocean landing was jarring, the procedure worked so well that they impacted just three miles from the recovery carrier the USS *Wasp*.

With the landing of *Gemini 12*, Project Gemini officially came to a close. President Lyndon Johnson issued a statement commending NASA for its successes and gave some perspective to the progress that had been made,

Ten times in this program of the last 20 months we have placed two men in orbit about the Earth in the world's most advanced spacecraft. Ten times we have brought them home. Today's flight was the culmination of a great team effort, stretching back to 1961, and directly involving more than 25,000 people in the National Aeronautics and Space Administration, the Department of Defense, and other Government agencies; in the universities and other research centers; and in American industry. Early in 1962 John Glenn made his historic orbital flight and America was in space. Now nearly five years later, we have completed Gemini and we know America is in space to stay.

Table 16-1. The Gemini Flights

Mission	Launch Date	Crew	Orbits	Duration	Notes
Gemini 1	April 8, 1964	Unmanned			Compatibility test
Gemini 2	January 19, 1965	Unmanned			Reentry test
Gemini 3	March 23, 1965	Grissom, Young	3	4 h 52 min	Test flight
Gemini 4	June 3, 1965	McDivitt, White	62	97 h 56 min	First EVA
Gemini 5	August 21, 1965	Cooper, Conrad	120	190 h 55 min	Long duration flight
Gemini 6	December 15, 1965	Schirra, Stafford	16	25 h 51 min	Rendezvous with *Gemini 7*
Gemini 7	December 4, 1965	Borman, Lovell	206	330 h 35 min	Long duration flight
Gemini 8	March 16, 1966	Armstrong, Scott	6	10 h 41 min	Docking, terminated
Gemini 9	June 3, 1966	Stafford, Cernan	45	72 h 21 min	Rendezvous, EVA
Gemini 10	July 18, 1966	Young, Collins	43	70 h 47 min	Docking, two EVAs
Gemini 11	September 12, 1966	Conrad, Gordon	44	71 h 17 min	Docking, EVA
Gemini 12	November 11, 1966	Lovell, Aldrin	59	94 h 34 min	Docking, three EVAs

Despite the impressive results of Projects Mercury and Gemini, just a little more than three years remained to fulfill President Kennedy's pledge to place a man on the Moon and return him safely to Earth. Although NASA was pleased with its accomplishments up to this point, time was running short.

A Postscript

The twelve Gemini launches occurred over a period of 31 months, from April 8, 1964 to November 11, 1966. The total cost of the program was around $1.2 billion. During this period Gemini accumulated 974 hours of flight time, 969 hours of which were manned, and nine EVAs lasting a little over 12 hours were completed. Statistics about these flights are presented in Table 16-1.

Project Gemini established a number of American and world milestones, including: (1) the first American to walk in space (*Gemini 4*), (2) the first orbital rendezvous (*Gemini 6*), (3) the longest duration spaceflight, 330 hours 35 minutes or 206 orbits (*Gemini 7*), (4) the first orbital docking (*Gemini 8*), and (5) the highest altitude (853 miles) reached by a manned craft (*Gemini 11*). Orbital space rendezvous were accomplished ten times, and nine dockings were completed.

In the mid-1960s public interest in manned spaceflight began to wane. Michael Collins offered an explanation for this change while defending Gemini's record, "Gemini's accomplishments were not perceived as being dramatic improvements, yet these ten manned flights were absolutely es-

sential in proving the technology of the machines and the desirability of the humans." Frank Borman, crewman of *Gemini 7,* concurred, "[Gemini] was a period of our space history that many Americans may have forgotten. . . . Perhaps only those who participated in Gemini can understand its importance to the achievements of Apollo."

NASA's string of space successes also played a role in the decline of public interest. In all, sixteen manned flights had been completed without the loss of a single astronaut. This record made it difficult to view spaceflight as a high-risk, suspense-filled endeavor. Those who flew these "successful" missions, however, knew better; they were not fooled. Frank Borman noted,

> Even in Gemini, where NASA consistently demonstrated an admirable commitment to safety, risk still couldn't be eliminated; there always was the unplanned incident, the unexpected malfunction, the uninvited emergency that popped up without warning. We almost lost *Gemini 8* and there wasn't a single mission in which something didn't go wrong. Every flight was an attempt to get answers to questions, but some questions remained unanswered.

Although the nation was beginning to view spaceflight as routine, the astronauts knew that as their missions became more complex, the likelihood that a serious mishap would occur was increasing.

17   The Flying Machines of Apollo

No single space project in this period will be more exciting, or more impressive to mankind, or more important for the long-range exploration of space; and none will be so difficult or expensive to accomplish.

—President John F. Kennedy

Apollo was one of the principle deities in ancient Greek mythology. He was the god of light and purity and was known for his faithfulness and dependability. His father, Zeus, was the powerful king of the gods, and his sister, Artemis or Diana, was the goddess of the Moon. The association of Apollo with light and dependability and his sister's realm of the Moon, made him a fitting symbol for America's lunar program.

The American space program had come a long way by the mid-1960s. The Explorer probes had discovered and mapped the radiation belts surrounding the Earth. Since the spatial extent limited, they did not pose a threat to a lunar mission. Project Mercury had shown that an astronaut could function in space and Project Gemini had developed rendezvous and docking techniques while demonstrating that astronauts could spend many days in space without becoming incapacitated. The unmanned lunar flights of Ranger, Lunar Orbiter, and Surveyor had allowed scientists to examine closely the Moon's space environment and surface properties. Many of the important technical issues associated with a lunar landing had been confronted and satisfactorily resolved. However, the vehicles that would carry a crew to the Moon had yet to be built and tested.

In September 1963, George Mueller was appointed head of the Office of Manned Spaceflight. Coming from the Air Force ballistic missile program he was a strong-willed person and a skilled administrator. Shortly after arriving at NASA headquarters he ordered that a realistic time-line for the nation's lunar program be constructed. He was shocked to learn that the earliest projected landing was in late 1971, well beyond the time period established by President Kennedy. On October 29 he therefore directed that a new development procedure called "all-up" testing be implemented. Simply put, this meant all three stages of the Moon rocket would be simultaneously tested during a flight. To von Braun's Huntsville group this approach was somewhere between "heresy" and "insanity." At Peenemuende each component of the V-2 had been verified before moving on to the next component. The Huntsville engineers planned to undertake a similar incremental approach with America's Moon rocket. Mueller realized this approach was too time consuming. He pointed out that to flight-test the lower stages of the rocket, a weight approximat-

Figure 17-1. George Mueller became the respected head of the office of Manned Space Flight in late 1963.

ing that of the upper stages would have to be attached anyway. Why take the time to fabricate dummy stages when real ones could be used instead? Secondly, in an actual flight the upper stages would have to ignite when the rocket was traveling at several thousand miles per hour and under severe aerodynamic stresses. These conditions could not be easily simulated on the ground. In his all-up procedure these engines could be tested once the lower stage was jettisoned. His most telling argument, however, and the one that convinced von Braun to adopt this approach, was not based on engineering but political considerations. Without an "all-up" approach, it would not be possible to reach the Moon before the end of the decade. If the president's commitment was abandoned, the political will to continue funding Apollo might evaporate.

When President Kennedy was assassinated in Dallas just a few weeks after Mueller's October 29 directive was issued, America's view of the lunar mission took on an added dimension. A Moon landing was now not just a contest between differing political systems, it was the final wish of the nation's popular fallen leader. To renege on Kennedy's pledge would dishonor his memory. This was unthinkable to the average citizen.

To make a lunar mission a reality, however, a more powerful launch rocket and larger spaceship had to be constructed. The American lunar program called for a three-

Figure 17-2. The Apollo command module (cone shaped object at the right end of the vehicle) and service module.

person craft. Two crewmen would descend to the Moon's surface while a third astronaut remained in orbit to monitor the condition of the mother ship.

The craft that the astronauts would travel to the Moon in consisted of three separate modules: the command module, the service module, and the lunar module. The command module served as the nerve center of the vessel. The service module carried the crew's life support-system, the propulsion systems, and the ship's maneuvering fuel. The lunar module would land the astronauts on the Moon and then return them to the orbiting spacecraft.

The Command Module

The command module (CM) was shaped like the head of a bullet (see Figure 17-2). As its name implies, it was the operational center of the Apollo spacecraft. It stood 9.8 feet high, had a maximum diameter of 12.5 feet, and could easily fit into a modest-size room. The CM was divided into three compartments—the forward compartment, the crew area, and the rear compartment.

Located at the top of the forward compartment was a docking probe that joined the CM and lunar module together. Beneath this fixture was the capsule's landing parachute. In the center of this compartment was a tunnel the astronauts used in traveling from the CM to the lunar module. Along its outside wall were a pair of reaction control engines that oriented the capsule for reentry into the Earth's atmosphere (see Figure 17-3).

Below this region of the craft was the crew area. It contained the ship's instrument panels, storage bays, guidance and navigational systems, space suits, food, and personal hygiene equipment. The craft's electrical power, life-support, and communication systems were also controlled from this part of the module. The crew compartment was cramped, having a volume of about 6 cubic meters, roughly equivalent to the interior room in a midsize car. The astronauts entered the CM through a hatch located above their seats. Along the outside wall were five windows, two faced forward and were used for visual sightings during a docking. The other three windows served as observation ports. The rear compartment of the CM contained an ablative heat shield, made of the same phenolic epoxy resin used on the Gemini capsule. The rest of the capsule was covered with a thin coating of this material.

The Service Module

The service module (SM) was connected to the bottom of the CM. To maintain symmetry, its 12.5-foot diameter matched that of the CM. It was 24.6 feet long (about the size of a small bus) and had a cylindrical shape. Except during the reentry phase of the mission, the CM could not operate without the SM.

The SM carried the ship's fuel, water, oxygen, and hydrogen supplies. Fixed and steerable radio antenna extended outward from the bottom of the SM and were used to communicate with the Earth. The Apollo craft's main rocket engine was located at the rear of this module. Its fuel consisted of hydrazine, unsymmetrical dimethylhydrazine, and nitrogen tetroxide.

Along the interior wall of the SM were several equipment bays, four of which contained fuel tanks and two others which contained electrical equipment, life support systems, and the astronauts' oxygen supply. An additional compart-

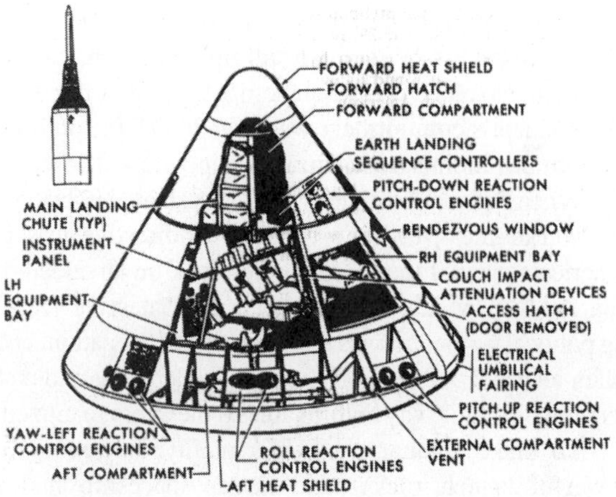

Figure 17-3. The Apollo command module (CM).

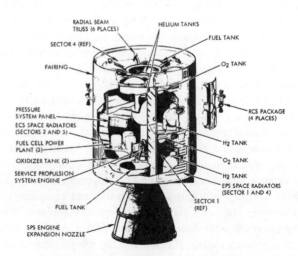

Figure 17-4. The Apollo service module (SM).

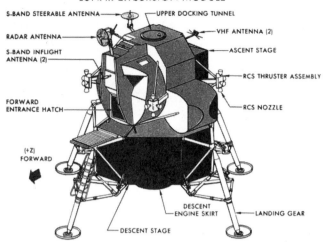

LUNAR EXCURSION MODULE

S-BAND STEERABLE ANTENNA — UPPER DOCKING TUNNEL

RADAR ANTENNA — VHF ANTENNA (2)

S-BAND INFLIGHT ANTENNA (2) — ASCENT STAGE

RCS THRUSTER ASSEMBLY

FORWARD ENTRANCE HATCH — RCS NOZZLE

(+Z) FORWARD

DESCENT ENGINE SKIRT — LANDING GEAR

DESCENT STAGE

Figure 17-5. The lunar module (LM) did not look like a moon craft, but the moon's lack of an atmosphere allowed it to have a nonaerodynamic shape.

Figure 17-6. When the probes at the base of the landing legs touched down, the LM's engine would be turned off.

ment carried instruments and sensors. Four sets of four thrusters were located symmetrically around the exterior wall of the upper half of the SM. In each cluster, the four thrusters were oriented at 90 degrees from each other. By selectively firing them, the craft could be moved in any direction.

The Lunar Module

The lunar module (LM) was built by Grumman Aircraft Corporation and was about the size of a two-level motel room. Standing 23 feet high, it had a maximum diameter across its landing legs of 31 feet. During the descent to the Moon it would be manned by two astronauts. The LM was a self-contained, two-stage craft. The upper stage was pressurized and housed the astronauts. It also carried their equipment, fuel tanks, communication equipment, as well as an ascent rocket. Two triangular windows were located on either side of the exit hatch and were used by the astronauts when flying the LM.

The lower stage of the LM had an octagonal shape consisting of large storage compartments. Joined to the sides of this base unit were four landing legs (see Figure 17-6). At the end of each leg was a 6.2-foot diameter, trash-can cover–looking fixture that was designed to prevent the craft from sinking into the lunar soil. A 5.6-foot-long probe extended from the bottom of three of the landing feet. When these sensors touched the lunar surface, the astronauts turned off the descent engine to avoid kicking up a dust cloud that might clog the rocket's engine. The descent engine was located in the center portion of the bottom stage. Upon separation from the CM/SM, this rocket was fired to slow the LM's velocity and begin the descent to the Moon's surface.

As Figure 17-6 shows, the LM resembled an enormous insect. An aerodynamic shape was unnecessary because the Moon lacks an atmosphere. Despite its peculiar appearance,

the LM was nevertheless capable of flight. Jim McDivitt, an astronaut from *Apollo 9*, described it in the following manner: "It sure flies better than it looks."

The Saturn Rockets

In 1961, when the decision was made to land on the Moon, America had only one large rocket under development, the *Juno 5*. Beginning in 1957, Wernher von Braun and his team worked on this booster at the Army Ballistic Missile Agency in Huntsville, Alabama. It employed a cluster of five modest-sized engines, that when fired together generated an impressive lift capability.

When von Braun's group was reassigned to NASA in 1959, the *Juno 5* was augmented by adding powerful upper stages. This rocket, although not employed in the Apollo program, served as a prototype for NASA's Saturn rockets. The Saturn series of rockets consisted of three successively more powerful and larger rockets, the Saturn I, the Saturn IB, and the Saturn V.

The Saturn I

The two-stage Saturn I rocket was designed to place an unmanned prototype of the combined CM and SM into Earth's orbit. Its first stage, the S-1, consisted of a cluster of eight H-1 engines (see Figure 17-7), each of which produced 94 tons of thrust for a total thrust of 752 tons. Its second stage, the S-IV, employed six RL-10A engines and produced a combined 45 tons of thrust. With an Apollo craft on top, the vehicle stood 163 feet tall and weighed 583 tons when fully fueled.

The first flight of a Saturn I, the *SA-101*, took place on October 27, 1961. Since its goal was to test the first stage of the rocket, the second stage was flown unfueled. This mission

Figure 17-7. The Saturn I first stage consisted of a cluster of eight substantial engines.

Figure 17-8. The Saturn I was the smallest of the Saturn series of rockets but was a very large booster.

was a success, as were its next three test flights. The first complete Saturn I flight, the *SA-105*, occurred on January 29, 1964, and placed an 18.7-ton payload into orbit. The next two Saturn I flights, *SA-106* and *SA-107,* occurred in May and September of 1964 and placed development versions of the Apollo craft in orbit.

The final three test flights of the Saturn I, *SA-108, SA-109,* and *SA-110,* were also successful and were used to place meteoroid detection satellites (the Pegasus series) into orbit. In order to accelerate development of the next version of the Saturn rocket, the last three scheduled Saturn I launches were cancelled.

The Saturn IB

The Saturn I could place a 10-ton payload around the Earth, but this was not sufficient for all three Apollo modules. Since it was vital to verify the Apollo spacecraft worked in orbit prior to proceeding to the Moon, a powerful new rocket, the Saturn IB, was pressed into service.

The Saturn IB combined a modified Saturn I first stage with a new and more powerful second stage. The thrust of the modified first stage, renamed the S-IB, was 820 tons. The second stage, called the S-IVB, used a single J-2 engine that produced 2.5 times the thrust of the S-IV, or 113 tons. This new second stage also carried computers and guidance equipment to control the rocket's flight. The Saturn IB with its attached Apollo spacecraft stood 224 feet tall and weighed 645 tons.

On February 26, 1966, about two and a half months after Borman and Lovell completed their fourteen-day mission aboard *Gemini 7*, a series of unmanned Saturn IB launches began. They had three objectives: (1) to test the behavior of

Figure 17-9. Launch of a Saturn IB.

its liquid fuel while in orbit, (2) to test the Apollo command and service module systems, and (3) to test the lunar module.

These unmanned test flights were successful and the first manned Saturn IB launch took place on October 11, 1968. This flight, designated *Apollo 7*, was flown by astronauts Schirra, Eisele, and Cunningham, and was the only manned Saturn IB launched during the Apollo program. This rocket, however, would later be used in the Skylab and Apollo-Soyuz programs.

Saturn V

For the manned Moon mission, NASA needed a still more powerful rocket, so a three-stage rocket, the Saturn V, was constructed. In addition to having a third stage, it employed improved versions of the engines used in prior rockets of this series. For example, the first stage of the Saturn I used eight engines to produce 752 tons of thrust. The first stage of the Saturn IB also employed eight engines to produce a larger thrust of 820 tons. In contrast, the first stage of the Saturn V, called the S-IC, employed only five engines but produced an astonishing 3,840 tons of thrust. This thrust is equivalent to that produced by 20,000 heavy locomotives or 200 aircraft carriers. During flight, the S-IC fired for about 2 minutes and was then jettisoned.

The S-IC, built by the Boeing company, was 33 feet in diameter and stood 138 feet tall. Its five F-1 engines burned liquid oxygen and kerosene and its outer four engines could swivel to provide the rocket with some directional control during flight. The Saturn V's second stage, the S-II, built by North American Aviation, also had a diameter of 33 feet and was 82 feet tall. Powered by a cluster of five J-2 engines, it produced a maximum combined thrust of 575 tons. While in flight, it fired for about 6 minutes, increasing the craft's velocity to 15,260 miles per hour. At this speed, the Saturn V could fly across the entire United States in about 12 minutes. The third stage of the Saturn V, the S-IVB, built by Douglas Aircraft Corporation, was essentially identical to the second stage of the Saturn 1B. This final stage produced 115 tons of thrust, had a length of 59 feet, a diameter of 21.6 feet, and was fired twice; once to reach orbit and a second time to propel the craft beyond the Earth's escape velocity and to the Moon.

Above the S-IVB was a cylindrical-shaped unit that housed the LM (see Figure 17-10). Once on the way to the Moon, the CM would dock with the LM and tow it away from this enclosure. This procedure is illustrated in Figure 17-13. The S-IVB would then be sent into orbit around the Sun or to crash into the Moon.

The Saturn V was an enormous rocket and demonstrated the extent to which the United States was prepared to go to honor its commitment to land a man on the Moon. The cost of its construction, $9.3 billion, limited government spending in other areas and required the talents of an army of engineers. Never before in the history of the world had such a powerful spacefaring vehicle been built. With all four of its stages attached, the Saturn V stood 363 feet tall (longer than a football field) and weighed an astonishing 3,196 tons. It was capable of placing 130 tons into Earth's orbit or sending

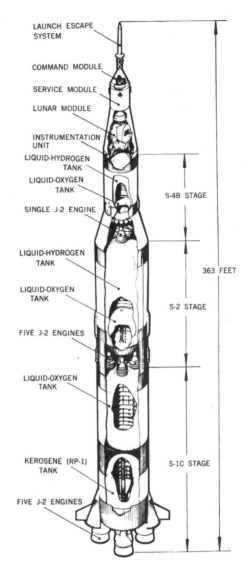

Figure 17-10. The components of America's monster Moon rocket, the Saturn V.

Figure 17-11. The Saturn V was so large it was transported to the Cape in parts. There it was assembled and housed in an enormous building prior to launch.

Figure 17-12. The relative size of the capsules and rockets used in Projects Mercury, Gemini, and Apollo demonstrate the tremendous progress that was being made in the American space program.

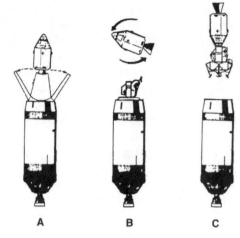

Figure 17-13. Once on the way to the Moon the astronauts would retrieve the LM from the S-4B.

a payload of 43 tons to the Moon. When fired, it produced seismic shock waves that could be detected at a distance of 1,000 miles. The noise from its engines was as loud as an active volcano, and its ascent could be seen in Jacksonville, Florida, located 150 miles from its launch site at Cape Kennedy. When the famous journalist Walter Cronkite reported on a test flight of the Saturn V in November 1967, he was amazed at the rocket's power, As it lifted-off Cronkite shouted, "Oh, my God, our building is shaking! Part of the roof has come in here!" Once this rocket and the Apollo spacecraft had been thoroughly tested, America would be ready to go to the Moon.

On the Path toward the Moon

The design of the three-crew Apollo spacecraft was based on information gathered from Projects Mercury and Gemini. The journey to the Moon would be completed using successively smaller subcraft. Although this plan was complicated by the need for multiple rendezvous and docking maneuvers, NASA realized it was the only way to reach the Moon by the end of the 1960s. It was not possible, given the state of technology at that time, to build a single rocket capable of traveling all the way to the Moon and back. In this respect, America's Moon craft bore little resemblance to the vehicle envisioned by Tsiolkovsky, Oberth, and Goddard. The time for that type of vessel had not yet come.

Table 17-1. The Saturn Family of Rockets

	Saturn 1	Saturn 1B	Saturn V
Height (ft)	187.9	224	363
Lift-off weight (tons)	582	645	3,195
First Stage			
Height (ft)	80.3	80.3	138
Diameter (ft)	21.4	21.4	33
Engines	8 H-1	8 H-1	5 F-1
Thrust (tons)	752	820	3,840
Fuel	Liq O, Kerosene	Liq O, Kerosene	Liq O, Kerosene
Second Stage			
Height (ft)	41.4	58.4	81.6
Diameter (ft)	18.3	22.7	33
Engines	6 RL-10A	1 J-1	5 J-2
Thrust (tons)	45	113	575
Fuel	Liq O and H	Liq O and H	Liq O and H
Third Stage			
Height (ft)	-	-	58.6
Diameter (ft)	-	-	21.7
Engines	-	-	1 J-2
Thrust (tons)	-	-	115
Fuel	-	-	Liq O, Kerosene

18 | *Apollo 1*: **The High Cost of Bravery**

We're in a risky business, and we hope if anything happens to us, it will not delay the program. The conquest of space is worth the risk of life.

—Virgil Grissom, astronaut

By the middle of 1966, a total of nineteen American astronauts had flown in space. Seven of these men, including Grissom, had flown on two missions. Since the beginning of Project Mercury, not a single astronaut had lost his life or was even seriously injured during a spaceflight. Three astronauts—Elliot See, Charles Bassett, and Theodore Freeman—had died in airplane crashes, but these accidents were not associated with NASA's hardware. NASA was proud of this record and the safety of the astronauts remained a high priority as work on Project Apollo accelerated.

Up to this time, all of the Apollo program flights had been unmanned. The first test flight, which took place in November 1963, was designed to identify problems in the capsule escape system. Between 1964 and 1965 additional solid-fueled Little Joe flights were conducted to validate this system. Unmanned Apollo Saturn I rocket flights also took place between July 30, 1965 and February 26, 1966, culminating in a 39-minute suborbital mission that lifted a 37,500-pound payload to a height of 5,500 miles.

On August 25, 1966, NASA moved closer to its goal of a manned flight when it successfully launched its first Saturn IB. This booster lifted an unmanned production model of a Command and Service Module to a height of 18,000 miles and paved the way for the construction of more powerful rockets. The Apollo program was obviously making steady progress and seemed on the right track. Confidence grew as technical obstacles were overcome by NASA's engineering skills. Barring any unforeseen setbacks, a lunar landing before the end of the decade was doable.

Apollo 1

On the morning of January 27, 1967, astronauts Virgil Grissom, Edward White, and Roger Chaffee prepared to conduct a launch-simulation test in preparation for their upcoming Saturn IB flight. A small army of technicians, over 1,000 in all, busily checked and rechecked the spacecraft and rocket to ensure that everything was functioning properly. The goal of this exercise was to duplicate all aspects of a launch in order to identify and correct any potential problems prior to the actual manned flight.

Throughout the morning work proceeded smoothly, and after an early lunch, the three astronauts suited up and rode to Launch Complex 34 to continue their tests. The capsule sat atop its Saturn IB. Grissom was troubled by the numerous problems his capsule was experiencing. He had worked for months on these problems and thought that his suggestions were being ignored. This made him so mad, he told an interviewer, that at times he couldn't see straight. Finally, a few days before the test, he walked up to the capsule and hung a big Texas lemon on it.

Immediately adjacent to the capsule was a clean room manned by employees of North American under the direction of Donald Babbitt. When sealed inside the capsule, the astronauts purged the cabin air and replaced it with pure oxygen, the atmosphere it would have in space. Work proceeded in a routine fashion but was frequently interrupted by communication problems between the capsule and support personnel. An exasperated Grissom radioed from inside, "I can't hear a thing you're saying. Jesus Christ! . . . How are we going to get to the Moon if we can't talk between two or three buildings!?" This delayed the simula-

Figure 18-1. From left to right, Edward White II, Virgil Grissom, and Roger Chaffee.

Figure 18-2. Prior to the fire, the astronauts would have been seated in their capsule in the fashion shown above.

tion countdown, and it was nearly sunset before the rocket and capsule were ready to be disconnected from their external power sources.

At 6:31 p.m., Babbitt and his employees were located directly outside the capsule performing their normal tasks when suddenly, over their radio speaker, they heard Roger Chaffee cry out, "Hey!" Then, after a scuffling sound, they heard, "We've got a fire in the cockpit." As the flames fed on the capsule's pure oxygen atmosphere, the cabin's air pressure increased to such an extent that the external wall burst open. Before Babbitt and his men could react, a sheet of flame shot from the broken craft, knocking them to the floor. A thick, acrid smoke filled the surrounding room.

Elsewhere in the launch complex, workers quickly realized a serious problem had arisen. On the first floor of the launch pad, Gary Propst watched in horror as his monitor showed flames within the cabin flickering around the capsule's window. He then saw an astronaut's arm grappling at the hatch mechanism. Instinctively Propst yelled, "Blow the hatch! Why don't they blow the hatch!?" From inside the capsule Chaffee yelled, "We've got a bad fire. . . . Let's get out. We're burning up. . . ." This was followed by a scream of pain and then silence.

After recovering from the shock of the sudden fire, Babbitt and three other technicians fought the flames engulfing the craft with fire extinguishers and worked furiously to remove the capsule's inner and outer hatches. It was difficult work. Over the communications net listeners could hear their pained voices as they were driven back, "It's too hot!" Despite their best efforts, it took a crucial 5 1/2 minutes to finally reach the inner hatch.

As Babbitt and his crew pried the inner hatch open, they were faced with a terrible sight. Grissom and White lay unmoving and tangled together at the base of the cabin. The heat had been so great the material in their suits fused with the craft's plastic and metal fixtures. Chaffee lay burnt on his couch, the restraining straps and oxygen hoses removed from his space suit. Even though he was not a doctor, Babbit knew the astronauts were dead.

Launch physicians Fred Kelly and Alan Harter arrived at the smoldering capsule about five minutes after the hatches had been opened. They tried to examine the astronauts for signs of life, but the fire had sealed their space suits shut with molten nylon. Finally they confirmed what everyone viewing the scene suspected, the astronauts were dead. Firemen arrived shortly later and with great difficulty, removed the astronauts' dead bodies from the charred craft.

Astronaut Frank Borman, a close friend of Ed White, visited the scene of the fire the next day and was shocked by the sight,

> As the first one who entered it [the capsule], I can't even begin to describe that chamber of horror. To me, the interior of a spacecraft had always provided a reassuring sight of gleaming instruments and spotless equipment, creating the illusion of total indestructibility. *AS-204* was a fire-blackened charnel house, a charred shell that wasn't even a recognizable facsimile of a spacecraft.

The loss of the astronauts was keenly felt at Cape Kennedy, and the emotional reaction was intense. Borman later wrote,

> I was no stranger to violent death and its aftermath, having occasionally been emotionally and professionally close to a victim or victims. But nothing I had witnessed in the past came close to what I saw at Cape Kennedy. The giant complex was in a mass state of shock.

Part of this reaction was due to the manner in which the astronauts had died. This was even felt by NASA's director, James Webb, "We've always known that something like this would happen sooner or later, [but] who would have thought the first tragedy would be on the ground?" Borman agreed,

> He was right—that's what made the Pad 34 disaster so hard to take. Three superbly trained pilots had died, trapped during a supposedly routine ground test that shouldn't have been any more dangerous than taking a bath. I don't think the grief would have been any less if they had perished in space, but at least it would have been more logical and half-expected.

An autopsy found that despite the intense nature of the fire, the astronauts would probably have survived it, albeit with extensive burns. Their death was caused not by burns but by inhaling toxic gases created by the fire. Had an explosive hatch, of the type used in the Mercury capsules, been installed in the Apollo command module, the astro-

Figure 18-3. The charred interior of the capsule demonstrates the fierce nature of the fire.

Figure 18-4. The above picture shows a portion of the burnt exterior of the Apollo 1 capsule.

nauts might have survived. Ironically, it was Grissom's own *MR-4* flight, where the capsule hatch unexpectedly blew open after landing, that had caused NASA to discard this hatch feature.

A few days after the accident, the astronauts were buried with full honors. Grissom and Chaffee were buried at Arlington National Cemetery and White was buried at the West Point Military Academy.

Post *Apollo 1* Changes

A NASA review board, headed by Floyd Thompson, director of the Langley Research Center, was convened to determine the cause of this tragedy. Frank Borman represented the astronauts. After an extensive investigation, they issued a 3,000-page report of their findings concluding that the fire was caused by an electrical arc within the command module. The extensive damage within the capsule, however, prevented them from precisely determining where the fire had started.

The Thompson Board found that despite the intense design scrutiny given to the construction of the Apollo Command Module, no one had realized the fire danger that a sealed capsule with an overpressurized atmosphere of pure oxygen presented during ground tests. Although the board's report contained words such as "ignorance," sloth," and "carelessness," the key descriptive term that described this accident was "oversight."

The panel recommended several changes to prevent a similar disaster in the future. It recommended that the air in the capsule during launches be changed from 100 percent oxygen to a less flammable mixture of 60 percent oxygen and 40 percent nitrogen. Also, wherever possible, the cabin materials should be rebuilt out of inflammable materials. The panel similarly recommended that material in the astronaut's space suits consist of an inflammable glass fabric. Finally, it recommended a new hatch design that could be opened by technicians within 10 seconds.

In the aftermath of the *Apollo 1* accident, NASA managers wondered if they had missed some warning signs that might have prevented this tragedy. Nearly two years earlier, in March 1965 the issue of a one- or two-gas capsule atmosphere and the possibility of a fire had been discussed. At that time, most of the studies dealt with a fire in space rather than on the ground and these reports concluded that a pure oxygen atmosphere was safer and less complicated than a multigas atmosphere. Other more recent warnings, however, existed. Hilliard Paige of General Electric alerted Apollo Spacecraft Manager Joseph Shea about the possibility of a ground fire just three weeks prior to the accident. Unfortunately no action was taken.

The tragedy of *Apollo 1* went beyond the deaths of the three astronauts. Several workers at Cape Kennedy were deeply disturbed by it, and some required psychological treatment. The astronauts and their wives were severely shaken. Astronaut Frank Borman and his wife Susan were particularly hard hit by this tragedy because they had grown quite close to the Whites. Borman viewed Ed White as an "astronaut's astronaut" and as the "brother I never had." One of the overlooked victims of *Apollo 1* was Ed White's wife, Pat. Never able to recover from the loss of her husband, she committed suicide in 1984. Naturally NASA would take the necessary actions to ensure that a similar fire never again occurred. This provided some conciliation but not enough, the cost of this lesson was too high, much too high.

19   Apollo Missions 4–10: From Earth to Lunar Orbit

I think the Russians would have gone all out to orbit the Moon if we hadn't done it when we did. We preempted that. If they had orbited the Moon, our press would have said they had won.

—Robert Gilruth, chief of the Manned Spacecraft Center

Just a little over two months after the tragic deaths of astronauts Chaffee, Grissom, and White, the unmanned portion of the Apollo program continued. While modifications were being made to the command module (CM), other important hardware issues had to be resolved. NASA engineers still didn't know if the powerful, new three-stage Saturn V rocket was reliable enough for manned flight and if the heat shield of the CM would be able to withstand the high temperatures generated when a craft returning from the Moon entered the atmosphere at nearly 25,000 miles per hour. They also had to prove that the service module (SM) engine could be repeatedly fired and restarted in space.

In order to find answers to these questions, NASA planned several unmanned missions that built upon each other. With each success, subsequent flights would test additional procedures and hardware until all of the problems associated with a lunar mission were resolved.

Apollo Missions 4–6

On November 9, 1967, the unmanned *Apollo 4* mission took off from the Kennedy Space Center. This flight was the first to use the new Saturn V rocket and its objectives were (1) to test the three stages of the Saturn V rocket, (2) to test the structural integrity and compatibility of the Saturn V and Apollo spacecraft, and (3) to determine if the CM's heat shield could withstand the heat generated by a craft returning to the Earth from the Moon.

As the Saturn V burst to life, its 7.5 million pounds of thrust sent shock waves through the launch site. It was an awesome event. The ground at the site shook so severely that an eyewitness later jokingly remarked that the question was not whether the Saturn V had risen, but whether Florida had sunk.

The accumulated experience NASA had gained from Projects Mercury and Gemini was evident in the lift-off of *Apollo 4*, it occurred within one second of its planned time. The launch was successful and about 10 minutes later, the Saturn V's third stage, the S-IVB, carrying the CM and SM and a mock-up of the lunar module (LM), was in orbit. The total weight of this craft was a remarkable 278,699 pounds.

Less than ten years earlier, the United States had struggled just to place the 31-pound *Explorer 1* into orbit.

After completing two orbits, the S-IVB again ignited and propelled the craft to a height of 11,000 miles. The CSM was then automatically ejected from the SIV-B and sensors within the capsule indicated that it was functioning properly. The SM engine then fired, sending the craft speeding back to the Earth where it reentered the atmosphere traveling at 24,911 miles per hour. Reentry proceeded normally and the capsule settled into the ocean within 5 miles of its intended landing site. Once retrieved by the USS *Bennington* and inspected, it was clear that the heat shield had protected the capsule's interior as hoped. The objectives of *Apollo 4* had been met.

Figure 19-1. The first launch of the Saturn V was an important milestone in America's quest to place a man on the Moon.

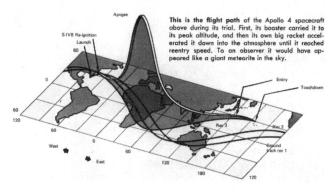

Figure 19-2. The flight path of Apollo 4.

Apollo 5 lifted off on January 22, 1968, at 5:48 p.m., using a Saturn IB rocket. The goal of this unmanned mission was to test the one piece of Apollo hardware that had yet to be flown in space, the LM. Specifically, *Apollo 5* was to determine if the LM's ascent and descent engines worked in the vacuum of space. Ten minutes after lift-off, the S-IVB and LM, called the LM-1, were in orbit. The LM-1 separated from the S-IVB, and its systems were remotely operated by ground controllers during the course of two orbits. Its descent engine was fired a total of three times and commands to fire the ascent engine were sent twice. In each case, the engines fired when commanded to do so. After these tests were completed, LM-1 was instructed to reenter the atmosphere where it was destroyed over the Pacific ocean by the heat of reentry. NASA officials judged the *Apollo 5* mission to be so successful that they cancelled a second unmanned LM flight.

After the success of the prior two flights, expectations were high among NASA engineers that *Apollo 6* would be similarly successful. *Apollo 6* lifted off on April 4, 1968. Its goals were similar to those of *Apollo 4*—to test the reliability of the Saturn V and the Apollo CSM. The results of this flight, however, were very troubling. All three stages of the Saturn V malfunctioned. The thrust from the Saturn V's first stage,

Figure 19-3. The Apollo 4 *capsule returned to earth severely charred but flightworthy.*

the S-IC, was uneven, causing the rocket to bounce like a giant pogo stick for about 30 seconds. This pogo motion had been identified and fixed during the early stages of the Gemini program so it was discouraging to see it reappear.

As *Apollo 6* continued upward, the S-IC was jettisoned and the second stage, the S-II, ignited. Shortly after ignition, however, two of its five J-2 engines prematurely shut down. To compensate for the loss of thrust produced by the failures in its lower two stages, the Saturn's third stage, the S-IVB, automatically fired for a longer duration. This increased the craft's velocity enough for the craft to reach orbit. Once there, ground controllers sent instructions to the S-IVB to restart its engines, but the booster failed to respond.

The CSM then separated from the nonfunctional S-IVB, and the SM engine fired, lifting the craft to a height of 10,000 miles. The SM engine then fired a second time and sent the vehicle racing back to the Earth. Unfortunately there was only enough fuel to increase the craft's speed to 21,600 miles per hour, a value less than what a returning craft from the Moon would have. The SM was jettisoned, and the CM landed in the Pacific Ocean where it was retrieved by the USS *Okinawa*. A postflight inspection of the CM showed it had survived the reentry and landed in good condition.

Despite the unsatisfactory performance of the Saturn V during the *Apollo 6* mission, there was little public criticism of NASA. The nation was preoccupied with other, more weighty, issues. On March 31, President Johnson made the surprising announcement that he would not seek reelection, but would instead devote his remaining time in office to ending the Vietnam war. On April 4, the same day as the launch of *Apollo 6*, civil rights leader Martin Luther King, Jr., was assassinated in Memphis, Tennessee. This terrible event was made even sadder when racially inspired riots erupted across the country. Political and social conflicts seemed to be tearing the country apart. In this setting, the partial failure of an unmanned Saturn V was not a matter of great concern.

In view of the mixed success of *Apollo 6*, NASA accelerated ground testing of the Saturn V. Up to 1,000 engineers worked on a solution for the S-IC's pogo problem, while others reexamined both the S-II and S-IVB to determine why they had failed to function properly. On June 24, 1968, after the successful ground simulation of a 167-hour Apollo flight, NASA felt confident enough that the engine problems had been solved to declare the unmanned portion of its test program was complete. It was ready to begin manned flights.

The Manned Flight of *Apollo 7*

The first manned orbital Apollo flight, *Apollo 7*, took place on October 11, 1968, twenty-one months after "the fire" on Launch Pad 34. Its crew consisted of astronauts Walter Schirra (veteran of *Mercury 8* and *Gemini 6*) and rookies Donn Eisele and Walter Cunningham. This would be the first and last manned Saturn IB flown in the Apollo program. The goals of *Apollo 7* focused primarily on engineering issues. It was also the first mission to use the modified CM.

Figure 19-4. The crew of Apollo 7. *From left to right, R. Walter Cunningham, Walter Schirra, Jr., and Donn F. Eisele.*

Figure 19-5. During Apollo 7, *Schirra demonstrated his sense of humor by displaying comical cards before the television camera.*

During this flight, the Apollo spacecraft's life-support and operating systems would be tested for 10.8 days, a time similar to that required for a lunar mission. This flight would also test the ability of the SM to perform critical maneuvers in space and rendezvous with the S-IVB stage of the Saturn IB. In a spaceflight first, live television pictures of this rendezvous were taken by the astronauts and broadcast to Earth. This activity was included with an eye toward the future Moon landing. William A. Lee, a management-level NASA engineer at Houston explained,

> One [objective] of the Apollo Program is to impress the world with our space supremacy. It may be assumed that the first attempt to land on the Moon will have generated a high degree of interest around the world. . . . A large portion of the civilized world will be at their TV sets wondering whether the attempt will succeed or fail. The question before the house is whether the public will receive their report of this climatic moment visually or by voice alone.

It was undeniable that television pictures would have a larger impact than radio. To increase public awareness of this mission and to better prepare for broadcasts from the Moon, live pictures became an important part of this and all subsequent Apollo flights.

The launch of *Apollo 7* proceeded smoothly, and once in orbit, the astronauts ignited the SM engine to begin their rendezvous with the S-IVB. The kick of this engine was stronger than the astronauts expected, causing a surprised Schirra to shout, "Yabadobadoo!" After several more firings of the SM engine, they rendezvoused with the S-IVB and conducted a mock docking.

As hoped, the live television coverage provided by the astronauts was a public relations hit. Although Schirra was initially against using valuable flight time for this purpose, he seemed to truly enjoy these broadcasts. During the start of one broadcast, he and Cunningham held a professionally printed sign up to the camera that read, "Hello From The Lovely Apollo Room High Atop Everything."

After completing their assignments and, more importantly, again verifying that the SM engine could be reliably restarted in space, the astronauts returned to Earth. *Apollo 7* completed 163 orbits and was recovered on October 22, 1968, by the USS *Essex* in rough seas. NASA considered the mission "101 percent successful."

To the Moon and Back: The Mission of *Apollo 8*

In August 1968, George Low, manager of Houston's Apollo Spacecraft Program Office, met with Christopher Kraft, assistant director of Marshall Space Center for flight operations, to discuss the upcoming *Apollo 8* mission. The original goal of this flight called for the CSM and LM to be tested in Earth's orbit. It was becoming increasingly clear, however, that the LM would not be ready for this mission. Low therefore proposed to maintain the Apollo program's momentum by changing the goal of *Apollo 8* to that of a circumlunar flight.

Kraft was enthusiastic about this proposal and gave his endorsement of it. Low soon found that a circumlunar mission had widespread support within NASA, extending from Frank Borman, the commander of *Apollo 8*, to von Braun, the builder of the Saturn V. Whether to proceed with this ambitious flight hinged on the outcome of *Apollo 7*.

With the success of *Apollo 7*, George Mueller, associate administrator for Manned Space Flight, called a confidential meeting on November 10, to finalize the goals of *Apollo 8*. Everyone at the meeting realized a lunar mission was risky. The Saturn V had performed poorly on *Apollo 4,* and despite von Braun's assurance that its problems had been fixed, the fact remained that the rocket had not been retested in an actual flight.

Figure 19-6. Low (left) and Kraft (right) were the initial forces behind the change of Apollo 8 *to a lunar mission.*

Apollo 7 employed a Saturn IB. There was an additional source of danger with a lunar mission. If the SM engine failed when the craft was on its way to the Moon, there was no way for the astronauts to return to Earth; they would die in space.

Mueller asked each person, in turn, for his opinion. Hilliard Paige of General Electric responded, "The checkout equipment is doing the same thing it has done before; there are no reservations from a reliability standpoint; and NASA should go, and is ready to go, into lunar orbit." Leland Atwood of North American seemed to express the view of the group when he said, "This is what we came to the party for." Based on these responses, Mueller decided to recommend that *Apollo 8* undertake a lunar mission. Two days after Mueller's meeting, on November 12, Tom Paine, the new acting chief administrator of NASA, announced at a press conference that *Apollo 8* would fly around the Moon.

The Flight of *Apollo 8*

At 3 a.m. on the morning of December 21, 1968, the crew of *Apollo 8,* Frank Borman, Jim Lovell, and Bill Anders, were awakened for a traditional preflight breakfast of steak and eggs. Preparations for their lift-off moved quickly, and by 7 a.m., they were strapped into their capsule awaiting launch.

Figure 19-7. The mission patch for Apollo 8 *depicted a flight path that circled the moon and then returned to Earth.*

The beaches and highways leading to the Cape were crammed by a record number of people who had come to see the launch of mankind's first mission to the Moon. Among NASA's guests was Charles Lindbergh, the famous aviator who had helped Robert Goddard secure funding for his early rocket research.

The *Apollo 8* crew had met Charles Lindbergh two weeks earlier while at a White House dinner hosted by President Johnson. Frank Borman was particularly pleased to meet Lindbergh because he was one of his childhood heroes. Prior to the launch, Lindbergh again met Borman, this time at the Cape. Lindbergh asked how much fuel the Saturn V used. When told the answer, he made a brief calculation, grinned and said, "In the first second of your flight tomorrow, you'll burn ten times more fuel than I did all the way to Paris."

At 7:51 a.m., *Apollo 8* lifted off its launch pad on its two-and-a-half-day journey to the Moon. Future *Apollo 11* astronaut "Buzz" Aldrin, who was among the spectators, said, "I had the best seat in the house, the roof of the launch control center. As always, the noise was overpowering and unEarthly. The thick concrete roof was bouncing like a trampoline. It was hard to believe that three tiny humans were strapped into the tip of that monster."

From a nearby underground control center, Wernher von Braun and his team of Huntsville experts anxiously scanned telemetry from the rocket for problems. The first stage of the Saturn operated normally, the pogo motion did not reoccur. After the first stage was jettisoned, all engines on the second stage ignited properly. Once this stage was released, the third stage engine also ignited and worked flawlessly. About 12 minutes after lift-off, the astronauts and over 100 tons of equipment were in orbit.

Two and half hours passed as the astronauts and ground controllers checked the condition of the Apollo craft. Finding no problems, the crew fired the S-IVB engine, raising the craft's velocity to over 25,000 miles per hour, and the astronauts proceeded on their way to the Moon. Once on a lunar trajectory, Borman turned a T-shaped handle in the CM that triggered explosive devices that released their craft from the booster. He then maneuvered the CSM away from the S-IVB.

About five hours after lift-off, *Apollo 8* passed through the van Allen radiation belt, discovered nearly ten years earlier by *Explorer 1*. Some scientists had warned NASA that the astronauts might encounter harmful, perhaps even deadly, levels of radiation here. Anders was glad to report that the radiation level within the craft was well within acceptable limits. The spacecraft provided sufficient shielding, so the radiation the astronauts were exposed to never exceeded that which a person received in a routine chest x-ray.

The first mid-course correction using the SM engine took place when the Apollo craft was 60,000 miles from Earth. This was a tense moment. If a malfunction occurred the mission would have to be aborted and it was far from certain the astronauts could return to Earth. The SM engine, however, performed flawlessly.

Figure 19-8. The crew of Apollo 8. *From left to right, James A. Lovell, William A. Anders, and Frank Borman.*

Figure 19-9. This view of the Earth rising above the lunar landscape was taken by Frank Borman while in lunar orbit.

Fourteen hours into the flight, Borman experienced some mild space sickness, he became nauseous and vomited twice. By the next day, however, his space sickness had passed. During the flight to the Moon, both Lovell and Anders also experienced similar, though less severe episodes of space sickness. Gemini flights had found that space sickness frequently occurred early in a flight as the body adjusted to weightlessness. After spending a day or two in space, it passed.

Halfway to the Moon, the astronauts conducted the first of six live television broadcasts. They commented about how beautiful the Earth looked from space. Borman was struck by the fact that although nearly 3 billion people lived on Earth, no great cities or engineering works were visible, the planet looked uninhabited. The CSM approached the Moon with the SM in front so the astronauts were unable to see the approaching Moon through the capsule's windows. When the spacecraft came closer to the Moon, an on-board computer would query the astronauts about whether to commit the craft to a lunar orbit. If instructed to proceed, the computer issued commands that fired the SM engine and slowed the craft's velocity. As the CSM came within 500 miles of the Moon, the computer asked for these instructions, and without hesitation, Borman punched the PROCEED button. The computer instructed the SM engine to fire and the craft's speed decreased from 5,700 to 3,700 miles per hour. They were now in lunar orbit. Almost immediately after the SM fired, the astronauts saw the Moon swing into view. On Earth, it was Christmas Eve.

Apollo 8 entered into an elliptical orbit with maximum and minimum heights of 169 and 61 miles. After completing two orbits, the SM engine again fired, placing the CSM into a nearly circular 70-mile-high orbit.

Once these maneuvers were complete, the astronauts looked at the Moon through the CM windows. Borman was struck by the bleakness of the Moon, "I felt as if *Apollo 8* had been transported into a world of science fiction, with incredible lighting and awesome, furlong beauty—desolate beyond belief." Lovell described the Moon's surface to mission control, "The Moon is essentially gray, no color. Looks like plas-

ter of Paris. Sort of grayish sand. . . . It makes us realize what you have back on Earth." During their second orbit, live television pictures of the Moon were transmitted back to Earth. The lunar surface reminded Anders of a beach scene, to him the Moon looked like "dirty beach sands with lots of footprints in it."

The astronauts spent much of their time photographing Apollo landing sites and practicing navigational tasks. During the third orbit Borman happened to glance out the window and saw the distant disk of the Earth floating above the lunar terrain. He was captivated by this view, "It was the most beautiful, heart-catching sight of my life, one that sent a torrent of nostalgia, of sheer homesickness, surging through me." He took the on-board camera from Anders and snapped a picture. Once developed back on Earth, this scene became one of the most famous pictures of the space age.

The astronauts were greatly moved by the sights they had seen and at the urging of Borman concluded their activities by reading a portion of the Bible, Genesis 1:1–10, for their audience back on Earth,

In the beginning God created the heaven and the Earth. And the Earth was without form, and void; and darkness was upon the face of the deep. And the spirit of God moved upon the face of the waters. And God said, Let there be light: and there was light. And God saw the light, that it was good: and God divided the light from the darkness. And God called the light Day and the darkness he called Night. And the evening and the morning were the first day. And God said, Let there be a firmament in the midst of the waters, and let it divide the waters from the waters. And God made the firmament, and divided the waters which were under the firmament from the waters which were above the firmament: and it was so. And God called the firmament Heaven. And the evening and the morning were the second day. And God said, Let the waters under the heaven be

Figure 19-10. The crew of Apollo 9 *(left to right), James A. McDivitt, David R. Scott, and Russell L. Schweickart.*

gathered together unto one place, and let the dry land appear: and it was so. And God called the dry land Earth; and the gathering together of the waters called he Seas: and God saw that it was good.

After orbiting the Moon ten more times, they fired the main SM engine for the 304 seconds needed to lift them out of lunar orbit and onto their return path to Earth. Lovell's joy and relief was evident in his transmission to ground controllers, "Please be informed there is a Santa Claus. The burn was good." They had been in lunar orbit for about 24 hours.

After a trip of nearly 500,000 miles, *Apollo 8* landed about 1,000 miles southwest of Hawaii in the predawn darkness on December 27. The capsule touched down within 3 miles of the recovery ship the USS *Yorktown*. As the rescue helicopter circled overhead, the helicopter pilot inquired, "Hey, *Apollo 8*, is the Moon made of limburger cheese?" A beaming Anders answered, "No, it's made of American cheese."

The flights of *Apollo 4* through *8* had successfully tested much of the hardware and many of the procedures needed for the upcoming lunar landing. The capsule had been altered so that another tragedy like that of *Apollo 1* would not happen. The Saturn V had been shown to provide a reliable, albeit, rough ride into orbit. The SM engine had been successfully fired and restarted on several occasions. The CSM had been flown to the Moon and back. The only piece of hardware not yet extensively tested in a manned flight was the LM.

Apollo Missions 9 and 10

The objectives of *Apollo 9* were to test the LM in space and simulate maneuvers needed for a lunar landing. Since this was the first mission where all of the Apollo hardware was simultaneously tested, the flight took place in Earth's orbit. Its ten-day duration was, however, long enough to accommodate a lunar mission. The astronaut crew consisted of

Figure 19-11. Apollo 9 *CM pilot Dave Scott was photographed by Schweickart as he stood in the hatch of the LM.*

James McDivitt, David Scott, and Russell Schweickart. As commander of *Gemini 4*, McDivitt was a space veteran. Scott had also flown in space aboard *Gemini 8*. This would be Schweickart's rookie flight.

Apollo 9 lifted off from the Kennedy Space Center just one second behind schedule on March 3, 1969. The Saturn V performed well, but during the second-stage burn, the pogo problem returned. Shortly after reaching orbit, the CSM separated from the S-IVB and then docked with the LM. With a combined mass of 91,058 pounds, the CM, SM, and LM constituted the most massive vehicle ever assembled in space. The SM engine was then fired to test the structural integrity of the combined vehicle and to raise the orbit of the combined craft. No problems developed during this maneuver.

On the second day of the mission Schweickart and then McDivitt entered the LM to test its systems. They fired the LM descent engine and were impressed with how well it performed. Both the LM and CM were then depressurized, and Schweickart and Scott opened their hatches, photographed each other, and retrieved experimental packages.

On the fifth day of their mission, Schweickart and McDivitt detached the LM and allowed it to drift away from the CM. They then rotated the LM so that Scott could see if the LM's landing legs were fully deployed and if the vehicle had suffered any damage. All looked well so the LM was moved 3 miles away, and for 6 hours, its crew performed lunar approach and descent maneuvers. They then jettisoned the LM's descent stage. From a distance of about 100 miles, they fired the LM's ascent engine and flew back to the CM, simu-

Figure 19-12. *Dave Scott photographed the LM as it moved away from the CM.*

Figure 19-13. *The crew of* Apollo 10. *From left to right, Eugene A. Cernan, John W. Young, and Thomas P. Stafford.*

lating a return from the Moon's surface. Smaller thrusters were then used to dock.

For the remaining five days of their flight, the astronauts continued to test the SM rockets and obtained photographs of the Earth in a variety of colors. Having completed their mission objectives, the LM was jettisoned and the astronauts returned to Earth.

Apollo 9 splashed down northeast of Puerto Rico on March 13,1969, and it was recovered by the USS *Guadalcanal*. The astronauts had spent 10.1 days in space, fired the SM main engine a total of eight times, tested the LM's propulsion and operating systems, and conducted a stand-up Extra Vehicular Activity (EVA). Only a final full-up mission remained before NASA was ready to attempt a lunar landing. This final checkout mission was assigned to *Apollo 10*.

Apollo 10 was a dress rehearsal for the upcoming lunar landing. It had three primary objectives: (1) to determine how the LM's guidance and navigation systems behaved in the lunar environment, (2) to conduct additional tests of the LM's engines, and (3) to practice tracking and communications procedures with ground controllers.

The crew of *Apollo 10* consisted of three experienced astronauts, Thomas Stafford, a veteran of *Gemini 6* and *9*, John Young, a crew member of *Gemini 3* and *10*, and Eugene Cernan, Stafford's crewmate on *Gemini 9*. This was the first NASA crew where each member had previously flown in space.

Apollo 10 lifted off on May 18, 1969. Over 100,000 people watched its lift-off from locations around the Cape. The ride up was rough. During the S-II burn, the pogo motion reappeared and the astronauts wondered whether the CSM and LM might somehow break apart as a result of this jostling. The vibration was so strong that Stafford could barely read the settings on the instrument panel. He momentarily thought about shutting down the booster and aborting the mission but decided, "No way! We've come this far—if she blows, then she

blows." When the third stage ignited, it made loud growling noises and shook the entire ship. Despite the tense moments associated with these events, the S-IVB safely carried the astronauts into orbit. After an instrument check, they fired the S-IVB rocket a second time. It growled and shook again, but it raised their velocity to the escape velocity of the Earth and placed them into the proper lunar trajectory.

Once on the way to the Moon, the astronauts released the CSM from the S-IVB. They then docked with the LM and retrieved it from its storage position atop the booster. The S-IVB was then allowed to continue in a solar orbit. The trajectory used for *Apollo 10* closely followed that planned for

Figure 19-14. *The CSM as seen by Cernan and Stafford in the LM during the mission of* Apollo 10.

Figure 19-15. NASA officials Low (sitting) and Kraft (standing) watch a television transmission from Apollo 10. *These two individuals allowed the Apollo program to leap forward because of their daring idea to make* Apollo 8 *a lunar mission.*

actly on time, bringing Stafford and Cernan to within 8.4 miles of the lunar surface. Here they took pictures and described the scene to mission control. The landing radar of the LM worked perfectly and they could clearly see the Moon's surface features. No attempt, however, was made to land. It was not part of their mission, and even if they had wanted to, the LM was too heavy to lift-off from the Moon.

Just before Stafford and Cernan were to jettison the descent stage of the LM, the craft began lurching wildly. The descent stage was quickly jettisoned, but the craft continued to move erratically. Stafford took control of the LM and brought it under control within a few minutes. Ground controllers later notified the crew that the problem had arisen because the astronauts had accidently switched the craft's guidance system to an improper setting.

Their mission complete, Stafford fired the LM's ascent engine. As they climbed back into orbit, the LM radar system locked onto the CM and maneuvered the craft toward the CSM. When within 26 feet, the astronauts manually docked the two craft. After Stafford and Cernan reentered the CM, the LM was jettisoned. As the two craft separated, ground controllers fired the LM engine and sent it into a solar orbit. The three astronauts then gathered further data about the lunar surface that would be of use to future lunar missions.

Apollo 10 completed 31 orbits of the Moon and returned to Earth on May 26. The CM landed in the Pacific Ocean about 415 miles from Samoa. After a flight time of 192 hours, the astronauts were safely back on the recovery ship, the USS *Princeton.*

NASA was very pleased with the success of *Apollo 10.* Except for some noncritical problems, all of the Apollo hardware and systems had worked properly. The goal of the next mission, *Apollo 11,* would be to land a man on the Moon.

Apollo 11. When the astronauts reached the vicinity of the Moon on May 21, they entered an elliptical orbit of 196 × 70 miles. Adjustments were then made to circularize the orbit to a height of about 60 miles.

After checking the LM systems and finding them all working properly, astronauts Stafford and Cernan fired its descent engine and proceeded toward the Moon's Sea of Tranquility. This was a critical burn. If the engine fired for just two seconds longer than its planned 30 seconds, they would crash into the Moon. After the allotted time the engine stopped ex-

20 *Apollo 11*: The First Manned Lunar Landing

We came in peace for all mankind.

—**From the plaque left on the Moon by the *Apollo 11* astronauts**

After years of strenuous efforts, America was ready to send a manned expedition to land on the Moon. If this mission was a success, it would elate the world and potentially usher in an entirely new chapter in the annals of human exploration. If it failed, it would be nearly impossible for NASA to fulfill America's long standing commitment to place a man on the Moon by the end of the decade. Considering the vast wealth and effort that had been expended toward this goal, failure was almost unthinkable.

NASA methodically worked toward the goal of a Moon landing for nearly a decade. This journey began with Projects Mercury and Gemini and included the unmanned flights of Ranger, Lunar Orbiter, and Surveyor. Ten unmanned launches of the Saturn I were conducted to test this rocket's aerodynamic properties, propulsion systems, launch facilities, and ground support equipment. They were followed by six unmanned Saturn IB and Saturn V missions that had three goals: (1) to test the increasingly powerful engines, (2) to verify that the command module (CM) and lunar module (LM) propulsion systems would work in space, and (3) to confirm that the capsule's ablation heat shield would protect a crew returning from the Moon.

The first manned Apollo flight, *Apollo 7*, was launched in October 1968 and proved the command service module (CMS) was a spaceworthy vehicle. *Apollo 8, 9,* and *10* followed in quick succession. *Apollo 8* was America's first manned circumlunar mission, while *Apollo 9* completed the first manned test flight of the LM. The pathway to the Moon

Figure 20-1. Neil Armstrong, commander; Michael Collins, CM pilot; and Edwin Aldrin, Jr., LM pilot.

was finally cleared when *Apollo 10* flew to the Moon and performed all the tasks required for a landing except touching down on the dusty lunar surface.

Despite this careful work, as the nation prepared to send *Apollo 11* on its mission to fulfill President Kennedy's pledge, nagging doubts remained. Had enough tests been completed? Had something been overlooked? Was our faith in the Apollo hardware justified or had we somehow been fooled by a string of good luck? Three men would now stake their lives on the belief that NASA's preparatory work had been properly and thoroughly completed.

Three American Heroes

The crew of *Apollo 11* consisted of astronauts Neil Armstrong (commander), Edwin "Buzz" Aldrin, Jr. (lunar module pilot), and Michael Collins (command module pilot). Each of these men had a distinctively different personality and none had flown together on a prior mission.

Armstrong grew up in middle America during the Depression and developed an appreciation for hard work,

Table 20-1. Steps to the Moon

Mission	Period
Project Mercury	1959–1963
Project Gemini	1964–1966
Ranger	1961–1965
Lunar Orbiter	1966–1967
Surveyor	1966–1968
Saturn 1	1961–1965
Saturn IB–Saturn V	1966–1968
Apollo 7–10	1968–1969

Focus Box: Symbolic Representations in *Apollo 11*

In the book, *First on the Moon*, Neil Armstrong describes the symbolism depicted in the *Apollo 11* patch: "We were all of us, interested in a number of small peripheral elements which go along with a flight. Things like the patch which we wear on our suits, and the names we select for in-flight communication between the two vehicles and between our vehicles and the ground. We were very conscious of the symbolism of our exploration, and we wanted the small things to reflect our very serious approach to the business of flying a lunar mission. The patch we designed was not intended to imitate the Great Seal of the United States; it was meant simply to symbolize a peaceful American attempt at a lunar landing. All of us in the crew particularly liked the idea of the olive branch, and as time went on we began to attach more and more importance to its inclusion in the patch design.

"The name *Eagle* was adopted subsequent to the selection of the patch design and was intended both to reflect the theme of the patch and also a degree of national pride in the enterprise. . . . None of us wanted our names on the patch. Adding them would have made it too crowded. So much for the department of aesthetics. More important than that, we were three individuals who had drawn, in a kind of lottery, a momentous opportunity and a momentous responsibility."

---

sacrifice, and a belief in the power of prayer. Addicted to flying at the age of 9, he later attended Purdue University on a U.S. Navy scholarship and majored in aeronautical engineering. Before graduating he was called to active duty, flew compact missions in Korea, "bridge breaking, train stopping, tank shooting and that sort of thing," was shot down in enemy territory, and rescued. After leaving the military, he completed his bachelor's degree and then obtained a master's degree

top

right

Figure 20-2. The landing site of Apollo 11 was in the southern end of the Sea of Tranquility, a large maria located on the western, front side of the Moon.

from the University of Southern California. Prior to joining NASA he worked for the National Advisory Committee for Aeronautics (NACA) as a research pilot and flew in the X-15 program. Selected as an astronaut in 1962, flight simulator operators recalled that no matter how difficult the task given to him, he never became angry or said a cross word.

Armstrong was a highly intense and focused individual who was also difficult to read. At times he seemed almost shy, but during some parties he played ragtime on a piano and was the last to leave. His wife, Janet, described him as highly nonconfrontational. She would later say, "Silence *is* a Neil Armstrong answer. The word 'no' is an argument."

Armstrong earned his position as commander on *Apollo 11* by serving as the commander of *Gemini 8*. He looked forward to the flight and was well aware of the dangers it held,

I have been in relatively high-risk businesses all of my adult life. Few of the others, however, had the possibility of direct gains in knowledge which this one has. I have confidence in the equipment, the planning, the training. I suspect that on a risk-gain ratio, this project would compare very, very favorably with those to which I've been accustomed in the past 20 years.

The phases used most often to describe Buzz Aldrin were "perfectionist," "magnificent confidence bordering on conceit," "humility," and "the best scientific mind we have sent into space." A professional, he had high expectations of himself and those around him and was intolerant of people not willing to give their best effort. Although he possessed a Ph.D. in engineering from the prestigious Massachusetts Institute of Technology, his single-minded determination to excel was not rooted in his educational background. It was a part of his being. His wife Joan would later state, "If Buzz were a trash man and collected trash, he would be the best trash collector in the United States."

Although possessing enormous ability and talent, he was uncomfortable in public settings and did not like to participate in idle conversation. Not given to outward displays of emotion, he was nevertheless self-confident. His passion for spaceflight was perhaps inherited from his father, who was also an avid flier, a former associate of Orville Wright, a student and friend of Robert Goddard, and on such close terms with Charles Lindbergh that he called him "Slim."

Michael Collins was a veteran of *Gemini 10*, and the youngest member of the crew. After his Gemini flight he had been assigned to fly aboard *Apollo 8*, but prior to lift-off he had to undergo a potentially career-ending operation. Removed from the *Apollo 8* crew he was assigned to *Apollo 11*. When asked if he were an optimist, he replied, "Not a perennial optimist, anyway. I think I'm more the fatalist—what's written is written, and I don't know what it is. I'm not always convinced that everything is going to work out well. On the other hand, there's nothing wrong in acting as if things will work out."

Easygoing and accessible, Collins liked fine wines, a good book, working in his rose garden, and dabbling in oil painting. Although a skilled engineer, he was not fond of computers and did not consider himself to be mechanically skilled. He was philosophical and humble about the upcoming mission and his role in it,

> I think man has always gone where he could, he has always been an explorer. There's a fascination in exploring and thrusting out into new places, but I can't say there is anything special about me—that I'm a natural explorer or I've always been bent in this direction—because it isn't true. I'll be a good explorer, but so would other people. I really think the key is that man has gone where he could and he must continue. He would lose something terribly important by having that option and not taking it.

Preflight Thoughts

The mission plan of *Apollo 11* called for Armstrong and Aldrin to land on the Moon's Sea of Tranquility (see Figure 20-2) in the lunar module, the *Eagle*, while Collins orbited above in the command module, *Columbia*. The launch date was set for July 16, 1969. Armstrong was pleased to be participating in this historic event, but was somewhat surprised to be doing so. When assigned to command *Apollo 11*, a little over a year and a half earlier, he did not believe his flight would land on the Moon, too many problems remained to be overcome. He thought the actual landing attempt would be made by a later crew,

> Our goal, when we were assigned to this flight last January, seemed almost impossible. There were a lot

of unknowns, unproved ideas, unproved hardware. The LM had never flown. There were many things about the lunar surface we did not know. . . . I honestly suspected, at the time, that it was unlikely that *Apollo 11* would make the first lunar landing flight. There was just too much to learn—too many chances for problems.

With the flight now imminent, Armstrong felt *Apollo 11* carried with it the hopes and dreams of a multitude of NASA engineers and contractors. He also realized his crew's performance would be judged by hundreds of millions of people around the world who were anxious to see if America could make good on its promise to land a man on the Moon by the end of the decade. Armstrong later wrote,

> This nation was depending on the NASA–industry team to do this job, and that team was staking its reputation on *Apollo 11*. A lot of necks had been put voluntarily on the chopping block, and as more and more attention was focused on the flight it became perfectly evident that any failure would bring a certain tarnish to the U.S. image.

On the day of the flight, weather conditions at Cape Canaveral were ideal. After an uneasy night's rest Armstrong, Aldrin, and Collins proceeded to the Saturn V and took the elevator up to the *Columbia*. As Collins was preparing to enter the capsule, his mind raced through a series of thoughts. He was struck by the historical significance of this flight and by an uneasy feeling about the dangers his crewmates faced. Later he recalled thinking,

> I am everlastingly thankful that I have flown before, and that this period of waiting atop is nothing new. I am just as tense this time, but the tenseness comes mostly from an appreciation of the enormity of our undertaking rather than from the unfamiliarity of the situation. I am far from certain that we will be able to fly the mission as planned. I think we will escape with our skins, or at least I will escape with mine, but I wouldn't give better than even odds on a successful landing and return. There are just too many things that can go wrong. . . . Here I am, a white male, age thirty-eight, height 5 feet 11 inches, weight 165 pounds, salary $17,000 per annum, resident of a Texas suburb, with black spots on my roses, state of mind unsettled, about to be shot off to the Moon. Yes, to the Moon.

As Aldrin waited for his fellow crewmates to be strapped into the craft, he was struck by the enormous public interest in this flight. Recalling his thoughts while viewing the scene surrounding their launch pad, he wrote,

> While Mike and Neil were going through the complicated business of being strapped in and connected

Figure 20-3. Apollo 11 *begins its journey to the Moon riding atop a Saturn V rocket.*

to the spacecraft's life-support system, I waited near the elevator on the floor below. I waited alone for fifteen minutes in a sort of serene limbo. As far as I could see there were people and cars lining the beaches and highways. . . . I savored the wait and marked the minutes in my mind as something I would always want to remember.

The Journey

The lift-off of *Apollo 11* occurred at 9:32 a.m. Eastern time from Launch Complex 39A at the Kennedy Space Center. The ascent proceeded as planned, but Collins remembers it as being a little shaky,

I was a little surprised by the initial ride on [the] Saturn V. It was rough for the first 15 seconds or so. . . . It was like a nervous lady driving her car down a narrow alleyway, unable to decide whether she's too far to the left or too far to the right, but she knows she's one and maybe the other. So she keeps jerking the wheel back and forth. . . . After 15 seconds it quieted down and the second stage was absolutely as smooth as glass. It had a sort of ethereal quality. You couldn't believe you had those big engines burning behind you.

Soon after reaching orbit and checking out the craft's systems, the astronauts reignited the Saturn's S-IVB rocket, increasing their velocity sufficiently to escape the Earth's gravitational pull. They were now beyond the point where an easy return to Earth was possible, but were untroubled by this fact. Armstrong later wrote,

After all the preflight preparation, there was actually somewhat less pressure on me during the flight itself. I no longer had a choice, an option, as to how I might best spend my time in training. There was one job to do and, just as with most jobs, once you're involved in it you feel more at ease.

The next phase of their flight called for the astronauts to detach the CSM from the S-IVB and retrieve the LM. This maneuver was completed without any problem and the S-IVB booster was released and sent into a solar orbit. The trip proceeded in an uneventful fashion and three days after leaving the Kennedy Space Center, *Apollo 11* entered into a lunar orbit. Again Collins described the scene,

Despite our concentrated effort to conserve our energy on the way to the Moon, the pressure is overtaking us (or me at least), and I feel that all of us are aware that the honeymoon is over and we are about to put our little pink bodies on the line. Our first shock comes as we stop our spinning motion and swing ourselves around so as to bring the Moon into view. We have not been able to see the Moon for nearly a day now, and the change is electrifying. The Moon I have known all my life, that two-dimensional small yellow disk in the sky, has gone away somewhere, to be replaced by the most awesome sphere I have ever seen.

The next morning astronauts Aldrin and Armstrong awoke early and entered the LM to check its systems. Finding everything in order, they detached their vehicle from the CM and prepared to descend to the lunar surface. As planned, Collins remained behind to pilot the CM. With a statement that concealed any hint of his earlier apprehension for his crewmates' safety, he tells them, "You cats take it easy on the lunar surface. . . ." As the LM drifted away from the CM, the astronauts exchanged brief barbs,

COLLINS: I think you've got a fine-looking flying machine there, *Eagle*, despite the fact you're upside down.
ARMSTRONG: Somebody's upside down.
COLLINS: You guys take care.
ARMSTRONG: See you later.

Collins would later be asked if he wasn't disappointed at having to remain in orbit while his companions journeyed to the Moon's surface. In his book *Liftoff*, he responds,

Figure 20-4. As the LM began its descent to the Moon with astronauts Armstrong and Aldrin, Collins inspected it from the CSM.

Figure 20-5. View from the Eagle *as it approached its landing site in the Sea of Tranquility. The ship is at a height of 63 miles.*

It is certainly true that I did not have the best seat on *Apollo 11*, and I would be a liar or a fool to say otherwise. But it is also true that I am perfectly satisfied with the seat I did occupy. I had always considered the first landing flight to be the premier one of the series, and I was proud to be a part of it—any part of it. By the time I joined Armstrong and Aldrin it had been clear to me for over a year that I had become a CSM specialist . . . and it was equally clear to me that of the three of us, I was the logical choice to stay in the *Columbia*. So I had no quarrel with either of my crewmates. I respected and admired them both.

The Landing

As Aldrin and Armstrong continued their descent to the Moon's Sea of Tranquility, they were busy working the LM controls. Aldrin read computer readout sequences while Armstrong piloted the craft.

At a height of about 6,000 feet above the maria floor, Armstrong reported to Collins that a yellow caution light, number "1202," had turned on. Neither Armstrong nor Aldrin knew if the caution light signaled a serious or a minor problem. Houston ground controller Steve Bales was also aware that this caution light had come on from telemetry data displayed on his console. He knew it was informing the pilot that the on-board computer system was trying to do too many tasks at the same time and that the computer would simply postpone less critical functions until later. The situation posed no danger to the astronauts and he promptly relayed this news to the nervous crew.

Shortly after this first warning light appeared, a second yellow LM light came on. Both Armstrong and Aldrin were again concerned about this sign of potential trouble. Aldrin later recalled that "our hearts shot up into our throats while we waited to learn what would happen. . . ." Steve Bales

again realized it was not a concern and reassured the crew. The descent continued.

As Armstrong and Aldrin neared the lunar surface, Armstrong noticed that the automatic landing system was about to set them down in an area strewn with boulders the size of a "volkswagen." Rather than risk damaging or even overturning the LM, he took control and flew to a nearby, smoother site. During this time, Armstrong and Aldrin were too busy to communicate the reason for this movement to Mission Control. Armstrong later recalled,

> We looked at the landing area and found a very large crater. This is the area we decided not to go into, we extended the range down-range. The exhaust dust was kicked up by the engine and this caused some concern in that it degraded our ability to determine not only our altitude in the final phase but also our translational velocities over the ground. It's quite important not to stub your toe during the final phases of touchdown.

Telemetry received at the Houston Control Center revealed that the LM was moving horizontally and pushing the limit of the available fuel. Just as controllers were about to transmit an abort signal, Armstrong set the craft down on the lunar soil. At 4:17 p.m. EDT on July 20, 1969, Armstrong reported back to Houston, "Tranquility Base here, the *Eagle* has landed." Houston replied, "Roger Tranquility. Be advised there's lots of smiling faces in this room, and all over the world." Armstrong replied, "There are two of them up here."

The initial concerns about the amount of fuel the LM had used during its landing were later found to be overblown, controllers determined the LM had at least 45 seconds of additional fuel, more than enough to provide an ample safety margin.

Figure 20-6. The Apollo 11 *landing site was flat and desolate. An 80-foot diameter crater was located about 100 yards from the* Eagle *on the opposite side of the craft from where this picture was taken.*

Both Armstrong and Aldrin were relieved by their successful landing. After a flight in which many things could have gone wrong, they were now on the Moon! Although they had faith in their abilities and in the LM, they could not totally withhold their relief. Aldrin recounted the scene,

> Neil and I are both fairly reticent people, and we don't go in for free exchanges of sentiment. Even during our long training we didn't have many free exchanges. But there was that moment on the Moon, a brief moment, in which we sort of looked at each other and slapped each other on the shoulder—that was about the space available—and said, "We made it. Good show." Or something like that.

Both men were surprised by the Moon's stark beauty (see Figure 20-6). Its surface was pockmarked by a multitude of small craters and a field of small rocks was located just beyond the landing site. Surface colors varied from a light gray-

tan, as Lovell had reported on *Apollo 8*, to an ashen gray. Because of Armstrong's decision to land at a smoother location, the LM touched down about four miles from its intended site.

During the next 6 1/2 hours, the astronauts checked out their craft's control systems, prepared for an emergency launch, sent data back to Earth, and put on their space suits and backpacks in preparation for mankinds' first exploratory journey on another world. Before stepping from the LM, Aldrin took time to give thanks to God and took private communion. Armstrong then cautiously left the *Eagle* and proceeded down the attached ladder. Pausing at the last step, he placed his footprint into the lunar soil and said,

> That's one small step for man, one giant leap for mankind.

Armstrong later wrote about the process leading to the choice of these words,

> I had thought about [what to say] a little before the flight, mainly because so many people had made such a big point of it. I had also thought about it a little on the way to the Moon, but not much. It wasn't until after landing that I made up my mind. . . .

The spacesuit the astronauts wore weighed 348 pounds on Earth but because of the Moon's lower gravity, it weighed only 53 pounds on the lunar surface. After convincing himself the Moon's surface was firm enough to support his weight, Armstrong took a few steps away and found it was surprisingly easy to move around. He quickly took some pictures of his surroundings and then collected a contingency sample of lunar soil and rocks. The soil was very dark, almost like powdered graphite. On the Earth an estimated 600 million people, about one-fifth of the world's population, watched history unfold on television screens. When Armstrong spoke his first words everyone realized the human race

Focus Box: American Astronaut Aldrin's Descent to the Lunar Surface

This dramatic sequence of pictures was obtained by Armstrong as Aldrin became the second person to set his footprint in the lunar soil.

had finally "left the cradle" and had entered a new era. It was a joyous moment that was shared by all people regardless of their political persuasion.

About 18 minutes after setting foot on the Moon's surface, Aldrin joined Armstrong. Aldrin described the surface as composed of light, rather powdery material that compacted easily and had considerable cohesion. He later wrote,

> In many places we sank only a fraction of an inch, but the rims of some small depression craters seemed to have a deeper soft layer. Our boots actually went in three or four inches. This created a tendency for slipping sideways when the boot finally hit something hard, and we tried to move around as much as possible on level areas, avoiding the little depressions.

Together the astronauts set up a TV camera a short distance from the LM that sent live pictures of their activities back to Earth. Ghostly black-and-white images of the scene showed the astronauts in what looked like deep-sea outfits, making abrupt movements in an invisible ocean. The sky was dark yet the landscape was brilliantly illuminated. They did not walk from place to place but moved by taking short hops.

Armstrong and Aldrin on the Moon

Because little was known about how the astronauts' bodies would react to conditions on the lunar surface, NASA planned a conservative exploration program. The astronaut's main objectives were to deploy an experimental package and collect rock samples. They would then return to Earth and report on the lunar environment and subsequent expeditions would be assigned more complex tasks based on their experience. As Armstrong explained,

> There were two viewpoints; whether we should concentrate on what's best for the results of this expedition, or whether it's best to prepare so that later expeditions may be even more fruitful. It's a matter of balancing two objectives and coming up with the operational restrictions.

For safety reasons, the astronauts were instructed to stay within 50 to 100 feet of the LM. If a problem unexpectedly arose this would give them plenty of time to get back to the relative safety of the LM. The Extra Vehicular Activity (EVA) was to be brief. Episodes of space sickness on earlier Apollo flights and unknown problems that might arise when working on the Moon's surface, caused NASA doctors to limit this time to under two hours. Until more was known, they felt it was foolish to tempt fate by over-exerting the astronauts. Despite the sound reasoning behind this decision, it was difficult for the astronauts to restrain their natural curiosity. Armstrong later wrote, "My only real problem on the surface was that there were so many places that I would like to have investigated, to find out just what was beyond the next hill, so to speak."

Figure 20-7. The footprints made by the astronauts were crisp. The surface had the consistency of wet sand.

The instrument package Armstrong and Aldrin placed on the Moon was called the Early Apollo Scientific Experiments Payload, or EASEP. It was a smaller version of a larger instrument package that would be deployed during future Apollo landings.

The EASEP consisted of three experiments. The first was called the Passive Seismic Experiment. It was designed to measure lunar vibrations and seismic activity on the Moon and to radio this information back to Earth. The second experiment was called the Laser Ranging Retro-Reflector Experiment. Once deployed a laser beam would be shot toward it from the

Figure 20-8. Experimental packages were set up on the Moon's surface. Aldrin leaves the laser reflector to set up the seismic experiment.

Earth and the reflected light would be used to determine the Moon's distance to within a few inches. This data would then be used to measure irregularities in the Moon's orbit around the Earth. The third experiment consisted of a sheet of aluminum foil, about the size of a window pane. Attached to a pole and facing the Sun, it would collect samples of the material bombarding the Moon's surface. At the end of the mission it would be rolled up and brought back to Earth for analysis.

The highest scientific priority of the expedition was to collect a sample of Moon rocks and core samples. The Moon rocks would allow the composition, geologic history, and age of material located at the landing site to be determined. Core samples, obtained by driving a hollow cylindrical tube into the lunar soil, would allow scientists to determine how the soil properties changed with depth. The rock and core samples were to be placed in two specially designed carrying cases, with a combined capacity of 50 pounds.

As the astronauts collected the lunar rocks, their conversation was monitored by mission controllers.

ALDRIN: . . . The rocks are rather slippery. . . Very powdery surface. . . . The powder fills up all the very fine porouses. My boot tends to slide over it rather easily. . . Say Neil, didn't I say we might see some purple rocks?

ARMSTRONG: Find the purple rocks?

ALDRIN: Yes, they are small, sparkly. . . I would make a first guess, some sort of biotite . . .

Aldrin found it surprisingly difficult to collect the core samples. He later wrote,

Technically the most difficult task I performed on the surface was driving those core samples into the ground to get little tubes of lunar material for study. There was significant and surprising resistance just a few inches

Figure 20-9. Aldrin stands at attention before the American flag. The wave in the flag was produced when wires supporting it in a horizontal direction bent during deployment.

down. . . . I had to hold on to the top of the core extension while I was hitting it with the hammer to drive it into the ground. I actually missed once or twice.

In fact, the surface was so hard the core sample tube was dented while being driven into the soil.

The astronauts also had some symbolic activities to perform. They took pride in the peaceful nature of their mission. This was reflected by the olive branch held by the eagle on their mission patch and by the plaque attached to the front landing gear of the LM. While on the surface, Armstrong unveiled the plaque and read its inscription, "Here men from the planet Earth first set foot upon the Moon, July 1969, A.D. We came in peace for all mankind." As Armstrong read these words, Aldrin used a video camera to provide the watching television audience with a view of the lunar landscape.

Back in Nassau Bay, Texas, a number of the Aldrin family members were watching the astronauts on television. Buzz Aldrin's uncle, Bob Moon, remarked that the landscape looked like west Texas. Joan Aldrin said aloud, "I've never seen anything like that before." To which Audrey Moon replied, "Who has?"

The astronauts also had another task they looked forward to completing, planting the American flag into the lunar soil. This task actually proved to be somewhat difficult. Aldrin explained,

It took both of us to set it up and it was nearly a disaster. . . . A small telescoping arm was attached to the flag pole to keep the flag extended and perpendicular. As hard as we tried, the telescope wouldn't fully extend. Thus the flag, which should have been flat, had its own unique permanent wave. To our dismay the staff of the pole wouldn't go far enough into the lunar surface to support itself in an upright position. After much struggling we finally coaxed it to remain upright, but in a most precarious position. I dreaded the possibility of the American flag collapsing into the lunar dust in front of the television camera.

Aldrin saluted the flag and then took a moment to survey the region surrounding the landing site. He recalled,

I looked high above the dome of the LM, Earth hung in the black sky, a disc cut in half by the day-night terminator. It was mostly blue, with swirling white clouds, and I could make out a brown landmass, North Africa and the Middle East. Glancing down at my boots, I realized that the soil Neil and I had stomped through had been here longer than any of those brown continents. Earth was a dynamic planet of tectonic plates, churning oceans, and a changing atmosphere. The Moon was dead, a relic of the early solar system.

Upon completing their explorations, Armstrong and Aldrin received a brief congratulatory call from President Nixon, who told them, "For one priceless moment, in the whole history of man, all the people of Earth are truly one—one in their pride in what you have done. And one in our prayers, that you will return safely to Earth."

On a personal note, the astronauts were pleased to leave medals commemorating the Soviet cosmonauts who had lost their lives in the quest to land on the Moon. As a final good-will gesture, they left a silicon disk containing messages from seventy-three heads of state. Other small symbolic tokens brought to the Moon included a piece of fabric from the Wright brother's plane. Once these duties were complete, the astronauts placed the lunar samples aboard the LM, reentered the craft, and rested. After spending about 9 hours on the Moon, of which about 2 hours and 21 minutes was EVA time, they prepared for lift-off.

The Return Home

The astronauts used the ascent stage of the LM to blast off from the Moon. This was a busy time for Armstrong and Aldrin. As they climbed away from the Moon, Aldrin's attention was drawn back toward the surface. As he glanced at their landing site, he saw the American flag they had struggled to place into the lunar surface fall over.

During the time that Armstrong and Aldrin were on the Moon, Collins was busy in the orbiting CM. He spent much of his time on the near side of the Moon trying to locate the LM. He also prepared the CSM for docking in the event his crewmates had to unexpectedly leave. While Armstrong was placing his foot into the lunar soil and making his historic statement, Collins was out of radio contact on the Moon's far side. Although hundreds of millions of people heard and witnessed this event, he would have to see it replayed on televi-sion once back on Earth. When the CSM finally emerged, Armstrong and Aldrin were setting up the flag. Collins never-theless enjoyed his time in the CM, especially when it was on the Moon's far side and out of radio contact. He later recalled his feelings during these times,

I am alone now, truly alone, and absolutely isolated from any known life. I am it. If a count were taken, the score would be three billion plus two over on the other side of the Moon, and one plus God knows what on this side. I feel this powerfully—not as fear or loneli-ness—but as awareness, anticipation, satisfaction, confidence, almost exhilaration. I like the feeling.

About 4 hours after leaving the Moon Armstrong and Aldrin docked with the *Columbia* in its 69-mile-high orbit. Aldrin was the first to reenter the CM. Collins was so happy and relieved to see him that his immediate inclination was to give him a big smooch on the forehead. Realizing Aldrin prob-ably would not appreciate this, he gave him a brisk handshake instead. After the astronauts transferred the lunar samples to the CM and Armstrong entered the capsule, the LM was jetti-soned. The main engine of the service module (SM) was then fired to increase the craft's speed to the value needed to escape from the Moon. They were on their way back to the Earth.

On July 23, one day before landing, the astronauts trans-mitted their final telecast from space. Each gave his impres-sion of the mission. Collins thanked the many thousands of dedicated people who had made their flight possible. Aldrin, still in awe of what he had seen, quoted a passage from the Bible, "When I consider the heavens, the work of Thy fingers, the Moon and the stars, which Thou hast ordained; What is man that Thou art mindful of him." Armstrong gave credit to those giants of science who had laid the foundations neces-sary for spaceflight and also thanked the many thousands of people who had "put their hearts and all their abilities in those craft" that had so ably served them on their lunar mission.

Figure 20-10. The Eagle *returns to the Columbia. Note that the bot-tom stage of the* LM *is gone. It was left on the Moon.*

Figure 20-11. Glad to be back on Earth, astronauts (left to right) Armstrong, Collins, and Aldrin grin from within their isolation chamber on the USS Hornet.

The *Apollo 11* CM reentered the atmosphere and splashed
down within 11 miles of its intended impact area in the Pacific
Ocean on July 24, 1969. The astronauts were then taken aboard
the USS *Hornet* where they were placed in an isolation chamber to guard against the possibility they had inadvertently been
contaminated by lunar microbes. Although it was considered
unlikely lunar microbes existed, if they did, Earth organisms
would have no natural defense against them. It was far better
under these circumstances to err on the side of caution. Using a
microphone connected to this chamber, the astronauts were
greeted in person by President Nixon who expressed the thanks
of a grateful nation for their safe return and successful trip.

The Moon Rocks

One of the major scientific objectives of the *Apollo 11*
mission was to retrieve a sample of lunar rocks. Great care
was taken to ensure that these samples were not exposed to
contamination by the Earth's environment. Once the *Columbia* was brought on board the recovery carrier, the CM was
connected to the quarantine trailer by a plastic tunnel. The astronauts then retrieved the sample containers and placed them
in a decontamination chamber. After their exterior surfaces
were decontaminated, the boxes were passed outside and
placed in special shipping packages.

The lunar samples were immediately flown to the Lunar
Receiving Laboratory at Johnson Space Center in Houston,
arriving there on July 25. Once the containers had again been
sterilized, they were placed inside the laboratories main vacuum chamber. Here they were weighed and found to contain
33 pounds and 6 ounces of material. The first sample con-
tainer was opened the next day at 3:45 p.m. while geologists
crowded around the observation port. Unfortunately the samples were covered by a layer of lunar dust so all their visual
observation revealed was that the rocks were irregular in
shape and their edges were slightly rounded.

Three days after the samples arrived in Houston, geologists working in the laboratory unexpectedly had an opportunity to view a lunar rock when much of the dust covering it
fell off. Numerous small, shallow pits were seen and three
common minerals—feldspar, pyroxene, and olivine—were
identified as the rock's major constituents. Based on its appearance, the rock was classified as igneous, a rock that had
condensed from a molten state. The gathered geologists speculated it could have been formed as part of a lava flow or
through local heating, perhaps caused by a meteor impact.

The following week was devoted to determining how to
distribute the material among the 142 scientific investigators
chosen to study it. A small sample of lunar dust was also mi-
croscopically examined, revealing that about half the material
was composed of a glassy substance having spherical to
needlelike shapes. Spectroscopic analysis of the lunar dust
showed its composition was very similar to that found by *Surveyor 5*. This craft had touched down about 15.5 miles from
the *Apollo 11* landing site and had the ability to perform a
chemical analysis of the material close to the craft. The similar composition at these two locations suggested the material
retrieved by the *Apollo 11* astronauts was representative of
that present over a much larger region on the Moon.

The lunar dust possessed abnormally high concentrations of titanium and small amounts of volatile elements.
These latter elements vaporize at relatively low temperatures
and would be rare in rocks formed at a high temperature. Although no one knew it at that time, this finding would prove
crucial in discussions about the validity of the model then in
use to explain the origin of the Moon.

21   The Missions of *Apollo 12* and *13*: A Rough Start to Lunar Exploration

Fellows, we're home.

—**James Lovell, speaking to his *Apollo 13* crewmates**

With the success of *Apollo 11*, NASA eagerly looked for ways to maximize the scientific return of its subsequent manned lunar missions. Projects Mercury and Gemini, as well as the prior Apollo flights, had provided information in three vital areas: (1) hardware development, (2) rendezvous and docking techniques, and (3) the human body's reaction to prolonged weightlessness. The flight of *Apollo 11* had contributed additional knowledge about these areas, but went beyond them by providing unique new data about the nature of the Moon's structure and surface.

Each of the obstacles standing in the way of a lunar exploratory program had now been overcome. Based on NASA's cumulative flight experience, officials believed the agency had done all it could to minimize the dangers faced by the astronauts. With safety concerns seemingly behind them, officials began to change their focus from the mechanics of the trip, to the scientific questions that could be addressed. Actually, a considerable amount of thought had already been given to this matter. In March of 1962, seven years before the launch of *Apollo 11*, an ad hoc group, the Sonett Committee, had been established by NASA's Space Sciences Steering Committee to identify which research objectives held the most promise. In their judgment these included studies about the Moon's geology, interior, and atmosphere. In order to take advantage of the unique circumstances offered by Apollo, the committee recommended that experiments aimed at providing information in these areas meet several criteria. They had to (1) address scientifically important questions, (2) provide data that could *only* be obtained from a manned lunar landing, (3) require the presence of an astronaut, and (4) serve as the basis for additional scientific or technological progress.

The Sonett Committee also suggested that NASA provide its astronauts with training in geology. This training would allow them to collect the samples required to construct a geologic map of their landing site and to determine the origin of its prominent features. Additionally, the committee recommended the astronauts deploy long-lived instruments on the Moon's surface that could monitor lunar activities and automatically transmit this data back to Earth. After these ideas received favorable reviews from independent experts, they were adopted as the scientific foundation for future lunar investigations. In order to ensure this research was conducted in a thoughtful and orderly manner, NASA established the Manned Space Science Division in July 1963.

The Crew and Objectives of *Apollo 12*

Apollo 11 was a transitional mission, concerned with both hardware issues and scientific exploration. *Apollo 12* was the first mission to focus equally on scientific and engineering questions. It was manned by a crew of three former Navy test pilots. Charles Conrad, the commander, was a veteran of *Gemini 5* and *Gemini 11*. He was a short and balding man who loved humorous jabs and jokes. Richard Gordon, pilot of the control module (CM), the *Yankee Clipper*, had been Conrad's crewmate on *Gemini 11*. He had the build of a boxer and was a man of strong emotions. Like Conrad, Gordon was a gifted pilot. The last member of the trio, Alan Bean, was to pilot the lunar module (LM), the *Intrepid*. He had been a student of Conrad's at the military's Pax River Flight Training School and was a stickler for details. Unlike his crewmates, he enjoyed constructing thorough checklists of their mission tasks.

Figure 21-1. The crew of Apollo 12, *from left to right, Charles Conrad, Jr., Richard F. Gordon, and Alan L. Bean.*

The three men had developed close friendships in the Navy and worked especially well as a team. Simulator instructors considered them one of the quickest learning and most competent teams they had ever trained.

There were seven primary objectives of *Apollo 12*: (1) to test the techniques for achieving pinpoint navigation and landings on the Moon, (2) to deploy a nuclear-powered scientific station called the Apollo Lunar Surface Experiments Package (ALSEP), (3) to perform geological inspections, (4) to survey an extended area of the surface of the Moon, (5) to test man's ability to work in the lunar environment, (6) to retrieve parts from the *Surveyor 3* craft that had landed on the Moon 31 months earlier, and (7) to photograph potential landing sites for future missions. Pinpoint landings were important because they allowed geologists to preselect precise regions of the Moon for investigation. *Apollo 11* had overshot its intended landing site by four miles, partly to avoid coming down in a boulder field. NASA's new goal was to reduce this landing error to several hundred feet.

The Flight of *Apollo 12*

Apollo 12 was launched from the Kennedy Space Center at 11:12 a.m. on November 14, 1969. Its destination was the Moon's Ocean of Storms, located about 950 miles west of the *Apollo 11* landing site and about 250 miles southwest of the large crater Copernicus (see Figure 21-2). The weather on the launch day was nasty. Heavy rain fell intermittently and sky conditions were highly variable, changing from partly cloudy to totally overcast. As the launch time approached, a circling air force weather plane reported acceptable cloud ceilings and wind conditions with no lightning present for 19 miles. Based on this information, NASA decided to proceed with the launch.

Apollo 12 lifted off from Launch Pad 39A, but only a half a minute into the flight, Conrad saw a bright flash of light through the capsule's window. Suddenly the master alarm within the ship went off and warning lights flashed all over the instrument panel. Never in any of their numerous training sessions had the astronauts seen anything like it. Conrad radioed ground controllers, "Okay, we just lost the platform, gang. I don't know what happened here; we had everything in the world drop out." Flight personnel were of little help since they were equally perplexed by this situation.

Luckily, John Aaron, an expert flight controller specializing in the capsule's electrical system, had seen a similar failure during practice sessions about a year earlier. He relayed a corrective action, which was quickly implemented by Bean. By the time the first stage of the Saturn V was jettisoned, the capsule's fuel cells, units that supplied electrical power to the CM, were brought back on-line. During this period, the Saturn V continued to ascend as though nothing had happened. Apparently, whatever had caused the capsule's electrical system to fail had not interfered with the rocket. Houston reassured the astronauts, "You're right smack-dab on the trajectory." Despite the anxiety the emergency had

top

right

Figure 21-2. The landing site of Apollo 12 *was located far from that of* Apollo 11 *but both were in lunar maria.*

caused, the astronauts entered orbit and began to relax. Automatic cameras near the launch pad later revealed that the bright flash Conrad had seen was caused by lightning striking the craft. A second strike was also recorded.

Once in orbit, an equipment check revealed that the only casualty of the lightning strikes was an on-board clock. This loss was a minor concern, and the three astronauts began to joke about their adventure. Speaking about the press representatives covering their flight, Conrad said, "That'll give them something to write about tonight," and then, thinking of how their wives must have reacted to the news of their emergency, he added, "I'll bet your wife, my wife, and Al's wife fainted dead away."

After ground controllers determined that the CM systems were working properly, they instructed the crew to fire the S-IVB to take them out of Earth's orbit. When ignited, the booster gave the astronauts the expected reassuring jolt as they pulled away from Earth's gravity. Conrad turned to his crewmate, "Al Bean, you're on your way to the Moon!" Bean replied, "Yeah, y'all can come along if you like."

Apollo 12 on the Moon

Apollo 12 successfully entered lunar orbit on November 17. As with earlier missions, the initial orbit was highly elliptical, varying in height from 194 miles to 72 miles. The service module (SM) engine was then fired to adjust the orbit to 70×62 miles. After sending a telecast showing the Moon's surface to viewers on Earth and completing 12 orbits, Conrad and Bean proceeded down to the surface in the *Intrepid*. They landed one and a half hours after leaving the *Yankee Clipper* at 1:54 a.m. on November 19. The *Surveyor 3* craft, one of the objects they wished to visit, was only about 600 feet from where they had settled down. After landing, a smiling Bean congratulated Conrad, "Good landing, Pete! Out-*stand*-ing, man!" Conrad later recalled this moment,

> Our first important task was the precision landing near *Surveyor 3*. When we pitched over just before the landing phase, there it was, looking as if we would land practically on target. The targeting data were just perfect, but I maneuvered around the crater, landing at a slightly different spot than the one we had planned. In my judgment, the place we had pre-picked was a little too rough. We touched down about 600 feet from the Surveyor. They didn't want us to be nearer than 500 feet because of the risk that the descent engine might blow dust over the spacecraft.

Conrad and Bean were scheduled to spend 31 1/2 hours on the Moon, 10 hours longer then the crew of *Apollo 11*. In keeping with NASA's stepwise approach to space exploration, two Extra Vehicular Activities (EVAs) lasting 3 1/2 hours each were planned, roughly tripling the amount of time allowed the crew of *Apollo 11*.

The first EVA began at 6:44 a.m. on November 19. Conrad was the first to leave the LM. As he stepped off the *Intrepid's* ladder and placed his foot onto the Moon, he jokingly alluded to his short stature, "Whoopee man, that may have been a small one for Neil, but its a long one for me." A half hour after exiting the *Intrepid*, he was joined by Bean. Play-

Figure 21-4. The ALSEP was deployed near its central power supply. Its instruments were not separated by much more than 30 feet.

ing the part of a happy tourist, Conrad approached him with his camera and said, "Okay, turn around and give me a big smile. Atta, boy. You look great. Welcome aboard."

The Apollo Lunar Surface Experiment Package

Conrad and Bean had a full schedule of activities jammed into their EVA. They set up an antenna for communication with Earth, deployed an apparatus called the Solar Wind Composition Experiment, and erected an American flag. They then set up the Apollo Lunar Surface Experiment Package. This package consisted of geophysical instruments powered by a radioactive fuel cell. During the Apollo program, each ALSEP employed a different set of instruments, but they all had one thing in common, each was a minor engineering miracle.

The *Apollo 12* ALSEP consisted of four primary instruments: a solar wind spectrometer, a suprathermal ion detector, a passive seismic experiment, and a lunar surface magnetometer. The solar wind spectrometer measured the energy spectrum of charged particles in the solar wind. Because the Moon does not have a significant atmosphere or magnetic field, particles emitted by the Sun directly strike the Moon. Therefore, the energy spectrum of charged particles detected on the Moon is the same as that emitted at the Sun's surface. The goal of the ion detector was to search for a tenuous lunar atmosphere and positively charged ions created by the impact of the Sun's ultraviolet light on the Moon's surface. The passive seismic experiment was designed to collect data about the internal structure of the Moon. It employed four extremely sensitive seismometers to measure lunar vibrations and was covered by a 5-foot-diameter thermal shroud to shield the sensors from day–night temperature variations. The last ALSEP experiment, the lunar surface magnetometer, detected fluctuations in the solar and lunar magnetic field by measuring the velocity and direction of electrons and protons hitting the lunar surface.

Figure 21-3. Alan Bean's first step on the moon.

Focus Box: The Apollo Lunar Surface Experiments Package (ALSEP)

On order to maximize the scientific return of the Apollo missions, each flight carried a set of experiments to obtain data about the Moon and its environment. The modular design of these experiments allowed them to be interchanged with others that were tailored to a landing site. While astronauts could only be on the Moon's surface for a matter of hours, the ALSEP's instruments were designed to transmit daily reports back to the Earth for an extended period of time. The experiments were powered by a centrally located radioisotope thermoelectric generator. Each ALSEP transmitted about 9 million readings to Earth per day.

Aldrin deploys the Apollo 11 *seismic experiment*

The ALSEP was deployed about 500 feet from *Intrepid* so it would not be covered by dust when the astronauts blasted off from the Moon. Its instruments were designed to transmit data back to Earth for a year, but five years after the return of *Apollo 12,* they were still faithfully operating. After completing these activities, the astronauts returned to the LM and rested. Conrad recalled the joy the astronauts felt while undertaking their tasks,

> Both of us really enjoyed working on the surface; we took a lot of kidding later about the way we reacted. But it was exciting; there we were, the third and fourth people on the Moon, doing what we were supposed to do, what we had planned to do, and keeping within schedule. Add to that the excitement of just being there, and I think we could be forgiven for reacting with enthusiasm.

Figure 21-6. The Intrepid *can be seen in the background as Alan Bean examines Surveyor 3's television camera.*

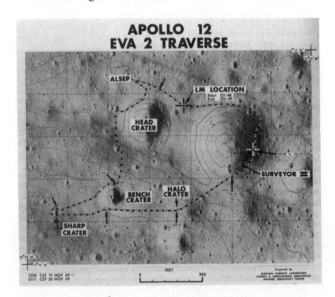

Figure 21-5. During their second EVA the astronauts explored several craters and visited the Surveyor 3 *landing site.*

Conrad and Bean began their second EVA at 10:55 p.m. It was scheduled to last 3 hours and 49 minutes. First, they verified that the seismometer deployed earlier was operating properly by walking toward the ALSEP. The vibrations created by their steps were instantly recorded and transmitted to controllers in Houston. They then made a long circuit around their landing site, ending at *Surveyor 3* (see Figure 21-5). During their walk to the spacecraft, Conrad and Bean took photographs and collected samples of lunar soil and rocks. When they reached the Surveyor, they took parts from it so that engineers could later determine how well it had fared in the harsh lunar environment. Conrad later recalled,

We cut samples of the aluminum tubing, which seemed more brittle than the same material on Earth, and some electrical cables. Their insulation seemed to have gotten dry, hard, and brittle. We managed to break off a piece of glass, and we unbolted the Surveyor TV camera. Then Al suggested that we cut off and take back the sampling scoop, and so we added that to the collection.

Before reentering the LM, they recovered the aluminum sheet from the Solar Wind Composition Experiment set up during their first EVA. While Conrad and Bean were engaged with these activities, Gordon conducted a variety of experiments aboard the *Yankee Clipper* and photographed the Moon's surface through blue, green, red, and infrared filters to help geologists identify various lunar mineral deposits.

Apollo 12 Returns to Earth

The LM lifted off from the Moon on November 20, 1969. Conrad and Bean then rendezvoused and docked with the orbiting CM without any problem. In consultation with Houston, the crew then decided to stay in lunar orbit for one extra day to complete their photographic assignment. It was particularly important to obtain high-resolution images of the Fra Mauro region, the intended landing site of *Apollo 13*. The astronauts took hundreds of individual pictures and thousands of feet of motion picture film of this area as well as the Descartes and Lalande craters, two other proposed Apollo landing sites.

With their tasks completed, the *Intrepid* was jettisoned and directed to crash into the Moon. It would strike the Moon about 45 miles from their landing site at a speed of 5,000 miles per hour. In its final contribution to the mission, the *Intrepid's* impact produced 30 minutes of shock waves that provided useful seismic information about the material located deep within the Moon.

Conrad, Bean, and Gordon then proceeded back to Earth in the CSM and splashed down in the Pacific Ocean on November 24 at 3:58 p.m., 3 miles from the recovery ship, the USS *Hornet*. In all, the astronauts returned with nearly 75 pounds of lunar soil and rocks and 15 pounds of *Surveyor 3* parts.

Apollo 13

The lightning blots that had struck the *Apollo 12* craft during its launch had nearly resulted in NASA's first aborted mission. Although they produced some heart-stopping moments, this situation was caused by adverse weather conditions, not by an engineering problem with the Saturn V or Apollo spacecraft. Given these factors, there was not a great deal of concern about the reliability of the Apollo hardware as NASA prepared to send its next mission, *Apollo 13*, to the Moon.

Assigning the number "13" to a spaceflight concerned some people, but the astronauts and NASA were not inclined to rename the mission because of superstitious fears. Black cats were not prohibited from the Kennedy Space Center nor were the guards on the lookout for witches or goblins. Yet as if to punish the agency for its arrogance, *Apollo 13* would experience an incredible string of misfortune.

A significant problem arose even before the launch. Ken Mattingly, a valued member of the *Apollo 13* crew and its CM pilot, was inadvertently exposed to the German measles just days prior to launch. Even though Lovell tried to keep him on the flight, NASA doctors felt the risk of having a sick astronaut in space was too great. Mattingly was replaced by John Swigert, Jr., the backup CM pilot. Never before had a crew member been dropped from a mission so close to a launch. NASA's decision was a terrible blow to Mattingly. He had devoted himself entirely to the Apollo program and had spent countless hours preparing for an opportunity to fly. Now just two days before lift-off, he was grounded.

The Crew of *Apollo 13*

Apollo 13 would be manned by astronauts James Lovell, Jr., Fred Haise, Jr., and John Swigert, Jr. Lovell, the mission commander, had the most spaceflight experience of any NASA astronaut, having logged 572 hours aboard *Gemini 7, Gemini 12,* and *Apollo 8*. Despite his impressive record, Lovell still hungered to place his footprints into the lunar soil. *Apollo 13* would finally allow him to achieve this goal. Privately, he had already told his wife this would be his last mission.

Figure 21-7. The crew of Apollo 13 *consisted of James A. Lovell (left), John L. Swigert, Jr. (middle), and Fred W. Haise, Jr. (right).*

In some ways, Swigert's personality was just the opposite of that possessed by the man he replaced. Swigert enjoyed an active social life, and his dating success had reached almost legendary stature among his fellow astronauts. Whereas Mattingly took pride in being meticulously dressed, Swigert could appear in a suit and tie while wearing white socks. In other important respects, though, they were strikingly similar. Both men worked tirelessly and conducted themselves as professionals who could be depended upon to do their job well.

Lovell and Haise would be America's third pair of astronauts to land on the Moon. Counting circumlunar voyages, *Apollo 13* would be the fifth mission to fly around the Moon in a little over one year. Perhaps because of the quick succession of flights, interest in NASA's lunar program began to decline. The public was enthralled by *Apollo 11*, modestly intrigued by *Apollo 12*, but almost indifferent to *Apollo 13*.

The Launch of *Apollo 13*

Apollo 13 lifted off from the Kennedy Space Center and had an uneventful trip to orbit. The S-IVB was then reignited, propelling the astronauts on their way to the Moon. As was now common practice, the CSM detached itself from the booster, turned around, and retrieved the LM, the *Aquarius*. To this point, *Apollo 13* had been a textbook perfect flight.

Once these tasks were completed, the astronauts sent a live television show back to Earth from the CM, the *Odyssey*. However, none of the network television stations chose to broadcast this event because they did not believe the public would find it significant enough to warrant interrupting their prime-time shows. After what the astronauts believed to be a successful broadcast, Swigert returned to the CM instrument panel, while Lovell and Haise prepared to take pictures of Comet Bennett.

The First Sign of Trouble

While Lovell and Haise busied themselves with their photographic task, Mission Control asked Swigert to stir the liquid oxygen and hydrogen in the SM's cryogenic tanks. In space, these gases become stratified making it difficult to measure their quantities accurately. An electric fan had therefore been installed to stir their contents. The oxygen tanks supplied the crew with their breathable air and the craft's three fuel cells used this oxygen along with the stored hydrogen to produce the ship's electrical power. Swigert flipped the appropriate switches and, after several seconds, turned the fans off. Moments later the astronauts heard a loud bang and felt the spacecraft shudder.

At first Lovell thought the bang was another one of Haise's pranks. In jest, he would occasionally hit the CM's repressurization switch just to see the surprised reaction of his crewmates. It was a harmless joke but it made a similarly loud noise. Lovell asked Haise, "Fred, do you know what that noise was?" But when he saw Haise's face, it was clear he

was likewise baffled. Something else must have happened. Moments later the master alarm in the capsule went off and the astronauts scrambled to reach the CM instrument panel. As Lovell and Haise hurried toward the instrument panel, Swigert was already staring at it in disbelief. Numerous warning lights were glowing. He radioed Mission Control, "Okay, Houston, we've had a problem." At that moment, Lovell joined him and looking at the display added, "We've had a Main B Bus undervolt."

The Developing Emergency

The electrical power for the CSM was supplied by three fuel cells located in the SM. These cells used the liquid oxygen and hydrogen from the cryogenic tanks to generate electricity. Current from these cells flowed to various instruments on the ship through two junctions, called Main Bus A and Main Bus B. Haise, an expert on the craft's electrical system, checked the status of fuel cell #3, the cell supplying power to Main Bus B. He was shocked to find that it was dead.

As the three astronauts attempted to identify what had gone wrong, the situation continued to deteriorate. Haise flipped switches to allow the capsule's instruments to draw power through Main Bus A, but found that its fuel cell was also dead. Two of the craft's three fuel cells had inexplicably died. As Haise tried to make sense out of what was happening, Lovell checked the gauges for the oxygen tanks that fed these two fuel cells. To his astonishment, he found tank #2 was empty and the amount of gas in tank #1 was rapidly declining.

Flight regulations required that all three fuel cells had to be functioning before a craft would be allowed to enter lunar

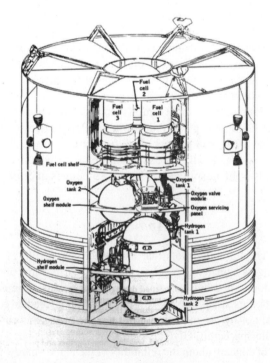

Figure 21-8. A view of the location of the three fuel cells (top), oxygen (middle), and hydrogen (bottom) tanks in the SM bay.

Figure 21-9. Concern over the fate of Apollo 13 *is evident in the faces of ground controllers as they discuss the situation.*

The Situation Worsens

Never before had NASA encountered such a dire emergency in space. By contrast, the problems experienced during the launch of *Apollo 12* were short lived, and the astronauts could have returned from Earth orbit. The situation for *Apollo 13* was an entirely different matter. Their ship had been crippled, and they were still traveling toward the Moon!

Ground controllers monitoring the telemetry from *Apollo 13* were perplexed. It seemed impossible for so many problems to arise simultaneously, they suspected the spacecraft was transmitting false status reports. For the astronauts aboard *Apollo 13*, however, there was no doubt a serious malfunction had occurred. They had heard the boom with their own ears and felt the ship shake.

The first thought Lovell, Haise, and Swigert had about their situation was that their craft had been hit by a meteoroid, but after a few moments consideration, this possibility was dismissed. If the ship had been struck, the cabin would have depressurized quickly and this obviously had not happened. As they frantically pondered various possibilities, Lovell glanced out a capsule window and was shocked to see a cloud of gas streaming away from them. He radioed Houston, "It looks to me that we are venting something. We're venting

orbit. It was clear there would be no lunar landing for *Apollo 13*. This was not, however, the worst of their problems. Without electrical power, the SM engine could not be fired and this engine was needed to return them to Earth. The astronauts were nearly 200,000 miles away from home, and it was beginning to look like they might not be able to get back.

Focus Box: The Explosion on *Apollo 13*

The failure of oxygen tank #2 in the service module of *Apollo 13* caused this mission to be terminated and nearly cost Lovell, Haise, and Swigert their lives. What caused this explosion? In 1965, the Apollo spacecraft was upgraded to accept 65-volt inputs largely to accommodate ground test equipment. Up to that time, a 28-volt input was used. During this upgrade, the thermostat inside the oxygen tanks was overlooked. This thermostat turned off a heater inside the tank to prevent it from overheating. During a test, tank #2 was filled with oxygen. Technicians found that they could not empty it using the normal procedure. To remove the oxygen the tank's internal heater was turned on. During this time, the excess voltage created an electrical arc that welded the thermostat contacts closed. Tests later showed that spots along the heating tube reached a temperature of nearly 1,000° F, damaging the teflon coating on the fan motor wires and exposing bare wire. The high temperature within the tank and exposed wires led to the explosion. When Swigert turned on the fan in tank #2, an electrical arc occurred, causing the oxygen inside the tank to explode. This explosion ruptured the tank, blew out the panel covering this section of the SM, and damaged oxygen tank #1, allowing its contents to also escape.

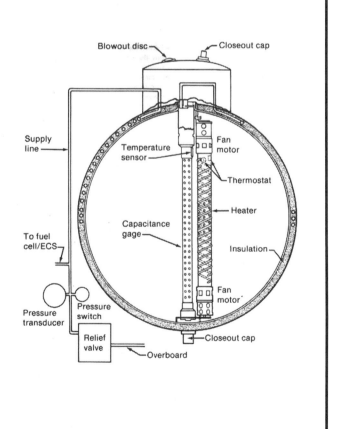

something into space. It's a gas of some sort." The cloud of escaping gas was their oxygen supply.

Although the cause of this failure was unknown, its consequences were frighteningly clear. The *Odyssey* was dying, and without electricity, the SM engine would be useless. The situation for Lovell, Haise, and Swigert looked bleak. If *Apollo 13* continued on its current path, the craft would go around the Moon and head back toward Earth. Unfortunately, unless some corrective action were taken, it would miss Earth by 45,000 miles. They would then be carried out into space and die when their oxygen supply ran out.

As Lovell surveyed his crippled ship he realized his dream of setting foot on the Moon was gone. But there was little time for reflection, he later recalled, "The knot tightened in my stomach, and all regrets about not landing on the Moon vanished. Now it was strictly a case of survival."

NASA Acts

About the only option left to save the crew was to employ the LM as a "lifeboat" and use its descent engine to return to Earth. As an expert on the LM, Haise realized this posed all sorts of questions,

> I never heard of the LM being used in the sense that we used it. We had procedures, and we had trained to use it as a backup propulsion device, the rationale being that the thing we were really covering was the failure of the command module's main engine. . . . In that case, we would have used combinations of the LM descent engine, and in some cases for some lunar aborts, the ascent engine as well. But we never really thought and planned, and obviously didn't have procedures to cover a case where the command module would end up fully powered down.

The LM contained enough fuel to get the crew back to the Earth, but before its rocket was fired it was essential that the *Aquarius* be correctly oriented. When the guidance system within the LM was powered up, it required input values from the computer in the CM. These numbers would then be manually entered by one of the astronauts. The problem with this plan was that the fuel cells that powered the CM computer were dead. However, Swigert had a solution. He directed the *Odyssey's* computer to use power from the CM's reentry batteries. The power from these batteries would later be needed when the ship reached Earth, but if the astronauts worked quickly, the amount of power consumed would be small.

During ground tests, Lovell had practiced converting the CM coordinates into the slightly different system for the LM. Now under tremendous pressure he asked Houston to check his values before he entered them into the *Odyssey's* computer. Haise recognized the crucial importance of this matter and cautioned, "Do it right." Then attempting to lessen the pressure, added, "take your time." Once Lovell's values

were verified, Haise entered them into the computer on the *Aquarius*. This procedure complete, Swigert turned off the *Odyssey's* computer and all other remaining CM systems to save as much power for reentry as possible. He had done all he could. Upon returning to the *Aquarius* from the *Odyssey*, he told Lovell and Haise, "It's up to you now."

Preparations to Return Home

As they continued to approach the Moon, Lovell fired the descent engine of the *Aquarius* to change their trajectory. He was relieved to see the computer terminated the burn exactly on schedule. The burn was good. At least they were now on a path that would bring them back to Earth.

After discussions with Mission Control, it was decided that after the spacecraft had swung around the Moon, the *Aquarius'* engine would be fired again to decrease their return time from four to two-and-a-half days. To align the ship, Lovell was instructed to use the LM's heavily filtered navigation telescope to view the Sun. The explosion of oxygen from tank #2 had released thousands of small pieces of debris that resembled stars through this instrument so the Sun was the only celestial object that Lovell could clearly identify.

As they passed behind the Moon, Swigert and Haise took time to admire it through the capsule's windows and take pictures. Lovell, worried about the upcoming burn, reminded his crewmates, "If we don't get this burn off, you won't get your pictures developed." Haise replied, "Relax Jim, you've been here before, and we haven't." Realizing there was plenty of time to prepare for the next LM firing, he joined his crewmates and watched the Moon pass majestically below them.

On the Way Home

Lovell and Haise were ready a full hour ahead of the scheduled time to reignite the LM's descent engine. Because the software in the LM was not written to handle the unique situation they faced, Lovell manually fired the engine. It went perfectly. In 62 hours *Apollo 13* would reenter the Earth's atmosphere.

Although encouraged by their steady progress, Swigert was still worried about the condition of the *Odyssey*. A CM had never been turned off for days in space and then restarted. Would it work? The *Odyssey* had to be operational since the *Aquarius* did not have a heat shield and would be destroyed if it attempted to return to Earth. He also knew that the CM reentry batteries only contained a small amount of power. An entirely new CM reentry procedure would have to be developed, tested, and written. These tasks normally required at least three months of intense work, but *Apollo 13* would need this script in just two days.

Luckily, the crew of *Apollo 13* had an expert working on their behalf in *Odyssey's* ground simulator—Ken Mattingly. At that very moment he was constructing a reentry procedure and was determined not to let his crewmates down.

Figure 21-10. The improvised carbon dioxide remover (taped box near center) was made from materials found within the capsule and from filters in the CM.

The Carbon Dioxide Problem

The LM was designed to accommodate two astronauts during a short two-day journey to and from the Moon's surface. It contained filters to remove the carbon dioxide, exhaled by the astronauts, from the cabin's air. The LM was now occupied by three people, and as time passed, the amount of carbon dioxide increased. If this situation were left unchecked, the astronauts would develop bad headaches, their hearts would beat faster, they would then become drowsy, fall into a deep sleep, and die.

The CM had plenty of carbon dioxide filters but they had a square shape, and the LM was designed to accommodate round filters. A way had to be found to fit a square filter into a round hole. If this were not done, the astronauts would suffocate from their own breath before they ever reached the Earth.

NASA engineers realized all the materials required to make the adapter had to be available in the Apollo craft. In an impressive display of ingenuity, a working device was constructed from such items as the cardboard back of the flight book, plastic storage bags, tape, and parts of a sock. Lovell and Swigert were given instructions on how to build this mailbox-shaped apparatus, and although it looked strange, when installed (see Figure 21-10), it worked. The carbon dioxide problem had been solved.

Another LM Burn

It was beginning to look as though all the barriers to a safe return to Earth had been overcome when another, life-threatening problem was discovered. If *Apollo 13* continued on its present course, it would not reenter the Earth's atmosphere but would skip off it, much like a rock thrown at a steep angle skips across a body of water. An additional LM burn was needed to correct their trajectory.

Lovell was instructed to first use the Sun to align the craft and then rotate the ship until the crescent Earth appeared in the gunsight of the LM, normally used for rendezvous maneuvers. A proper alignment was absolutely critical for this burn, and as Lovell maneuvered the craft, his apprehension grew, "I rotated the spacecraft to the attitude Houston had requested. If our alignment was accurate, the Sun would be centered in the sextant. When I looked through the AOT (Alignment Optical Telescope), the Sun just had to be there. It really had to be." After completing these adjustments, and looking through the device, he shouted to controllers, "We've got it!" Once this alignment was obtained, the LM descent engine was to be fired. Lovell and Haise would control the motion of the craft and Swigert would call out the start and stop times for the burn. To say the least, this was an unorthodox maneuver, but they had little choice but to try. Without this correction, *Apollo 13* would be lost.

Although the engine was fired for only 14 seconds, it was an incredibly tense time. After shutting down the engine, the crew had no idea how successful it had been. Mission controllers radioed the astronauts, "Nice work." Lovell replied, "Let's hope it was." As additional telemetry was received by ground stations, it was clear the proper correction had been made. It was now time to restart their frigid CM and prepare for reentry.

Ken Mattingly had methodically worked out a way to bring the *Odyssey's* systems back to life within the power limitations imposed by the on-board batteries. For 2 hours he dictated each step of this process to Swigert and then tried to reassure him, "We think we've got all the little surprises ironed out for you." Swigert replied, "I hope so because tomorrow is examination time."

The Last Obstacle

As Swigert turned on the systems within the *Odyssey*, he carefully placed a piece of tape over the switch labelled "LM

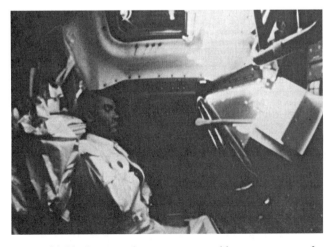

Figure 21-11. Because the astronauts could not use power for heaters, the temperature within the LM became very cold. Fred Haise developed a high fever on the journey back to the Earth.

Figure 21-12. *The extensive damage done to the service module of* Apollo 13 *is evident in this picture. The panel that normally covered the oxygen tank had been blown away.*

Figure 21-13. *Haise, Lovell, and Swigert are welcomed aboard the recovery ship the* Iwo Jima. *They were dehydrated and exhausted and had lost a combined 31 pounds during their flight.*

Jettison." He was exhausted and did not want to make a careless mistake by jettisoning the LM with his crewmates inside.

Once the explosive bolts holding the SM to the CM fired and the two modules separated, Lovell turned the *Odyssey* around to get a good look at the departing SM. He wanted to examine it for any sign of the problem that had caused his mission to fail. He was astonished by what he saw and reported to Houston, "There's one whole side of that spacecraft missing!" Shreds of Mylar tape and torn wires could be seen in the exposed compartment. The astronauts took pictures of the damaged SM until it was just a distant point. Haise summed up the crew's feelings about the extent of the damage the LM had suffered, "Man, that's unbelievable."

Mission controllers were also surprised by Lovell's description of the crippled SM. Had the *Odyssey's* heat shield been damaged by the explosion? It was almost unthinkable that after all of the troubles the astronauts had overcome, they might now be killed by the heat of reentry. However, there was little time to dwell on this matter since only two and a half hours remained before *Apollo 13* would reenter the atmosphere.

Despite Mattingly's superb effort at developing a reentry procedure, it was far from certain all of the CM systems would work. It was extremely cold inside the *Odyssey*, colder than some of the components' manufacturing specifications allowed. In addition the capsule was literally dripping with condensed water vapor from the astronauts' breath. There was a strong possibility this moisture would cause an electrical short as power once again flowed through the craft's systems.

As Swigert and Haise turned on the CM instruments, they often had to wipe water away from the dials just to see if the instruments were working. Despite the high potential for problems, the process was completed without a single failure. The fact this was successful was almost surely due to the extensive redesigns and safety measures put into the capsule as a result of the tragic fire on *Apollo 1*.

With less than two hours left before splashdown, it was now time to jettison the *Aquarius*. Lovell made some final small maneuvers and radioed, "OK Houston, *Aquarius*. I'm at the lunar excursion module (LEM) separation attitude and I'm planning on bailing out." After the *Odyssey* and *Aquarius* hatches were sealed, Swigert flipped the LEM JETT switch and radioed, "Houston, LEM jettison complete." Kerwin responded form Mission Control, "OK, copy that. Farewell Aquarius, and we thank you." The crew now prepared to reenter the atmosphere. Lovell smiled to his crewmates, "Gentlemen, we're about to reenter. I suggest you get ready for a ride."

As the heat of reentry built, the normal communication blackout between the capsule and ground occurred. As the capsule sped through the atmosphere, no one at Mission Control knew whether the heat shield had done its job or whether the electronic system used to release the parachutes had worked. Finally, live television pictures from the recovery site showed the *Odyssey* descending with all three of its orange parachutes deployed. Lovell, Haise, and Swigert were safely home. As they settled in the ocean, the astronauts could see water splashing onto the capsule's windows. Their ordeal finally over, a relieved Lovell turned to his fellow astronauts and said, "Fellows, we're home."

No matter what else he might be, Shepard in his heart and mind was and always would be an eagle born to fly. Even all the way to the Moon and back.

—*Moon Shot*, p. 250

Despite the numerous problems experienced during the flight of *Apollo 13*, the mission was considered by NASA to be a "successful failure." The circumstances the astronauts had faced were so formidable that many people initially thought the crew had little hope of surviving. The professionalism displayed in bringing this mission to a successful end deflected criticism of the agency and increased the public perception that NASA could solve any problem it might encounter during a spaceflight.

Apollo 13 also produced some useful scientific data. On April 14, 1970, the empty third stage of its Saturn V rocket crashed into the Moon about 87 miles from the landing site of *Apollo 12*. The energy released in this impact was equivalent to the detonation of 11 tons of TNT and a series of seismic waves were created. These signals were detected by the seismographs left by *Apollo 11* and *12* and provided geologists with additional information about the Moon's structure.

Although the joy over the safe return of the *Apollo 13* astronauts was genuine, public interest in additional lunar missions continued to decline. Seizing on this lack of support, opponents of the space program slashed NASA's budget, causing the flights of *Apollo 18, 19,* and *20* to be canceled. Mindful of the mounting political resistance and the risks inherent in a lunar landing, high-ranking NASA officials began to wonder whether it was wiser to continue its Moon program or to move aggressively into other areas. In the book *Moon Shot*, Alan Shepard and Deke Slayton wrote about Robert Gilruth's (director of the Manned Spacecraft Center in Houston) position on this matter,

> Gilruth offered the theory that by landing on the Moon the country had achieved the political goal laid down by President Kennedy. So why risk any more lives? Forget the Moon, he urged top government officials, and get cracking with a manned space station in Earth orbit. That way, if any serious problems occurred, returning astronauts to Earth would be simpler and safer.

Still troubled by internal and external discord over their direction, NASA prepared for the launch of *Apollo 14*.

Figure 22-1. Shepard was happy to return to space after having served in administrative positions for several years.

Alan Shepard, one of the nation's best-known astronauts, was selected to command this expedition. Almost ten years had passed since his historic suborbital flight in May of 1961, but the memory of how it had inspired the nation was still fresh in everyone's mind. Less than three weeks after its completion, President Kennedy committed the United States to a manned lunar landing and the Apollo program was started. His new mission might once again play an important role in defining future space policy. If he was not able to land on the Moon, it was entirely possible that the remaining three Apollo flights might be cancelled. On the other hand, if his mission were successful, it might give the Apollo program a much needed boost.

Shepard's Long Effort to Return to Space

After his suborbital mission, Shepard had hoped to complete a second Mercury flight and was disappointed when the program ended before this could occur. He was, nevertheless, pleased to be assigned, along with Tom Stafford to the crew of

Gemini 1. Six weeks into his training program, Shepard awoke feeling nauseated and dizzy. When he met Deke Slayton later that morning he told him about this unexpected illness.

> All of a sudden, Deke, I fell. I was so dizzy! The room was spinning around and suddenly I'm on the floor. I got up holding onto the wall, and right away, I got so sick I vomited. I thought, Jesus, what the hell did I have to drink last night? It must have been one hell of a hoorah, but, well, that just wasn't the case.

Over the next several days Shepard's dizziness continued and a consultation with the flight surgeons was arranged. After a thorough exam they concluded he had a serious inner ear problem, called Meniere's syndrome. There was no cure for this disorder in which fluid builds up in the semicircular canals of the ear causing a person to lose balance and experience attacks of vertigo, but sometimes it simply vanished with time. Shepard hoped this would be the case, but until that time arrived, he was not permitted to fly. In frustration, he approached Deke and asked for his advice,

> Deke, what the hell do I do? Should I just hang it up? Quit? I've had my moment of glory, if you want to call it that, but it's a hell of a long way from being enough. Something burns inside me, deep down and bright, and it tells me I've got to fly again, to keep flying, and one day go all the way to the Moon.

Slayton empathized with his friend because he too had been grounded for medical reasons. He recommended that Shepard remain with NASA until the situation resolved itself. In the mean time, he assured his friend, "I've got a job for you." He then tried to encourage him, "I'm keeping my foot in the door, and yours should be right there next to mine."

By 1968 Shepard's condition had declined to the extent that he was nearly deaf in one ear—his flying career seemed over. At this low point, Tom Stafford told him about a physician, Dr. William House, who had devised a new surgical procedure for those who suffered from this condition. Faced with the prospect of never flying again, Shepard realized that at the age of 47 time was running out. He flew to Los Angeles and after discussing the procedure with Dr. House made his decision, "Let's go for it." To avoid publicity, Shepard admitted himself to the hospital under the bogus name of Victor Poulos. Eight months later, in a meeting with Dr. House, he received the desperately hoped for words, "My friend, you're cured."

With his medical problem solved, Shepard was eager to return to space. Deke Slayton was pleased to accommodate his friend's wish and assigned him, along with Stuart Roosa and Edgar Mitchell, to the crew of *Apollo 14.* In terms of prior mission selections, the composition of this crew was unusual. Shepard was the only member who had any prior spaceflight experience and that was limited to a 15-minute suborbital mission several years earlier. To make matters worse, neither Roosa nor Mitchell had ever served on an Apollo backup crew—the normal path an astronaut took before being assigned to a flight.

In view of these circumstances, George Meuller, NASA associate director for Manned Space Flight, vetoed Slayton's decision and reassigned Shepard's crew to *Apollo 14.* This would allow them more time to train. To accommodate this change, the prime crew of *Apollo 14* was moved up to the *Apollo 13* mission.

Shepard's *Apollo 14* Crewmates

Shepard played an important role in the selection of his crewmates. He considered both of them to be outstanding pilots. When grumbling arose over their selection, he quickly put a stop to it, telling the other astronauts, "I picked these guys for one reason only. They're tops. They can do the job. I want them with me."

Stu Roosa was born in Durango, Colorado, on August 6, 1933. Raised in Claremore, Oklahoma, he developed a conservative philosophy and viewed being in the military as an honor rather than as an obligation. In 1953 he joined the air force and was trained as a fighter pilot, logging 5,000 hours of jet flying time. Later he attended the University of Colorado and received a B.S. in aeronautical engineering. His careful work during the flight of *Apollo 9* had impressed both Christopher Kraft and Deke Slayton and probably earned him his spot on *Apollo 14* as pilot of the control module (CM), the *Kitty Hawk.* Prior to his selection, Roosa had seen Shepard as an intimidating and mysterious figure. However, as their training progressed, this view faded and Roosa often referred to him as the mission's "fearless leader."

Ed Mitchell had earned a Ph.D. from the Massachusetts Institute of Technology in 1964 in aeronautics and astronautics and was referred to by many astronauts as "The Brain." Like Shepard and Roosa, he was married. An experienced test pilot, Mitchell could become impatient with others, especially when they could not supply him with the information he needed. He would pilot the lunar module (LM) (the *Antares*) and accompany Shepard to the Moon's surface. Like

Figure 22-2. The crew of Apollo 14*: (left to right) Stuart Roosa, Alan Shepard, and Edgar Mitchell.*

Haise, the CM pilot on *Apollo 13*, he was not only an expert on the LM systems, but also possessed a deep understanding of the operation of the CM as well.

For nineteen months, longer than any previous crew, Shepard, Roosa, and Mitchell practiced for their upcoming flight. A lot had changed since Shepard's flight aboard *MR-3*. Perhaps the most obvious difference was in the power of the rockets. For example, the escape rocket on the top of the Saturn V produced twice the power of the Redstone rocket used in his ballistic flight.

The Flight of *Apollo 14*

The night before the launch of *Apollo 14* was an anxious one for Deke Slayton. He rode the launch tower elevator to the top of the Saturn V to conduct an inspection of the vehicle his friend would ride into space,

> It wasn't quite the same as kicking the tires of a car, but the sense of inspection was the same. . . . Memories of Gus Grissom flooded his mind. It had taken Slayton a long time to accept losing that man. Gus had been closer than a brother. Now Shepard would ride an even bigger dragon. . . . He shook his head. He didn't know if he had it in him to bear the loss of Alan Shepard. He wasn't at all sure he could get over the loss of another so close. Alan filled that special space reserved for best friend. . . .

When the astronauts arrived at the pad the next afternoon, Slayton was with them. He watched as they rode the elevator to the top of the gantry and walked across the catwalk to the capsule. Shepard paused and then looked around,

top

right

Figure 22-3. The landing site of Apollo 14. *This was also the proposed landing site for* Apollo 13.

Deke saw Shepard stop and stare down at him. He had the feeling of many years compressed to this single moment. Up there was the only one of the original seven Mercury astronauts who was about to realize the dream of one day going all the way out. Shepard, in a very real sense, was going for all seven of them. Deke gave him a thumbs-up.

Lift-off occurred at 4:03 p.m. on January 21, 1971. Shepard and his crew were surprised at how little noise penetrated the capsule, the firing engines sounded like distant thunder. Ten seconds later the Saturn cleared the launch tower and was heading skyward. After two and a half minutes they were 40 miles high and traveling at the speed of 6,000 miles per hour. Approximately 11 minutes into the flight, the second stage of the Saturn fell away and the third stage fired. In two more minutes they were in orbit! After checking the spacecraft's systems and finding everything in order, the Saturn S-IVB fired and propelled the astronauts toward their destination, the Fra Mauro, the aborted landing site of *Apollo 13*.

As was now standard practice, once out of Earth's orbit, the CMS was decoupled from the S-IVB. Roosa then brought this vehicle around to retrieve the *Antares*. However, as he piloted the *Kitty Hawk* into the *Antares's* docking fixture, the mechanism failed to lock the two vehicles together. Over the next hour and a half, two additional unsuccessful docking attempts were made. Within the control room at Houston, an official was overheard whispering to a colleague, "If they can't dock, there goes the farm." Everyone, including the crew, knew *Apollo 14* was in serious trouble.

Shepard and his crew were committed to doing everything possible to solve the docking problem. In their minds, there was "no way" this failure was going to jeopardize their mission. They quickly devised a plan to retrieve the *Antares*. The capsule would first be depressurized, the hatch would then be opened, and Shepard would physically pull the two craft together. Ground controllers listened carefully to this plan, but NASA engineers had developed a less risky alternative. They asked Roosa to perform a hard docking, to ram the CM into the *Antares* docking mechanism, and to then keep the two craft firmly pressed together. This procedure would drive the *Kitty Hawk* past the outer latches to the twelve latches on the docking ring itself. As Roosa prepared to try this, Shepard encouraged him, "juice it." Their craft slammed into the probe with such force that the impact rocked both vehicles. Moments later, the capture light came on and the three men shouted "Got it!" Shepard radioed Houston "We have a hard dock." He then added, "Houston, we're ready for business up here." It was a frustrating episode, but after six attempts the docking had been completed and the astronauts were on their way.

The trip to the Moon proceeded without further incident. Once in lunar orbit, the crew took a moment to play tourist. Shepard radioed Houston, "This is really a wild place." Roosa jokingly added, "Fantastic! You're not going to believe this, but it looks just like the map." Mitchell, struck by the Moon's barren surface told controllers, "That's the most stark and desolate looking piece of country I've ever seen."

On the 12th orbit, Shepard and Mitchell separated from the *Kitty Hawk* and began to check out the *Antares's* systems. All was going well and as a final test they instructed the computer to conduct a simulated descent. Moments later Shepard exclaimed, "Hey, our abort program has kicked in!" A second simulation was run and again the computer detected an abort signal. If the on-board computer detected this false signal during the actual descent to the Moon, it would automatically fire the ascent engine, and the mission would be over. In frustration, Mitchell turned to Shepard and asked, "Al, are we snakebit?"

Ground controllers were perplexed by this problem and telephoned the computer's programmer, Donald Eyles of M.I.T., for help. He immediately wrote some corrective code that was verified in simulators and then transmitted to the *Antares*. Mitchell finished entering these instructions just 15 minutes before the landing attempt would have to be scrubbed. Shepard then began the descent, and although the astronauts half expected the worse, the fixes worked.

The relief Shepard and Mitchell felt at overcoming this potentially mission ending problem was short lived. As they approached the lunar surface, the *Antares's* landing radar failed to lock-on to the Moon's surface. Mitchell informed his crewmate, "Al, I'm not getting a landing radar update." Without this data, the astronauts could not determine their distance from the ground. This was more than an inconvenience because the astronauts could not see the surface from the windows of the *Antares*. As a result, they were flying blind. Regulations required that the mission be aborted if the radar system did not operate by the time they were within 10,000 feet of the surface. As the astronauts descended to 22,000

feet, the radar was still not working, and Mitchell could be heard by ground controllers encouraging the apparatus, "Come on, Come on!"

In his mind, Shepard was running over the possibility of reorienting the *Antares* so he could see the lunar surface and then manually landing the craft. This very dangerous maneuver was forbidden by mission rules, but Shepard and Mitchell were both experienced test pilots and they privately discussed this option. Back at Mission Control, Deke Slayton had noted a change in the tone in Shepard's voice and thought, "By God, he's gonna take her all the way down with or without his radar." Almost at the last possible moment, as Mitchell was trying various fixes, he reset the radar's circuit breakers and suddenly the instrument panel came to life and the *Antaras'* caution lights blinked off. The astronauts were elated and briefly celebrated as Shepard piloted their craft toward the surface. Minutes later, the *Antares* landed softly on an undulating portion of the Fra Mauro highlands, about 110 miles east of the *Apollo 12* landing site and within 87 feet of their planned destination. After this harrowing descent, Mitchell turned to Shepard and asked, "The truth now. Just between you and me. Would you have really flown us down without the radar?" With a mischievous smile on his face, Shepard answered, "You'll never know, Ed, you'll never know."

Shepard and Mitchell Explore Fra Mauro

Shepard and Mitchell began their first Extra Vehicular Activity (EVA) at approximately 10:18 a.m. on February 5. As Shepard placed his foot into the lunar soil, he thought about his ten-year struggle to reach the Moon and the many circumstances that had almost ended his dream. He radioed Mission Control, "It's been a long way, but we're here." They replied, "Not bad for an old man." Surrounded by the stark beauty of the Fra Mauro and with the blue and white crescent Earth overhead, Shepard was overcome with emotion and for several moments cried. Those who knew of his difficult, long-term struggle

Figure 22-4. A distant view of the Apollo 14 *landing site in the Fra Mauro highlands.*

Figure 22-5. Happy to at last be on the Moon, Shepard poses next to the American flag.

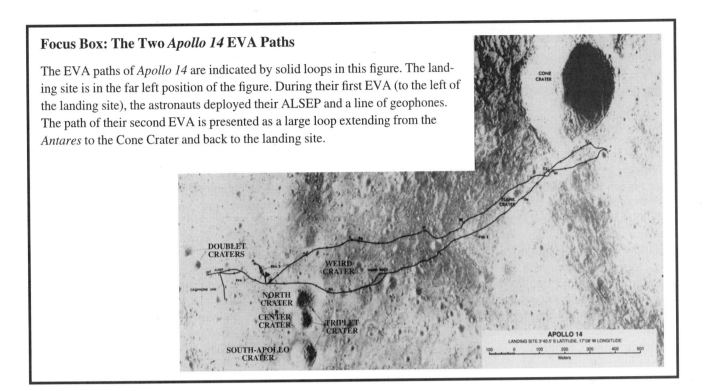
odyssey understood the depth of his feelings. Shepard's joy and relief at the successful landing was shared by his wife, Louise. When learning that her husband was safely on the Moon, she jokingly told the family, "We can't call him Old Man Moses anymore. He's reached his Promised Land."

During their first EVA, a contingency sample of soil and rocks was collected and an American flag was planted into the lunar soil. Shepard radioed his impressions of their landing site back to Earth, "Gazing around at the bleak landscape, it certainly is a stark place here at Fra Mauro. It's made all the more stark by the fact that the sky is completely dark. This is a very tough place. . . . The soil is so soft it comes all the way to the top of the LM's footpad." Shepard and Mitchell then unloaded a wheeled equipment transporter called the Modularized Equipment Transporter or MET and deployed their Apollo Lunar Surface Experiment Package (ALSEP).

The *Apollo 14* ALSEP contained two new experiments. The first consisted of an array of geophones and a "thumper" to create impacts of known size. When the thumper was pressed against the ground and triggered, an explosive charge was fired into the ground. Mitchell fired a series of these charges along a 360-foot-long track to gather information about the density of the subsurface material. The second experiment consisted of an apparatus that launched grenades that exploded on the lunar surface at greater distances. After the astronauts left the Moon, this device would be activated by ground controllers, and the explosions would provide additional seismic data. The astronauts adjusted to the Moon's gravity so well that their EVA was extended to 4 hours and 50 minutes, 40 minutes longer than originally planned.

The destination of the second EVA was the 1,000-foot diameter Cone Crater, located about 3,300 feet from their land-

Figure 22-6. Shepard and the MET, or "lunar rickshaw." This wheeled vehicle was hard to drag along the Moon's dusty surface.

ing site. To carry their tools and equipment, the astronauts used a cart they called the "lunar rickshaw" (see Figure 22-6). The term was appropriate since it was designed to be pulled by the astronauts. It carried maps, core tubes, a scoop, camera, hammer, and numbered sample bags. The rickshaw was also used to transport the lunar samples retrieved during the EVA.

The primary objective of the second EVA was to obtain rock samples from the rim of the Cone Crater (see the Focus Box above). During their training sessions, scientists had explained the reason behind this task,

It's our belief that Cone was carved out of the Moon's surface more than four billion years ago. One hell of a meteoric impact. By the way that crater is gouged, there's every chance it ripped away rocks from maybe 300 feet down and that these rocks were tossed about along the crater rim. If that's what you bring back with you, then we'll be able to study material that came into existence about the same time the planets and moons of the solar system were still forming out of dust and gas.

On their way to the crater, Shepard and Mitchell collected samples at two intermediate sites. As they walked, Mitchell jokingly mimicked the military cry to charge: "To the top of the Cone Crater!" The terrain steepened as they continued on their journey and it was actually faster to carry the MET rather than to drag it through the lunar dust. Shepard found this situation humorous, and as they trudged along, he called out the cadence of left, right, left, right, left, right.

Despite the lighthearted nature of this scene, the astronauts were becoming worried. The uneven lunar terrain made it difficult to determine the direction to the Cone Crater, distances were difficult to judge, and after walking for what seemed to be a long time, the crater was still nowhere in sight. Shepard commented on this,

> Ed and I had difficulty in agreeing on the way to Cone, just how far we had traveled, and where we were. We did some more sampling, and then moved toward Cone, into terrain that had almost continuous undulations, and very small flat areas. Soon after that, the surface began to slope upward even more steeply, and it gave us the feeling that we were starting the last climb to the rim of Cone.

After proceeding for a while longer, a tired and frustrated Shepard grunted to Mission Control, "Damn crater. It's like it's challenging us to make it to the top." The climb up the side of the crater proved to be too difficult. Shepard reported that it was like climbing up a sandy hill. "You take one step

up and you slip back half a step." After seeing that the astronauts were becoming exhausted by this effort, Mission Control instructed them to abandon the climb and collect samples from the nearby boulders. Mitchell was chagrined by this decision and responded, "I think you're finks."

Both astronauts, however, realized that their EVA time was running out and they had several other activities to complete. Later it was determined that they had come within 30 feet of the crater's rim. After collecting samples from their surroundings, the astronauts headed back to the *Antares*, stopping along the way at Weird Crater and others to obtain additional rock and core samples. They also dug a small trench to determine the rigidity of the soil and to see if different layers were present. Shepard and Mitchell then proceeded back to the *Antares*, retrieved the solar wind experiment, aligned the ALSEP antenna, and collected some final samples. This second and last EVA lasted 4 hours and 35 minutes and covered a distance of about 2 miles.

In a final playful gesture before leaving the lunar surface, Shepard threw a makeshift javelin into the sky, and being an avid golfer, hit a couple of golf balls with a golf club (see Figure 22-8) he had fashioned from their tools. He then joined Mitchell in the *Antares*. In all, they had gathered 94 pounds of lunar samples.

After 33.5 hours on the lunar surface, Shepard and Mitchell lifted off to rejoin the *Kitty Hawk*. This time the docking went perfectly. While his companions were on the Moon, Roosa had been busy conducting a photographic mineralogical survey of the Moon from orbit. He also obtained detailed pictures of the Descartes Crater, the proposed land-

Figure 22-7. Ed Mitchell uses a map to direct him toward the Cone Crater. The lunar dust was very sticky and coated the astronauts' suits all the way to their knees.

Figure 22-8. Shepard hits a nice golf shot with a makeshift club just prior to leaving the Fra Mauro.

Figure 22-9. Mitchell (left) and Shepard (right) view some of the rocks they returned to Earth.

ing site of *Apollo 15,* and photographed a 200-foot crater that had been produced by the impact of *Apollo 13's* S-IVB booster.

During the return journey to Earth, the astronauts conducted experiments to determine the effect of zero gravity on welding, liquid and heat flows, and the separation of molecules. They landed on target in the Pacific Ocean and were recovered by the USS *Orleans* on February 9, 1971.

Shepard later offered this assessment of the *Apollo 11, 12* and *14* missions,

> I look back now on the flight carrying Pete's crew [*Apollo 12*] and my crew as the real pioneering explorations of the Moon. Neil, Buzz, and Mike in *Apollo 11* proved that man could get to the Moon and do useful scientific work, once he was there. Our two flights—*Apollo 12* and *14*—proved that scientists could select a target area and define a series of objectives, and that man could get there with precision and carry out the objectives with relative ease and a very high degree of success.

There was little doubt that the Apollo program was evolving from a technological demonstration into a real scientific program. The emphasis placed on acquiring data concerning the origin and evolution of the Moon would become even more pronounced in the upcoming flights of *Apollo 15, 16*, and *17.*

23 *Apollo 15–17*: Lunar Exploration Takes Priority—the J Missions

Guess what we just found? I think we found what we came for.

—David Scott, astronaut

The success of Alan Shepard's flight to the Fra Mauro largely removed the lingering doubts NASA felt about the reliability of the rockets and spacecraft used in the Apollo program. The agency was now intent on addressing questions dealing with the origin and evolution of the Moon rather than hardware issues. To highlight this new focus the remaining flights were referred to as J missions. *Apollo 11, 12,* and *14* had visited the lunar maria, *Apollo 15, 16,* and *17* would land in the unexplored maria/highland interface regions and the rugged lunar highlands.

Apollo 15

NASA's first J mission, *Apollo 15,* was to visit a maria/highland interface region. It carried an all-air force crew commanded by David Scott, a veteran of *Gemini 8* and *Apollo 9,* and rookies James Irwin and Alfred Worden. Worden would pilot the control module (CM) *Endeavor,* named after the famous 18th-century sailing ship commanded by Captain James Cook, and Irwin would fly the lunar module (LM) *Falcon*, named after the mascot of the Air Force Academy.

Scott was born on June 6, 1932, in San Antonio, Texas. The son of an air force general, he graduated from the U.S. Military Academy at West Point and received an M.S. degree in aeronautics and astronautics from M.I.T. in 1962. After graduating from the USAF Experimental Test Pilot School at Edwards AFB he was nearly killed when a jet he was flying crashed during simulated X-15 landings. Selected as an astronaut by NASA in October 1963, he specialized in spacecraft guidance and navigation systems. James Irwin, was born in Pittsburgh, Pennsylvania, on March 17, 1930. He graduated from the Naval Academy at Annapolis in 1951 and earned an M.S. degree in aeronautical and instrumentation engineering from the University of Michigan in 1957. He elected to serve in the air force and completed several advanced piloting schools. Selected for astronaut training in 1966, he specialized in Apollo lunar module systems. The last member of the crew Alfred Worden was born on February 7, 1932 in Jackson, Michigan. Like Scott he was a West Point graduate and served in an air force fighter squadron. Prior to joining the astronaut corps in 1966, he was an instructor at the aerospace school at Edwards.

Figure 23-1. The crew of Apollo 15*: (left to right) David Scott, Alfred Worden, and James Irwin.*

A New Era of Apollo Exploration

The destination of *Apollo 15* was a fascinating, but geologically complex region located near the Moon's Hadley Rille. Named after an 18th-century mathematician, the Hadley Rille is large enough to be seen from Earth using a modest-size telescope. Situated along the edge of the Imbrium Basin, it meanders over a hundred miles of the lunar surface, reaching a width of one mile and a depth of 1,200 feet. A second interesting object geologists wished to investigate was Mt. Hadley. Located about two miles north of the landing site, this mountain is a member of the 12,000-foot-high Apennine Mountain chain (see Figure 23-2). Finally, three dome-like features believed to have been formed by volcanic action or by the impact of a huge meteorite were close enough to the landing site to be closely examined. A major goal of *Apollo 15* was to retrieve rock and soil samples to determine the origin of all of these objects.

Prior to the flight, the geology of the Hadley-Apennine landing site was described to news reporters by astronaut/geologist Harrison Schmitt,

The Imbrium Basin formed very early in the geologic history of the Moon. The Apennine Front represents one of the upthrown rims, one of several rings of mountains that were created by [the] Imbrium impact event. By being upthrown—that is, the outer part thrown up and the inner part of the basin dropped down—it exposed a section of lunar crust . . . some 25,000 feet [deep].

NASA scientists hoped that this vertical rock wall would reveal the past history of the Moon just as similar features on Earth reveal our planet's early history.

To prepare for this expedition, Scott, Irwin, and Worden were trained in field geology in the gorges of the Rio Grande River in New Mexico and the Little Colorado River in Arizona. These regions were believed to be similar to those the astronauts would encounter on the Moon. In all, the astronauts participated in fifteen different field trips to prepare for their upcoming mission.

Apollo 15 on the Moon

Apollo 15 was launched from the Kennedy Space Center without any problems on July 26, 1971. Unlike Shepard's *Apollo 14* mission, the crew had no trouble retrieving the LM from the S-IVB. The journey to the Moon proceeded according to plan except for a troublesome water leak in the *Endeavor* that occurred during the third day of the flight. *Apollo 15* went into lunar orbit on July 29 and on the next day Scott and Irwin landed 1,500 feet northeast of their planned destination. The touchdown was rougher than the astronauts expected and caused them some anxious moments. Irwin recalled,

> Man, we hit hard. Then we started pitching and rolling to the side. My first thought—and I'm sure Dave's, too—was, "Man, are we going to have to abort?" I thought surely we had ruptured something, that something might be leaking. . . . We just held our breath for about ten seconds.

Finally, the craft stopped shaking and shortly afterward they heard a reassuring message from Houston: "You have a STAY." Relieved, Scott and Irwin hugged and patted each other on the shoulder, they were on the Hadley Delta!

Scott and Irwin Explore the Moon

The astronauts on *Apollo 15* conducted a record three Extra Vehicular Activities (EVAs) lasting a total of 19 hours, about twice the time allowed the crew of *Apollo 14*. To maximize the results from these EVAs, a novel new tool would be employed, an electrically powered vehicle called the "Lunar Rover." This four-wheeled vehicle looked like something built from a giant erector set and it was hard to believe this "car" would actually work. Despite its unsophisticated

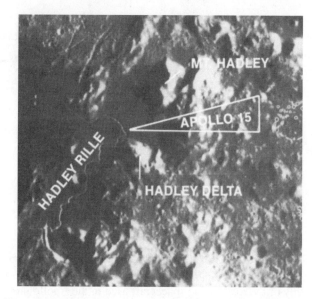

Figure 23-2. The landing site of Apollo 15. *The Hadley Rille is seen as a curving feature starting in the left center and continuing to the lower left.*

appearance, the rover performed admirably and permitted the astronauts to explore regions located miles from their landing site.

Scott and Irwin began their first EVA at 9:30 a.m. on July 31. They gathered a contingency sample of lunar soil and rocks and tested the rover by driving around the landing site. The vehicle's front wheels would not turn properly, but since the rover could be steered using its independently controllable rear wheels, this was not a great concern.

As Scott and Irwin proceeded on their first EVA, they found that riding the rover was a little like riding a "bucking bronco." In *A Man on the Moon*, Andrew Chaikin writes about the astronaut's experience,

Figure 23-3. The Lunar Rover was first used on Apollo 15 *and greatly extended to an area the astronauts could explore.*

Focus Box: The Lunar Rover

The Lunar Rover was jointly built by Boeing and General Motors at a cost of $12 million per vehicle. It weighed 484 pounds, was 9.8 feet long, and 6.6 feet wide. Built to accommodate two astronauts, it had a cruising speed of 7.8 miles/hour. Power was provided by two silver zinc batteries that gave the vehicle a total range of 54 miles. An umbrella-shaped antenna sent data and images to the Earth. An on-board navigation system allowed the astronauts to drive easily to their destinations. Three rovers were carried to the moon.

Aboard Rover 1, the ride was more exciting than anyone, including Scott and Irwin, had banked on. Though the Rover averaged only 5 to 7 miles an hour, it seemed more than fast enough to the two men perched atop it. Every time the Rover hit a bump—and the 'plain' at Hadley was all bumps and hollows—it took flight. Each new obstacle set one or two wheels off the ground for a long moment. . . . The ride was especially harrowing for Jim Irwin . . . especially when Scott had to swerve to avoid a crater. The transmissions from the Rover were punctuated every so often with a warning from Scott—"Hang on!"—followed by Irwin's grin-and-bear-it laughter.

Figure 23-4. The Hadley Rille as seen from the slopes of the St. George Crater. Rock fragments can be seen at the bottom of the rille.

All-in-all, though, the astronauts were delighted with how well rover handled. They drove along the rim of the Hadley Rille, stopping at Elbow Crater to collect samples, and then proceeding to St. George Crater located a short distance away, near a bend in the rille. From this point they had a spectacular view of the rille and transmitted pictures of this scene back to Earth. After collecting soil and rock specimens as well as two core tube samples, they returned to the *Falcon* and deployed their Apollo Lunar Surface Experiment Package (ALSEP) experiments.

While Irwin was busy deploying the ALSEP, Scott attempted to drill two 10-foot-deep holes into the lunar soil to measure the heat emitted from the radioactive decay of elements within the Moon's interior. This task proved to be so difficult that its completion was postponed until the next EVA. Their first EVA had lasted 6 hours and 33 minutes.

Scott and Irwin's second EVA began at 6:47 a.m. on August 1 and proved to be the most scientifically rewarding of the mission. They drove almost 8 miles toward the Apennine Front, stopping and collecting samples at small craters along the way. They knew the material along the rim of these craters had been blasted from deep below the lunar surface and was, therefore, of great interest to geologists.

As they examined ejecta along the rim of a small crater called the Spur Crater, Irwin suddenly called to Scott, "Oh, man, look at that!" Scott stared at the rock Irwin had noted and replied, "Guess what we found? I think we found what we came for." Before them was a little white anorthosite rock sitting on a pedestal-shaped boulder. Lunar scientists believed the primordial lunar crust was composed of this type of material so this rock was thought to be very old. Indeed, subsequent tests revealed it was at least 4.5 billion years old, nearly a billion years older than the oldest rock found on the Earth. Press representatives at Houston later dubbed this specimen (lunar sample 15415) the "Genesis Rock." Irwin was struck by the peculiar manner in which it was discovered,

Figure 23-5. The "Genesis Rock" called attention to itself by sitting on top of a mound of soil. Scott suspected that this rock was part of the Moon's primordial crust.

That was remarkable to see that little white rock sitting on that pedestal almost free from dust. Most of the rocks up there are covered with dust, dust that has been moved about the surface of the Moon for millions and millions of years. Here was that rock clearly displayed for us . . . like it was just there for us to pick it up.

Scott and Irwin concluded their second EVA by drilling a second hole for the heat flow experiment, excavating a narrow trench from which samples were collected, and erecting an American flag. Tired but happy with their effort, they returned to the *Falcon* for a well-deserved rest. Their EVA had lasted a record 7 hours and 12 minutes. NASA was elated by this excursion, and at a hastily called press conference, Flight Director Gerald Griffin pronounced it "the greatest day of scientific exploration that we've seen in the space program."

The third and last EVA of *Apollo 15* started early on August 2. Scott and Irwin drove to the east rim of the Hadley Rille and obtained a 10-foot-deep core sample. While describing the appearance of this region to ground controllers, Scott mentioned that he could see several well-defined layers in the west wall of the rille. This seemingly off-hand remark had enormous implications for the various theories about the origin of the maria. A layered structure implied the maria were not formed from a single cataclysmic event, such as the impact of a large meteor or an enormous volcanic eruption, but by multiple lava flows extending over millions of years. In addition to this tantalizing observation, rocks collected from this area were later dated to be surprisingly young, only 3.3 billion years old, more than a billion years younger than the Genesis Rock.

When their work at the rille was complete, Scott and Irwin returned to the *Falcon*. Once there, Scott performed a short experiment to demonstrate that all objects, regardless of their mass, fall at the same rate in a gravitational field (see the Focus Box below). Before reentering the LM for the last time, the astronauts adjusted the rover's TV camera to record their lift-off. As he gazed at the surrounding lunar mountains, Irwin said, "I am reminded of my favorite biblical passage from Psalms, 'I'll look into the hills from whence cometh my help . . . ,'" and then remembering his audience back home, he added, "But of course, we get quite a bit from Houston, too." Scott and Irwin blasted off the Moon at 9:42 a.m. on August 2 and rejoined the *Endeavor* without difficulty.

While Scott and Irwin were working on the lunar surface, Worden was conducting his own set of experiments using a new set of instruments, called the Scientific Instrument Module (SIM), in one of the bays of the service module (SM). The

---

Focus Box: Falling Objects on the Moon

Galileo said that in the absence of air, all objects fall at the same rate, regardless of their mass. *Apollo 15* astronaut Scott performed an experiment on the Moon using a hammer and a feather to show that was this true. Below is Scott's narrative of this experiment:

"In my left hand, I have a feather; in my right hand, a hammer. I guess one of the reasons we got here today was because of a gentleman named Gálileo a long time ago who made a rather significant discovery about falling objects in a gravity field, and we thought where would be a better place to confirm his findings than on the moon. . . . So we thought we'd try it here for you. . . . The feather happens to be, appropriately, a falcon feather for our *Falcon*, and I'll drop the two of them here, and hopefully, they'll hit the ground at the same time. . . . How about that! That proves Mr. Galileo was correct about his findings."

Galileo Galilei (1564–1642)

panel concealing this package was opened just prior to entering lunar orbit. Special cameras in the SIM were designed to identify mineral deposits and to allow geologists to construct high-quality topographic maps. Objects as small as a desk could be seen on these photographs.

Other instruments in the SIM measured the chemical composition of the lunar soil, emission from radioactive atoms, and the amount of fluorescent x-rays emitted from the lunar soil. Ultraviolet pictures of star fields were acquired and the composition of the Moon's tenuous atmosphere was measured. Finally, another device in the SM searched for volcanic activity as the craft passed over deep fissures in the lunar surface.

After Scott and Irwin rejoined Worden, they remained in lunar orbit for two more days completing experiments and photographic tasks. Shortly before leaving orbit, they released a 78.5-pound satellite to measure the Moon's gravitational field and to monitor solar flare activity. This was the first time a satellite had been released from an orbiting spacecraft.

The mission of *Apollo 15* ended when the *Endeavor* splashed down in the Pacific Ocean on August 7, 1971. The crew was recovered by the USS *Okinawa*. The accomplishments of this mission were impressive. The astronauts had conducted three recordbreaking EVAs, explored 17.5 miles of the Moon's surface, returned 169 pounds of samples and additional measurements about the Moon's surface, and successfully deployed a lunar satellite.

A Review of Lunar Exploration

Including *Apollo 15*, four teams of astronauts had set foot on the Moon. *Apollo 11* and *12* had established that the maria consisted of giant beds of basaltic rock and that at least two episodes of maria formation had taken place. Rocks from the *Apollo 14* landing site predated the formation of the maria and provided information about the Moon's early history. They consisted of breccias, or congealed rocks, and suggested the Moon was intensely bombarded about 3.5 billion years ago. During this era, infalling meteorites formed craters all over the surface and the huge basins were blasted out by the impact of asteroid-sized objects. The shock wave created by these gigantic collisions piled up enormous amounts of material, forming features such as the Fra Mauro hills.

Prior to *Apollo 15* lunar geologists hotly debated the nature of the Moon's original crust. By dating the "Genesis Rock" scientists were able to determine that the Moon's primordial crust was made of *anorthosite*, an igneous rock composed of feldspar. Additional observations from the orbiting (CSM) of *Apollo 15* settled this argument by showing that this type of rock covered the ancient lunar highlands.

The seismographs left behind by *Apollo 12, 14,* and *15* continued to transmit information about the Moon's interior. These data suggested the lunar crust was approximately 40 miles thick and that a hot, though not molten mantle, rich in iron and magnesium existed. Other Apollo instruments provided data about the Moon's core. The absence of a detectable magnetic field implied the core was not molten. Magnetic fields are believed to arise from circulatory motions within a liquid core so the Moon's core was apparently too rigid for these motions to arise.

These findings were impressive, but the Moon's surface is also covered by highland regions. Before the Moon's geologic history could be reconstructed more data was needed about them. Unfortunately, this knowledge would not come easily. The highland terrain is rough and landing on it carried a considerable amount of risk.

The Exploration of the Lunar Highlands

Apollo 16 was the first mission to target the lunar highlands. Its goal was to explore the 8,000-foot-high plateau located about 150 miles northeast of Tranquility Base near the Descartes Crater. To help the crew prepare for this mission, this region had been extensively photographed by Stu Roosa during the flight of *Apollo 14*.

NASA geologists were attracted to the Descartes Plateau because it contained two types of geologic terrain: the Cayley Plains and the Descartes Formation (see Figure 23-6). Cayley Plains are abbreviated highland versions of the vast low-lying lunar maria. They are flat and were believed to have been created by volcanic activity. The Descartes Formation was also thought to have a volcanic origin and to contain rocks that were both older and younger than the lunar maria.

Scientists from the U.S. Geological Survey had carefully studied Roosa's *Apollo 14* photographs and had identified numerous objects that looked like lava flows and cinder cones. The color and shape of these features suggested its lava was richer in silicates than in the lowlying maria and represented a different era in the Moon's geologic history.

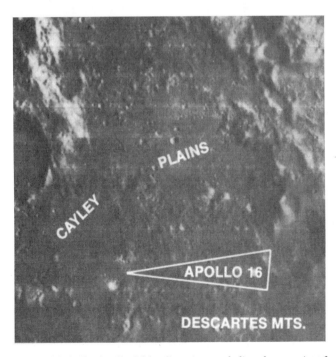

Figure 23-6. The Apollo 16 *landing site was believed to consist of volcanic material like that found on the lunar maria.*

A final inducement to explore the Descartes region was the presence of two medium-sized craters called the North and South Ray Craters. Scientists believed the light-colored rays emanating from these objects consisted of material blasted to the surface from deep within the Moon. Therefore, samples collected from these rays might provide information about the Moon's internal composition.

The Crew of *Apollo 16*

The crew of *Apollo 16* consisted of John Young, mission commander, Thomas Mattingly II, pilot of the CM *Casper*, and Charles Duke, Jr., pilot of the LM *Orion*. Next to Jim Lovell, Young had flown in space more than any other astronaut. He was a veteran of three missions, *Gemini 3, Gemini 10,* and *Apollo 10.* A tough-minded individual, he possessed a sharp mind and an intuitive feel for the solution of complex engineering problems. Thomas Mattingly was originally assigned to be the CM pilot of *Apollo 13,* but was replaced just days prior to launch when he was inadvertently exposed to the German measles. This would be his first spaceflight. This would also be the first spaceflight for Charles Duke. Duke was a former navy pilot and held a master's degree from M.I.T. in aeronautics and astronautics. He was married and had two children. A born-again Christian, he would later leave NASA to become a full-time minister.

The Flight of *Apollo 16*

Apollo 16 lifted off from the Kennedy Space Center on April 16, 1972. On the way to the Moon, Mattingly carried out an experiment to determine if liquids could be purified under conditions of weightlessness, but otherwise, the trip was routine. On April 19 their craft entered an elliptical 69 × 12 mile-high orbit. After checking out the *Orion's* systems and finding them in working order, Young and Duke separated from *Casper* and began preparations to descend to the lunar surface.

Figure 23-7. The crew of Apollo 16 *(left to right), Thomas Mattingly II, John Young, and Charles Duke.*

As these two astronauts worked in the LM, Mattingly prepared to pilot *Casper* back into a 70-mile-high circular orbit, but a problem arose with the engine control system. As he tested the secondary control systems and touched the yaw thumbwheel, the *Casper* began to shake violently. He radioed Houston, "I don't know what's wrong with this thing. It feels like it's going to shake the spacecraft apart." Listening to this conversation, the spirits of Young and Duke sank. Considering the seriousness of the problem, Duke felt certain Houston would cancel the lunar landing. Later he recalled,

> The way [Mattingly] described it, it sounded really bad to me. I thought, what we're going to do is end up joining back up again and coming home. That was a real low point. Not because of any worry about not getting home because you could use the lunar module to burn out of orbit. But it was the fact that you'd trained all that time and been up there [in lunar orbit] for a day. And you can see your landing spot every time you come around—there it is, 50,000 feet below you, 8 miles down—and they were going to say no.

Although Duke was pessimistic about their chances of landing, Houston still held out hope and frantically worked to save the mission. They asked Mattingly to switch to the *Casper's* backup guidance system. Once this was done, the spacecraft stopped shaking. After a delay of six hours, Houston gave its OK for Duke and Young to proceed to the surface.

Relieved by NASA's decision, Young and Duke began their descent sequence and landed within 500 feet of their intended touchdown point. Once on the surface, Duke could not contain his joy and shouted, "Wowww! Whoa man! Old *Orion* is finally here Houston! Fantastic!" A more reserved Young glanced out the LM window, and noting the roughness of the highland terrain radioed, "We don't have to walk far to pick up rocks Houston, we're among 'em."

Young and Duke Explore the Lunar Surface

The first EVA of *Apollo 16* began at 11:58 p.m. EST on April 21. About 30 percent of the ground was covered by rocks and 70 percent of it was pockmarked by craters. They collected a contingency sample of rocks and to their surprise found all of them to be *breccias*, coarse-grained rocks fused together by an impact event. According to the lunar scientists who had planned the mission, the landing area should have been covered by volcanic rocks or basalts.

Young and Duke soon made a second, unexpected discovery. Scattered around their landing site were magnetized rocks. Since geologists believed the Moon had never possessed a molten core, it should not have developed a magnetic field. Not only did these rocks disprove this idea, it was later found that the primitive magnetic field on the Moon, like that of the Earth, had changed its polarity. These two findings had enormous implications. The prevailing theories of the origin of the Descartes Plateau and the Moon itself would have to be substantially revised.

Having concluded their preliminary tasks, Young and Duke unloaded their lunar rover and travelled 200 to 300 feet to set up their ALSEP. While deploying these instruments, Duke accidently broke the cable leading to the heat flow experiment. Both astronauts were disheartened by this accident, but there was no time to fix it. They also set up a semiautomatic observatory to obtain ultraviolet photographic and spectroscopic observations of stars. The Earth's atmosphere absorbs this light but the Moon provides a natural platform for its detection.

Figure 23-8. Lunar geologists expected Young and Duke to find basaltic rocks of the type shown above over much of their landing site.

Figure 23-9. Charles Duke drills a core sample at the edge of Plum Crater (located near Flag Crater). The Lunar Rover is seen in the upper left.

After deploying the ALSEP instruments, Young and Duke drove their rover toward a 1,000-foot-wide crater called Flag (see Focus Box, p. 180). As they rode toward this site, Duke described some of the boulders they passed and noted they also looked like breccias. After hearing Young and Duke classify rock after rock as a breccia, the lunar geologists back on Earth realized their prediction that the Descartes Plateau had a volcanic origin was probably wrong. Revealing both their frustration and surprise at this unexpected finding, they asked the astronauts, "Have you seen any rocks that you're certain aren't breccias?" Duke replied, "Negative."

The *Apollo 16* EVAs

Once Young and Duke arrived at Flag, they collected a number of rocks, all of which appeared to be breccias. One of these rocks, a 26-pound, football-shaped specimen, would be the largest sample brought back during the Apollo program. It was even given a name, Big Muley, in honor of William Muehlberger, head of the *Apollo 16* Surface Geology Team. On the way back to the *Orion*, Young and Duke made one additional stop to collect samples and to take a magnetometer reading to search for any remnant lunar magnetic field. They then returned to the *Orion* and rested.

Their second EVA was scheduled to last over 7 hours and cover 6 miles. The first site they visited was Survey Ridge, a broad, rock-strewn, elevated region located north of Crater Stubby. It was thought that rocks from this area might have been ejected from nearby South Ray Crater. They then drove to Stone Mountain and climbed nearly 500 feet up its side. Looking back at the *Orion* the astronauts were surprised to see many jagged-looking rocks. Because the region was assumed to be old, geologists had predicted that the boulders would be more rounded. Once again, it looked like the geologists were wrong.

From their relatively high position on Stone Mountain, Young and Duke had an excellent view of the South Ray Crater. Because of the lighting conditions at that time of day, the edges of this 1,500-foot-wide crater appeared to be a dazzling white. Ejecta from it was visible for miles in all directions. Young had wanted to visit this site, but Earth-based radar measurements had suggested its terrain was too rough for the rover. Now that he could see it, Young felt they could have safely driven to it, but this was no longer a possibility— their schedule was simply too tight. Instead, they collected samples from a small trench dug into the side of Stone Mountain, hoping these samples would reveal how the mountain had been formed. Geologists thought it was formed by volcanic activity, but again the rocks looked like breccias.

Young and Duke collected additional samples from a nearby crater and then shuffled back down the mountain. They stopped at another crater to collect samples, and as they walked to the rover, noticed an unusual shoe-sized white rock. "That's the first one I've seen here that I really believe is a crystalline rock," Young said. It did not have the darkened appearance of a volcanic rock, but like the Genesis Rock of *Apollo 15*, it seemed to contain anorthosite material.

The third and final EVA of *Apollo 16* began on April 23rd at 10:25 a.m. and lasted 5 hours and 40 minutes. Young and Duke first stopped at the rim of the 3,000-foot-wide and 650-foot-deep North Ray Crater, located about three miles from the *Orion*. This was the first time astronauts had visited a large, deep crater, and they hoped samples from its rim would reveal the composition of the material located far beneath the plateau. Once they reached the crater, they were immediately impressed by the sight and radioed Houston, "Oh, spectacular! Just spectacular." For 40 minutes they collected samples and took pictures. They then decided to investigate a large, distant rock. As they walked further and further from the rover's TV camera, their image became increasingly smaller. Back at Houston, geologists watching them on mon-itors broke into laughter when astronaut Jack Schmidt said, "And as our crew sinks slowly in the west. . . ."

When Young and Duke finally reached their destination, what had seemed like a modest-size boulder turned out to be a colossal, four-story-high rock. It was so big they named it "House Rock." After a brief examination, they declared it too was a breccia. This finding coupled with the lack of basalts found during their previous two EVAs made the volcanic origin hypothesis of the region untenable. They collected samples from House Rock and returned to the *Orion*.

The Conclusion of *Apollo 16*

After spending 71 hours on the Moon, Young and Duke blasted off and rejoined the orbiting *Casper* with 207 pounds of samples. During their final hours in orbit, the astronauts released an 85-pound satellite from a bay of the SM. Like the prior one left during *Apollo 15*, it was designed to collect data about the Moon and its near space environment.

The astronauts jettisoned the *Orion* on April 24, but it tumbled out of control. As the *Orion's* orbit decayed, this craft eventually impacted the Moon and sent shock waves through the lunar soil that were detected by the Apollo ALSEPs.

During the return journey to Earth, Mattingly, like Worden before him, conducted an EVA to retrieve a film cassette from the panoramic and mapping cameras in an SM bay. These cameras had obtained over 12,000 feet of high-resolution photographs of the lunar surface. Once back in the *Casper*, he also conducted a brief experiment on the effects of unfiltered sunlight on microbes. The astronauts landed in the Pacific Ocean at 2:45 p.m. EST on April 27, 1972, and were recovered a short time later by helicopters from the USS *Ticonderoga*.

Apollo 17: The Taurus-Littrow Valley

NASA geologists had selected a 4.5-mile-wide, box-shaped valley called the Taurus-Littrow as the final destination of the Apollo program. Located near the southeastern shore of the Sea of Serenity, this valley is covered by small cone-shaped hills geologist believed had been formed by volcanos. Another set of objects, called the Sculptured Hills, bordered the valley and looked like volcanic features. Following the surprising results from the Descartes Plateau, geologists looked forward to an expedition of a volcanic area where material from deep within the Moon could be found.

Other interesting features also attracted the interest of geologists. Along the sides of the valley floor were two rocky prominences called the North and South Massifs. These objects appeared to be old, and it was hoped that rocks from them would be among the oldest recovered from the Moon. A landslide at the base of the South Massif also offered the opportunity of collecting a variety of different rocks. The valley also possessed a lunar scarp, a narrow, winding uplifted ridge. Scarps are fairly common on the lunar maria but none had been explored. A final inducement to land in this valley was the fact that it was covered by some of the darkest material found on the

Focus Box: The *Apollo 16* EVA Paths

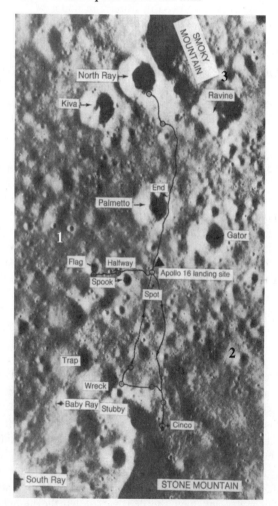

Three lunar EVAs were made by the astronauts of *Apollo 16*. The first was to the Flag Crater. The second was to the base of Stone Mountain (bottom of picture) and the third was to the North Ray Crater (top of picture). The EVAs are indicated by solid lines.

Figure 23-10. The target of Apollo 17 *was the Taurus-Littrow Valley, a region surrounded by a ring of mountains. Its floor is part of the Serenitatis Basin. South Massif is located in the center of this photograph.*

Figure 23-11. The crew of Apollo 17: *(left to right) Harrison Schmitt, Eugene Cernan (sitting), and Ronald Evans.*

Moon. Geologists hoped it was composed of relatively young volcanic ash, perhaps only half a billion years old.

NASA officials conceded that the Taurus-Littrow Valley possessed a large number of interesting geological sites but they were also concerned about the safety of the crew. Landing here was even riskier than on the Descartes Plateau. The Taurus-Littrow floor was only 4.5 miles wide, smaller than the amount by which *Apollo 11* had overshot its landing site. To insure a safe landing, the precise landing techniques used by *Apollo 12, 14, 15,* and *16* would have to work perfectly. Because of the added danger this flight entailed and some reservations about its need, Chris Kraft, director of the Manned Spacecraft Center, cautioned its commander, "Don't take any chances out there; just get home alive."

The Crew of *Apollo 17*

The crew of *Apollo 17* was commanded by Eugene Cernan, a veteran of *Gemini 9* and *Apollo 10*. Harrison Schmitt served as the pilot of the LM (*Challenger*), and Ronald Evans served as the pilot of the CM (*America*).

At the time of the launch Cernan was 38 years old. He was a charismatic leader who looked forward to the challenge this mission presented. Unlike some of his fellow astronauts, who were disheartened by the conclusion of the Apollo program, he was still excited about America's manned space program. He viewed the end of Apollo simply as the "end of the beginning" and looked forward to even greater space adventures.

Harrison Schmitt was a trained geologist and the only professional scientist to fly to the Moon. The scientific com-

munity had urged NASA to include a scientist on at least one Apollo mission, so he was their standard bearer. Not only did he feel great pressure to meet the expectations of the scientific community, he also had to prove to NASA that a scientist/astronaut could contribute to the flight and landing. He and mission commander Cernan got on well together even though their personalities were quite different. Cernan was among the most talkative of the astronauts, whereas Schmitt was one of the most quiet and introspective.

The third crew member of *Apollo 17* was Ronald Evans. He had watched most of the Apollo flights carry his fellow astronauts to the Moon from a grassy spot along Cape Kennedy with his wife, Jan. After each Saturn V rose into the sky, he would turn and say to her, "One day, that's going to be me." He was absolutely thrilled to have been selected to fly aboard *Apollo 17* and considered opponents of the space program to be kooks. He liked his fellow crewmates and referred to Schmitt as "Dr. Rocks."

Apollo 17 on the Moon

Apollo 17 lifted off from the Kennedy Space Center on December 7, 1972, in a spectacular night launch, a first for a Saturn V. Four days later, Cernan and Schmitt landed on the Moon.

Once on the surface, Cernan and Schmitt quickly left the *Challenger*. Schmitt surveyed the lunar landscape and remarked, "A geologist's paradise if I ever saw one." He then threw himself into his assigned tasks. Cernan also took a moment to view the scene and was struck by the sight of the brilliant blue Earth above the lunar landscape. He called to Schmitt, "Oh man. Hey Jack, just stop. You owe yourself 30 seconds to look up over the South Massif and look at the Earth." But Schmitt, now focused on his work, could not be persuaded. He replied, "What? The Earth? Aaah! You seen one Earth, you've seen them all."

Cernan's and Schmitt's initial tasks included the deployment of their ALSEP and the collection of a contingency sample

Each of the three paths taken by the astronauts of *Apollo 17* are shown in this picture of their landing site. Their first EVA is to the lower right, their second EVA is to the lower left, and their third EVA is toward the upper right. Numbers in the picture are referred to in the text.

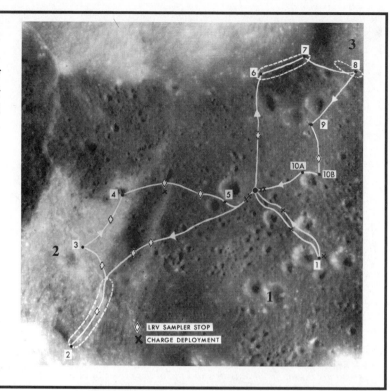

of lunar rocks and soil. These tasks proved more difficult and time consuming than they had hoped. The instrument to measure gravity waves failed to work and the core tubes proved hard to obtain. Next they conducted a brief geology traverse to a medium-sized crater called Steno to collect additional samples. These tasks completed, they returned to the *Challenger* and rested. Schmitt was somewhat disappointed after completing this 7 hour excursion. He had not been able to identify the

Figure 23-13. Harrison Schmitt examines Split Rock at the base of the North Massif.

source of the mysterious dark soil that covered the landscape. Rather than being of volcanic origin, he wondered if it might consist of the shattered remains of boulders that littered the area. He wished he could have devoted more time to this puzzle.

The next day the astronauts continued their exploration by driving 5 miles to the 7,500-foot South Massif. Cernan guided the rover across the rock-strewn terrain while Schmitt analyzed the surface and radioed his impressions back to Houston. Once they arrived at the foot of the massif, they entered a crescent-shaped crater Schmitt had named Nansen, after the Norwegian arctic explorer Fridtjof Nansen. They began an hour-long investigation and collected numerous

Figure 23-12. Apollo 17 astronaut Harrison Schmitt explores the region around Camelot Crater (Stop 5 labelled in the Focus Box above).

rock and soil samples, including a white fragment that would later be dated to have an age of 4.5 billion years.

After leaving this site, they stopped at the Lara Crater to collect additional samples and then proceeded to what appeared to be a landslide at the foot of the massif. This feature was thought to have been created when ejecta from the distant crater Tycho rained down on this region. By dating the landslide, geologists hoped to indirectly determine the age of Tycho.

The next stop on their EVA was a 360-foot-wide crater called Shorty. Orbital pictures showed its rim was covered by a very dark material that might be composed of volcanic ash. Once at Shorty, Schmitt radioed to Houston, "Okay, Houston, Shorty is clearly a darker-rimmed crater. The inner wall is quite blocky . . . and the impression I have of the mounds in the bottom is that they look like slump masses that may have come off the side." To his surprise he noticed the material beneath the gray top soil near his foot was orange. He called Cernan's attention to this and then attempted to determine how far the colored soil extended. He found it was distributed in a symmetrical pattern around the rim of the crater and also found zones of yellow and red material. Based upon their training at volcanic craters in Hawaii, the astronauts guessed that the different colors were produced when the soil was oxidized by volcanic gases. Schmitt radioed mission control, "If I ever saw a classic alteration halo around a volcanic crater, this is it."

Cernan and Schmitt collected soil samples and obtained a three-foot-long core tube. When these samples were later analyzed, it was found that Shorty Crater was not a volcanic crater at all but a normal impact crater. The orange material consisted of tiny beads of molten glass whose color resulted from low concentrations of silica mixed with high amounts of iron and titanium. This glassy material was thought to have been created when molten rock was blasted high above the surface by the explosion that formed Shorty, cooled, and fell in a rain of small droplets. The beads ranged in color from orange to black.

Cernan and Schmitt then returned to the *Challenger* for a well-deserved eight hours of sleep. Before climbing into his hammock, Schmitt contemplated their next EVA—the last of the Apollo program—and radioed mission control, "Tomorrow we answer all the unanswered questions. Right?"

Their third EVA began at 4:30 p.m. and was planned to last more than 7 hours. The astronauts' primary destinations were the North Massif and the Sculptured Hills. They first drove to the foot of the North Massif and examined a large boulder that had slid 1,500 feet down to the valley floor. They were surprised to find this rock was composed of breccias and volcanic rock, a testament to the great force of the impact which had carved out the vast lunar maria. After collecting samples from it they proceeded to the Sculptured Hills and took specimens from a boulder on a mountain side. Along their way back to the *Challenger* Schmitt noticed a layer of bright white soil and collected a few additional samples.

The lunar exploratory work of Cernan and Schmitt was now nearly complete. Back at the landing site, Cernan selected a rock to be broken into small pieces and distributed to

Figure 23-14. The crew of Apollo 17 *left this plaque on their Lunar Rover to commenerate the final Apollo mission to the Moon and to reemphasize the peaceful nature of the program.*

museums around the world. He then pointed the TV camera to a plaque attached to the *Challenger's* front landing gear and read its inscription,

Here man completed his first exploration of the Moon, December 1972 A.D. May the spirit of peace in which we came be reflected in the lives of all mankind.

It bore the signatures of the three astronauts and President Nixon. He then added his own thoughts about the future, "This is our commemoration that will be here until someone like us, until some of you who are out there, who are the promise of the future, come back to read it again and to further the exploration and meaning of Apollo."

Figure 23-15. Cernan and Schmitt lift off from the moon on their journey back to Earth.

Figure 23-16. A Navy helicopter airlifts the crew of Apollo 17 *to the awaiting USS* Ticonderoga.

As Cernan prepared to climb up the LM's ladder, he took a final look at the Earth shining over the lunar landscape, and thinking of their return trip to Earth said, "Godspeed the crew of *Apollo 17*." After a stay of 75 hours on the Moon, Cernan and Schmitt lifted off with 243 pounds of lunar samples and rejoined the CM, *America.*

While his fellow astronauts were on the lunar surface, Evans explored the Moon from orbit using sensing instruments, cameras, and a radio sounding device that could penetrate three-quarters of a mile into the Moon's surface. On the return voyage to Earth, he performed a spacewalk to retrieve this data. The astronauts splashed down in the Pacific Ocean on December 19, 1972. Less than one hour later they were aboard the recovery ship the USS *Ticonderoga.*

Closing Thoughts

Apollo 17 was to be the last American manned flight to the Moon. As with the end of the Gemini program, numerous people were either let go by NASA or were reassigned to other projects, in this case to the upcoming *Skylab* and *Apollo-Soyuz* missions.

The end of *Apollo 17* was a bittersweet time at the Kennedy Space Center. A unique period in the history of human exploration had come to a close—a victim of public apathy, political indifference, and pressures to reallocate money to other pressing national needs. In retrospect, the change in public attitude from excitement to indifference should have been expected. NASA had portrayed Apollo as a race to beat the Soviet Union to the Moon. America had won that race and had gone far beyond its original commitment. It was impossible to argue that the goals of Apollo had not been met. Six teams of astronauts had landed on the Moon and explored its surface. They had visited the lunar maria, the highland/maria interface, the highlands, and numerous craters. Hundreds of pounds of lunar samples had been gathered and their analysis would revolutionize our scientific knowledge about the Moon's origin and geologic history.

The American political leadership, rightfully or not, saw little to be gained from additional lunar landings. The political climate had changed and the superpowers had moved into an era of peaceful coexistence. There was a strong feeling that the time was ripe to give increased attention to global political issues. Nevertheless, there was a noble side to America's lunar adventure, a side that would have appealed to Tsiolkowsky, Goddard, and Oberth. Mankind had finally and irreversibly escaped the confines of the Earth. Harrison Schmitt expressed the legacy of Apollo in the following way,

What will historians write many years from now about the Apollo expeditions to the Moon? Perhaps they will note that it was a technological leap not undertaken under the threat of war; competition, yes, but not war. Surely they will say that Apollo marked man's evolution into the solar system, an evolution no longer marked by the slow rates of biological change but from then paced only by his intellect and collective will. Finally, I believe that they will record that it was then that men first acquired an understanding of a second planet.

Introduction to Section IV
The Early Russian Manned Space Program

I have come to the conclusion that some of our traditional ideas of international sportsmanship are scarcely applicable in the morass in which the world now flounders. Truth, honesty, justice, consideration for others, liberty for all—the problem is how to preserve them . . . when we are opposed by people who scorn these values.

—President Dwight Eisenhower, 1955

The early rocket programs in Russia, Germany, and the United States were led by Tsiolkovsky, Oberth, and Goddard. These talented, independent thinkers thrived on the intellectual challenges presented by spaceflight, and believed the rocket was the vehicle humans would ultimately use to journey into space. Konstantin Tsiolkovsky developed the theoretical basis of spaceflight and wrote about the possible uses of satellites decades before these craft were actually built. Hermann Oberth also worked on the physics of rocketry, and then attempted to transform his design into a functioning device. His efforts would inspire the next generation of German rocket engineers. Robert Goddard's dream was to construct a vehicle capable of interplanetary flight. He was a talented physicist who possessed the skills of an engineer. He not only constructed the first liquid-fueled rocket, he designed and built much of the hardware needed to control its flight.

The era of the "lone-wolf" rocket researcher, typified by these men, gave way to a period where groups of rocket enthusiasts assumed the lead in this field. In this regard, Oberth was a transitional figure. Starting as an isolated researcher, he then helped create the German rocket club the *VfR*. The *VfR*, and groups elsewhere, moved the field of rocketry forward, but were soon overwhelmed by the formidable technological and financial demands of the field. These obstacles were addressed in the next era in the development of rocketry. Starting in the years prior to World War II, national governments began to devote larger and larger sums of money to rocketry. Motivated by the desire to strike at a distant enemy, powerful rockets were built in Germany. These weapons had no effect on the outcome of the war, but their development solved many of the problems hindering the construction of large, operational rockets.

At the end of World War II two politically opposed military superpowers, the United States and Soviet Union, existed. As these countries competed for the allegiance of the nonaligned nations of the world, a deep sense of mutual distrust arose, and enormous sums of money were spent on

Figure Intro. 4-1. Robert Goddard (1934) holding a centrifugal pump and turbine he had built for his rocket.

Figure Intro. 4-2. Soviet Premier Khrushchev berated the United States for sending spy planes deep within the Soviet Union. This event increased cold-war tensions.

Figure Intro. 4-3. In October 1961, American (foreground) and Russian (background) tanks faced each other in a tense situation at Check Point Charlie in Berlin.

Figure Intro. 4-4. A group of GIRD members pose in front of one of their early rockets.

armaments. Nuclear weapons were refined and rockets were developed that were capable of carrying warheads to targets thousands of miles away. Space programs, whose goals were tied to scientific exploration, also grew out of these military efforts. By the middle of the 1950s, the world came to view rocket technology as a way of measuring the relative strengths of the communist and capitalistic systems.

In the early 1960s, the American space program lagged behind that of the Soviet Union. All of the publicly acclaimed successes, the launch of the world's first satellite, the first living creature to be placed into space, and the first man in orbit, belonged to the Soviets. In the cold-war climate prevailing at that time, each of these triumphs were viewed by American policy makers with a sense of foreboding. They feared the Soviets intended to use this technology to gain a military advantage. Much to the chagrin of the American leadership, the world increasingly saw the Soviet space successes as evidence of a growing Soviet superiority over the West.

Much of the credit for these milestones was attributed by the communist press to an unnamed figure called the "Chief Designer." The Soviets steadfastly refused to reveal the identity of this person, claiming he would then become the target of western assassination attempts. The mystery surrounding this individual further unsettled an American society unaccustomed to secrecy.

Development of the Soviet Union Rocket Program

The space program in the Soviet Union developed much like those in America and Germany. Rocket research was ini-

tially conducted by amateurs, who in order to share the expense of this hobby pooled their expertise and created clubs. In Russia two groups formed, one supported by the state, the *Gas Dynamics Laboratory* (GDL), and the other, the *Group for Studying Reaction Propulsion* (GIRD), composed of amateur rocket enthusiasts.

During World War II the American and Soviet rocket programs floundered because they lacked a powerful and knowledgeable advocate. However, after the end of the war both of these countries dramatically expanded their work in this area. The American effort was spearheaded by Wernher von Braun and his German coworkers. The Soviet Union also employed German scientists, but their program was soon taken over by Soviet rocket scientists. Chief among this new group of Soviet engineers was Sergei Korolev.

After the devastation caused by World War II, it seemed that the Soviet Union would have to spend years rebuilding their country's infrastructure. Because of this, few people thought they would be able to pose a challenge to America's leadership in technology. If the Soviet Union was to have any chance at this, they would have to find a man who possessed extraordinary organization skills and engineering talent. Fortunately for the Soviets, Korolev possessed all of these qualities.

In many respects Korolev's career paralleled that of von Braun. The roles played by Oberth and Dornberger in von Braun's life were played by Tsander and Tukhachevsky in Korolev's life. Like von Braun, Korolev's initial postwar focus was on the development of military rockets.

24   Sergei Pavlovich Korolev (1907–1966)

Like Napoleon he sometimes made decisions without full information. If you work in a field where there is no experience you are forced to make decisions without full information. Therefore you are liable to make mistakes. The good commander makes the right decisions instinctively.

—Boris Rauschenbakh, a worker for Korolev, in *Korolev*

Sergei Pavlovich Korolev was born in the small Ukranian town of Zhitomir on January 12, 1907. His mother, Maria Nikolaevna Moskalenko, was a Cossack, a proud people renowned for their bravery and independent spirit. His father, Pavel Yakovlevich Korolev, was a literature teacher at the local high school in Zhitomir. The marriage of Pavel and Maria was arranged by their parents, and initially, the couple seemed to be happy. However, problems soon arose. Still a teenager herself, Maria found it difficult to deal with the financial stresses placed on her by the family's small income. This situation became worse when Pavels parents unexpectedly died and he adopted two of his younger sisters. In addition to this problem, Maria longed to continue her education, as her sister Anna was doing, in Kiev. These troubles and others finally caused the marriage to break apart and Maria took Sergei to her parents' home in Nezhin, about 100 miles northeast of Kiev. After settling into his new house, Sergei asked about his father, and in an attempt to put an end to this matter and to spare further worry, his mother said he had died during the move.

Maria was still less than twenty years old when she returned to her parents' home. She was anxious to complete some of her earlier goals and soon left Nezhin to join her sister in Kiev. This left Sergei in the care of his grandparents. Maria made frequent weekend trips back to Nezhin to visit Sergei and later wrote that this arrangement had little effect on his childhood. However, Yaroslav Golovanov, Korolev's Russian biographer, painted a different picture, portraying Sergei as a sad child who felt a deep loss, "Everyone loved him, but he missed the love of his parents at a time when he needed it most. He was always neatly dressed, always excellently fed, and always lonely, nearly always sad. . . ."

A similar assessment of Korolev's early childhood days was given by his Nezhin school teacher, Lidia Grinfeld. She recalled that his grandparents did not encourage him to make friends and restricted his range of activities,

He didn't have any child acquaintances of his own age, and never knew child's games with little friends. He was often completely alone at home. . . . They didn't allow him to run in the street. The gate was always

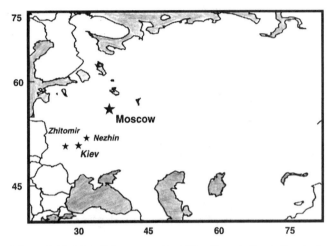

Figure 24-1. Map of the area around Kiev. Stars indicate Korolev's birthplace of Zhitomir and his mother's birthplace of Nezhin.

bolted. He would sit a long while on the upper cellar door, and watch what was happening in the street.

In some respects Sergei's childhood isolation resembled that of Tsiolkovsky. However, Tsiolkovsky's loneliness was caused by a physical condition, a hearing loss that was beyond his control, whereas Korolev was forced to pay the price for his mother's premature marriage. Tsiolkovsky described his childhood as the darkest period of his life, while Korolev would later tell one of his relatives, "I did not have any childhood."

Sergei liked school and according to his teacher was an "attentive, industrious, and capable" student. He was an avid learner and especially loved arithmetic and reading. At the age of six he was taken to the nearby fairgrounds in Nezhin to watch an airplane demonstration. Only ten years had passed since the Wright brother's first flight at Kitty Hawk, so the sight of an airplane was still a novelty. People from all walks of life came to see the airplane performance. The atmosphere at the Nezhin fairground was filled with excitement and the whole town had gathered to watch the flight. This experience, like Goddard's early vision at his Aunt Ward's farm in Auburn,

Massachusetts, had a profound effect on Sergei. At that moment, he decided to devote his life to the study of flight.

In November 1916 Maria married Grigory Balanin, a young electrical engineer, whom she had met in Kiev. Shortly afterward, Balanin was offered a job with the railroad in the coastal city of Odessa. He moved the family and soon his fortune improved when he was appointed head of the port's power station. Life for Sergei was finally settling into an enjoyable routine. His mother was content, he got along well with his

Figure 24-4. British marines were sent to Odessa to help evacuate people trying to escape the Russian civil war.

Figure 24-2. From left to right, Maria Nikolaevna, (Korolev's grandmother), Sergei Korolev, and Maria Matveevna, his mother.

stepfather, and the family income was sufficient to provide for their needs. However, this tranquil period was short lived. In 1918 the Russian civil war broke out and Odessa was filled with gunfire as one faction battled against the other.

With the outbreak of fighting, the Soviets gained control of Odessa but after only a few weeks they were driven off by German and Austrian troops who occupied the port. By November these troops left, but were followed by contingents of British, Serbian, and French soldiers. Korolev's stepfather was not impressed by any of these troops and described the situation as, "New bandits come to take the place of the old ones."

By February 1920 the civil war had largely ended and the Soviets were in charge of the city. Years of turmoil had taken their toll. Odessa had never been an elegant city and after years of conflict conditions had worsened. N.Y. Aberzguz, a schoolmate of Korolev, described life in Odessa at this time,

> Apartments were heated with cast iron stoves. Often they were fueled simply with household furniture. . . . The water lines were always breaking. On the streets you could hear the click of the "woodies" of the Odessa residents—sandals with soles made out of two boards held together with strips of leather. . . .

Aberzguz also described the educational changes imposed by the Soviets,

> Reform in the system of public education was being implemented. Pay and private gymnasiums [high schools] and commercial schools were eliminated, and the separate education of boys and girls was abolished. The schools were undergoing polytechnical transformation and vocational instruction was introduced.

Korolev was trained as a roofer and tile maker and found he had a flair for carpentry. He was a good student, but his real interest remained in aircraft. He often became impatient

Figure 24-3. Sergei Korolev at the age of four. At this time he was living with his grandparents in Nezhin.

when school activities took time away from his private studies. In 1923 he joined the Society of Aviation and Aerial Navigation of the Ukraine and the Crimea. He also volunteered to perform routine chores for a military seaplane detachment stationed in Odessa.

One day a pilot friend asked if he wanted to accompany him for a short ride. Korolev eagerly climbed aboard the craft and they soared into the sky. While over the bay, Sergei maneuvered onto the wing to check the plane's oil level. At this moment, the seaplane's engine suddenly stalled and the craft shook violently. Sergei was sent tumbling into the air and would have been killed if the plane had not been just a few tens of feet above the sea. Thoroughly drenched and somewhat embarrassed, Sergei swam back to shore. His desire to fly, however, was undiminished.

In 1924 Korolev graduated from high school and asked his school sweetheart, Xenia Vincentini, to marry him. She was intent on continuing her own studies and declined his proposal. He then left Odessa to enroll in the aviation program at the Kiev Polytechnic Institute. Sergei and Xenia parted on good terms and continued to write to each other.

Korolev's Technical Training

Korolev spent two years in Kiev and then transferred to the Moscow Higher Technical School. Andrei Tupolev, a noted aeronautical engineer who would later play an important role in Korolev's life, served as one of his instructors. Korolev's aviation training now progressed rapidly. In the fall of 1927, he entered a design in the first Koktebel glider competition, and in 1929 he made his first solo flight in an Avro 504k biplane.

In 1930 Korolev was awarded a degree in aeronautical engineering based in part on the design of a plane he had begun in the spring of 1927. Tupolev served as his advisor and was impressed by his student. He later wrote,

> . . . it was sufficient just to help him a little, to make some minor corrections here and there. . . . He already made an excellent impression on me then as regards both his personality and his talent for design. I would say that he was a man with unlimited devotion to his job and his ideas.

These impressions would prove to be right on the mark. After completing his studies, Korolev was assigned to Factory 22 in Fili and designed military seaplanes. Korolev was a good worker, but did not stand out among the other engineers. His real talent, that of coordinating the efforts of others, had not yet been discovered.

That same year, Korolev flew his own entry in the Koktebel competition and earned high praise for his design work. In another competition, he submitted a novel glider design that allowed the craft to complete acrobatic maneuvers. Unfortunately, during the competition, he was suddenly taken ill and hospitalized with typhoid fever. When this disease led

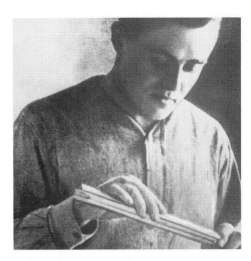

Figure 24-5. A young Sergei Korolev using a slide ruler. He enjoyed his technical education in aeronautics.

to a serious ear infection, his mother rushed him back to Moscow for additional medical treatment.

While recuperating from his illness at home, Korolev worked on the design of a new two-seater light plane, the SK-4. A prototype of this craft was built, but performed poorly. The aircraft was plagued with mechanical problems caused by poor machine work. This experience instilled in him the belief that the fabrication process had to be closely monitored if a quality product was to be built, a lesson that would prove invaluable in his later position in the Soviet space program.

Rocketry in the Soviet Union

Beginning in the 1920s, the pace of rocket research in the Soviet Union began to accelerate and two groups came into existence. One of these, the Gas Dynamical Laboratory (GDL), was located in Leningrad. It was supported by the government and was headed by Valentin Glushko (1908–1989). The second group, the Group for Studying Reaction Propulsion (GIRD),

Figure 24-6. Valentin Glushko would become one of the Soviet Union's leading rocket engine designers.

Figure 24-7. Friedrikh Tsander was an early Russia rocket enthusiast. He helped foster Korolev's love of rocketry.

was located in Moscow and was led by Friedrikh Tsander (1887–1933). It consisted of amateur rocket enthusiasts.

Tsander's dream, like that of Goddard, was to build a rocket capable of traveling to the planet Mars. While working with Tsander at the Central Aerohydrodynamical Institute in Moscow, Korolev became enthralled by the rocket and joined GIRD. During his free evenings and weekends he enthusiastically worked on various GIRD projects.

The problems that Tsander's group faced were difficult and their experimental program progressed slowly. Tsander had designed a liquid-fueled rocket, but before it could be tested, he died. The honor of launching the Soviet Union's first liquid-fueled rocket would belong to another GIRD member, Mikhail Tikhonravov. Launched on August 17, 1933, about 20 miles west of Moscow in the woods of Nakhabino, this rocket weighed 40 pounds and used gasoline and liquid oxygen. The flight lasted 18 seconds, about seven times longer than Goddard's historic first flight in 1926.

The Rise of Soviet Military Interest in Rockets

In the 1930s, the work of GIRD drew the attention of the Soviet military. Marshal Mikhail Tukhachevsky, chief of Armaments for the Red Army, became interested in the military application of rockets as early as 1932 and provided monetary support to both GRL and GIRD. In order to make the funding process easier, these two groups merged into a single organization, called the Reaction Propulsion Institute (RNII). Tukhachevsky named Ivan Kleimenov, a military engineer, to be its director. Korolev was appointed deputy chief engineer. The aim of the RNII was to develop weapons for the Soviet armed forces.

Korolev's main task at RNII was to design long-range rockets and guidance and control systems. His work progressed well and by the summer of 1934 he was promoted to chief engineer. This was a happy time in Korolev's life. He

Figure 24-8. The launch of the first liquid-fueled rocket in the Soviet Union. Korolev is seen standing to the left.

Figure 24-9. Left to right, Xenia, Natasha, and Sergei Korolev. Sergei was 29 years old.

loved his work and had married his high school sweetheart Xenia. Their first child, Natasha, was born on April 10, 1935.

Stalin's Purges and Their Effect on the Soviet Rocket Program

In the early morning hours of June 27, 1938, Korolev was taken from his home by the Soviet secret police, the NKVD, and charged with "subversion in a new field of technology." Specifically, he was arrested on charges of collaborating with anti-Soviet elements in Germany. This was a brutal era in Soviet history. Mass imprisonments and executions of people considered to be a threat to the government were common. Not realizing that these arrests were planned by Stalin himself, Korolev's mother sent a telegram asking for his help. It read: "For the sake of my sole son, a young, talented rocket expert and pilot, I beg you to take the necessary steps to resume the investigation." Not surprisingly, her appeal went unanswered.

Many high-ranking people in the Soviet rocket program experienced an even worse fate. Marshal Tukhachevsky, the military leader of the Soviet rocket program, was accused of collaborating with the Germans. He was given a token trial and then executed along with several officers from his staff. The brutality of this period had few bounds. Tukhachevsky's wife and two brothers were also arrested and shot. Even his daughter, when she came of age, was sent to prison. His mother and sister were forced to leave the country and would later die in exile. Kleimenov, the head of RNII, and fellow worker Gyorgi Langemark, were imprisoned and then shot after being found guilty of collaborating with the Germans. Three months before Korolev's arrest, the former head of GRL, Glushko, was arrested by the NKVD and sentenced to eight years in prison.

Korolev was always reluctant to talk about this awful period in his life, but a fellow RNII worker, Arvid Pallo, described what it was like to live in those times: "That was a peculiar period. When two guys would say a third guy was bad, that was enough. We lost the gift of speech in those days." He later spoke about the circumstances leading to Korolev's arrest,

He had been demoted about a year before his arrest, from chief of a group to a senior engineer. Kleimenov in that time wanted to focus on the development of solid rockets. Korolev wanted to work on a winged vehicle powered by liquid rockets. Some workers wrote to the NKVD that Korolev was holding back progress. . . . When Korolev didn't show up for work one day the other workers knew that he had been arrested.

The arrests and executions extended across Soviet society and included party officials, plant managers, army officers, and ordinary citizens. Between 1937 and 1938 an estimated 4.5 to 5.5 million Russians found themselves falsely accused and placed in prison. Of this number 800,000 to 900,000 received death sentences. The wholesale arrests during the purges were particularly destructive to the Soviet rocket program. Gregory Tokaty-Tokaev, a Soviet rocket expert who defected to England, later wrote,

> . . . the political arrests and murders of the 1935–40 period caused much greater damage than was realized abroad. Far too many scientists, technologists and managers were destroyed, humiliated or disheartened. And rocket experts were no exception. . . . Almost all who worked on a project with . . . [Marshal Tukhachevsky], or who were in contact with him—as all leading rocket specialists were—had now to face the danger of being an "accomplice of a spy."

Korolev would later talk about his brief trial with cosmonauts Gagarin and Leonov. He told them he was beaten, led into a room, and then asked by the judge, "Do you agree with the accusations?" "I didn't commit any crime," he replied. The judge answered, "None of you swine have committed a crime. Ten years hard labor. Go! Next!"

Even though he was not sentenced to be shot, Korolev still faced a potentially deadly situation because conditions in the prison camps were terrible. The inmates lacked food and shelter and suffered from all sorts of diseases. Minor offenses were dealt with in a pitiless and swift manner. Because of these harsh conditions, about 10 percent of the prisoners died each year. The camp that Korolev was sent to was particularly notorious. About 30 percent of its population of 500,000 prisoners died annually. Although he survived the experience, his health was permanently impaired and he lost all of his teeth.

Andrie Tupolev, Korolev's former teacher, was also arrested in 1937. However, instead of being interred in a regular prison, he was sent to a special facility for people who possessed skills needed by the Soviet military. Here the attitude of the guards was very different. What mattered were results, the political views of the prisoners were irrelevant. Probably, at the request of Tupolev, Korolev joined this group in September 1940. In an incredible display of loyalty to a regime that had deprived them of their freedom, the captives worked

Figure 24-10. Marshal Mikhail Tukhachevsky was an early supporter of the Soviet rocket program. He provided funding to both GDL and GIRD.

Figure 24-11. Walking toward the camera, from left to right, Nikita Khrushchev, Joseph Stalin, and Georgy Malenkov in 1945.

Figure 24-12. In a dramatic change of fortunes, Korolev was appointed a colonel in the Soviet Army and was sent to Germany to gather materials from their V-2 program.

hard to help the war effort of their country. Two of the Soviet's most potent military aircraft, the Tu-2 light bomber and Ilyushin-2 fighter, were designed by Tupolev's group.

Conditions at this new site were far better than in other Soviet prisons. The food and housing were good and after a day's work, the inmates were given free time. More importantly, their families were allowed to visit. Xenia and Natasha met with him in a prison located close to his work site. Xenia tried to shield Natasha from the fact that her father was a convict by fabricating stories about his situation,

> I remember when I was five or six we were allowed to visit for half an hour. My mother had told me that father was a pilot, and I asked him how he could land in such a small yard. We visited him several times. . . .

In 1942 Korolev was moved to another facility operated by the NKVD in Kazan, a town about 400 miles west of Moscow. Here he joined Glushko as chief designer to work on jet engines and rockets. By October 1943 a rocket engine was built and attached to the Pe-2 bomber.

At the end of the war the Kazan inmates were released by the NKVD and the facility was turned over to the People's Commissariat of Aviation Industry. Many members of the research group decided to stay and continue their wartime research. Glushko was appointed chief designer and Korolev was assigned to be his deputy. In the summer of 1945 Korolev was commissioned a colonel in the Red Army and was sent to Germany to gather information about their former enemy's V-2 program.

The German Impact on the Russian Rocket Program

Like the Americans, the Soviets were anxious to acquire the German rocket technology. They were aware of Operation Paperclip, the American effort to capture the key German rocket scientists, and agents were dispatched to grab von Braun from the Americans. However, they found that he was too closely guarded for their schemes to succeed. Other Soviet agents and military personnel were given the task of gathering V-2 test equipment, drawings, and any partial or fully assembled rockets. They were aided in this effort by sympathetic Germans. Helmut Grottrup, a senior engineer who had worked for von Braun at Pennemunde, was particularly helpful. This work suited Korolev and he began to demonstrate the managerial skills that would quickly cause him to advance to the top ranks of the Soviet rocket program.

Yuri Mozzhorin, a senior expert in the Soviet rocket program recalled this time,

> You could talk to him about simple as well as complicated things. You'd think his time in prison would have broken his spirit, but on the contrary, when I first met him in Germany when we were investigating the V-2 weapons, he was a king, a strong-willed purposeful person who knew exactly what he wanted. By the way, he was very strict, very demanding, and he swore at you, but he never insulted you. He would always listen to what you had to say.

In the fall of 1946, 5,000 German weapon experts were kidnapped from their homes and sent, along with their families, to locations outside of Moscow. Strict orders were given that the Germans were to be treated kindly and were to be allowed to take whatever possessions they wished with them. Grottrup's wife, believing a famine existed in Russia, tested the limits of this directive by insisting that the cows on the farm where she and her husband were staying also be brought along. The animals were dutifully collected and shipped to Moscow.

Many of the German rocket engineers were initially sent to a newly formed research group called the Scientific Research Institute (NII-88) located northeast of Moscow. The institute was housed in an abandoned artillery factory whose buildings were in poor condition. When it rained, puddles formed on the floors of the production plant. The heating system did not work and repairs were needed everywhere. Circumstances were so primitive that the engineers lacked drafting tables and instead used large boxes as desks. The institute consisted of three divisions. The first dealt with experimentation, the second with the fabrication of components, and the third, with missile design. Korolev was placed in charge of this latter unit.

The German workers assigned to NII-88 were soon moved to the island of Gorodomliya to work on problems dealing with ballistics and aerodynamics. A rocket launch facility was built in a remote section of southern Russian called Kapustin Yar and the first reconstructed V-2s lifted-off from this site in October 1947. Of the eleven V-2s fired, five landed on target. This was about the same success rate that the Germans experienced during the war. The first Russian copies of the V-2, the R-1, were successfully tested a year later.

While the R-1 tests were being conducted, Korolev's group worked on its more powerful successor, the R-2. The Germans worked on a similar rocket called the G-1. Unlike the U.S. effort where von Braun's group worked closely with their American colleagues, German and Russian engineers rarely worked together. Advances made by the Germans were communicated to their Soviet counterparts through written reports. The German engineers were viewed as teachers rather than as long-term participants in the Soviet rocket program. They were to be sent back to Germany when their job was done.

The R-2 and G-1 were intended to have twice the range of the V-2 and to be much more accurate. It was soon clear, however, that it was too expensive to support both of these programs. In accord with the Soviet plan to gradually become self-sufficient, the G-1 was phased out and the Germans were sent back to their homeland. The first group left the Soviet Union in January 1952 and the final group departed in 1954. From that time onward, the Soviets pursued their work without outside assistance.

Early Soviet Military Rockets

The Soviet weapons program progressed rapidly in the postwar era. In 1949 the first Soviet atomic bomb was detonated and in 1953 they exploded their first hydrogen bomb. To exploit fully the potential of these weapons, Korolev was instructed to begin work on a rocket that could carry these multiton bombs to targets located thousands of miles away.

The 20-ton, intermediate-range missile, the R-2, was declared operational in 1951. Work then proceeded on its more powerful successor, the R-3. This rocket weighed 75 tons and was intended to have a range of 1,800 miles. Unfortunately, its construction required an unachievable technological leap, and the program was terminated in 1952. Korolev's group then focused on the more modest R-5. This rocket, which weighed 30 tons and had a range of 720 miles, was successfully launched from Kapustin Yar on March 15, 1953.

Stalin was intimately involved in the Soviet rocket program and personally approved many of its projects. Having such a powerful patron ensured that all of the rocket program's vital needs were met, but it had its downside as well. Major decisions were never debated, so few members of the political leadership had a good grasp of the technical state of the Soviet program. In his memoirs, Nikita Khrushchev, a former head of the Soviet Union, wrote about the impact of Stalin's heavy oversight,

> . . . he completely monopolized all decisions about our defenses, including—I'd say especially—those involving nuclear weapons and delivery systems. We were sometimes present when such matters were discussed, but we weren't allowed to ask questions. Therefore when Stalin died, we weren't really ready to carry the burden which fell on our shoulders.

Stalin was a strong advocate of the liquid-fueled rocket and saw it as a way of countering America's growing military power. The fact that Korolev could provide him with needed expertise made Korolev one of "Stalin's fairhaired boys." Despite his wartime treatment, Korolev seems to have personally liked Stalin. In an interview with Korolev's daughter, Natasha, she said her father, "thought that Stalin had no knowledge of the arrests and purges. He thought that Stalin's death was a tragedy. He had respect for him. . . ."

Stalin's death bequeathed immense power to Korolev. His frank and forthright reports impressed the new political

Figure 24-13. The R-7 rocket was designed by Korolev's group. It was powered by twenty engines distributed into five clusters. The outer four clusters were ejected after burning for 140 seconds.

leaders and he quickly gained their confidence. They considered Korolev the country's premier rocket expert and grew to rely on his opinion. Khrushchev wrote of the Soviet leadership's profound ignorance about the state of the country's program at that time,

> Not too long after Stalin's death, Korolev came to a Politburo meeting to report on his work. I don't want to exaggerate, but I'd say we gawked at what he showed us as if we were a bunch of sheep seeing a new gate for the first time. . . . Korolev took us on a tour of a launching pad. . . . We were like peasants in a marketplace. We walked around and around the rocket, touching it, tapping it to see if it was strong enough.

The relationship Korolev enjoyed with the Soviet leadership gave him substantial power to direct the nation's rocket program. Nevertheless, he avoided pushing the limits of his authority. Sergei Belotserkovsky, head of cosmonaut academic training, later wrote,

> The top people's attitude to Korolev was purely that of consumers. For as long as he was indispensable, for as long as they needed him to develop missiles as a shield for the Motherland, he was allowed to do whatever was necessary. . . .

The Development of the Soviet ICBM

The next rocket that Korolev worked on was the R-7. This was a uniquely Soviet rocket that employed some of the basic ideas formulated by Tsiolkovsky decades earlier. Attached to the rocket's exterior walls were four boosters that ignited on lift-off and burned for 140 seconds. After their fuel was exhausted, they were jettisoned and a central cluster fired for an additional 180 seconds (see Figure 24-13). It had a range of 4,800 miles and a lift capability of 5.4 tons.

Test flights of the R-7 were conducted at a new launch facility called Tyuratam (later renamed Baikonur). Five launch attempts were made between May and July 1957, but all failed. The first successful flight took place just a month later on August 21, 1957 and marked the birth of the world's first Intercontinental Ballistic Missile (ICBM). The American counterpart, the Atlas, would not be successfully flown for another fifteen months.

A Victorious Defeat

Korolev's next project, the R-9, was intended to have a range of 7,500 miles and a payload of 1.7 to 2.2 tons. However, from a military prospective the R-9 had a major flaw. It had to be fuelled just prior to launch and this process took nearly a day to complete. This limitation made it impossible to respond quickly to a developing international situation.

An alternative rocket, the R-16, that relied on propellants stored within the rocket itself, was developed in parallel with the R-9 by a group lead by Mikhail Yangel, a former colleague and rival of Korolev. The R-16 could be launched in 30 minutes and only Korolev's influence with the government prevented an immediate decision in its favor. Mindful of Korolev's power, the military wisely decided to postpone the final rocket selection until their respective test programs were complete.

The military's enthusiasm for the R-16 proved to be premature. Its first demonstration launch resulted in one of the greatest rocket disasters the Soviet Union would experience. As the R-16 was being readied for flight, it exploded on the pad, killing about 165 people. Among the dead were high government officials, members of the launch and design teams, and Marshal Mitrofan Nedelin, head of the Soviet strategic rocket forces. S. Averkov, an eyewitness to the nightmarish scene described the disaster,

> . . . a flash of fire erupted from the second stage engine nozzle. The powerful jet immediately ruptured the oxidizer tank. Nitric acid gushed out onto the concrete. Both the rocket and the launch structures were engulfed in a firestorm. . . . The rocket broke in half and fell on the launch pad, crushing those who were still alive. . . . Some people were devoured by fire; others, still running, were overcome by poison gases. . . .

Despite this tragedy, the quick launch capability of the R-16 was too important to the military to be ignored. They recommended that the R-16 be selected as the principle Soviet ICBM. In a final effort, Korolev countered this recommendation by suggesting that high-speed pumps be used to fuel the R-9, but this solution was rejected. Even though the R-16 was adopted, Korolev's influence in the Soviet rocket program was so great that a few R-9 rockets were ultimately deployed.

Korolev took the selection of the R-16 over the R-9 philosophically. It was perhaps easier for him to accept this defeat because his interests had now changed from using the rocket as a military weapon to a vehicle to explore space. He did not resist efforts to transfer the ballistic missile program to other facilities and viewed his "defeat" as a victory because it freed him to focus on the challenges posed by spaceflight.

Korolev Brings the Soviet Union into the Space Age

In 1948, Mikhail Tikhonravov suggested to Korolev that his group launch the world's first satellite. Captivated by this idea, Korolev sought approval for this project from his supervisory group, the Academy of Artillery Sciences. At first this idea was rebuffed because it seemed to have little connection with the design of ICBMs for the Red Army. The academy's chairman stated, "The topic is interesting. But we cannot include your report. Nobody would understand why. . . They would accuse us of getting involved in things we do not need to get involved in. . . ." However upon further reflection, the chairman changed his mind, and risking the criticism of the other committee members, he allowed design work to proceed.

After five years of work, Korolev presented a draft decree to the Central Committee of the Communist Party authorizing a satellite. However, as the document moved through channels, the section dealing with the satellite was removed. Undeterred by this setback, Korolev proposed this idea directly to Dimitri Ustinov, the Soviet Minister of Armaments. In order to make it more palatable, Korolev shrewdly emphasized that the technology developed for this effort could be used to improve the Soviet ICBMs. Phrased in this context, the project was allowed to proceed.

Ironically, it was the United States that finally created the conditions that allowed Korolev to accelerate his satellite program. On July 29, 1955, the Eisenhower administration announced that it would place a satellite in orbit as part of its contribution to the International Geophysical Year. The Soviets viewed this as a technological challenge, so on January 30, 1956, the Council of Ministers responded by issuing a decree authorizing the construction of a Soviet satellite.

Despite having all the necessary permissions, the Soviet satellite project proceeded at a frustratingly slow pace. Korolev felt the R-7 was ready, but the satellite's instrument package continued to fall behind schedule. Gyorgi Grechko, one of the engineers working on this program, later recalled that these ongoing delays forced Korolev to act,

Figure 24-14. A statue was erected in Moscow to honor Korolev, the Soviet Union's "Chief Designer."

. . . these devices were not reliable enough, so the scientists who created them asked us to delay the launch month by month. We thought that if we postponed and postponed we would be second to the U.S. in the space race, so we made the simplest satellite, called just that—*Prostreihiy Sputnik,* or 'PS.' We made it in one month, with only one reason, to be first in space.

The delayed scientific package would eventually be launched as *Sputnik 3.*

Having solved one problem, Korolev now encountered another. When he asked the State Commission for permission to use one of the R-7s, they hesitated. The first five test flights of the R-7 had failed and only one flight, its last one, had been successful. Sharp arguments arose among the committee members regarding the need for additional tests. Korolev, sensing that time was running out, and willing to accept the risk of a launch failure, threatened to take the matter to the Central Commission of the Communist Party. Faced with the possibility of being charged with delaying an approved project, the committee reluctantly allowed him to proceed.

Korolev now rushed to make the final preparations for a launch. The 174-pound *Sputnik 1* was made as simple as possible, containing only a radio transmitter and a temperature sensor. For aesthetic and engineering reasons, he decided it should have a spherical shape. Remembering his earlier problems with quality control, he personally supervised the satellite's construction, making sure no detail was overlooked.

Launch day was a proud moment for Korolev. He appreciated, certainly more than the Soviet political establishment, the historic nature of this moment. Colonel Mikhail Rebrov, who was at Baikonur, recalled the scene and Korolev's reaction to the launch,

> . . . The carrier rocket was rolled out to the launch pad in the early morning of October 2, 1957. Korolev walked in front, together with all the other chief designers. They walked in silence the entire 1.5 kilometer long way from the assembly-testing building to the pad. No one will ever know what was going through Sergei Korolev's mind at the time. Later on, when the sputnik was installed in orbit, and its call sign was heard over the globe, he said: "I've been waiting all my life for this day!"

In a glorious fashion, Korolev had brought the Soviet rocket program into the space age. He was confident *Sputnik 1* was only the beginning. Using the powerful rockets developed for the Soviet military, he believed that a manned orbital flight was within the Soviet's grasp.

25   The Soviet Launch Facilities and Vehicles

Whatever the obstacles on his path, man will make his way to the stars.

—Sergei Korolev

Following the end of the second world war, the United States expanded its rocket program. New facilities were built and existing sites were enlarged in California (Vandenberg Air Force Base), New Mexico (White Sands Missile Range), Virginia (Wallops Flight Center), and Florida (Cape Canaveral). Likewise, the Soviet Union accelerated its program by building three launch sites, one for small rockets and satellites (Kapustin Yar), one for manned flights and planetary probes (Tyuratam), and a third for military projects and the launch of polar orbiting satellites (Plesetsk).

Kapustin Yar

The first of the new Soviet facilities was built about 580 miles southeast of Moscow near the town of Kapustin Yar. The fact that this region was sparsely populated and relatively inaccessible made it an excellent site for test flights, but also made it a difficult place in which to work. Gyorgi Tyulin, head of the Soviet Technical Commission, described this region as, "... bare, lifeless steppe with dry sage brush gray from dust, camel's thorn, and sparse little islands of spurge. There was essentially no water. The hot wind chased the swirling dust and balls of tumbleweed."

Figure 25-1. In this figure the location of the Soviet Union's three rocket launch sites are shown. The directions at which rockets are launched are shown by arrows.

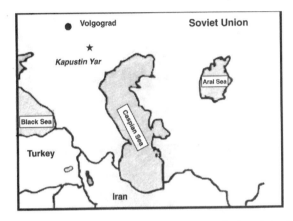

Figure 25-2. The star in the above map shows the location of the Soviet missile base, Kapustin Yar. Turkey is located in the lower left-hand corner.

Anxious to begin test flights as soon as possible, a base was bulldozed out of the barren desert. Little thought was given to the construction of living quarters, so conditions were primitive. Soldiers lived in tents and dugouts and technicians stayed in hastily constructed wooden shacks. Military officers and government officials commuted from nearby villages where accommodations were far below the standards of Moscow. Planes landed and departed on a freshly cut dirt runway that dissolved into an impassable quagmire of slippery mud during rain storms. Rockets were transported to the site by rail and were then assembled in wooden barracks located close to the launch pads. These buildings were poorly constructed and provided little protection from the region's freezing winter winds.

The first rocket launched from Kapustin Yar, a captured German V-2, lifted-off on October 18, 1947. Its flight was only a partial success. As the V-2 ascended into the sky, it veered off course and crashed about 100 miles from its intended impact site. Puzzled by this failure, the Russians asked their German specialists to determine what had gone wrong. After analyzing the flight data they made some changes to the rocket's guidance system and the next flight was successful. The Russian engineers closely studied each

V-2 flight, noting where problems existed and improvements could be made.

In a demonstration of the high priority given to the rocket program, a Soviet copy of the V-2, the R-1, was tested less than one year later on October 10, 1948. This rocket looked almost identical to the V-2, but was heavier by 12,420 pounds. Remarkably, of the nine R-1 rockets fired during these tests, seven reached their target. Russian rocket pioneer, Boris Chertok, later described the building of the R-1 as "unbelievably difficult," but he also took enormous pride in this accomplishment,

> We had to develop a lot of technologies from scratch. Furthermore where to get the materials . . . absolutely new materials . . . never produced by our industry before? . . . Just try now, in our great shops with wonderful equipment, to produce a new missile in a year. Give the personnel good rations and bonuses and you still won't be able to do it. Yet we did it.

Despite this successful start, Kapustin Yar was largely abandoned only a few years after becoming operational. American intelligence agencies had placed sensitive listening devices near the Turkish border to gather information about this Soviet effort. Since most of the rockets being tested had military applications, the listening posts presented a serious security problem to the Soviets. As work progressed on more powerful rockets, officials decided to build new launch facilities at a site beyond the range of the American listening posts.

The move from Kapustin Yar was also motivated by a practical consideration. The Earth's rotational velocity is greatest at the equator. By launching a rocket from this location, the Earth's maximum rotational velocity is imparted to the rocket's velocity, making it easier to place an object in orbit.

Tyuratam

The new Soviet site chosen to replace Kapustin Yar was a few degrees to the south and farther to the east. Located near the town of Tyuratam in the Soviet state of Kazakhstan and the Aral Sea, it was accessible using the Moscow-Tashkent rail-line and was beyond the range of the American installations. Construction work began in the mid-1950s.

In public announcements, the Soviets gave the location of this facility as close to the small mining town of Baikonur. This announcement was made to mislead western analysts, but spy satellites soon revealed that the spaceport was actually about 222 miles to the southwest (see Figures 25-3 and 25-4). The nearby city of Lenisky (now also named Baikonur) grew in size as work at the complex expanded.

By 1955 construction at Tyuratam was well underway and by 1957 its first launch pad was complete. The complex covers 60 miles in the east/west direction and 30 miles in the north/south direction. At least eighty pads have been built at this site. The early Sputnik flights blasted off from launch

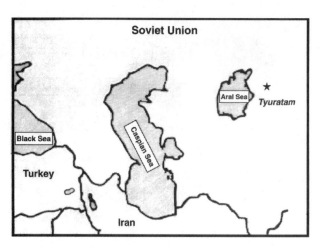

Figure 25-3. The Soviet launch facilty of Tyuratam is shown by a star in the above picture.

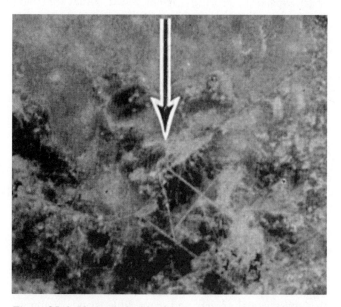

Figure 25-4. Photographs like this Landsat view revealed that the Soviet rocket base of Baikonur was actually located near the town of Tyuratam.

pads that are still employed today. Smaller rockets are generally launched from the eastern portion of the complex. Following the procedure adopted at Kapustin Yar, each rocket arrives by train in sections and is assembled on-site. It is then transported horizontally on rail cars, as shown in Figure 25-5, to a launch pad where it is lifted into a vertical position. Support towers are then placed around the rocket and technicians work on attached platforms to complete final status checks.

The Tyuratam cosmodrome continues to play an analogous role to that of the Kennedy Space Center at Cape Canaveral, Florida. All of the Soviet manned flights and planetary missions lift-off from this facility. The launch pad used by Yuri Gagarin is located about 19 miles from Leninsky. The small houses occupied by Gagarin and Korolev are also preserved as memorials and are frequently visited by foreign dignitaries and cosmonauts.

Figure 25-5. A Soviet rocket is conveyed to its launch pad in a horizontal fashion along rail lines.

Because of Tyuratam's northerly location relative to the Kennedy Space Center, almost all of its payloads are launched to the northeast, into orbits that are inclined between 50 and 72 degrees with respect to the Earth's equator. This launch pattern is shown in Figure 25-1.

Plesetsk

The third and newest Russian cosmodrome, Plesetsk, is located about 108 miles south of the Russian city of Archangel, just outside the arctic circle. The base surrounds the towns of Mirny (located on the shore of Lake Plestsy) and Kochmas. The Emtsa River passes through it. Military surveillance, meteorology, navigation, and electronic intelligence gathering satellites are launched from this site. It is also used for long-range ballistic missile tests and for the launch of polar orbiting civilian satellites. Plesetsk serves a similar

Figure 25-6. The Soviet launch complex of Plesetsk is located south of Archangel near the arctic circle. This site is used by the Soviet military to place satellites into a polar orbit.

role for Russia as Vandenberg Air Force Base in California does for the United States.

The first launch from Plesetsk (Cosmos 112) occurred on March 17, 1966. By the late 1960s more rockets were being fired from this cosmodrome than from Kapustin Yar and Tyuratam combined. Because most of the activities at Plesetsk are related to military projects, the Soviets did not even acknowledge the existence of this facility until 1983.

Rockets Used in the Soviet Manned Space Program

To meet the challenge of placing heavy satellites into orbit, the Soviets built a series of powerful rockets. Unlike their American counterparts, these rockets evolved from a single basic design. Therefore, each successive rocket looked like a larger version of its predecessor. This system had the advantage of decreasing the development time needed to produce a more powerful booster, but this process could only be carried so far. When an entirely new rocket was needed for the Moon project, the Soviets found it difficult to make the required design changes.

In the late 1950s, President Eisenhower made a policy decision to avoid using military rockets in the civilian space program. The Soviets viewed this type of thinking as foolish since it would require work to be needlessly duplicated. Their program relied on modified versions of their ICBMs. This difference in philosophy allowed the Soviets to gain an initial lead in the space race, but it also imposed a veil of secrecy surrounding their launches. The Soviets did not want outside observers watching their launches because this would reveal sensitive military information. As in America, the capability of military rockets was a closely guarded secret.

The R-7: The Sputnik Rocket

The R-7 was a descendent of the Soviet R-1 to R-5 medium-range ballistic missiles. It was built as a cold-war weapon to deliver nuclear bombs to targets in the United States. The U.S. Department of Defense refers to this rocket as the SL-1. The R-7 was propelled by a single liquid oxygen/kerosene burning engine, called the RD-107. Designed by Glushkov, this engine contains four combustion chambers and four smaller verniers that produce 102 tons of thrust. When used in the Soviet space program, the R-7 was augmented by four attached RD-108 booster engines (see Figure 25-7). Each of these engines possesses four combustion chambers and two vernier rockets powered by kerosene and liquid oxygen and produces 96 tons of thrust. In this configuration the rocket is referred to as the SL-2 in the United States. This rocket stands 95 feet high, has a diameter of 34 feet, and uses 20 combustion chambers and 12 vernier rockets.

During lift-off, the SL-2 produces nearly 500 tons of thrust. Once the boosters have exhausted their fuel supply, they are jettisoned, but the core engine continues to fire. The

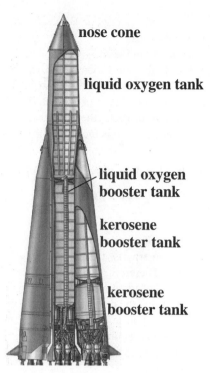

Figure 25-7. A diagram of the augmented R-7 rocket. The basic R-7 (SL-1) did not have the attached booster tanks.

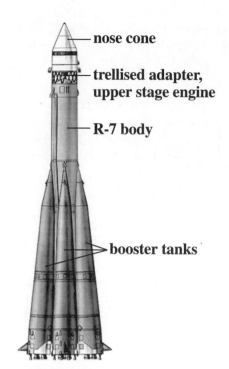

Figure 25-8. The A-1 shown here is configured for a lunar satellite launch. For Vostok launches the nose cone assembly was larger.

SL-2 is capable of placing a 3,300-pound satellite into a low Earth orbit and was used to launch *Sputnik 1, 2,* and *3.* The largest payload of these three satellites, *Sputnik 3,* had a mass of 2,920 pounds.

The A-1 (SL-3)a: The Vostok Rocket

The A-1 was developed in 1958 and consists of an SL-2 with an additional upper third stage. This stage uses a single RO-7 engine that produces 5.6 tons of thrust. The Vostok capsule containing Yuri Gagarin was placed into orbit by an A-1. In this respect, the A-1 played a similar role to the American Atlas rocket that placed *Friendship 7* and John Glenn in orbit during the Mercury program. The A-1, however, produces about twice the thrust of an Atlas. At lift-off, the A-1 stands 131 feet high, weighs 638,000 pounds, and is capable of placing an 11,000-pound object into a low Earth orbit.

The first test flights of the A-1 were conducted in late 1959 and early 1960 and quickly drew the attention of the American Central Intelligence Agency (CIA). On May 1, 1960, Lieutenant Gary Powers flew the high-altitude U-2 jet aircraft from a base in Afghanistan on a mission to photograph this rocket as it sat on its launch pad at Tyuratam. During Power's flight his U-2 was detected by Soviet radar and was damaged by a surface-to-air missile. Powers ejected from his crippled plane and parachuted to Earth, landing in a wooded region of the Ural mountains. Captured soon after landing, he was convicted by a Soviet court in an internationally televised trial and imprisoned as a spy.

The A-2: The Voskhod/Soyuz Rocket

The A-2, or SL-4, is similar to the A-1 but employs a more powerful upper stage. Its improved third stage is 26 feet long, 8 feet in diameter, and uses a RD-461 engine to produce five times the thrust of the RO-7.

The A-2 can place a 16,500-pound payload into a low Earth orbit and was employed in the Voskhod and Soyuz programs. It was also used to send Soyuz and Progress spacecraft to the Soviet Mir space station. At lift-off the weight of the A-2 is about 682,000 pounds. If an escape tower is attached to the top of the rocket, the rocket is called an A-2e. The height of an A-2e is 162 feet.

The Proton Rocket

The two-stage Proton rocket was constructed exclusively for the Soviet Union's space program. It was built by the design bureau headed by Vladimir Chelomei and was a direct competitor to Korolev's N-1 Moon rocket. Early two-stage versions of the Proton were able to place a 26,000-pound satellite into Earth orbit. A four-stage version of this rocket was able to place 66,000 pounds into orbit. The Proton was the Soviet analogue to the American Saturn IB. Unfortunately for the Soviets, the Proton did not become fully operational until after *Apollo 11* had landed astronauts on the Moon's surface.

The first stage of the Proton is powered by six RD-253 engines that use nitrogen tetroxide and unsymmetrical dimethyl hydrazine for fuel. The second stage consists of four smaller rockets that also burn this fuel. Later, a third and

Focus Box: The A Series of Soviet Rockets

The Soviet A rockets used the R-7 as a lower stage. The A-1, called the SL-1 in the west, consists of an R-7 with an attached upper stage. This three-stage rocket was used to launch *Luna 1,* the first craft to exceed the Earth's escape velocity. The A-2 (Voshkod) employed a R-7 with a more powerful upper stage. In the west this rocket is called the SL-4. In the A-2 (Soyuz), also referred to as the A-2e or Molniya, an escape stage is attached to the top of the rocket. The A-2e is frequently used to launch planetary probes.

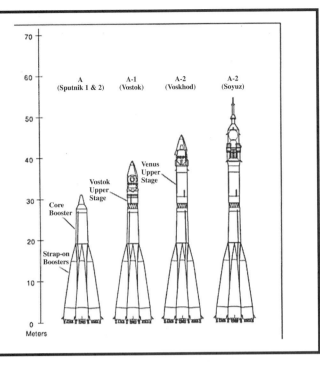

fourth stage were added to the Proton. The third stage uses a single engine and four small vernier rockets. The fourth stage is restartable, and like the Agena, is fired to propel the craft out of Earth orbit. The first successful two-stage Proton launch placed a 26,800-pound satellite into orbit on July 16, 1965. On September 14, 1968 a four-stage Proton sent a capsule carrying plants and small animals around the Moon. A week later this capsule splashed down in the Indian Ocean and was recovered.

The N-1

The N-1 was originally intended as a multiple-use rocket. Korolev thought it could be employed as a military rocket, as a launch vehicle for communication and meteorological satellites, and as a vehicle to propel probes to the nearby planets. As the race to the Moon intensified, Korolev saw the N-1 as the Soviet answer to the American Saturn V.

The selection of the N-1 as the Soviet's Moon rocket was challenged by Chelomei, a prominent figure in the Soviet cruise missile program. Chelomei argued that his rocket, the UR-500 (later to become the Proton), should be used for this task. Chelomei was a worthy opponent to Korolev. Not only was he a good rocket designer, he was politically shrewd. In an effort to gain support for his projects, he hired Khrushchev's son, Sergei, as his deputy. One reporter claimed that this gave Chelomei so much influence that, "if he had wanted the Bolshoi theatre for his enterprise he could have gotten it."

The competition between the N-1 and the UR-500 diluted the Soviet space effort at a critical time and slowed progress on these rockets. After Korolev's untimely death in

1966, the N-1 project was further hurt when the flow of materials to the UR-500 was increased. Vasili Mishin, Korolev's successor, fought this move but lacked Korolev's influence.

The first three test flights of the N-1 were conducted between February 21, 1969 and June 27, 1970, but each failed when the rocket experienced difficulties soon after launch. By this time the Americans had already landed *Apollo 11* and *12* on the Moon, and with the Moon race lost, Soviet support for the N-1 began to decline. Nevertheless, a fourth test was carried out on November 23, 1972. During this flight an N-1 carried a full-scale mock-up of a lunar orbiter and lander. Unfortunately, this attempt also ended in disaster when the N-1 exploded in a tremendous fireball after a flight of only 107 seconds. A fifth test was planned for August 1974 but by this time the political will to continue had disappeared. The N-1 was cancelled by Soviet Premier Leonid Brezhnev and Mishin was relieved of his post. Valentin Glushko was appointed as his successor.

Soviet officials were anxious to avoid the loss of prestige that would occur if the failure of the N-1 program became known. Their claim that the communist system could produce superior technological products compared to capitalism was hard to accept after the loss of the Moon race. It was therefore imperative to hide this latest failure. Sergei Kryukov, a Soviet rocket engineer later wrote, "For many years visitors to the museum in the Korolev design bureau in Kaliningrad were unable to see even a spot of evidence that there ever was an N-1 program." Glushko, Mishin's replacement, was especially careful to cover up this Soviet failure. Kryukov wrote that Glushko attacked this task with a passion and, "incinerated every notion of the N-1 with a hot iron."

26 Vostok and Voskhod: The Soviet Union's Early Manned Missions

One person alone, in a single, seated spacecraft, will never undertake interplanetary travel such as flight to the moon.

—Tikhonravov, et al., *Ten Years of Soviet Space Research*

Within a year of the triumphant launch of *Sputnik 1* on October 4, 1957, the Soviets began work on a manned space program. The goal of placing a man in space was officially authorized in 1958 and work on a single-person spacecraft commenced in 1959.

Unlike their American counterparts, Soviet officials saw little value in conducting a series of manned suborbital flights. They felt this activity would hamper their program by diverting funds and engineering time to an enterprise that would produce little useful data. The different approaches used by the United States and Soviet Union all but guaranteed that the initial Soviet lead in the space race would grow.

President Kennedy's announcement that the United States would land a man on the Moon by the end of the 1960s was viewed by the Soviet Union as a political as well as a technological challenge. Buoyed by their recent space successes, they were eager to once again beat the Americans and adopted a three-stage program to reach the Moon. The first step, named Vostok, closely paralleled Project Mercury. Its objectives were to place a single manned capsule into orbit around the Earth, to determine the effects of weightlessness on the human body, and to test the newly constructed space hardware. The next program, called Voskhod, was similar to Project Gemini. Its goals were to develop the techniques needed for rendezvous and docking, determine the effect of weightlessness on the body during longer duration flights, and to increase the crew size from one to three cosmonauts. The third stage of the Soviet program was named Soyuz and was analogous to the American Project Apollo. The Soyuz capsule would transport three cosmonauts to the Moon. Its role was similar to that of the combined Apollo command and service modules. Considering the vast differences between the Soviet and American systems, it is remarkable that the plans adopted by each country to reach the Moon were so similar.

The Original Soviet Cosmonauts

America's astronauts were selected from military test pilots to insure that sensitive data about the state of the U.S. rocket program would not be revealed. The selection of experienced pilots also guaranteed that if problems arose, people

trained to cope with tense situations would be at the controls of the spacecraft. For similar reasons, the Soviets also recruited military pilots for their program.

The Soviet selection process began in the fall of 1959, and by March 14, 1960, a total of twenty men, called cosmonauts, had been chosen. These men were I.N. Anikeyev, P.I. Belyayev, V.V. Bondorenko, V.F. Bykovsky, V.I. Filatyev, Y.A. Gagarin, V.V. Gorbatko, A.Y. Kartashov, Y.V. Khrunov, V.M. Komarov, A.A. Leonov, G.G. Nelyubov, A.G. Nikolayev, P.R. Popovich, M.A. Rafikov, G.S. Shonin, G.S. Titov, V.S. Varlamov, B.V. Volynov, and D.A. Zaikin (see the Focus Box on p. 204). Colonel-General N.O. Kamanin was placed in charge of their training and Colonel E.A. Karpov served as his deputy.

Of this group, only twelve cosmonauts would eventually fly in space. The fate of the remaining eight demonstrated the tremendous pressure they were under and the high standards demanded by Kamanin. Cosmonaut Varlamov left the corps in 1960 after breaking his vertebra, Kartashov retired after becoming ill during centrifuge training, Anikeyev was dismissed in 1961 after getting drunk and clashing with a military patrol, and Filatyev and Nelyubov were asked to leave based on their role in this incident. Bondarenko, the youngest of the cosmonauts (not included in the following picture), was killed in a fire at the end of a ten-day isolation test on March 23, 1961. Cosmonaut Rafikov left the program in 1962 due to training problems, and Zaikin resigned in 1968 when he developed an ulcer. The remaining twelve cosmonauts came to be known as the Star City 12.

The Vostok Cosmonauts

Six of these twenty cosmonauts (Gagarin, Kartashov, Nikolayev, Popovich, Titov, and Varlamov) were selected to participate in the Soviet Vostok program, an effort to place a single man in orbit. Shortly after beginning this effort, Kartashov was involved in a centrifuge accident and was replaced by Nelyubov. Nelyubov was then dismissed for his role in the episode involving Anikeyev. A talented but proud man, Nelyubov never recovered from this disgrace. On February 22, 1966 he got drunk and killed himself by jumping

Focus Box: The Original Soviet Cosmonauts and Korolev

Front row (left to right): P.R. Popovich, Gorbatko, Khrunov, Gagarin, Korolev, Korolev's wife holding Popovish's daughter, Karpov (training director), Nikitin (parachute instructor), Fedorov (doctor).
Middle row (left to right): Leonov, Nikolayev, Rafikov, Zaikin, Volynov, Titov, Nelyubov, Bykovsky, Shonin. Back row (left to right): Filatyev, Anikeyev, and Belyayev.

under a moving train. An additional change occurred when Varlamov fractured a vertebra during a swimming pool accident and was replaced on the Vostok team by Bykovsky.

The Vostok Capsule

As shown on the next page, the Vostok capsule consisted of two main parts: an upper reentry module and a lower instrument/retrorocket module. The weight of the craft was 4.7 tons, and it had a length of 14.4 feet and maximum diameter of 7.9 feet. The craft was about the size of a family van.

The reentry module housed the cosmonaut, and to simplify its engineering Korolev decided to give it a spherical shape. The module had a diameter of 7.5 feet and a mass of 2.5 tons. By carefully positioning heavy objects along its walls, the module would automatically assume the correct orientation during reentry. This provided an additional measure of safety that was absent in the design of the American Mercury capsule. The cosmonaut was not expected to remain in this compartment all the way from space to the Earth's surface. At a height of around 4.2 miles the pilot ejected and descended to the ground by parachute.

The instrument/retrorocket module of Vostok was 7.5 feet long, had a maximum diameter of 7.9 feet, and weighed 2.3 tons. This part of the craft was not recovered from space. It was jettisoned just before the cosmonaut began his reentry and subsequently destroyed by frictional heating in the Earth's atmosphere.

Although the Soviets did not employ suborbital missions to test their equipment, several flights were conducted to insure that the Vostok functioned properly before manned flights were undertaken. Many of these flights, called Korabl-Sputniks, carried live dogs and other small animals. The launch of Laika aboard *Sputnik 2* was a precursor to these

Figure 26-1. The left-hand part of the diagram shows a Vostok launch. The right side shows how the capsule and cosmonaut descend separately to earth.

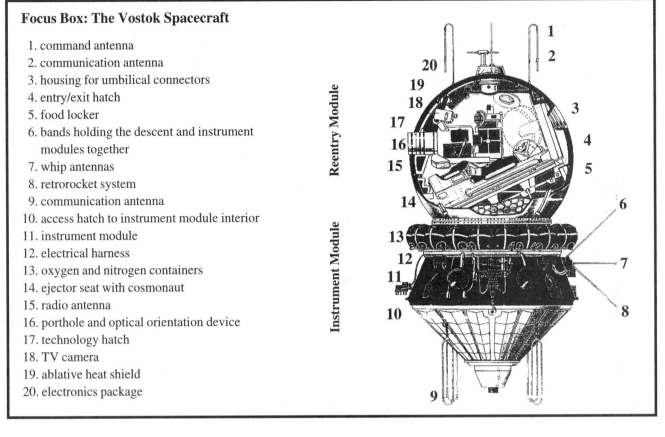

missions. In the last two flights of this series, *Korabl-Sputniks 4B* and *5*, dogs were also used to test the capsule's ejection system. By March 1961 the Soviet hardware was deemed reliable enough to proceed with a manned flight.

Of the six Vostok cosmonauts, Gagarin and Titov were considered to be the most likely candidates to fly this mission. On April 3, 1961, the Soviet government gave its official go-ahead for a manned flight. Seven days later General Kamanin privately informed the cosmonauts that Gagarin would fly this mission and Titov would serve as his backup. This first manned Vostok flight was discussed in a previous chapter, so only a brief review of it will be presented here.

The Flight of *Vostok 1*: The First Man in Space

After a short delay caused by a problem with the capsule's main hatch cover, the historic flight of *Vostok 1* blasted-off from the Soviet Space Center at Tyuratam at 06:07 GMT on April 12, 1961. Little was known about how a man would react in space, so Gagarin was given relatively few tasks. He was not even allowed to pilot the craft. Under the orders of General Kamanin, the Vostok's controls were locked prior to launch. Only in the event that ground communications with the capsule were lost or an emergency arose would Gagarin be allowed to fly the craft.

The capsule worked as expected and Gagarin conducted a short experiment to determine if a person could eat and drink in space. Live television pictures from within the capsule were sent to Soviet ground stations and then transmitted to a worldwide audience. Gagarin sent greetings as the Vostok flew over friendly communist countries. After completing one orbit, the reentry system was automatically fired and the module began its descent.

As a planned part of the landing, Gagarin was ejected from the orbital module and parachuted to Earth. International rules at that time required that the pilot return with his craft in order to establish a new aviation record. Aware of this

Figure 26-2. Yuri Gagarin and Sergei Korolev not only held very responsible positions within the Soviet space program, they were also good friends.

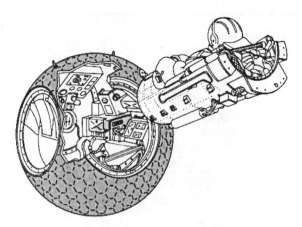

Figure 26-3. A cosmonaut was ejected from the Vostok reentry module as shown above. Once free from the supporting apparatus, the cosmonaut parachuted to Earth.

Figure 26-4. Gherman Titov relaxed while awaiting medical tests at a facility in Moscow as part of the cosmonaut selection process.

rule, the Soviets were very evasive in describing Gagarin's descent. It was not until 1978 that they openly admitted he had not landed in his capsule.

The successful 108-minute flight of *Vostok 1* made space history. Gagarin became an instant celebrity and international respect for the technological sophistication of the Soviet Union, already enhanced by the Sputnik launches, skyrocketed. This latest Soviet triumph provided another shock to the still struggling American space program. The United States responded by launching astronauts Alan Shepard's and Virgil Grissom's suborbital flights. Shepard's flight was successful, but at the end of Grissom's flight the hatch of *Liberty Bell 7* unexpectedly blew open and the capsule sank. Grissom's helmet was later recovered from the ocean as a 10-foot-long shark circled the area.

Vostok 2: The First Daylong Mission

Vostok 2 was originally planned as a three-orbit flight, but in order to capture yet another spectacular space first, Khrushchev directed that it be extended to a full day. This flight would be flown by Gherman Titov with Andrian Nikolayev serving as his backup. The method the Soviets used for crew selections was identical to that employed in the United States, backup pilots flew the next scheduled flight. As preparation for the lift-off of *Vostok 2* moved forward, the spirits of both cosmonauts were high.

Vostok 2 was launched at 9 a.m. on August 6, 1961 and flawlessly entered a 110 × 146 mile orbit. Once in space, Titov tested the Vostok's attitude control systems, ate, and completed some physical exercises. When asked for a status report by ground controllers he replied, "I feel splendid. Just splendid . . . Everything is going fine. Everything is ship-shape. . . . The Earth is very beautiful. . . . Right now I'm looking at a mountainous region. . . . The view is splendid! The peaks are in the clouds."

During his sixth orbit, Titov began to experience the effects of space sickness and became nauseous. After sleeping for a little more than eight hours, he awoke refreshed and

without any signs of his former illness. He then prepared the capsule for reentry. On the 17th orbit the retrorockets fired and the capsule began its descent. He later described this part of his flight in his book, *I am Eagle*,

> *Vostok 2* fell like a stone through the air, stabilized in an arrow-like drop toward the Earth. I activated several switches and controls, gripped the contour seat carefully, and glued my eyes on the chronometer. The sweep hand came around, moving steadily closer to the moment when explosive charges would go off, and I would be hurled with the seat away from the falling spaceship. Ten—then only three more seconds to go. A red light flashed on the indicator panel. . . . Thunder crashed into my ears at the same time I felt a tremendous force beneath me. The ejection shell exploded exactly on schedule, and in a blur I saw daylight flash before my eyes as the entire seat burst away from the spaceship.

Shortly after parachuting to Earth near the village of Krasny Kut, Titov was greeted by cosmonaut Nikolayev who asked, "Well, how was it up there?" With a big smile Titov replied, "Very interesting!" Following a debriefing and a parade in Moscow, he proudly presented his flight logbook to Korolev at a reception in the rocket assembly plant.

Despite the successful conclusion of this mission, Soviet doctors were alarmed by Titov's space sickness. Gagarin had not become similarly ill during his relatively short flight. Would this illness incapacitate cosmonauts on long-duration flights? Unable to answer this question, Titov was permanently forbidden from making another flight. Vladimir Yazdovsky, a medical expert assigned to this flight, sadly noted that Titov's reaction to weightlessness would require much more study. In response to a question about the physiological problems posed by this flight a doctor replied, "they will compel us to make substantial changes in the general training program and figure out how best to condition the vestibular apparatus." Concerns about space sickness caused a delay in subsequent Vostok flights. More than a year would pass before the next mission lifted-off from Tyuratam.

Vostok 3 and *4*: A Surprise Dual Flight

On February 20, 1962 the United States finally matched the flight of *Vostok 1* when astronaut John Glenn orbited the Earth aboard *Friendship 7*. During his 5-hour, 3-orbit mission, Glenn tested the Mercury capsule's maneuvering system and conducted a series of experiments. After the completion of this flight, a duplicate mission was flown by astronaut Carpenter. The United States finally was becoming a real competitor in the space race.

On August 11, 1962, about 10 weeks after Carpenter's Mercury flight, Andrian Nikolayev was launched aboard *Vostok 3*. Soon after reaching orbit he released the straps holding him to his couch and floated in the capsule. He did not experience the space sickness that had afflicted Titov and enjoyed the sensation, he later reported,

> It was an amazingly pleasant state of both body and soul, not to be compared to anything else. You don't weigh anything, you aren't supported by anything, and yet you can do everything. Your mind is clear, your thoughts precise. All your movements are coordinated. Both vision and hearing are perfect. You see everything and hear everything transmitted from the ground.

Western analysts expected Nikolayev's flight would be longer, but otherwise similar to that flown by Titov. They were completely caught off guard when *Vostok 4*, carrying Pavel Popovich, was launched one day later. The Soviets announced that this joint mission would gather data needed for a future rendezvous in space.

Vostok 3 and *4* were placed into such precise orbits by their booster rockets that the cosmonauts came within 4 miles of each other and Nikolayev was able to see sunlight gleaming off *Vostok 4*. Since neither craft was equipped with maneuvering engines, they gradually drifted apart. Nevertheless, the fact that the Soviets were able to launch two craft with such accuracy struck many observers as amazing. Sir Bernard Lovell, a noted British radio astronomer, described the flight as, "the most remarkable development man has ever seen." President Kennedy was reported to have been "severely jolted" by this news.

As if to show off their new dexterity in space, the two Vostoks reentered the Earth's atmosphere at nearly the same time and landed within minutes of each other near the city of Karaganda. Soviet doctors were relieved that neither Nikolayev nor Popovich was seriously affected by space sickness. They had been in space for 94 and 71 hours, respectively, compared to Titov's 25 hours. Nikolayev seemed completely unaffected by weightlessness and Popovich experienced only a brief period of disorientation. Soviet doctors correctly deduced that space sickness was not an inherent consequence of space flight, but was dependent on the individual cosmonaut.

The First Woman in Space: *Vostok 5* and *6*

The last Vostok mission would provide the Soviets with another space first. On June 14 and 16, 1963, *Vostok 5* and *6* were launched into similar orbits in what appeared to be a repeat of their previous dual expedition. *Vostok 5* was manned by Valeri Bykovsky, but *Vostok 6* carried the first woman into space.

Valentina Tereshkova seemed unlikely to make history as the first female in space. She was born in the ancient Russian town of Taroalavl, located on the Volga River, and both of her parents were peasants. Her father was killed by the Germans during World War II. Prior to joining the cosmonaut corps she worked as a loom operator in a textile mill.

Tereshkova's primary qualifications for selection as a cosmonaut seems to have been her parachuting experience and her fearless nature. A member of a flying club, by the time she applied to the space program she had made over 120 jumps. Not all of these were totally successful. On one occasion she landed in the Volga River and nearly drowned. Over 400 women applied for admission into the program, but only four others were selected, Tanya Kuznetsova, Valantina Ponomareva, Irina Solovyeva, and Zanna Yorkina.

The idea of placing a woman in space came directly from Khrushchev who saw it as an opportunity to make a political statement about the equality of women under the Soviet system. He selected Tereshkova to fly *Vostok 6* from the list of female cosmonauts and was influenced by the fact that she was a member of the working class. Yuri Gagarin closely monitored the progress of the female cosmonauts and agreed with Khrushchev's choice. When asked to describe Tereshkova's ability by a reporter, Gagarin replied, "She was born for space."

Vostok 5 and *6* were placed in similar orbits and came within three miles of each other. The two cosmonauts engaged in several inter-ship discussions and both seemed at

Figure 26-5. Andrian Nikolayev (left) piloted Vostok 3. *Pavel Popovich piloted* Vostok 4.

Figure 26-6. Tanya Kuznetsova (right) and fellow cosmonaut Valentina Tereshkova (left) undergoing training for the Vostok program.

Figure 26-7. As the pilot of Vostok 6, Tereshkova would be the first woman to fly in space.

ease. Tereshkova reportedly slept so soundly that controllers had difficulty waking her. With all going well, her flight was extended beyond its originally scheduled 17 orbits. By June 19 she had been in orbit for three days, longer than all of the American astronauts combined.

The *Vostok 5* and *6* missions had an experimental component as well as a political one, to observe the Earth and stars from space, but these objectives were completely overshadowed by Tereshkova's participation in the mission. After completing 48 orbits she landed 375 miles northeast of Karaganda. Bykovsky landed a few hours later, 324 miles northwest of Karaganda. His 5-day, 81-orbit flight established a new space endurance record.

Both cosmonauts were treated to a reception in Moscow. About four months later, on November 3, 1963, Tereshkova married fellow cosmonaut Andrian Nikolayev. Gagarin and Bykovsky served as witnesses and Khrushchev was the toastmaster. The couple had a daughter named Yelena.

With no other space flights planned for her, Tereshkova resigned from the cosmonaut corps to raise her daughter. In 1967 she was elected to the Supreme Soviet and in 1974 she was elected president of this body. Not surprisingly, she also became a prominent figure in women's groups and peace organizations and traveled extensively on behalf of the Soviet government. Sadly Tereshkova and Nikolayev divorced in 1980 because of marital problems.

Despite Khrushchev's claim that men and women possessed equal status in the Soviet system, none of the other female cosmonauts were assigned to a mission. With some bitterness, they left the space program in 1969 when it became clear that their chances of flying were minimal. This was unfortunate because they were exceptional individuals. Kuznetsova went on to train crews for geophysical expeditions, Pononareva wrote a book on space rendezvous procedures and then went to work for the Soviet Academy of Sciences. Solovieva took part in Antarctic expeditions, and Yorkina wrote a book on the history of cosmonautics.

The Voskhod: The First Multiple Crew Spacecraft

The Soviet manned spaceflight program was now ready to move on to the multimanned Soyuz craft. However, the construction of this craft was behind schedule, so Korolev decided to use this extra time to launch a series of flights to bridge the

---

Focus Box: The Vostok Cosmonauts

Flight Record

Vostok 1	Gagarin	Apr. 12, 1961
Vostok 2	Titov	Aug. 6, 1961
Vostok 3	Nikolayev	Aug. 11, 1962
Vostok 4	Popovich	Aug. 12, 1962
Vostok 5	Bykovsky	June 14, 1963
Vostok 6	Tereshkova	June 16, 1963

Pictured from left to right are Titov, Gagarin, Tereshkova, Bykovsky, Nikolayev, and Popovich.

gap between these two programs. The craft employed for these flights was a modified Vostok called the Voskhod.

Although the Voskhod was outwardly similar to the Vostok, there were four major differences. The Voskhod could (1) carry a crew of up to three cosmonauts, (2) possessed a backup retrorocket for possible use during reentry, (3) carried an extendable airlock that allowed a cosmonaut to conduct a space walk, and (4) used solid propellant rockets attached to its parachute system to slow the capsule's descent speed prior to touchdown. These differences are discussed in more detail below.

In the Vostok program the spaceship's orbit was low enough that the craft would naturally return to Earth within eight days. Even if the Vostok's retrorockets failed to operate, the cosmonauts could be safely recovered. However, the planned orbits for the Voskhod were so high that the cosmonauts' air and food supplies would be exhausted long before their craft would reenter the atmosphere. To prevent a potential disaster, Korolev insisted the Voskhod possess a backup retrorocket system that could be used if the primary system failed.

The second major difference between the Vostok and Voskhod programs dealt with the way in which the cosmonauts landed. In the Vostok program, the cosmonaut ejected from the capsule and parachuted to Earth. Since the Voskhod carried more than one cosmonaut this procedure was impractical. The Voskhod crew would land with the craft. To prevent injury, landing rockets were used, as shown in Figure 26-8.

A third difference was related to the small crew space available. The Voskhod had the same exterior dimensions as the Vostok. In order to accommodate three cosmonauts, the bulky spacesuits they wore were eliminated. The Soviet explanation for this dangerous procedure was that the craft was so reliable there was no need for the spacesuits. In reality, if a three-person crew were to fit inside this craft, there simply wasn't room for these suits.

The Voskhod flights had four objectives: (1) to test the multimanned spacecraft and measure the efficiencies that a larger crew made possible, (2) to develop the tools needed to perform a successful space walk, (3) to gather more data about weightlessness, and (4) to develop the procedures needed to dock multiple spacecraft together in space.

The launch of the first multimanned Voskhod was preceded by a single unmanned test flight (*Cosmos 47*) on October 6, 1964. After a flawless 16-orbit mission, the capsule returned to Earth as planned and preparations progressed quickly for a manned launch. This mission would be flown by cosmonauts Vladimir Komarov, Boris Yegorov, and Konstantin Feoktistov.

Komarov was the only member of this crew who belonged to the original group of twenty cosmonauts. Like the American astronaut Deke Slayton, doctors had earlier grounded him when it was found his heart possessed an irregular beat. After gathering statements from Moscow's leading heart specialists that this would not cause a problem in space, he was recertified for flight in 1963. Yegorov was a short,

Figure 26-8. During the landing of a Voskhod, rockets attached to the parachute fired just before the capsule touched down. This allowed the cosmonauts to escape injury.

Figure 26-9. The crew of Voskhod 1. From left to right, Feoktistov, Komarov, and Yegorov. There was insufficient room in the Voskhod capsule for them to wear spacesuits, so they wore ordinary clothing.

dark-haired doctor whose speciality was space medicine. He would be the first doctor to fly in space. Feoktestov was a trained engineer and had the most interesting background of any member of the cosmonaut corps. While serving as a scout during the second world war, he had been captured by the Germans, shot in the head, and left for dead. Remarkably he recovered from this wound and went on to earn an engineering degree from the Bauman Technical College in Moscow. Assigned to Korolev's engineering group, he was largely responsible for the redesigned Voskhod.

On October 12, 1964, five days after the return of *Cosmos 47*, the three cosmonauts walked to their awaiting R-7 rocket; their silver-gray and blue jackets standing in sharp contrast to the unearthly spacesuits worn by the American astronauts (see Figure 26-9). After entering the Voskhod capsule and attaching their headphones, they blasted off into a 107 × 245 mile orbit.

Each of the cosmonauts was assigned a unique set of duties. Komarov was to pilot the craft and test its attitude control system. Yegorov was to collect data about weightlessness and space sickness so doctors could develop exercises to lessen the impact of these illnesses on future cosmonauts. Feoktistov was to photograph the polar aurora and stars through the capsule's window and make detailed observations of the Earth's horizon to determine if it could be used for navigational purposes.

Worldwide reaction to this flight was immediate and positive. While overseas headlines were praising this latest Soviet accomplishment, Americans experienced another attack of self-doubt about their ability to catch the Soviets. *The New York Times* reported that there was "an air of resignation among U.S. officials." Astronaut Malcolm Scott Carpenter, pilot of Project Mercury's *Aurora 7*, expressed the view of many when he said, "I wouldn't have been surprised if two had been sent up. But three? The Russians seem to do this, just what you don't expect them to!"

After completing 16 orbits, the crew of *Voskhod 1* returned safely to Earth, landing near the Soviet city of Kustanai. Komarov was rumored to have argued for an extended flight but was denied permission. Instead Korolev simply radioed him a phrase from Hamlet, "there are more things in Heaven and Earth, Horatio, than are dreamt of in your philosophy." During their flight the Khrushchev regime had been overthrown. Leonid Brezhnev, Alexei Kosygin, and Nikolai Podgorny had seized control of the government and Khrushchev was "retired" to his home on the Black Sea. The first public appearance of the new leadership occurred a few days after the return of *Voskhod 1* when the cosmonauts were welcomed back in a Moscow ceremony.

The First Spacewalk: *Voskhod 2*

Like the prior Voskhod flight, *Voskhod 2* was preceded by an unmanned test flight (*Cosmos 57*). Shortly after this craft achieving orbit on February 23, 1965, observers were astonished to see *Cosmos 57* apparently self-destruct, break-

Figure 26-10. The crew of Voshkod 2 *on their way to the launch pad. Cosmonaut Belyayev is to the left and Leonov seems to be enjoying a good laugh.*

ing into at least 150 separate pieces. The mystery deepened when the Soviets did not offer any explanations for this failure. Several days later a second failure occurred when a Voskhod capsule was severely damaged during a drop test. Despite these setbacks, Korolev felt confident enough to recommend that *Voskhod 2* proceed.

On March 18, 1965, five days before the first manned Gemini flight, *Voskhod 2* blasted off carrying cosmonauts Pavel Belyayev and Alexei Leonov, both members of the original cosmonaut class. Belyayev had served as a pilot in the Soviet Air Force and had seen action against the Japanese during World War II. He felt fortunate to be on this mission. During a routine training exercise he had sustained an ankle injury in a parachute jump and was temporarily grounded. After relentless exercises and the strong personal recommendation of Gagarin, his flight status was restored just in time for the *Voskhod 2* mission.

Leonov was powerfully built and had a cheerful disposition, but he also possessed a fierce determination. One night his car went off the road and fell through the ice of a pond. Rather than panicking in this dire situation, he pulled himself free and then rescued his wife and driver. In a second incident, the ejector seat of his parachute escape system failed to detach during a practice jump. Held tightly by the chair's constraints, Leonov plummeted toward the ground and certain death. By bending the chair's, metal restrainers with his bare hands, he was finally able to free himself and land safely.

Leonov was not only an athletic man but he was an accomplished artist as well. He particularly enjoyed painting space scenes and futuristic missions to alien worlds. Along with Gagarin, Leonov would become an eloquent spokesperson for the Soviet program.

A little over one and a half hours after lift-off, Leonov left the Voskhod cabin and entered the attached airlock. The airlock was added to the Voskhod because the equipment inside the crew compartment was not designed to operate in a

Figure 26-11. Next to Yuri Gagarin, Alexei Leonov was the Soviet Union's best known cosmonaut.

While I was floating in space, over the Kuban steppes, I was in radio contact with Yuri Gagarin. He asked me anxiously how I was feeling. I replied that I could see an enormous amount, but that it was difficult for me to describe it all. . . . A fantastic sight! The stars appeared to be motionless. The sun

Figure 26-12. Leonov completed the world's first spacewalk during the Voskhod 2 mission. A trained observer, he nonetheless found it difficult to describe the beauty of the Earth from space.

vacuum. Once the airlock was depressurized, he opened the outer hatch and conducted mankind's first spacewalk. While floating in space he was attached to the capsule by a 16-foot umbilical cord that also contained a communication link to the capsule. His oxygen supply was carried in his backpack. Leonov floated in space for over 12 minutes. Recalling the wonder of this experience he later said,

Focus Box: Leonov's Spacewalk

During lift-off the two cosmonauts sat side by side with Leonov closest to the extendable airlock. Once in orbit the airlock was pressurized (1) and Leonov put on his spacesuit (2). He then entered the airlock (3), closed the interior hatch and depressurized it (4). Leonov then placed a movie camera on the outside of the airlock (5) and conducted his spacewalk (6). Once his spacewalk was complete, he retrieved the camera (7), reentered and pressurized the airlock (8). He then opened the hatch to the Voskhod (9) and rejoined Belyayev (10). Leonov's spacewalk lasted 12 minutes and occurred on March 18, 1965. It would be repeated by the American astronaut Edward White during the *Gemini 4* mission on June 7, 1965.

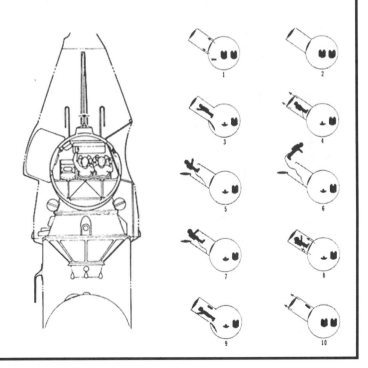

Table 26-1. Voskhod Flights

Mission	Launch Date	Orbit	Result
Cosmos 47	Oct. 6, 1964	16	Unmanned test
Voskhod 1	Oct. 12, 1964	16	First three-man crew
Cosmos 57	Feb. 22, 1965	—	Explodes in space
Voskhod 2	Mar. 18, 1965	17	First spacewalk
Cosmos 110	Feb .22, 1966	330	Test flight using two dogs

seemed as if sewn onto black velvet. We ourselves [the capsule] seemed not to be moving at all. I only felt a slight movement in the spaceship at the moment when I got out. When I pushed myself off, it moved for an instant in the opposite direction.

While in the vacuum of space, Leonov's spacesuit expanded slightly due to the lack of external pressure. At the end of his spacewalk, he placed his hand-held camera in the airlock and it lodged inside. With considerable difficulty he finally managed to squeeze back inside the airlock, repressurize the unit, and rejoin Belyoyev.

With the spacewalk completed, the cosmonauts obtained pictures through a porthole and conducted an experiment on the ability of the eye to determine colors while in space. With these activities completed, they settled down for a well-deserved rest. A tense moment occurred after the cosmonauts awoke and attempted to return to Earth. Much to their chagrin, the automatic orientation system failed to align the craft for reentry. Belyayev later recalled, "You can imagine that we were really quite excited when we saw that the sun-orientation system had failed. . . . Thirty seconds passed before a decision was taken . . . we received permission to make a manual landing in the eighteenth orbit."

On the next orbit Belyayev fired the backup solid propellant rockets, the first time this option had been tried in space, and the craft descended to Earth. Although the backup system worked well, the Voskhod landed off target in a densely wooded, snow-covered region of the Ural mountains, about 108 miles northeast of the city of Perm. The area was so remote recovery teams could not immediately reach the cosmonauts so food and clothing were dropped to them by air. Belyayev and Leonov spent the rest of that day and most of the next waiting for the recovery team to arrive. Finally, they were rescued and skied 12 miles to a firebreak where a helicopter was waiting to fly them back to Tyuratam.

Voskhod 2 would be the last Voskhod flight. Despite the fact that only two missions had been completed, the Soviets achieved two important space firsts: the launch of the first multimanned spacecraft and the first spacewalk. The near disastrous reentry problem during *Voskhod 2*, coupled with ongoing development delays with the Soyuz spacecraft, delayed further manned spaceflights for another two years.

The End of the Moon Race: A Deadly Dash to the Finish

Maybe for the same expense automatic stations might have done more. But the fact of a man being on the moon is great in itself and this represents the self-assertion of mankind.

—Georgi Petrov, director of the Soviet Institute of Space Research

Five months after the end of the Gemini program and two years after the launch of *Voskhod 2*, the Soviet Union resumed its manned space flight program using a new capsule called the Soyuz (meaning Union). This craft, designed by Sergei Korolev and his assistant, Leonid Voskresensky, was highly maneuverable, yet large enough to accommodate a crew of three cosmonauts. After being tested in Earth orbit, it was expected to carry cosmonauts to the Moon. In this respect the Soyuz was intended to fulfill the same role as the Apollo command module.

The Soyuz Spacecraft

The Soyuz spacecraft had an overall length of 23 feet, a diameter of 7.5 feet, two solar panels with a total collection area of nearly 150 square feet, and a launch weight of about 6.8 tons. As shown in Figure 27-1, the craft consisted of three principle compartments, the orbital module (A), the reentry module (B), and the instrument module (C).

The 7.5-foot diameter orbital module was located at the front of the craft and had a spherical shape. It served as the cosmonauts work, rest, and sleep area and housed the controls for the craft's communication systems. Cosmonauts could view the Earth and photograph celestial objects through four portholes located around its perimeter. A hatch was used to exit the craft and conduct spacewalks. Depending on the mission objectives, a docking mechanism (as seen in the Focus Box on p. 214) could be attached to the front of the module.

The Soyuz reentry module was located behind the orbital module and could be occupied by up to three cosmonauts. This dome-shaped unit was 7.2 feet long and contained the craft's orientation and maneuvering controls, an atmospheric regeneration system, the main and backup reentry parachutes, and TV cameras. A periscope was attached to its outer wall to aid the cosmonaut during docking maneuvers.

The third Soyuz module, the instrument compartment, had a cylindrical shape and was about 7.5 feet long. It contained the craft's electrical supply system, maneuvering and main engines, fuel, and long-range radio equipment. Two large solar panels were attached to its exterior wall to supply the craft with electrical power.

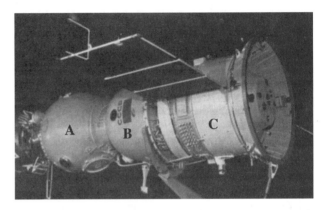

Figure 27-1. The Soyuz spacecraft with the three components discussed in the text labeled A, B, and C.

Flight Tests

Three test flights of the Soyuz were conducted prior to its first manned launch. The first test, named *Cosmos 133*, occurred on November 28, 1966. This capsule improperly reentered the atmosphere, and rather than land in China, the craft's self-destruct mechanism was triggered. On December 14, a second Soyuz test ended in a launch pad explosion. A third test was conducted on February 7, 1967, but as the Soyuz reentered the atmosphere a maintenance plug burned through the hull and the interior of the capsule was damaged.

Despite these setbacks, the Soviet government was determined to launch a manned Soyuz flight to commemorate the fiftieth anniversary of the 1917 Revolution that led to the birth of the communist system. As final preparations were made for this mission there was an air of foreboding among both the engineers and cosmonauts at Tyuratam. Based on its poor test record, they feared too many problems remained unsolved to risk a manned flight. It is probable that if Korolev had still been alive, he would have used his considerable power to cancel this flight. Rumors circulated that Vasily Mishin, his successor, had tried but failed to have this flight postponed. Despite all of the Soyuz mishaps, the launch was ordered to go forward.

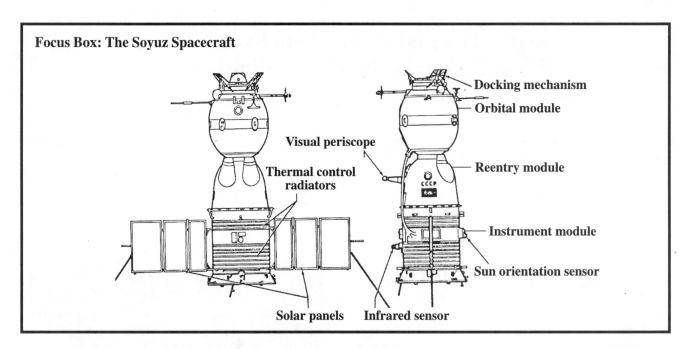

The Tragic End of *Soyuz 1*

On April 23, 1967, veteran cosmonaut Vladimir Komarov was driven in a transport van from his quarters at Tyuratam to the Soyuz launch pad. In contrast to the exuberant mood of prior flights, Komarov appeared dejected and projected an air of morbid resignation. The cosmonauts riding with him tried to lift his spirits by joking and singing songs, and by the time they arrived at the launch complex Komarov's mood seemed to improve.

As a veteran of two prior space flights, Komarov knew there was good reason to be concerned. His mission would be the most complex attempted by either the United States or So-

viet Union and the poor record of the Soyuz weighed heavily on his mind. When his Soyuz had arrived at Tyuratam, pre-launch checks found a record 203 defects. How many other problems, he wondered, had been missed? Extremely apprehensive about his safety, Komarov prophetically told a friend of Gagarin's, "I'm not going to make it back from this flight."

Komarov's scheduled 72-hour flight was packed with activities. He was to test the Soyuz systems, perform scientific and technical experiments, investigate the medical and biological effects of weightlessness, rendezvous with another Soyuz craft to be launched while he was in orbit, conduct a crew transfer, and then land at a precisely determined site within the Soviet Union. Many believed this list of objectives was unrealistic.

Despite these misgivings, Komarov's mission started off well. No problems occurred during the launch. Unfortunately, soon after reaching orbit difficulties began to occur. One of the craft's two solar panels failed to deploy, leaving the instruments with an inadequate supply of electricity. In particular, the capsule's automatic orientation system lacked the necessary power to operate. The Soyuz maneuvering system then malfunctioned, as did the craft's thermal controls. Komarov found it was difficult even to maintain radio contact with ground controllers.

In view of these mounting problems, officials decided not to proceed with the second Soyuz launch and to terminate Ko-

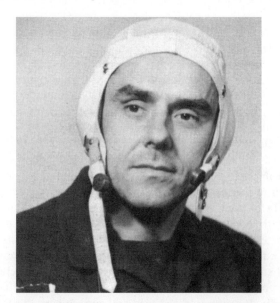

Figure 27-2. Vladimir Komarov (1927–1967) was the pilot of the ill-fated Soyuz 1 *flight. He was the first person to die during a space mission when his craft crashed on April 24, 1967.*

Table 27-1. Early Soyuz Flights		
Mission	**Launch Date**	**Result**
Cosmos 133	Nov. 28, 1966	Failure
Cosmos 140	Feb. 7, 1967	Failure
Soyuz 1	Apr. 23, 1967	Cosmonaut dies

marov's flight as soon as possible. On the 19th orbit he was instructed to fire the retro-rockets manually and reenter the atmosphere. A final devastating malfunction occurred when the capsule's main parachute failed to properly deploy and the backup chute became tangled. With neither of these parachutes fully inflated, the Soyuz crashed violently into the ground.

When the recovery helicopter arrived at the scene of the crash, the only part of the reentry module that could be identified was the rim of the hatch. The heat of reentry had been so great the Soyuz's metal hull was still burning. The horrified rescuers doused the fire by throwing dirt onto the crumbled vehicle. Gagarin, who had accompanied the team, helped in the gruesome task of removing his friend's broken body from the wreckage.

Komarov's death sent shock waves through the Soviet space program, but public announcement of this disaster was delayed for a full day. No mention was made of the haste with which the mission was launched nor of the prior Soyuz failures. Members of the space community were bitterly aware of the circumstances leading to this tragedy. Victor Yevsikov, a Soyuz engineer, later lamented,

> Some launches were made almost exclusively for propaganda purposes. . . . The management of the OKB-1 Design Bureau knew that the Soyuz had not been completely debugged, and more time was needed to make it operational, but the Communist Party ordered the launch, despite the fact that four preliminary unmanned tests had revealed faults. . . . The flight took place despite Vasily Mishin's refusal to sign the endorsement papers for the Soyuz reentry vehicle, which he considered unready.

Figure 27-3. Komarov's widow lays flowers at his grave site in the Kremlin wall.

Two days after the crash a somber cosmonaut corps looked on as Komarov's ashes were interred in a place of honor in the Kremlin wall. After thirty-five astronauts and cosmonauts had completed twenty-two missions consisting of 2,500 hours in orbit, spaceflight had claimed its first victim. In a newspaper article appearing shortly after the funeral, Gagarin wrote of the determination of the cosmonauts to continue their work despite this tragic loss, "Nothing will stop us. The road to the stars is steep and dangerous. But we're not afraid. Every one of us cosmonauts is ready to carry on the work of Vladimir Komarov—our good friend. . . ."

Despite Gagarin's call to proceed, further Soyuz launches were postponed until the multiple problems that plagued Komarov's flight could be studied and corrected. Only three months earlier, America had experienced a similar trauma when its three *Apollo 1* astronauts were killed during a ground test. In an eerie way, the admonition given to a Boston audience just months before these two tragedies by Dr. Eugene Konecci of the National Aeronautical and Space Council was fulfilled. He noted, "We intuitively know that some day, maybe in the not too distant future, we or the Soviets will suffer a manned space catastrophe." The sadness and shock of these deaths was felt around the world.

The Soviets valued the lives of their cosmonauts as much as America valued the lives of their astronauts. When the Apollo astronauts were killed, the American program paused for a long self-evaluation. William Shelton in *Soviet Space Exploration: The First Decade* wrote of the effect that Komarov's death had on the Soviet program, ". . . it ended whatever immediate hopes the Soviets had for manned rendezvous, docking, and crew exchange, and it stopped the Soviet manned spaceflight program in its tracks. . . ." A year and a half passed before the Soviet Union launched its next manned flight.

The Proton and Zond Flights

While preparations were progressing for the flight of *Soyuz 1*, a parallel program involving the Soviet's Moon rocket and capsule was nearing the point where flight tests were needed. The rocket and capsule were called the Proton and Zond, respectively. As seen in Figure 27-4, the Zond looked like a Soyuz without an orbital module. However, important differences existed. The Zond could accommodate two cosmonauts, possessed more thrusters than the Soyuz, had an umbrella-shaped long-distance radio antenna, and a thicker heat shield. Its launch weight was a little more than 11,000 pounds.

Two unmanned tests of the Proton/Zond combination were conducted on March 10 and April 8, 1967. The Soviets referred to these flights as *Cosmos 146* and *154*. *Cosmos 146* was intended to test the Zond during a simulated return from the Moon. Although the fourth stage of the Proton failed to lift the capsule as high as planned, the Zond still reentered the atmosphere with a lunar return velocity of 6.7 miles/second, making this flight a qualified success. Few details were pro-

The Zond

The Soyuz

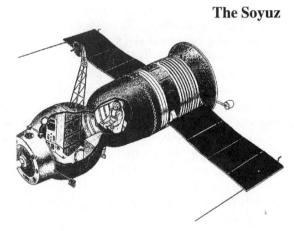

Figure 27-4. The Zond spacecraft was a modified version of the Soviet Soyuz. The Zond was intended to go to the Moon, whereas the Soyuz was used for missions in Earth orbit.

vided about *Cosmos 154,* but it seems to have had a similar objective. Once again, the fourth stage of the Proton failed to place the Zond in its intended orbit. An additional problem then arose that prevented the capsule from achieving the desired high-speed reentry.

Tests of the Redesigned Soyuz

Six months after the tragic loss of Komarov, an extensively reengineered Soyuz was ready to be tested in space. The first two of these unmanned flights, called *Cosmos 186* and *188,* were launched on October 27 and 30, 1967. Shortly after *Cosmos 188* achieved orbit, *Cosmos 186* automatically rendezvoused with it and the two craft docked. For 3.5 hours they remained joined together. The Soyuz then separated, reentered the atmosphere, and landed. This remarkable feat

was repeated in April 1968 during the flights of *Cosmos 212* and *213*.

A final four-day Soyuz test flight, *Cosmos 238*, would later be successfully launched on August 28, 1968. Before this mission could occurr the Soviet program would suffer another major blow.

The Death of a Hero

On March 29, 1968, Yuri Gagarin and Vladimir Seryogin climbed aboard a Soviet MIG-15Uti for a training mission. This flight was a normal part of Gagarin's routine and it received little public attention. After completing the planned flight, Gagarin and Seryogin were returning to base when a MIG-21 passed close to their plane, creating an invisible layer of unstable air referred to as jetwash. When Gagarin's jet entered this air, it went into a tailspin and crashed. Both Gagarin and Seryogin were killed instantly.

The emotional impact of Gagarin's death transcended the political differences between countries and expressions of shock and sympathy poured in from around the world. Gagarin's historic orbital flight had made him an international hero and his untimely death at the age of 34 saddened people everywhere. In a public ceremony his remains were placed alongside those of Komarov and Korolev in the Kremlin wall.

In the span of a little over three years, the Soviet space program had lost its chief rocket designer, Sergei Korolev, one of its most experienced cosmonauts, Vladimir Komarov, and its widely respected leader of the cosmonaut corps, Yuri Gagarin. Even the Soviet Union's most ardent supporters realized these losses would make it extremely difficult for the Soviets to beat America to the Moon.

The Space Race Is Resumed

After the successful flights of *Cosmos 186, 188, 212,* and *213,* the Soviet Union resumed its manned space flight program. On October 25, 1968, *Soyuz 2* was launched and *Soyuz 3* lifted off on the next day.

Soyuz 2 was unmanned, but *Soyuz 3* carried cosmonaut Georgi Beregovoi. At the age of 47, Beregovoi was the oldest man to fly in space. A veteran of the second world war who had flown 185 combat missions, his courage and expertise were unquestionable. These qualities made him an ideal pilot for this flight since the memory of Komarov was still fresh in most people's mind. Beregovoi's presence at the controls provided the much needed boost in confidence that this flight would succeed.

Beregovoi's objective was to rendezvous and then dock with *Soyuz 2*. The rendezvous process went smoothly and an on-board computer took his craft to within 600 feet of the orbiting *Soyuz 2*. At this point he took control of *Soyuz 3*, but for some unexplained reason did not dock. Soviet officials later claimed this was not part of the mission and that all of the prelaunch goals had been met. This explanation left many ob-

Figure 27-5. Georgi Beregovoi was the commander of Soyuz 3. This was the first manned Soviet mission after the disasterous flight of Soyuz 1.

servers unconvinced. After completing some additional maneuvers and obtaining observations of weather patterns and forest fires on Earth, Beregovoi landed near the Soviet town of Karaganda.

On November 10, 1968, the Soviet Union launched *Zond 6* into a lunar trajectory. As this craft sped by the Moon several photographs of the lunar surface were obtained. The spacecraft returned to Earth and its film was recovered. Once these pictures were developed, they were shown around the world and provided the Soviets with a propaganda bonanza. This flight also caused concern within NASA. Were the Soviets close to placing a man in orbit around the Moon?

Unknown to western space experts, *Zond 6* had actually encountered several problems that made a manned circumlunar trip far too risky. During atmospheric reentry a gasket failed, causing the cabin to decompress. If a crew of cosmonauts had been aboard, they would all have died. Additionally, as the Zond streaked through the atmosphere, its parachute system failed and the capsule crashed. None of the live specimens carried on this trip survived. I.B. Afanasiev, a colleague of Korolev, later wrote of the demoralizing effect this failure had on the cosmonauts,

> The hopes of the cosmonauts of the lunar detachment to be able to soon make a flyby of the Moon were rapidly dissipating under the onslaught of the ever newer emergencies that were occurring with each new launch.

Meanwhile, in December 1968, America made history by achieving what the Soviet Union had been unable to accomplish. Astronauts Borman, Lovell, and Anders guided their Apollo craft into an orbit around the Moon. NASA viewed *Apollo 8* as an important milestone, but realized further work would be needed before a lunar landing could be attempted. However, this flight seems to have had a profound impact on the Soviet leadership. Afanasiev wrote that after this mission, "the political sense of continuing the circumlunar program . . . had been lost."

Soyuz 4 and 5 Accomplish Another Space First

A second pair of Soyuz craft (*Soyuz 4* and *5*) were launched on January 14 and 15, 1969. Their objectives were similar to those of the ill fated *Soyuz 1* and *2* missions, to practice rendezvous and docking procedures and complete a crew exchange. *Soyuz 4* carried a single cosmonaut, Vladimir Shatalov, while the crew of *Soyuz 5* consisted of three cosmonauts, Boris Volynov, Yevgeni Khrunov, and Alexei Yeliseyev. This was the first three-cosmonaut crew the Soviets had launched since *Voshkod 2* and demonstrated they had finally recovered from the shock of Komarov's death.

The two Soyuz craft rendezvoused and then performed the first successful docking between two manned spacecraft.

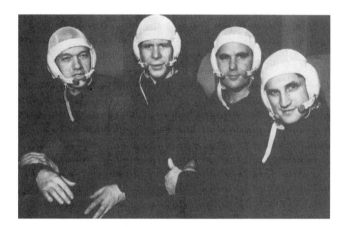

Figure 27-6. The crews of Soyuz 4 and 5. From left to right, Yeliseyev, Khrunov, Shatalov, and Volynov. During the flight Yeliseyev and Khrunov would leave Soyuz 5 and join Shatalov in Soyuz 4.

Figure 27-7. The docking of Soyuz 4 and 5 was the first time that two manned craft had been joined together in space.

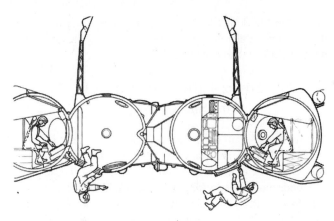

Figure 27-8. The crew transfer from Soyuz 5 *to* Soyuz 4 *would take place as shown in this sketch. As one cosmonaut entered* Soyuz 4, *the other would leave* Soyuz 5.

Two members of the *Soyuz 5* crew, Khrunov and Yeliseyev, left their craft through the orbital module and worked their way over to *Soyuz 4* by using exterior handholds (see Figure 27-8). Having completed this first ever crew exchange or simulated space rescue, the two craft separated and these crewmen returned to Earth with Shatalov. One day later, on January 18, 1969, *Soyuz 5* safely returned. Interestingly, the Soviets publicized this mission as a necessary step toward the creation of a space station. No mention was made of the Soviet–American Moon race.

A Change in the Direction of the Soviet Space Program

A surprising change of attitude about the Moon race now began to emerge from public announcements by the Soviet government. On January 1, 1969, a joint resolution was passed by the Council of Ministers authorizing the continuation of the lunar program, but also directing that an unmanned probe be sent to the Moon to collect and return samples to Earth. It also stated work on a Soviet space station should be accelerated.

Table 27-2. Early Soyuz Flights		
Cosmos 186	Oct. 27, 1967	Unmanned test flight
Cosmos 188	Oct. 30, 1967	Unmanned test flight
Cosmos 212	Apr. 14, 1968	Unmanned test flight
Cosmos 213	Apr. 15, 1968	Unmanned test flight
Cosmos 238	Aug. 28, 1968	Unmanned test flight
Soyuz 2	Oct. 25, 1968	Unmanned target
Soyuz 3	Oct. 26, 1968	G. Beregovoi
Soyuz 4	Jan. 14, 1969	V. Shatalov
Soyuz 5	Jan. 15, 1969	B. Volynov,
		Y. Khrunov
		A. Yeliseyev

The End of the Moon Race

As it became increasingly apparent that America would attempt a manned lunar landing around the middle of 1969, the Soviet Union frantically worked on the automated probe authorized by the Council of Ministers. To the Soviets the scientific mission of this probe was secondary to its political goal. If this probe could land on the Moon's surface and return samples of the soil, it would steal some of the thunder from the expected American landing. The political benefits of returning Moon rocks before the Americans was considered important enough for this mission to be given a high priority.

By the early spring of 1969 the Soviets had designed and constructed their automated Moon lander. It stood 13 feet tall and weighed 2 tons. The first attempt to launch this probe took place on April 15, 1969, but failed when its Proton rocket malfunctioned. Soviet anxiety rose when in May 1969 *Apollo 10* astronauts Tom Stafford, Eugene Cernan, and John Young circled the Moon and brought the lunar module to within 10 miles of the Moon's surface. All of the obstacles to the Moon had now been cleared. NASA announced that its long-awaited lunar landing mission, *Apollo 11*, would lift-off on July 16.

With time rapidly running out, the Soviet Union launched a second automated lunar probe on June 14, 1969, but as with the prior attempt, problems with the Proton rocket caused it to fail. While yet another probe was readied, the Soviets launched an N-1 rocket on July 3 with a mock-up of their manned lander. However, as the lower part of the booster cleared the tower, one of its engines blew up and the remaining engines automatically shut down. In a scene reminiscent of America's early Vanguard fiasco, but on a much grander scale, the rocket fell back to Earth, creating a fireball that incinerated the launch pad and surrounding structures.

In a last frantic gamble to beat America to the Moon, a third automated probe, *Luna 15*, was launched on July 13, 1969. This time the Proton worked well and the probe was sent on its way to the Moon. *Apollo 11* lifted off on its own journey three days later. The final leg of the race to the Moon was underway.

On July 17, *Luna 15* entered into an orbit around the Moon well ahead of the America spacecraft. As the Soviet leadership had hoped, public interest in this probe was high. Western news commentators helped deflect attention from *Apollo 11* by asking a stream of scientists and engineers to speculate about the mission of this Soviet probe. On July 20 *Luna 15* changed its orbit and came closer to the Moon's surface. While this maneuver was being carried out, the Apollo lunar module, the *Eagle*, began its descent. A short time later, Neil Armstrong reported to a listening world, "the Eagle has landed." At that moment the fate of the still orbiting *Luna 15* became irrelevant. For the first time in history, a human was on the surface of the Moon. Interest in the Soviet's unmanned probe vanished, and in an ironic twist of fate, *Luna 15* crashed into the Moon's Sea of Crises after a 4-minute descent.

28 The Soviet Salyut Space Station

The whole idea is to move from Earth to settlements in space.

—Konstantine Tsiolkovsky

With the loss of the Moon race, the Soviet Union turned its attention toward the construction of an Earth orbiting space station. Soviet rocket scientists had been interested in a station long before the start of the space race with the United States. Decades earlier, Tsiolkovsky had written about its benefits as a meteorological outpost and as a departure point for trips to nearby planets. His writings inspired Soviet engineers to explore this matter more fully.

As mentioned in the last chapter, in early 1969, the Soviet Union launched *Soyuz 4* and *5* to develop the necessary skills and test the hardware needed for rendezvous and docking. Although these flights were conducted as part of the Soviets' Moon program, they had applications for a space station. Even at this late date Soviet officials still felt there was a chance of beating America to the lunar surface. That hope was finally dashed by the flight of *Apollo 11* in July 1969. This American triumph brought an end to the Moon race, and gave the Soviets an impetus to set a new goal.

Figure 28-1. Konstatine Tsiolkovsky wrote about the advantages that a space station held for mankind in the early 1900s. Although this idea was far ahead of its time, it gained acceptance in the 1960s.

While the Soviets publicly dismissed the importance of a manned lunar program, an aggressive effort to build a permanently manned space station was begun. The resources devoted to this project were enormous compared to those allocated by the United States to its space program.

Two space station designs were pursued in the Soviet Union in the 1960s, one by Korolev's group, the OKB-1, and the other by Vladimir Chelomei, head of the OKB-52. Reflecting their duties within the Soviet rocket program, Korolev's station would provide the Soviets with a civilian orbiting scientific laboratory, whereas Chelomei's would serve as a military observation post.

Like Tsiolkovsky, Korolev was intrigued by the opportunities a space station offered. In 1962 he designed a facility consisting of a series of linked Vostok capsules. Cosmonauts could use the combined capsules as one large work and living area. Over the next few years this concept evolved into a plan for a 75-ton station called the *Orbital Station 1*. Powered by solar panels, it would consist of up to six individual sections that could be linked together. An engineering model of this station was actually built in 1965.

Vladimir Chelomei was a well-known figure in the Russian aerospace industry. He promoted his version of the space station, called the Almaz, as an observation post from which the Soviet military could monitor the movements of the American armed forces, particularly those at sea, where Soviet reconnaissance capabilities were limited. The Almaz consisted of a control room, a powerful television system, a rapid-fire cannon for defense against a potential attack by American space planes, and crew quarters. The station's personnel were to be ferried to the Almaz aboard a new vehicle called the Transport Logistics Spacecraft.

The American military had a similar program on the drawing boards. Their station would consist of a Gemini capsule with an attached laboratory and would be launched using a Titan 3 rocket from Vandenberg Air Force Base. Navy and air force pilots were recruited to occupy this station, but before work had progressed very far the escalating costs of the Vietnam war caused the entire project to be canceled in 1969. Many of these pilots would later fly in the American space shuttle program.

Soyuz 6, 7, and *8:* A Soviet Space Fleet

In the late 1960s, it was clear the Soviet Union was directing much of their effort toward an Earth-orbiting space station. But before a station could be launched, more data was needed about the prolonged effects of weightlessness on the human body. Procedures also had to be found to counteract its negative effects. In addition to these medical issues, techniques for building and maintaining equipment in space had to be developed and tested. Experiments in each of these areas were assigned to the upcoming Soyuz flights.

Soyuz 6, the Soviet's first post-*Apollo 11* manned spacecraft, lifted-off on October 11, 1969 carrying two cosmonauts, Georgi Shonin (commander) and Valeri Kubasov (flight engineer). This flight had an extensive list of objectives, including testing improved Soyuz electronics and hardware, validating the craft's maneuvering and navigation systems, obtaining photographs of geological structures on the Earth, conducting an investigation of the Earth's atmosphere, completing medical and biological experiments, and testing various welding techniques. In all, the cosmonauts were scheduled to conduct fifty separate experiments.

One day following the launch of *Soyuz 6,* *Soyuz 7* blasted-off from Tyuratam. It carried a crew of rookie cosmonauts: Anatoli Filipchenko (commander), Viktor Gorbatko (research engineer), and Vladislav Volkov (flight engineer). Gorbatko was a member of the original 1960 class of cosmonauts, and although he had not flown in space, he had served as a backup crewman on *Voskhod 2, Soyuz 2,* and *Soyuz 5.*

The stated objectives of this flight were to perform orbital maneuvers, conduct navigational exercises with *Soyuz 6,* and obtain scientific data on the brightness of stars and the Sun. They were also to conduct a series of Earth resource experiments centered on the area surrounding the Caspian Sea and the extensive forests of central Russia. As part of this

Figure 28-3. The crew of Soyuz 7. *From left to right, Anatoli Filipchenko, Vladislav Volkov, and Viktor Gorbatko. Gorbatko had been a cosmonaut for the longest time, but none of this crew had prior spaceflight experience.*

mission cloud formations over the Soviet Union would be photographed in a joint land and space meteorological study.

A day after the launch of *Soyuz 7, Soyuz 8* lifted-off carrying cosmonauts Vladimir Shatalov, overall commander of the orbiting Soviet space fleet, and flight engineer Alexei Yeliseyev. In all, seven cosmonauts were now in orbit aboard

Figure 28-2. The Soyuz spacecraft was the Soviet counterpart to the Apollo command module. Modified versions of this craft are still used today in the Soviet manned space program.

Figure 28-4. The lift-off of Soyuz 8 *from Tyuratam on October 13, 1969. By this time* Apollo 11 *had landed on the Moon.*

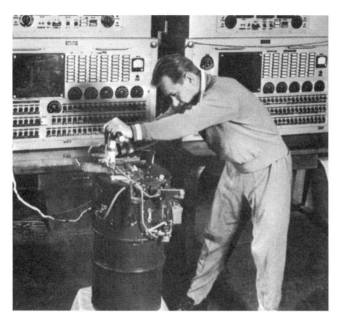

Figure 28-5. Valeri Kubasov training with the Vulcan, an experimental welding apparatus that he would use during the flight of Soyuz 6.

three separate ships, a new space record. The objectives of *Soyuz 8* were to investigate the polarization of sunlight reflected from the Earth's atmosphere, coordinate the flight of the Soyuz fleet, and perform joint orbital maneuvers with the other ships.

On October 15, the three Soyuz craft rendezvoused. *Soyuz 8* then approached *Soyuz 7,* but to the surprise of western observers did not dock. The cosmonauts obtained photographs of the other craft, separated, and practiced various joint maneuvers. On the following day, *Soyuz 6* cosmonaut Kubasov tested a new welding device. The orbital module of his craft was depressurized and he controlled the equipment from within the reentry module. After repressurizing the orbital module, the welded samples were collected and stored away for later analysis. The results of these tests would play a role in future Soviet space station work.

Soyuz 6 returned to Earth on October 16, and was followed by the other two craft at one-day intervals. Although the mission was pronounced a complete success, there was widespread speculation in the West that a docking had been planned between *Soyuz 7* and 8 but had failed. In a postflight interview, Shatalov denied this and stated that neither craft was even equipped for this task. He went on to explain, "It was not necessary; we had already done this in January, and before that it had been done, entirely by automatic control, with Cosmos satellites." Brain Harvey, in his book *The New Russian Space Programme*, disputes this statement and claims that the Soviets now admit that a docking had been planned.

At the conclusion of this mission, it was clear that the spaceships had performed most of their experiments in an independent fashion. Why then did the Soviets fly this as a three-craft mission? A possible explanation is that they wanted to direct world attention away from the lost Moon race and toward their space station efforts. For the Soviets the joint flight heralded the end of one program and signaled the start of a new era. Support for this assessment comes from the fact that soon after the conclusion of this flight, the Soviets publicly announced they were no longer interested in pursuing a manned lunar mission. Their goal was to construct an Earth- orbiting space station.

Figure 28-6. The return of the crews of Soyuz 6, 7, *and* 8 *was celebrated by an awards ceremony in Moscow. From left to right the crewmen are Shatalov, Shonin, Filipchenko, Kubasov, Gorbatko, Yeliseyev, and Volkov.*

Soyuz 9: Paving the Way for Long-Duration Spaceflight

The Soviets' first manned night launch, *Soyuz 9,* sped toward space on June 1, 1970. Its crew consisted of Andrian Nikolayev, commander and a veteran of *Vostok 3,* and Vitali Sevastyanov, flight engineer and a space rookie. Unlike other recent Soyuz flights, this mission involved only a single craft.

Soyuz 9 was the first mission whose goals dealt primarily with the Soviets' space station program. The main goal of this flight was to gather medical data about the long-term effects of weightlessness on the human body. The longest Soviet spaceflight to date, that of *Vostok 5,* had lasted five days, and the spaceflight record, held by *Gemini 7,* was only fourteen days. Since Soviet crews would man the space station for several weeks at a time, more data was urgently needed on the medical effects of prolonged weightlessness.

Like *Soyuz 7, Soyuz 9* also had secondary objectives related to Earth resources. One of these was to collect data about coastal currents and water temperatures to help Soviet trawlers develop better methods of locating large quantities of fish. A geological survey of Siberia, measurements of the snow coverage in the mountains of the southern republics, and meteorological readings that would complement simulta-

neously collected data by aircraft, sounding rockets, and the Meteor weather satellite would also be obtained.

To address the medical problems posed by weightlessness, the cosmonauts spent two hours each day exercising on a stationary bike or on other equipment to work their muscles. The cosmonauts disliked these sessions partly because there were no shower facilities on the cramped Soyuz. They had to wash up using only wet towels. The cosmonauts grudgingly continued this routine, but on at least one occasion attempted to "play hookie" by falsely reporting to ground controllers that they had completed their exercise routines.

By the twelfth day in space both men were showing signs of fatigue. They became sluggish and controllers found it difficult to wake them from their rest periods. Officials decided to end the mission before the situation grew worse. After spending 17 2/3 days in space and completing 285 orbits, Nikolayev fired the Soyuz retro-rockets and the cosmonauts returned to Earth.

A ground team soon reached the Soyuz recovery site, and as a precaution, lifted the cosmonauts out and placed them on stretchers. Nikolayev and Sevastyanov were airlifted back to Star Town and placed in a biological isolation chamber as a precaution against becoming sick while in their weakened state. The chamber was most likely a relic from their abandoned Moon program. After this mission, it was never used again. The cosmonauts' recovery from weightlessness was more difficult than publicly acknowledged. Medical records show they were slow to respond to questions, tired quickly, and their faces were pale and wrinkled. Neither cosmonaut could stand without assistance for three days, and their heartbeats were twice as fast as normal. Despite these difficulties, doctors were encouraged by the fact that these effects were only temporary. Both men regained their health within one month.

The medical data returned by *Soyuz 9* played an important role in the Soviet space program. Based upon its findings, doctors devised new exercises to counteract the body's adverse reactions to weightlessness. Special clothing was constructed to improve blood circulation and to exercise seldom used muscle groups. This work was later summarized by Nikolayev,

> In principle prolonged spaceflights present no problems. We now know that the barrier of weightlessness can be overcome. No radical measures are necessary for short flights of one or two weeks. On

Figure 28-7. Soyuz 9 *being prepared for launch at Tyuratam. At the top of the vehicle is the orbital module, the middle section is the descent module, and the lower section is the service module.*

Table 28-1. Transitional Lunar to Space Station Soyuz Flights

Mission	Launch Date	Crew
Soyuz 6	Oct. 11, 1969	Shonin, Kubasov
Soyuz 7	Oct. 12, 1969	Volkov, Gorbatko
Soyuz 8	Oct. 13, 1969	Shatalov, Yeliseyev
Soyuz 9	June 1, 1970	Nikolayev, Sevastyanov

flights lasting some months in zero-gravity we shall have to introduce strict programs for providing artificial stress for muscles, heart and circulation.

The World's First Space Station: *Salyut 1*

The Soviet government officially approved the construction of the first Earth-orbiting space station in late 1969, and by 1970 work was well underway. It was hoped that this station, called Salyut (translated Salute), would be ready for launch by April 12, 1971, the tenth anniversary of Yuri Gagarin's historic orbital flight.

Considering that *Salyut 1* represented mankind's first attempt to build an Earth-orbiting space station, it was an impressive start. The station was bigger than a school bus, having a length of 48 feet, a maximum diameter of a little over 13 feet, and a mass of about 19 tons (see the Focus Box below). The station was divided into four compartments: (1) the transfer module, (2) the work module, (3) the scientific instrument bay, and (4) the propulsion module. Each compartment had a cylindrical shape and, with the exception of the propulsion module, was pressurized.

The transfer compartment had dimensions of 6.6 × 9.8 feet with a docking fixture attached to its front. Once a Soyuz was docked, the cosmonauts entered the station through an air-tight hatch. The transfer compartment also housed the station's biology experiments, cameras, and a telescope. This latter instrument, called the Orion stellar telescope (identified as #11 in the Focus Box), was the station's primary optical astronomical instrument. It was designed to obtain ultraviolet stellar spectra in the wavelength range of 2000–3000A. A gamma-ray telescope was also carried by the Salyut that was capable of detecting photons 150,000 times more energetic than visual light.

Attached to the external wall of the transfer compartment were rendezvous antenna, docking lights, a TV camera, heat regulation panels, orientation sensors, micrometeor detection panels, and two large solar panels to provide power to the craft. An EVA hatch was located in the side of this unit.

Immediately following the transfer compartment was the work module. This larger cylindrical unit was 9.5 feet wide and 12.5 feet long. It was, in turn, connected to a larger 13.4 × 13.4 foot scientific instrument bay. These two modules contained the craft's communications and instrument control equipment, power supplies, and life-support equipment. Along its interior walls were small storage units.

As noted earlier, the Salyut space station was intended to be occupied by cosmonauts for extended periods of time. Considerable attention was given to maintaining the crew's physical condition. Salyut cosmonauts were required to exercise daily on a treadmill and data on physiological changes in their bodies were continuously collected and transmitted to doctors. The crew members were also required to wear a special rubberlike suit that exercised little used muscle groups.

A series of pioneering biology experiments to gather new information about the effect of weightlessness on the growth of living organisms was planned as part of the *Salyut 1* mission. Observations would be collected on the growth of frog eggs and other small plants and insects. Genetic changes of short-lived fruit flies would also be collected as well as data about the growth of higher-order plants.

The fourth Salyut compartment, the propulsion unit, was unpressurized and could not be entered by the cosmonauts. It had a length and maximum diameter of about 7.2 feet and contained the station's main rocket, stabilization and

---

Focus Box: Salyut 1 with Docked Soyuz

1. Docket Soyuz
2. Rendezvous radar
3. Solar panel
4. Main control console
5. Exercise treadmill
6. Scientific instrument bay
7. Main rocket fuel
8. Propulsion unit
9. Attitude thrusters
10. Movie camera porthole
11. Orion telescope

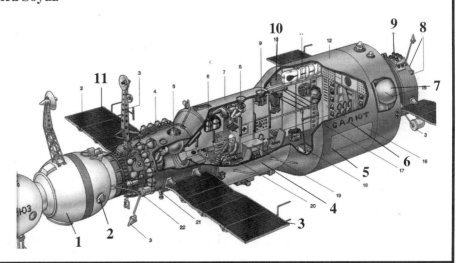

orientation jets. Two additional solar panels were attached to its exterior wall. The Salyut was viewed by observers around the world as an impressive accomplishment.

A Fitful Beginning to Space Station Operations

A new era of space exploration began on April 19, 1971 when the Soviet Union placed *Salyut 1* into orbit. Three days later cosmonauts Vladimir Shatalov (commander), Alexei Yeliseyev (flight engineer), and Nickolai Rukavishnikov (research engineer) lifted-off aboard *Soyuz 10*. Shatalov had commanded *Soyuz 4* and *8*, Yeliseyev had flown in *Soyuz 2* and *5*, and Rubavishnikov, although a rookie, had served on the backup crews of *Soyuz 6* and *7*. Soviet preparation for this mission had been intense. Prior to lift-off, Shatlov spoke to a press representative, "We prepared so painstakingly you may almost say that the ship is engraved on our eyeballs."

After an uneventful flight and rendezvous, *Soyuz 10* docked with the space station on April 24, 1971. This successful beginning was followed by one of those frustrating episodes that plague development programs. A failure in the docking mechanism made it impossible to enter the station. This mishap brought the mission to an embarrassingly abrupt end. The cosmonauts undocked, flew around the station to inspect its exterior fixtures, and sadly returned to Earth. It was little consolation that *Soyuz 10* made the Soviet program's first night landing. The cosmonauts had been in space for just 48 hours, the shortest mission since Komarov's tragic flight in *Soyuz 1*. The occupation of the space station was left to the crew of *Soyuz 11*.

29 The Flight of *Soyuz 11* and Its Aftermath

. . . feelings on the ground were close to exhilaration. The three cosmonauts had smashed every record in sight. The concept of an orbiting station had been vindicated. Salyut was the first space station beyond the Earth. The scientific haul from the mission would keep the scientists busy for years.

—**Brain Harvey in *The New Russian Space Programme***

Soyuz 11 lifted-off on June 6, 1971 with cosmonauts Georgi Dobrovolsky (commander), Viktor Patseyev (research engineer), and Vladislav Volkov (flight engineer) onboard. Dobrovolsky was a graduate of the Soviet Air Force Academy and had been in the space program since 1963. Patseyev had earned a master's degree from the Penze Industrial Institute and had worked as a design engineer at the Central Aerologicial Observatory. Volkov was an aircraft designer who had joined the cosmonaut corps in 1966. He was the only member of the crew with prior spaceflight experience, having served as the flight engineer on *Soyuz 7*.

The launch was routine and the Soyuz automatic systems brought the craft to within 300 feet of *Salyut 1*. At this point, Dobrovolsky took control of the Soyuz and maneuvered toward the station. The corrections made to the Soyuz docking mechanism worked perfectly. Exactly 24 hours after the launch of *Soyuz 11* the two craft were joined firmly together. First Patseyev and then Volkov crawled through the airlock connecting the two craft and entered the Salyut. This marked the first time an Earth-orbiting space station had been occupied. Dobrovolsky followed his crewmates a short time later. Having just left the cramped quarters of the Soyuz, he excitedly radioed to ground controllers, "the station's huge: there seems to be no

Figure 29-2. The launch of Soyuz 11 *from Baikonur on June 6, 1971.*

end to it!" The next few days were spent activating the Salyut instruments and using the station's propulsion system to lift its orbit from 127 × 149 miles to 155 × 169 miles.

The crew of *Soyuz 11* had an ambitious program of activities. During their stay, they were to conduct tests of the station's maneuvering, orientation, and navigation systems, study the Earth's atmosphere, and collect data in the fields of Earth resources, space physics, biology, and medicine.

On June 9, they began transmitting daily television "chat" sessions to Earth. Two days later, on June 11, they used the station's gamma-ray telescope, and on June 19, the Orion telescope was used to obtain ultraviolet spectra of a sample of stars. The cosmonauts attached multicolor filters to their cameras to map snow coverage, soil moisture content, crops, and tree disease within the Soviet Union. From their high vantage

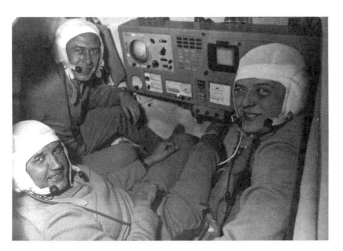

Figure 29-1. Soyuz 11 *crewmembers Viktor Patsayev (bottom left), Georgi Dobrovolsky (upper left), and Vladislav Volkov (right).*

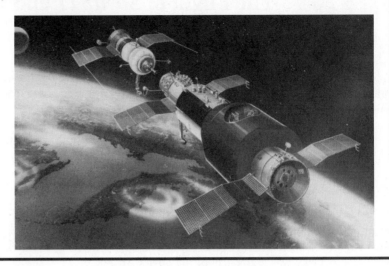
point, they also observed the motion of cyclones and typhoons and reported on the progress of a sand storm as it crossed the African continent. Few problems were encountered and the crew's time aboard the station seemed to pass quickly. On June 24, they broke the eighteen-day spaceflight record held by *Soyuz 9*.

By June 26, the cosmonauts had spent three weeks in space and their busy schedule of experiments and exposure to weightlessness finally began to take a toll. It was obvious to controllers the cosmonauts were becoming fatigued. Instruments on the station that measured the crew's blood pressure, heart rate, breathing rate, and air intake, suggested the crews' condition would continue to deteriorate. Doctors feared the cosmonauts' ability to respond to an unexpected emergency might be jeopardized. Already a small electrical fire had broken out on the station, but it had been quickly extinguished. If another situation arose, the results could be disastrous. Therefore, on June 28 the crew was instructed to mothball the station and return to Earth.

Up to this time, officials had every reason to be pleased with the mission results. In the book *The New Russian Space Programme*, Brian Harvey writes,

> . . . feelings on the ground were close to exhilaration. The three cosmonauts had smashed every record in sight. The concept of orbiting stations had been vindicated. Salyut was the first space base beyond the Earth. The scientific haul from the mission would keep scientists busy for years. . . . It was clear that this was what spaceflight was all about—not lunar stunt shows. And an American answer to Salyut was still some years off. It was the good old days all over again. A tremendous reception was being prepared for the three cosmonauts' return.

After loading logs and experimental data, the cosmonauts reentered their Soyuz and undocked from the Salyut on June 29th. They first flew around the station, inspecting its exterior for damage, and then moved away to begin preparations for reentry. Glancing back toward the station, controllers heard Volkov excitedly shout, "I can see the station. It's sparkling in the sunshine." Dobrovolsky then told controllers "So long, I am starting the landing procedure." Signals, automatically transmitted to ground controllers from their craft, indicated everything was normal.

As the capsule descended through the atmosphere no other communications with the ground took place. The module's parachutes opened at the proper time and the capsule landed just before sunrise on June 29th, undamaged and on target. Communications had not been reestablished, but officials assumed the craft's radio had simply malfunctioned. The mission appeared to have been a complete success. A recovery helicopter landed close to the capsule and its crew rushed to remove the Soyuz hatch and congratulate the cosmonauts. As the recovery team members looked into the capsule, they were stunned to find the cosmonauts were dead!

It was later discovered that an exhaust valve had accidently opened when the orbital and descent modules of the Soyuz separated just before reentry. Investigators suspected the crew realized what had gone wrong. Patsayev had unstrapped his safety belt, probably as part of an attempt to close the valve. Before this task could be completed, the capsule's air escaped and all three cosmonauts suffocated. Vasili Mishin later mournfully expressed the frustration and remorse of the space establishment, "That valve had been checked hundreds of times in test units and it had been used on all our previous craft. It had always worked fine. It never occurred to anyone that such a simple device could fail."

The unexpected death of the Soyuz crew sent shock waves through the Soviet Union. After establishing a multitude of space firsts and completing numerous difficult experiments, their death during the final moments of the mission seemed especially harsh. The daily televised reports

Date	Sample Activities	Date	Sample Activities
June 6	Launch from Tyurtam.	June 18	Orion telescope observations. Television broadcast.
June 7	Docking successful.		
June 8	*Salyut 1* systems activated.	June 19	Aerosol content of atmosphere measured. Medical tests. Meteorological observations.
June 9	Salyut engine used to change orbit.		
June 10	Physical exercises. Medical tests.	June 20	Television broadcast. Earth photography.
June 11	Spectroscopic Earth measurements. Gamma-ray telescope used to measure distribution of celestrial gamma-rays.	June 21	Orion Telescope studies. Gaama-ray radiation study. Space environment study.
		June 22	Polarization of reflected light from Earth. Meteorological study.
June 12	Television broadcast. Medical tests. Photography.	June 23	Photography of USSR Engine firings.
June 13	Earth Resources experiments. Measurements of cosmic radiation.	June 24	Space endurance record. Weather observations. Physical exercises. Earth photographs.
June 14	Navigation experiment. Coordinated weather observations.	June 25	Experimental program activities. Preparation for return to Earth. Television broadcast.
June 15	Simultaneous space and aircraft photographic studies. Joint Salyut and Meteor satellite cloud cover measurements.	June 26	Micrometeorite measurements. Increased exercises. Preparation for landing.
		June 27	*Soyuz 11* checks. Television broadcast. Weather observations.
June 16	Salyut systems evaluated. Atmospheric study.		
June 17	Spacecraft antenna tests. Medical tests. Space environment density measurement.	June 28	Medical tests. Soyuz system reactivation.
		June 29	Departure from *Salyut 1*.

broadcast from the station had made the cosmonauts nearly as well known as Gagarin and Leonov. The joyous reception that had been planned in Moscow was now replaced by yet another somber procession as the cremated remains of the cosmonauts were carried to the Kremlin wall and placed in a position of honor next to their comrades. Tributes flowed in from around the world, and in what seemed like an endless line, grief-stricken Soviet citizens walked by to pay their last respects to a crew they had grown to greatly admire.

Figure 29-3. The ashes of the crew of Soyuz 11 *were placed in the Kremlin wall in honor of their service to the Soviet state.*

Soviet space officials had hoped a successful Salyut mission would relieve some of the pain the nation still felt about the lost Moon race. The tragic end of *Soyuz 11* dashed those hopes and caused a further time-consuming reevaluation of the Soviet manned spaceflight program.

Postflight Changes

After the disaster of *Soyuz 11*, the Soyuz craft was redesigned. It was also decided that on future missions the cosmonauts would wear spacesuits during the critical launch, docking, and reentry phases of the flight. In order to accommodate the extra weight and room needed for the suits, the crew size was reduced from three to two cosmonauts. The external solar panels of the Soyuz were removed. Power would now be supplied by on-board batteries that could be recharged from the Salyut. As was the case for *Apollo 1*, officials wanted to make sure the lessons learned by the untimely death of their spacemen were fully incorporated into the redesigned craft. The alterations proceeded at a painstakingly slow pace, but no manned missions were permitted until the changes could be completed and tested. It would take two years before the Soviets placed another crew in space.

The End of *Salyut 1*

After spending several months in space, the orbit of *Salyut 1* was gradually decaying. To avoid the possibility that the station might crash into a populated area, ground con-

Table 29-1. Failed Salyut Launches

Mission	Launch Date	Reason for Failure
Salyut 2-1	July 29, 1972	Booster problem
Salyut 2	April 3, 1973	Station depressurization
Cosmos 557	May 11, 1973	Station thruster problems

trollers brought it down over the Pacific Ocean on October 11, 1971. In a sad ending for such a magnificent ship, the station broke apart as it streaked through the atmosphere. Its smaller parts were consumed in flames while larger pieces fell harmlessly into the ocean. It was a disappointing finish to a station that had seemed to offer such promise.

A String of Space Station Failures

Although *Salyut 1* had been in space for 175 days, it had only been occupied for 23 days. The crew of *Soyuz 10* had been unable to enter the station, and the next crew had manned it for only a relatively short amount of time. The space station program was proving to be dangerous and very costly. Given the latest disaster, the events of the next two years would severely test the Soviets' commitment.

The bad luck, begun with the launch of *Soyuz 10*, continued. On July 29, 1972, a new station, the *Salyut 2-1*, was launched, but failed to achieve orbit when the second stage of its booster rocket malfunctioned. Things seemed to improve when on April 3, 1973, a third station, *Salyut 2,* was successfully placed into a 129 × 156 mile orbit. Western observers

expected that after some orbital adjustments, a crew would be sent to occupy this station. Surprisingly, as the days passed, no launch occurred. It was later learned that shortly after reaching orbit, a fire in the engine compartment burned a hole in the station's wall causing the crew quarters and work areas to depressurize. Shortly after April 14, the station began to tumble, and on May 28, fifty-five days after its launch, *Salyut 2* reentered the atmosphere and was destroyed. Its larger parts fell harmlessly into the Indian Ocean.

The Soviets attempted to launch a fourth space station, *Cosmos 557*, on May 11, 1973. This was less than two weeks before the scheduled lift-off of *Skylab*, America's first space sta-tion. The timing of this Salyut launch was not a coincidence. After the loss of the Moon race and the public announcements that they would direct their efforts towards the construction of a space station, Soviet officials felt it was imperative to have a functioning station in orbit prior to the launch of Skylab. They were estatic when the launch of *Cosmos 557* was successful. Unfortunately, this joy soon disappeared. Shortly after reaching orbit, the station's orientation thrusters fired and then failed to shut down, causing the station to tumble end over end. This failure was so complete that the station reentered the atmosphere just eleven days later on May 22.

In a final humiliation, *Skylab* was successfully launched on May 23, 1973, one day after *Cosmos 557* fell from space. Although the American station immediately experienced problems (see the chapter titled "Skylab: America's Space Station and Its Early Problems"), the absence of a Soviet station in orbit was viewed as a political disaster. During the last two years, four stations had been launched but only *Salyut 1* had been occupied. At the time of the launch of Skylab, no Soviet stations were in orbit.

30 The Soviet Space Station Program Regains Its Momentum

It can be said with confidence that the 1970s will be the epoch of the development and use of long-term manned orbital stations with changing crews.

—Boris Petrov, Soviet space academician and *Voskhod 2* doctor

The next series of Soviet flights were designed to test the modified Soyuz craft. The first of these flights, *Cosmos 573*, was unmanned and was launched on June 15, 1973. It went perfectly, and a manned test flight, *Soyuz 12*, was scheduled for one month later.

Soyuz 12 carried cosmonauts Vasile Lazarev (commander) and Oleg Makarov (flight engineer) into orbit on September 27, 1973. Lazarev was born in 1928 in southern Siberia and was accepted into the cosmonaut program in 1964. He was a trained doctor and pilot. He had served as a backup doctor for *Voskhod 1* and as backup commander for *Soyuz 9*. Makarov was born in the Kalinin region of Russia (located near Moscow) in 1933 and was an engineer. Prior to his assignment to the cosmonaut corps in 1964 he had worked with the group that had built *Sputnik 1*. He had also served as the backup flight engineer for *Voshod 1*.

The goals of *Soyuz 12* were to test the craft's manual and automatic maneuvering systems. Secondary tasks included the acquisition of multicolor photographs of agricultural regions in the Soviet Union and the continuation of biological experiments begun on the previous flights. This two-day mission was not intended to produce a space spectacular, but was a normal part of a development program. Its two-day duration was considered an adequate amount of time for a craft to rendezvous and dock with a station. Lazarev accepted the lackluster nature of this flight in a professional manner. When asked to describe his mission he said, "Our tasks were purely technological, in other words, our everyday duties, which we simply happened to be doing in a different environment." Lazarev would later fly *Soyuz 18-1* and find that spaceflight was still an unpredictable and dangerous job.

After completing their planned 32 orbits, the cosmonauts landed 248 miles southwest of Karaganda. According to Soviet reports, the mission was textbook perfect, and Vladimir Shatalov (newly appointed to the post of chief of cosmonaut training) stated the results were, "to the complete satisfaction of all involved in this experiment." Although the Soviets were pleased with *Cosmos 573* and *Soyuz 12*, these flights were largely overshadowed by the American *Skylab* mission.

NASA was pleased to learn of this series of Soviet successes. The Soyuz would be employed in the upcoming

Figure 30-1. The crew of Soyuz 14 *consisted of cosmonauts Pavel Popovich (left) and Yuri Artyukhin.*

American/Soviet Apollo-Soyuz flight scheduled for 1975 and the recent string of Soviet space failures had begun to raise doubts about the safety of the Soviet equipment. These successes helped lessen those fears.

The third Soyuz test flight, *Soyuz 13,* was launched on December 18, 1973 and was manned by 31-year-old cosmonauts Pyotr Klimuk (commander) and Valentin Lebedev (flight engineer). Klimuk had been in the space program since 1965. At the age of 23, he was the youngest person to have entered the Soviet program. Lebedev had been in the cosmonaut corps for only one year. Trained as an engineer, he had worked on the design and testing of spacecraft systems.

Many of the activities undertaken on *Soyuz 12* were to be repeated on this flight, but the Soviet's growing confidence was demonstrated when an additional research component was added. Once in orbit, the crew would conduct a food production experiment using two types of bacteria and employ the Orion telescope to obtain ultraviolet spectra of stars about 100 times fainter than could be obtained with the Skylab equipment. No serious problems with the Soyuz were uncovered during this eight-day mission and the cosmonauts landed safely in a snow storm on December 26, 1973, about 124 miles southwest of Karaganda.

Unlike the *Soyuz 12* craft, which relied on batteries for power, *Soyuz 13* employed solar panels. Its configuration was similar to a pre-Salyut Soyuz and was flown because it would be used in the upcoming Apollo/Soyuz mission. A final unmanned test of the improved Soyuz, *Cosmos 656*, was launched on May 27, 1974. This mission was successfully completed two days later.

Salyut 3

The Soviets finally placed a new space station, *Salyut 3,* into orbit on June 25, 1974. *Salyut 3* included several improvements over earlier stations. Its solar panels could be rotated to face the Sun, a new interior floor plan was used, and it carried more sophisticated scientific instruments. Once in a stable orbit, Pavel Popovich (commander and a veteran of *Vostok 4*) and Yuri Artyukhin (flight engineer and 10-year member of the cosmonaut corps) were sent to man it. Both cosmonauts were active military officers.

The mission of *Soyuz 14* included military activities, so a complete list of their tasks was not made public. The main instrument on the Salyut, a folded 33-foot focal length telescope, was probably used to take pictures of U.S. sea forces. The announced scientific program consisted of studies in space medicine, biology, physics, geology, and engineering. During the flight the medical effects of weightlessness also received considerable attention. Electrocardiograms were routinely obtained to monitor the condition of the cosmonauts' cardiovascular systems. The flow of blood in their bodies and its composition were recorded as well as the quantity of air passing through their lungs. Biological specimens were also cultivated and their growth was monitored. Nonmedical tasks included observations of the Earth's atmospheric properties, photographs of geological formations, observations of glacier movements, and studies of cloud formations.

After completing a heavy load of experiments, the cosmonauts undocked from the Salyut and safely landed on July 19, completing a sixteen-day mission. Unlike the crew of *Soyuz 9*, who had spent eighteen days in space, they did not have trouble adjusting to Earth's gravity. Upon landing, they climbed out of their capsule and greeted the recovery team members.

After the return of *Soyuz 14*, a second attempt was made to occupy *Salyut 3*. *Soyuz 15* lifted off on August 26, 1974 carrying rookie cosmonauts Gennadi Sarafanov and Lev Demin. Both of these cosmonauts were military officers. After a successful launch and rendezvous, the Soyuz automatic docking system failed as they approached the station. Sarafanov attempted a manual docking, but had used so much fuel officials feared it would be exhausted before a docking could be

Figure 30-2. The crew of Soyuz 15*: (top) Lev Demin, (bottom) Gennadi Sarafanov.*

Table 30-1. Soyuz Missions *Salyut 1* through *3*			
Mission	**Launch Date**	**Days**	**Purpose**
Soyuz 10	April 22, 1971	2	Man Salyut 1
Soyuz 11	June 6, 1971	23	Man Salyut 1
Soyuz 12	Sept. 27, 1973	2	Soyuz test
Soyuz 13	Dec. 18, 1973	8	Soyuz test
Soyuz 14	July 3, 1974	16	Man Salyut 3
Soyuz 16	Aug. 26, 1974	2	Man Salyut 3

achieved. The crew was instructed to abort the mission and return to Earth. The failure of this mission again caused concern within NASA. The Apollo/Soyuz mission was less than a year away, yet the Soyuz still seemed to have problems.

Soyuz 15 was the last flight launched to *Salyut 3*. Like *Salyut 1*, only a single crew had ever occupied the station and their experimental program had lasted only sixteen days, about a week *less* than had been accomplished during *Soyuz 11*. On September 23 a module containing a film canister was automatically ejected from the space station, reentered the Earth's atmosphere, and was recovered. The station continued in an automatic mode and its systems were monitored by ground controllers to determine the manner in which its electronic equipment deteriorated with time. The station was commanded to reenter the atmosphere on January 24, 1975 and was destroyed during its passage through the atmosphere.

Salyut 4

The Soviets successfully launched their next space station, *Salyut 4,* on December 26, 1974, about a month before the reentry of *Salyut 3*. A few days later the station was raised to a higher orbit so less frequent reboosts would be required. Outwardly, this station looked very similar to *Salyut 1*. The pri-

This space station looked very similar to the *Salyut 1*. One obvious difference is the addition of the third large solar panel attached to the scientific instrument bay. The docked craft is a modified Soyuz. Its solar panels had been removed and the craft was powered by internal batteries.

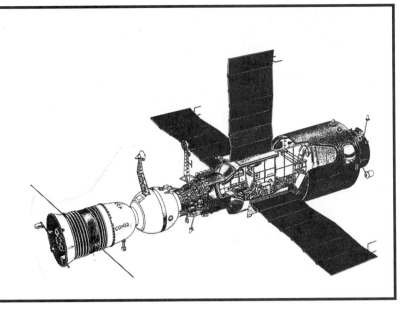

mary difference was a third large solar panel, and unlike *Salyut 1*, all of these panels could be rotated to maximize their exposure to the Sun. A cutaway of this station is shown above. The objectives of *Salyut 4* were purely scientific, including experiments in astronomy, biomedicine, and natural resources. The station carried X-ray and infrared telescopes and spectrometers, as well as micrometeorite and neutral particle detectors. Its medical equipment included an array of instruments: a blood analyzer, a bone tissue density monitor, a muscle stimulator, a pulmonary ventilation recorder, a physical conditioning suit, and a unit for monitoring various body parameters.

Figure 30-3. Cosmonauts Georgi Grechko (left) and Alexei Gubarev preparing for their upcoming Soyuz 17 *flight to* Salyut 4.

The first crew launched toward *Salyut 4* was that of *Soyuz 17*. Lifting-off on January 11, 1975, it carried cosmonauts Alexei Gubarev (commander) and Georgi Grechko (flight engineer), both were space rookies. Gubarev was born in Gvardeitsy, Borsky, in 1932. A veteran of the Soviet Army, he joined the cosmonaut corps in 1963. At the time of the launch, Grechko was 32 years old. He had been educated at the Leningrad Institute of Mechanics, was one of the Soviet's leading space craft designers, and had worked under Korolev.

The rendezvous and docking parts of this flight proceeded without incident. After entering the Salyut and turning on its systems, the cosmonauts quickly began working on a long list of experiments. During their thirty days in space (a new Soviet space duration record) they used the solar telescope to make observations of the Sun and the X-ray telescope to observe the Crab Nebula, the remnant of a supernova explosion first seen in 1054 A.D. Green peas were grown and a large number of biomedical experiments with fruit flies, bacteria, eggs, and frogs were completed. In all, the crew conducted 100 separate experiments.

A Near Catastrophe: *Soyuz 18-1*

The successful *Soyuz 17* mission was followed by the launch of *Soyuz 18-1* on April 5, 1975. It was manned by cosmonauts Vasily Lazarev (commander) and Oleg Makarov (flight engineer), both veterans of the engineering flight, *Soyuz 12*. Although the cosmonauts considered their previous flight unexciting, this mission would be anything but routine.

During lift-off Lazarev and Makarov experienced a life-threatening situation when the Soyuz craft and booster rocket failed to separate. The R-7 rocket went into an uncontrollable tumble and was being severely stressed by aerodynamic forces. The cosmonauts frantically asked ground controllers to blast their capsule free of the rocket, but telemetry

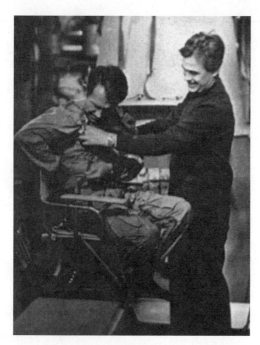

Figure 30-4. Cosmonauts Oleg Makarov (left) and Vasily Lazarev prepare for their flight to Salyut 4. Their mission would end when their R-7 booster rocket malfunctioned.

Figure 30-5. Flying after Soyuz 18-1 took real courage. Cosmonauts Vatili Sevastyanov (left) and Pyotr Klimuk (right) pose in front of a Salyut mock-up.

readings failed to show any problem. The cosmonauts knew their lives were in imminent danger and alternately pleaded and swore at the controllers to act. Still puzzled by the conflicting data, officials finally decided to rely upon the judgement of the cosmonauts and freed the Soyuz from its wildly flying rocket.

Although Lazarev and Makarov were relieved to be separated from the R-7, they soon found their adventure was far from over. As the Soyuz dropped to Earth, the g forces on their bodies rapidly increased, quickly passing the normal values experienced during reentry. At 18g the cosmonauts lost conscious. Instruments within their capsule later indicated the force on their bodies had reached the unheard of level of 20g. During this time they weighed an incredible one and a half tons.

The reentry module landed in near darkness on a steep hill in the Altai Mountains, but then started rolling down the mountain side. Inside the craft the cosmonauts had regained consciousness, but were being thrown against the interior Soyuz fixtures. The craft tumbled down the hill heading toward a steep cliff. If it had fallen over this ledge, the cosmonauts would have been killed. Luckily, the craft stopped when its parachute became tangled in nearby trees. Bleeding and badly shaken, the cosmonauts exited the craft in the freezing darkness and lit flares to direct the recovery helicopters to their location.

Lazarev was the most severely injured of the two, having sustained a concussion and broken ribs. Fortunately, a group of villagers saw the Soyuz's parachute as the capsule fell from the nearly dark sky and came to their aid. The next day, Lazarev and Makarov were retrieved by a Soviet recovery team.

The postflight review of the mission was unusually acrimonious. Officials at Tyuratam blamed the crew for the failed mission and refused to pay them the customary reward of 3,000 rubles for making a spaceflight! Supporters of the cosmonauts blamed the mishap on the support crew for not carefully inspecting the rocket prior to launch. This argument went all the way to the top of the Soviet government and was finally settled by Brezhnev who directed the crew be paid. Later it was found that the core stage of the R-7 did not properly jettison.

Only nineteen months had passed since the successful *Soyuz 12* mission that Lazarev had called unexciting. Now his ability and courage were being questioned. It was almost more than he could take. Despite Brezhnev's decision, the accusations that he had panicked tainted his career. He remained in the space program but never again flew in space. After drinking some improperly made home-brewed vodka, he died in 1990. His crewmate, Makarov, would not be assigned to a prime crew until 1980.

This mission ended just three months before the Apollo/Soyuz mission and caused NASA officials considerable concern. They were assured by their Soviet counterparts that a similar problem would not reoccur because *Soyuz 18-1* had been launched using an "old-fashioned" booster.

Soyuz 18

A third Soyuz mission to *Salyut 4, Soyuz 18,* blasted-off from Tyuratam on May 24, 1975. The mission cosmonauts were Pyotr Klimuk (commander) and Vitaly Sevastyanov (flight engineer). This time the R-7 booster behaved properly and placed them into the desired orbit. After making the required orbital adjustments, the crew docked with *Salyut 4* during its 2,379th orbit.

This was the first time a Soviet space station had been occupied by more than a single crew. The goals of this flight were to continue the experimental program started by the crew of *Soyuz 17*. They were to spend sixty-three days in space, breaking the Soviet record set by Gubarev and Grechko on the prior *Soyuz 17/Salyut 4* flight. During their stay the cosmonauts collected a wealth of astronomical, medical, and geological data. Over 600 pictures of the Sun and more than 2,000 Earth resources photographs covering 5 million square miles of the Soviet Union were obtained. The crew used a short-wave diffraction spectrometer, an x-ray telescope, an isotope spectrometer, and an experimental package to measure the emission of atomic oxygen high in the Earth's atmosphere. In all they spent thirteen days working in the area of geophysics, thirteen days on astrophysics, ten days on medical experiments, six days on technical studies, two days collecting atmospheric data, and two days on space station photography.

Near the end of their mission, mechanical problems began to arise. The Salyut life-support system, which had already operated two months beyond its design lifetime, began to deteriorate. Spots of green mold appeared on the station's interior walls and soon spread throughout the vessel. The humidity became so high that the windows fogged up. Officials, sensitive to the deteriorating situation, directed the cosmonauts to return to Earth. They landed on July 26 about 34 miles from the town of Arkalyk.

The Progress Resupply Vessel

One of the barriers to long-duration missions aboard the station was the small quantity of expendable supplies that could be transported along with the crew in the Soyuz craft. If truly long missions were to take place, a means of resupplying the station had to be developed. The Soviet answer to this need was the unmanned cargo ship, the Progress.

Following the Soviet scheme of modifying existing craft rather then designing new ones, the Progress was a stripped-down version of the Soyuz (see Figure 30-6). In this craft the orbital and reentry modules were used for storage and all life-support equipment was removed. The Progress was not intended to be a recoverable vehicle. After delivering supplies, this ship would be sent back into the atmosphere where it would be destroyed.

The Soviets conducted two unmanned tests, *Cosmos 772* and *Soyuz 20,* of the automatic systems that would control this craft. *Cosmos 772* lifted-off on September 29, 1975 on a three-day shake-down cruise. The autopilot worked as planned and the reentry module was recovered and examined for failures. *Soyuz 20* blasted off from Tyuratam on November 17, 1975 on a ninety-one-day mission. In addition to testing the docking system, this mission had two other objectives: (1) to verify that the Soyuz craft could be successfully

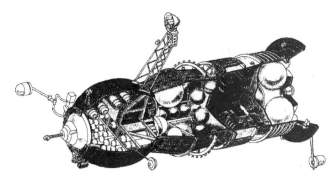

Figure 30-6. The Progress craft was used to ferry supplies to the space station. Based on the Soyuz design, the orbital compartment was used as a storage room and the inner reentry compartment was modified to transport fuel. The craft was totally automated and did not carry a crew.

flown after being deactivated for an extended period of time, and (2) to collect data on the effects of weightlessness on living organisms placed within the capsule. To meet the latter objective, *Soyuz 20* carried live turtles, plants, and seeds. These specimens would be recovered and examined at the conclusion of the ninety-day flight. This mission went well and was judged to be a complete success.

Salyut 5

Salyut 5 was to be the last of the early generation of Soviet space stations. Like *Salyut 3* it was a military space station, so public announcements about its goals were limited.

Salyut 5 was launched into orbit on June 22, 1976. About two weeks later, on July 6, *Soyuz 21* lifted off with cosmonauts Boris Volynov (commander) and Vitaly Zholobov (flight engineer). Volynov had formerly flown in *Soyuz 5* and *Voskhod 3,* but this was Zholobov's first spaceflight. The launch and rendezvous proceeded normally and the crew was soon aboard the Salyut performing experiments with pharmaceuticals, magnets, ceramics, and optical glasses. They also

Figure 30-7. Here cosmonauts Vitaly Zholobov (foreground) and Boris Volynov (background) are shown training for their Soyuz 21 *mission to* Salyut 5.

collected data on the hatching of fish eggs and how a guppy adjusted to weightlessness. Meteorological data and observations of air pollution were also obtained. Much of the crew's time was devoted to materials processing, an activity that would become increasingly important in later space station work. Four instruments to melt various types of materials were tested as part of this program.

Although the *Soyuz 21* mission was expected to last sixty-six days, the cosmonauts returned after spending only forty-eight days in space. The mission was prematurely terminated when Zholobov became ill and acutely fatigued. The air in the station had also developed an unbearable acrid odor.

A second flight, *Soyuz 23,* manned by commander Vyacheslav Zudov and flight engineer Valeri Rozhdestvensky, was launched to *Salyut 5* on October 14, 1976. The expedition got off to an ominous start when the bus taking them to the launch pad broke down. Soon after lift-off their rocket veered off-course and placed the Soyuz in too low of an orbit. The cosmonauts attempted to rendezvous with the space station, but used so much fuel that a docking attempt was scrubbed. The crew was instructed to return to Earth.

The cosmonauts' bad luck continued as they came down at night in terrible weather conditions. Snow was falling, winds were gusty, and the temperature was a frigid 1.4°F. Visibility was so poor the recovery helicopters were unable to locate the reentry module as it descended. This was particularly unfortunate, because instead of landing on the snow-packed ground, the capsule splashed down in a partially frozen salt water lake.

Figure 30-8. Alexei Leonov (left) talks with the crew of Soyuz 24, *Viktor Gorbatko (center) and Yuri Glazkov (right).*

Gusts of wind filled the Soyuz main parachute and pulled the capsule along the water's surface. The reserve chute then opened, became waterlogged, and dragged the craft under water. Since the Soyuz's exit hatch was completely submerged, the cosmonauts were trapped! The cabin was watertight so they were safe for the time being, but the capsule con-

Mission	Launched	Flight	Crew	Days Occupied
Table 30-2: Soviet Space Station Missions				
Salyut 1	April 19, 1971	*Soyuz 10*	Shatalov, Yesiseyev, Rukavishnikov	Failed
		Soyuz 11	Dobrovolsky, Volkov, Patsayev	23
Salyut 2-1	July 29, 1972		Not occupied	
Salyut 2	April 3, 1973		Not occupied	
Cosmos 557	May 11, 1973		Not occupied	
Salyut 3	June 25, 1974	*Soyuz 14*	Popovich, Artyukhim	16
		Soyuz 15	Sarafanov, Demin	Failed
Salyut 4	Dec. 26, 1974	*Soyuz 17*	Gubarev, Grechko	30
		Soyuz 18-1	Lazarev, Makarov	Failed
		Soyuz 18	Klimuk, Sevastyanov	63
Salyut 5	June 22, 1976	*Soyuz 21*	Volynov, Zholobov	49
		Soyuz 23	Zudov, Rozdestvensky	Failed
		Soyuz 24	Gorbatko, Glazko	18

tained only five hours of air. Unpleasant memories of the disastrous end of *Soyuz 11* filled them with apprehension. At daybreak one of the searching helicopters located the Soyuz and pulled it to shore. Once on dry ground, Zudov and Rozhdestvensky opened the hatch, gulped fresh air, and profusely thanked their surprised rescuers who thought the cosmonauts had perished.

On February 7, 1977, another attempt was made to man *Salyut 5*. *Soyuz 24* blasted off from Tyuratam carrying commander Viktor Gorbatko and flight engineer Yuri Glazhkov. The booster rocket performed well and the cosmonauts docked with the space station one day later. The acrid air in the Salyut was vented to space and the station was refilled with breathable air. When this operation was complete, Gorbatko and Glazhkov entered the station and continued the metallurgical work and air pollution observations begun by the prior crew. They also grew plants and fungus.

After completing their 17.7-day mission, Glazhkov and Gorbatko returned to Earth on February 25, 1977. A fourth flight to the station was considered but cancelled when analyses found the Salyut's fuel supply was too low to support another round of activities. After spending fourteen months in space, *Salyut 5* was instructed to reenter the atmosphere where it was consumed in flames on August 8, 1977.

Summary

After the loss of the Moon race, the Soviet Union directed its efforts towards constructing a permanently manned space station. Several initial setbacks showed this would not be an easy task. Even after the death of Komarov, the crew of *Soyuz 11*, and a series of unsuccessful space station launches, this effort persevered. One cannot help but wonder what the reaction would have been if a similar series of failures had occurred in the American program. Would the program have been allowed to continue? While the United States was cutting back on its space program, the Soviets seemed to have recaptured their former enthusiasm. They had a clearly defined objective, whereas the American effort seemed to lack a long-term goal.

After much work and heartache, the Soyuz spacecraft was proving to be a reliable vehicle and the medical problems associated with occupying a space station for extended periods of time appeared to have been solved. An automated cargo vessel, the Progress, had been tested and shown to perform as planned. After a harrowing start, the first generation of Soviet space stations had proven to be useful scientific laboratories. The Soviets were now ready to proceed to their next generation of sophisticated space stations while the American program struggled to make its new Space Shuttle operational.

Introduction to Section V
The Era of International Cooperation

Space policy is not above politics. . . . Chief executives cannot protect the civilian space agency from the forces that batter other discretionary spending programs. Space policy exceptionalism, as attractive as that notion continues to be, is not an appropriate view of reality.

—Roger Launius and Howard McCurdy in Spaceflight and the Myth of Presidential Leadership

The race to the Moon between America and the Soviet Union had produced a clear winner, the United States. Although the plaque attached to the lunar lander of *Apollo 11* read, "We came for all mankind," the Moon landing was widely seen as an American success. The Soviets had been beaten and polls showed the United States was viewed as "without peer in power and influence." However, as the previous chapters have discussed, had it not been for the untimely deaths of Korolev and Gagarin, the first flag to be raised on the Moon might well have been adorned by a hammer and sickle instead of stars and stripes.

The United States and the Soviet Union had adopted a similar approach to the Moon. The Soviet counterparts to projects Mercury, Gemini, and Apollo were Vostok, Voshkod, and Soyuz. During the early phase of the Moon race the Soviets' enjoyed a sizeable lead, but with the death of Korolev in 1966, they were unable to complete their Moon rocket.

NASA had begun planning for the post-Apollo era in 1964 when President Johnson instructed NASA to create a Future Program Task Force. Unfortunately NASA was unable to reach a consensus on how to proceed. George Mueller of the office of Manned Spaceflight supported a manned mission to Mars, but other powerful figures, including Wernher von Braun, favored the construction of an Earth-orbiting space station and a fleet of space shuttle craft. NASA's immediate objective, a scaled-down version of the Space Shuttle, was finally approved by President Nixon. This decision was influenced by cost and political considerations, but the question of NASA's long-term objective was left unresolved. Walker McDougall contrasts this situation with the one America adopted in its lunar program in his book . . . *the Heavens and the Earth: A Political History of the Space Age*, "So the Space Shuttle emerged, but no decision on the goals of future spaceflight. Apollo was a matter of going to the Moon and building whatever technology could get us there; the Space Shuttle was a matter of building a technology and going wherever it could take us."

In the 1970s the space programs in the United States and Soviet Union took separate paths. Work in the United States was directed toward Skylab, Apollo/Soyuz, and to the construction of a reusable space shuttle, while the Soviet Union remained true to its stated goal of perfecting its space station. As part of the Space Shuttle effort a consortium of European nations (the forerunner to the European Space Agency) would construct a "research and applications module" called Skyhab, and Canada would build a robotic arm to move objects to and from the shuttle's cargo bay.

By the late 1980s the costs of space exploration and a new spirit of East–West cooperation led to a new era of international cooperation. The end of national "lone-wolf" programs was signaled by the Shuttle/Mir program and the multinational agreement to build the International Space Station. The final section of this book deals with the events leading to this era and to the possible future international programs, Moon Base and a manned mission to Mars.

31

Skylab: America's Space Station and Its Early Problems

The legacy of the Skylab program to be passed on to planners and operators of future manned space programs is best stated in two words: "Can do!"

—**Rocco A. Petrone** in *Skylab: Our First Space Station*

As early as 1960 NASA and its subcontractors began to explore missions that could take advantage of the new technologies being developed for the Apollo program. In 1962 the Douglas Aircraft Company (later to become the McDonnell Douglas Corporation) submitted a proposal to NASA to build a space station. Their plan called for the use of the empty second stage of the Saturn 1B, the S IV-B, as an Earth-orbiting observatory. During lift-off the S IV-B would be fully fueled and its engines would provide the power necessary to place this stage and an Apollo CSM into orbit. A team of three astronauts would then be launched to dock with the empty tank and modify its interior into a working laboratory. NASA accepted this proposal on December 1, 1965 and named the project the orbital workshop (OWS).

Soon after work began on this station, problems arose when congressional support for the space program began to decline. NASA's budget requests were severely cut and it became impossible to support both the Apollo and space station programs simultaneously. Faced with this financial dilemma, NASA planners decided to postpone work on the orbital workshop until the Apollo program was completed.

Budgetary related delays in the station's construction actually led to some fortuitous benefits. With the cancellation of

Figure 31-2. The official Skylab patch shows how the space station was intended to look in space. The Sun (upper left) was one of its main research objectives.

the final three Apollo lunar missions, a Saturn V also became available to launch the space station instead of the smaller Saturn 1B. The use of this more powerful rocket (see Figure 31-1) allowed several improvements to be made. First, the S IV-B could be launched empty, avoiding the need to submerge the station's instruments and fixtures in highly corrosive fuel. Second, the station's interior structures could be completely installed prior to launch rather than being placed in orbit by a later flight and assembled during a series of difficult space walks. Third, the suite of scientific instruments could be upgraded by adding a complement of optical and x-ray telescopes called the Apollo telescope mount. Finally improvements could be made in the astronauts' living quarters to make their stay aboard the station more comfortable. NASA adopted this revised plan on February 20, 1970 and renamed the project Skylab.

The Skylab Components

Skylab consisted of four distinct units: (1) the multiple docking adapter, (2) the Apollo telescope mount, (3) an air-

Figure 31-1. The Saturn 1B (right) was used to ferry astronauts to the space station, whereas the much more powerful Saturn V (left) was used to place the station into orbit.

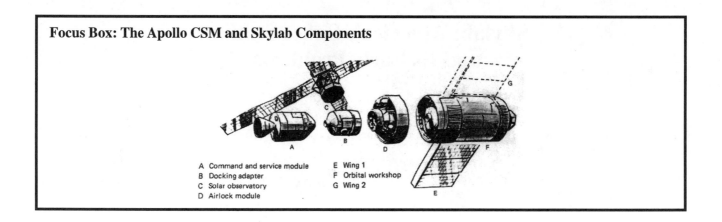

A Command and service module   E Wing 1
B Docking adapter   F Orbital workshop
C Solar observatory   G Wing 2
D Airlock module

lock module, and (4) a large compartment called the orbital workshop. These components are shown in the Focus Box above. The entire structure was over 86 feet long, weighed about 145,000 pounds, and had an interior volume of 13,000 cubic feet.

The multiple docking adapter contained two docking ports. The main port, which would be used by the Apollo CSM, was placed at the front, while an auxiliary port was located on one side. The multiple docking adapter also housed the Apollo telescope mount controls, Earth resource experiments, and various pieces of stored equipment. The second Skylab component, the Apollo telescope mount, was a converted lunar module and was located on top of the multiple docking adapter. Its telescopes were designed to observe the Sun and other celestial objects in the optical, ultraviolet, and x-ray regions of the spectrum.

Skylab's third component, the airlock module, connected the multiple docking adapter and the orbital workshop. It also contained the station's communication, electrical, and environmental control systems. Attached to its inner walls were cylinders of compressed gas that provided the station with air.

The largest Skylab module was the orbital workshop. It was divided into two sections by a wire-mesh floor. The bottom part (the part furthest from the airlock module) served as the astronauts' living area. It contained three bedrooms, an exercise area, restroom facilities, a kitchen, and a living room. The top section contained the station's smaller scientific instruments and two airlocks. Instruments placed in these airlocks allowed Skylab to observe the Earth and sky simultaneously. A third, much larger airlock was located at this end of the orbital workshop and was to be used by the astronauts for spacewalks.

Electrical power was to be provided by six solar panels. Two large panels were located on either side of the orbital workshop and four smaller collectors were attached to the top of the Apollo telescope mount.

The Mission of Skylab

Skylab's mission was very broad, including activities in the fields of astronomy, Earth resources, biology, medicine, meteorology, materials science, and manufacturing.

The astronomy program had two primary goals. The first was to measure the emission of stars and galaxies at ultraviolet and x-ray wavelengths. These forms of light are absorbed by atoms in the Earth's atmosphere and cannot be observed from the ground. Its second astronomical objective was to study the Sun to learn more about its energetic outbursts and how these events affect the Earth.

A second important focus of Skylab was to study the Earth. From its vantage point high above the atmosphere, the station's cameras could photograph 75 percent of the Earth's surface and 80 percent of its developed land. These images, obtained through a series of special filters, provided information about the negative impacts of agriculture, air and water pollution, and forestry on the environment. Additional photographic studies would provide information to researchers in the fields of geology, geography, oceanography, and meteorology.

The astronaut crew was expected to remain on Skylab for several weeks at a time so medical tests would be routinely conducted to monitor the influence weightlessness was having on their bodies. This data would be employed by NASA doctors to construct new exercises to counteract any negative effects. Doctors knew very little about the long-term medical problems posed by weightlessness and, therefore, wanted to proceed in a cautious manner. In agreement with these wishes, the initial Skylab crew would occupy the station for only twenty-eight days, twice the length of the American space endurance record set by *Gemini 7*.

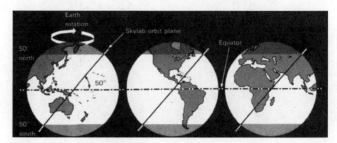

Figure 31-3. The orbit of Skylab was inclined 50° with respect to the Earth's equator. This allowed much of the Earth's surface to be studied from orbit.

During postflight exams doctors found that the *Gemini 7* astronauts had lost muscle tissue, red blood cells, and bone mass. Their cardiovascular systems had also been weakened and their body fluids had migrated from the lower to upper parts of their bodies. Doctors did not know whether these changes would become even more of a problem on a long duration mission or whether they stabilized at some safe level. Skylab astronaut Alan Bean considered the investigation of these medical issues to be one of the primary goals of the entire Skylab program, "We really don't know what's going to happen to the guys in the long term, and finding out is probably, in my mind, the single most important thing we've got to do in Skylab."

Skylab's materials science and manufacturing experiments were designed to determine whether more precisely formed crystals and purer alloys could be made in a weightless environment. If so, these substances might be of use in the construction of advanced electronics. NASA hoped Skylab's materials processing experiments might raise commercial interest in space-based manufacturing facilities.

Since Skylab was America's first space station, a final important objective was to determine how the station and its equipment would perform over extended time periods. Space stations were intended to have minimum operational lifetimes of several years. Therefore NASA was interested in finding ways in which future stations could be made more reliable and conducive to human habitation.

During the Skylab program, three astronaut crews would occupy the station for twenty-eight, fifty-nine, and eighty-four days, respectively. In all, the astronauts would conduct 270 experiments. The results of these studies would then be distributed to 182 scientists, many of whom were located in foreign countries.

The Launch of Skylab

The unmanned Skylab space station was placed into a nearly circular 270-mile-high orbit by a Saturn V on May 14, 1973. NASA officials were elated by this successful launch, but telemetry readings soon indicated that a serious problem existed. Sensors suggested that the station's external micrometeoroid/thermal shield had been torn away during the craft's journey into space.

Although scientists considered it very unlikely that this shield would be needed to protect the station from micrometeoroids, it had an important second role, to prevent sunlight from overheating the station. In the absence of this shield, the temperature within the station soon rose to 190°F, a level well above what an astronaut could tolerate. Officials also feared that the sensitive materials and instruments within the station were being destroyed by this intense heat.

Ground controllers discovered that by adjusting the spin and orientation of the station, the interior temperature could be reduced to around 105°F. However, this procedure was rapidly consuming the station's fuel, so an alternative method had to be worked out. Telemetry from Skylab also indicated that an additional serious problem existed, one of the station's two large solar panels was missing and the other had failed to deploy. Without these panels, Skylab would not have enough power to operate its instruments.

Plans to Repair Skylab

NASA officials realized that the loss of Skylab's micrometeoroid/thermal shield threatened this $2.5-billion program. Once controllers implemented procedures to temporarily stabilize the craft's temperature, round-the-clock sessions were held at Huntsville, Houston, and at other locations to find additional ways to lower the station's temperature.

Two possible methods were found; but both had serious drawbacks. One solution called for the astronauts to enter the lab and deploy a parasol-like device through the Sun-viewing airlock of the orbital workshop. This fixture would reduce the amount of solar heating, but would also permanently obstruct one of the station's two scientific ports. The second option required the astronauts to place a Sun shield over the craft manually. Although this seemed to be a straight forward task, it required a difficult Extra Vehicular Activity (EVA).

Officials also knew that the jammed solar panel had to be freed if the station was to become operational. After various options were discussed, officials decided to allow astronaut Charles Conrad to pilot the Apollo command module close enough to the stuck panel to permit his crewmate, Paul Weitz, to free it during a stand-up EVA.

In the days immediately following the launch of Skylab, procedures to accomplish both of these repairs were rehearsed in the neutral buoyancy water tank at Marshall Space Flight Center in Huntsville. After a week and a half of practice, the astronauts felt confident they would be able to make the required repairs.

The Repair Mission

Astronauts Charles Conrad, Joseph Kerwin, and Paul Weitz lifted-off from the Kennedy Space Center on May 25, 1973, eleven days after the launch of Skylab. Their objectives were to repair the station and then begin its scientific program. Their Saturn IB performed flawlessly and placed them into the desired orbit. Once the CSM arrived at the space station, Conrad confirmed what NASA engineers had feared, "Solar wing two is gone completely off the bird . . . solar wing one is . . . partially deployed." After completing an inspection of the jammed solar panel, the astronauts radioed Houston that they felt confident it could be freed.

Once the crew had eaten dinner, Weitz attempted to free the panel by standing in the hatch of the command module and pulling at the restraining debris. Despite using all his strength, he was unable to free the array. Conrad radioed the disappointing news back to Houston, "We ain't going to do it with the tools we got."

After this failure, controllers directed the astronauts to dock with the station while ground support teams tried to

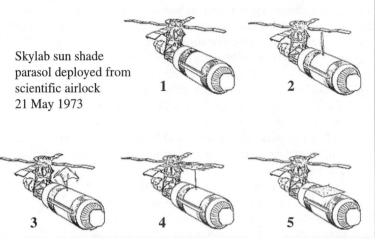
come up with another solution. As Conrad piloted the command module into the station's docking mechanism, he was shocked to find that the docking latches failed to close. Three alternative docking methods were tried, but none worked. As a last resort, the crew depressurized their capsule, opened the forward tunnel hatch to remove the probe's back plate to bypass some of its electrical connections. After centering the probe and drogue, they used the command module's thrusters to bring the craft together. This time, when the two craft met all twelve latches snapped shut. They were docked!

The next day astronauts Weitz and Kerwin entered the blistering station, describing the heat as "like the desert." They pushed an umbrella-like device through the Sun-facing airlock and then opened it, creating a parasol-shaped heat shield. This task took 2 hours to complete because the astronauts had to make frequent trips back to the command module to cool off. Despite these difficult working conditions, their efforts paid off. Over the next four days the station slowly cooled down to a comfortable 80°F (see the Focus Box above). On May 29th, the station was declared livable and the astronauts began conducting their planned medical tests, solar observations, and Earth resources experiments.

The solar panels on the Apollo telescope mount supplied the station with 4.5 kilowatts of power, but this was uncomfortably close to the minimum needed to operate the craft's instruments and scientific equipment. On May 30th, this problem was made worse when one of the station's batteries died. If the remaining large solar panel could not be freed, NASA officials felt the station would have to be abandoned.

Figure 31-4. The OWS solar panel (light-colored fixture) was held closed by a restraining strap (labeled A). The gray material (B) is part of the damaged micrometeoroid shield.

Figure 31-5. Skylab as photographed by the approaching crew of Skylab 3. *The docking port is located at the right end of the station.*

Freeing Skylab's Stuck Solar Panel

While the Skylab crew carried out their schedule of activities, Russell Schweickart, a veteran of *Apollo 9*, was hard at work in the water tank at Huntsville searching for a procedure to release the jammed solar panel. He found a way to do this, but it involved a hazardous EVA by two astronauts.

When Schweickart's scheme was radioed to the station, both Conrad and Kerwin were skeptical, but resolved to give it a try. After putting on their spacesuits and leaving the station they climbed toward the stuck panel carrying a nylon rope and a cable cutter. The plan called for them to cut through the obstructions holding the panel down and then use the rope to pull the panel into place. After working for half an hour, Kerwin made a slight modification to Schweickart's procedure that allowed him to cut the strap holding the panel. Together, the two astronauts then pulled the array into place. Conrad happily radioed Houston, ". . . those panels were out as far as they were going to go. . . ." A short time later the power within Skylab jumped to nearly 7 kilowatts, solving the station's power problem.

By the end of their twenty-eight day mission not only had the astronauts made Skylab operational, they completed 100 percent of their medical program, 80 percent of their solar observations, and 60 percent of their Earth resources experiments. The three astronauts were rightfully proud of their accomplishments and were pleased to leave Skylab in great shape for its next crew.

32

The Conduct of Long-Duration Missions: Lessons from Skylab

Skylab showed that man is a more adaptable creature, and space a more suitable home for him, than anyone had previously expected.

—Henry Cooper in *A House in Space*

The scientific program of Skylab focused on four primary areas: astronomy, biology, Earth resources, and space medicine. The results of these studies were important and they are discussed in the Appendix A-3. However, Skylab was also intended to answer additional questions. Can astronauts productively work in space and what types of activities can they complete? How should the procedures, developed for short-duration missions, be modified to accommodate long-term missions? How might future space stations be made more productive and conducive to human habitation? These questions looked to a future where extended duration missions would be more common. NASA engineers realized Skylab was America's first space station and that design flaws were to be expected. They eagerly looked forward to identifying these faults and finding solutions to make future stations more functional.

The Skylab Crew Accomplishments

The Skylab space station was occupied by three teams of astronauts. These missions were named *Skylab 2, 3,* and *4.* The first crew was commanded by Charles Conrad. Joseph Kerwin and Paul Weitz served as the mission's scientist/pilot and pilot, respectively. As discussed in the previous chapter, *Skylab 2* was launched on May 25, 1973, and had three main objectives: (1) to deploy a parasol-shaped Sun shield on the exterior of the station, (2) to free a jammed solar panel, and (3) to complete experiments in each of the fields listed above. During their twenty-eight-day flight, the crew met the first two objectives and completed forty-six of their planned fifty-five experiments.

The *Skylab 3* crew consisted of astronauts Alan Bean (commander), Owen Garriott (scientist/pilot), and Jack Lousma (pilot). Launched on July 28, 1973, these men extended the space endurance record to fifty-nine days, tested the astronaut maneuvering unit developed during the Gemini program, deployed a new Sun shield, repaired nine pieces of equipment, and replaced six of the station's gyros. They also obtained 305 hours of solar observations, 16,000 photographs and 18 miles of Earth Resources photographs, and completed 333 medical experiments.

Figure 32-1. Skylab's first crew was commanded by Charles Conrad (center) and was manned by Joseph Kerwin (left) and Paul Weitz (far right).

Gerald Carr commanded Skylab's third mission. His crewmates were Edward Gibson (scientist/pilot) and William Pogue (pilot). Like the prior crews, this astronaut team made repairs to the station and completed an extensive research program including 338 hours of solar observations, fifty-six experiments, and twenty-six science demonstrations. In all they spent 1,563 hours working on various research tasks.

Based on the above statistics, it is impossible to argue that the Skylab crews were not productive. They were able to complete experiments in a variety of fields as well as to perform complex repairs and routine maintenance tasks. Skylab proved that astronauts could work productively in space on a large variety of problems.

Short-Term and Long-Term Missions

Although the Skylab astronauts were obviously productive, it is also true that as the duration of the missions increased, so did the number of conflicts between the crews and ground

Figure 32-2. The second crew of Skylab (left to right), Owen Garriott, Alan Bean, and Jack Lousma.

personnel. Some of these conflicts resulted from the personalities of the crew members, but most were caused by the differing role perceptions of the ground teams and the crews.

Differences in the Skylab Crews

Charles Conrad and Alan Bean, the commanders of *Skylab 2* and *3*, were the only space veterans among the Skylab astronauts. Conrad had flown on *Gemini 5*, *Gemini 11*, and *Apollo 12,* and Bean had served as the lunar module pilot on *Apollo 12*. In contrast, Gerald Carr, Skylab's third commander, had served on the support teams for *Apollo 8* and *12*, but had never flown in space. As it turned out, the flight assignments given to the Skylab commanders were peculiar. The most experienced astronauts commanded the shorter missions, whereas the least experienced commander was placed in charge of the longest mission.

Conrad's crew also had a distinct advantage over the other two, since all of its members were former Navy officers

Figure 32-3. The third Skylab crew (left to right), Gerald Carr, Edward Gibson, and William Pogue.

who had previously served together. They shared a common background and worked exceptionally well together. During their mission they maintained cordial relations with ground controllers, and after returning from space, they had few complaints about the station.

Bean's crew members were the favorites of Mission Control. They cheerfully did everything asked of them, they were efficient, and even asked for more work. Like Conrad's crew, they enjoyed working together and quickly established good relations with ground controllers. After the flight, they had few suggestions on how the station might be improved.

Skylab's third crew was the "antithesis" of the earlier two crews. Soon after boarding Skylab and starting their experimental programs, poor relations developed between them and ground officials. The crew often felt frustrated by the station's organization and complained that numerous items were misplaced. They felt ground controllers were unduly rushing them to complete tasks and towards the end of their mission sharp exchanges took place.

Mission Parameters and the Role of Ground Controllers

Each Skylab flight would set a new American space endurance record. At the time of the launch of *Skylab 2*, the longest spaceflight was the twenty-four-day mission flown by the ill-fated crew of *Soyuz 11*. Skylab's first flight would beat this Soviet record by four days, but NASA wanted to test the boundaries of man's endurance, not merely increase the existing records by nominal amounts. If space stations were to become economical orbiting laboratories, procedures had to be established that allowed crews to spend months in orbit. It was simply impractical to shuttle crews continuously to and from the space station.

In keeping with NASA's cautious approach to manned spaceflight, the duration of each Skylab mission would not be dramatically increased. Each mission's length would be approximately twice as long as the former one. Doctors feared extended exposure to weightlessness might cause irreversible damage to an astronaut's body, so they insisted that mission lengths be restricted to gradual increases.

The Skylab flights were also a new experience for personnel at Mission Control. In the Mercury, Gemini, and Apollo flights, astronaut time was always carefully rationed. Since the primary goal of these short missions was to collect as much new data as possible, the comfort of the astronauts was not given a high priority.

Over the years an institutionalized desire had developed within NASA to exceed preflight expectations. In postflight press briefings, officials took pleasure in being able to say that mission objectives had not only been met, but additional tasks had been completed. By the time of Skylab, overachievement had become an unwritten objective that permeated the thinking at all levels of the manned space program. Because the astronaut's time in space was so limited, planners constantly sought ways for tasks to be more efficiently completed. Normally, a spaceflight is packed with activities, so the easiest way of accomplishing this was by having the astronaut work longer hours or at a faster rate. As a result, by the end of a mission the astronauts were often exhausted, but the additional achievements helped them quickly forget the strains under which they had been placed.

A different approach was adopted for NASA's long-duration missions. Dramatic differences exist between the relatively short Mercury, Gemini, and Apollo flights and space station missions. Time for rest periods was allowed and a less rigid work schedule was followed. Despite the wisdom inherent in this new approach, it collided with the NASA mindset of maximizing the use of an astronaut's time while in space.

NASA's Institutionalized Desire to Excel

Compared to the use of automated probes, manned space flight was enormously more costly. In order to justify these expenses, NASA officials constantly searched for examples to demonstrate how the presence of an astronaut provided significant benefits to the mission. One of the primary ways of achieving this goal was to exceed preflight expectations. As a result, when the astronauts were specifically told that once in space they were to "relax a bit," neither the astronauts nor ground controllers took the admonition very seriously. Henry Cooper, author of several books on the space program, succinctly described the extent of the ingrained desire to exceed expectations in his book *A House in Space*, "It was probably as impossible for Mission Control—or the astronauts, for that matter—to follow instructions to slow down a spaceflight as it would have been for the United States Marines to have followed an order to assault a beachhead gently."

Micromanaging Astronaut Time

As controllers monitored the activities of the *Skylab 2* and *3* astronauts, they learned how much time was required for crew members to complete even the smallest task. Armed with this information, they began to revert to their old ways of more closely scheduling activities. The Skylab astronauts themselves, particularly the second crew, inadvertently encouraged this process by asking for more work. In writing about this request, Cooper noted, "About the worst thing the flight controllers who were known around NASA as a 'highly motivated group' could imagine was having three astronauts up in space with nothing to do." Accordingly, timelines were compressed and additional experiments were added. By the time Carr's crew occupied the station, controllers felt they had accumulated enough information to schedule every minute of astronaut time efficiently. The initial response of Carr's crew to the burden this placed on them was not to complain, but to work longer hours and at a faster rate. Carr later

wrote that his crew offered only a modest amount of resistance to their accelerated schedule,

> We had told the people on the ground before we left that we were not going to allow ourselves to be rushed; yet we got up [there] and we let ourselves just get driven into the ground! We hollered a lot about being rushed too much, but we did not, ourselves, slow down . . . and do things just one after another like we said we were going to do.

As each mission progressed, controllers worked to maximize the number of completed experiments. Flight Director Neil Hutchinson clearly became caught up in this process. He described the rationale the controllers used to quicken the pace of the mission, "The initial purpose of Skylab may have been to explore simply how to live in space, but the cost of the program—two and a half billion dollars, caused us to change our minds." Unfortunately, the "us" Hutchinson referred to only included the controllers. In response to this change in attitude, more and more detailed instructions were sent to the astronauts and their schedules were tightened. With a sense of satisfaction, Hutchinson wrote,

> Back at the first mission, we weren't good enough to schedule the guys tight, but by the time the second mission ended, we knew exactly how long everything took. We knew how long it took to screw in each screw up there. We could have planned a guy's day without leaving a spare minute if we wanted to—we had that ability. We prided ourselves here that, from the time the men got up, to the time they went to bed, we had every minute programmed. The second crew made us think this way. You know, we really controlled their destiny.

Figure 32-4. Bean struggles with teletype instructions sent from the ground that insisted in staying rolled-up.

To Hutchinson and his fellow controllers this was a laudable situation and meant they were doing their jobs well. He continued,

> We sent up about 6 feet of instructions to the astronaut's teleprinter in the docking adapter every day—at least forty-two separate sets of instructions—telling them where to point the solar telescope, which scientific instruments to use, and which corollaries [other experiments] to do. We lay out the whole day for them, and the astronauts normally follow it to a T! What we've done is we've learned how to maximize what you can get out of a man in one day.

Despite the sense of accomplishment this procedure gave ground controllers, it was becoming harder and harder for the independently minded astronauts to bear. They resented being treated as robots, and as they worked faster, they made mistakes. Carr's crew especially Pogue, became frustrated by this situation and he radioed controllers,

> Now, I don't like being put in an incredible position where I'm taking somebody's expensive equipment and thrashing about wildly with it and trying to act like a one-armed paperhanger trying to get it started in insufficient time! . . . Somebody thinks something up in an office, it sounds good, and then all of a sudden you find yourself trying to do it for the first time [up here]; never having done it before, you're gonna take probably four or five times as much time to do the task than the man who has been needling the flight planners to have it included said it would [take].

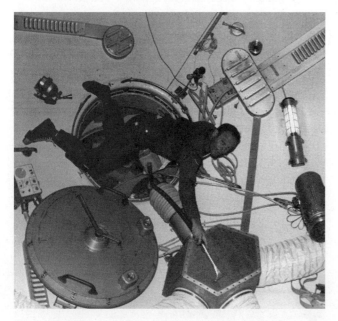

Figure 32-5. Because of the directional air flow within the station, Lousma (above) found that items tended to float up and adhere to the dust and filter screen on the workshop ceiling.

Realizing there was a limit to what he could handle, Pogue came to feel things were getting out of hand, he later wrote,

> "I came to realize, during Skylab, that what we were doing was taking a human and making him function in a way he was not designed to. . . . We were trying to function at a higher level of efficiency than we could. I then proceeded to make errors and berate myself. . . . When I tried to operate like a machine, I was a gross failure. . . . I think a person needs to more or less recreate himself, to pause and reflect occasionally. . . . We've got to appreciate a human being for what he is."

A third of the way through their mission, Carr realized that the crew and ground controllers "were not yet marching to the same drumbeat." The old ways of conducting flights at Mission Control were exhausting his crew. In assessing this situation Cooper wrote,

> Actually, everyone except the third crew had lost sight of what was to have been the chief purpose of Skylab, and that was to see whether men could really live in space for long periods of time; NASA had originally construed the word live to mean decently, as one might on Earth, with regular shifts and time off for relaxation.

By the halfway point of their eighty-four-day flight, Carr felt he had no choice but to take some action. On December 28 he transmitted a frank message to Mission Control, "We'd all kind of hoped before the mission . . . [that] everybody had the message that we did not plan to operate at the [previous crew's] pace. . . . Are we behind? . . . We'd like to have some straight words on just what the situation is right now." After this admonition, officials instructed the ground controllers to reduce the astronaut's work load. Upon reflection, Hutchinson later realized why his group had such a difficult time resisting the urge to assign Carr's crew more work, "Our system was designed to squeeze every minute out of an astronaut's day. . . . Suddenly the system is asked to stop for a few hours, or a day, to give a man some time off. The system doesn't want to do that!" He then added,

> . . . I saw we'd done a bad thing by forcing them. I saw they needed time to think about what they were doing and to reestablish themselves. They were not asking for time to read beddy-bye stories! . . . We now see that time off is mandatory. A man has to get mental enjoyment out of something other than his work. We now feel that an astronaut's time must be inviolate.

Carr later regretted he had not spoken up sooner. He and his crew knew a problem existed, but instead of objecting they chose to try to accommodate the additional requests. To their dismay, they found this attitude only encouraged the controllers to seek even more ways of compressing their schedule. Carr wrote, "We swallowed a lot of problems for a lot of days because we were reluctant to admit publicly that we were not getting things done right. That's ridiculous [but] that's human behavior."

Inter-team Communications

An additional problem encountered during Skylab was the poor manner in which ground support teams interacted with each other. In the short-duration Mercury, Gemini, and Apollo spaceflights, support teams had well-defined areas of responsibility and interactions between these groups were minimal. However, situations sometimes arose aboard Skylab when the expertise of more than one team was needed. In these cases decision-making processes did not always run smoothly. When NASA doctors realized Carr's crew was feeling stressed, the medical staff was not sure how to make their concerns known. Dr. Jerry Hordinsky, the crew surgeon for Carr's mission, described their dilemma,

At conferences, when we were on the side of easing up, of saying that the flight plans were too much, the engineers couldn't understand what we meant. We witnessed Mission Control getting off on the wrong foot, but there was no place to blow the whistle . . . communication between us and the flight planners was not good. They told us that "the flight schedule was a nonmedical duty" so there was a bad interface. It took us three weeks to see what was going on; then we went to bat.

One of the main lessons learned from Skylab is that a system to resolve the conflicting objectives of support groups must be in place. Teams must be willing to look at a matter from several different perspectives and make decisions based on several different sets of needs.

Making Space Stations More Livable

An additional lesson learned from Skylab was the need to construct schedules and space hardware that allow the astronauts to maintain, as much as possible, a sense of normalcy and structure.

No matter how much preflight planning might take place, situations will occur that cause the crew's anxiety level to rise. To minimize the cumulative effect of these episodes, it is important to establish an orderly daily routine. The more Earth-like this routine is, the easier it will be for the astronauts to maintain their composure. Even a small change can have an effect on an astronaut's temperament. This is illustrated by the following statement made by Bean in a review of the *Skylab 3* mission,

We found the first three or four days that we tended to let the meals move a little bit. . . . It became obvi-

ous after [a while] that there was enough work . . . to work all the time and not ever to eat. I think that kind of upset us a little bit and probably was responsible . . . for the fact that we were upset the first few days. So when we stopped letting the food move around, things kind of stabilized out for us.

This desire for order extends to the placement of fixtures in the station. The astronauts found it disorienting when the Skylab workstations lacked a single vertical axis. As Bean later told engineers, "If you want to put everything on the ceiling instead of on the floor, we can sure handle that. It's just that we don't want half and half."

Bean also pointed out that once the astronauts entered the station, they needed some time to learn where things were stored and to get accustomed to weightlessness. He estimated it would take an astronaut three times longer to set up an experiment at the beginning of a mission than near its end. Garriott basically agreed with this assessment, telling postflight debriefers, "Seems to me that after eight to ten days, I decided that it just seemed natural to be in zero gravity. . . . It's comfortable; it's pleasant."

The astronauts found that little organizational matters, if left unattended, could evolve into time-consuming annoyances. This situation was especially troublesome when different crews occupied the station. On Skylab there were 40,000 separate items stored in over a hundred cabinets. During preflight training sessions, each item was stored in its proper place. In space, however, the astronauts tended to let this housekeeping chore slip. Bean defended this practice as a time-saving measure. In a briefing from Skylab, he told Carr not to become overly concerned with this matter, "We just leave stuff around. We'll stow it before we go, otherwise you get in the business of stowing and unstowing if you're not careful."

Although Bean's suggestion seemed reasonable to Carr prior to his flight, immediate problems arose when his crew occupied the station and began their work. Carr later recalled,

Things were not stored where they were supposed to be stowed. We got ourselves into a mode of having to ask ground where everything was. In some cases ground pointed out proper places . . . and in other cases we just had to look for things till we found them. . . . Everything took two or three times as much time as we thought it would take.

Over time housekeeping tasks became less important to the crew occupying the station because they were familiar with where things were located. However, when a new crew came aboard, they found it took them a great deal of time to find objects that were not in their proper place.

A second item associated with the livability of the station dealt with preflight training. Jack Lousma, the pilot on Bean's crew, told officials that working in a weightless environment presented some problems their training sessions had not du-

Figure 32-6. Gibson particularly enjoyed monitoring the Sun with Skylab's array of solar telescopes.

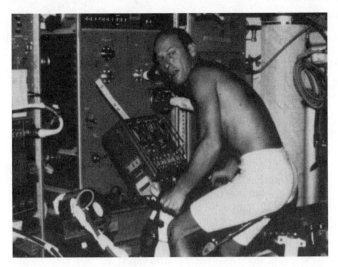
Figure 32-7. Conrad using the station's bicycle ergometer to exercise his body's cardiovascular system.

plicated very well. In particular, he found it difficult to fix instruments that required many small parts. He told debriefers,

> It's not difficult to handle one or two small items, but if you have many, many small items, it is difficult to handle them because they all want to float in their own different directions and all you have to do is blow on them or let them bump into each other and touch something, and they go off. You spend all of your time grabbing things to keep them from getting away.

Lousma contrasted this situation with that on Earth where a dropped object lands relatively nearby. In space, objects could travel all the way to the other end of the station and become lost. He also noted that if he placed small objects in his pockets, they tended to float out and drift away. In his opinion, the solution to this problem was to construct a workbench where objects could not escape.

A third problem area the Skylab astronauts identified was the way in which experiments were conducted. In long-duration missions they felt it was important for the astronauts to become more actively involved with their experiments. They argued this would produce several benefits: (1) important transient and unpredictable phenomenon could be recognized and studied, (2) peculiarities in the experiment could be noted and examined, (3) the causes of failures could be more easily documented, and (4) crew members could be kept mentally stimulated and challenged. Gibson wrote,

> I find that so many of the things we have on board, you do by rote, or by checklist, that you don't think about them; they're just push-the-button-and-make-sure-it-works type of experiments. There's nothing wrong with some of those; you can learn an awful lot from them. But they sure are hard on the operator

if you're going to do that all day. And I think it is the solar telescope, and the out-of the window observations of the Earth below, that keep us challenged and mentally awake. Without these, you'd be ready for the rubber room when they brought you back.

A final issue dealt with the human need not to feel excessively confined. Most of the astronauts recommended that a large part of the spacecraft be uncluttered and designated as a recreational area. In Gibson's view, "You will need [a] place where people can get away from any claustrophobia which they might get in small compartments. At least I feel that if I were penned up [in a small area] for months, it would begin to feel pretty much like a cell."

Extending the Skylab Mission

During Skylab's three manned missions, the astronaut crews spent a total of 171 days in space and the station circled the Earth 2,476 times. When the final crew landed on February 8, 1974, Skylab still contained a five-month supply of oxygen and a six-week supply of water. Computer models suggested the station would not reenter the Earth's atmosphere for another nine years.

Despite these facts, NASA had no plans to send another crew to occupy the station. Faced with a shrinking space budget, the decision was made to focus on the Shuttle program. Nevertheless, officials were hopeful Skylab would still be in orbit by the time the Shuttle was ready to fly. If a use could then be found, a propulsion unit could be carried to the station to lift it into a higher orbit. Unfortunately, delays in the Shuttle program and a more rapid decay of the station's orbit than predicted made this plan impossible. Skylab reentered the atmosphere and broke up over Australia on July 11, 1979.

The Unused Legacy of Skylab

In retrospect, NASA's Skylab program fell victim to the same shortcoming that had prematurely ended the Apollo program. The agency had not presented Skylab as a preliminary step leading to a permanent Earth-orbiting space station, but as an end in itself. In the public's mind Skylab's goal was to complete a limited number of scientific experiments. After acquiring 941 hours of solar observations, 46,000 Earth re- sources pictures, and 825 hours of life science data, this goal had been met. There was little support for extending a costly program whose main contribution would be to produce incremental results. Despite the insights Skylab provided for the construction of Earth-orbiting space stations, the application of this knowledge would have to wait for another time. A second fully functional copy of Skylab was decommissioned by NASA and placed on display at the National Air and Space Museum in Washington, DC.

33   The Apollo-Soyuz Test Project

Only a few years before, the idea of a joint American–Soviet space mission had been judged unthinkable. Now five space-men from both countries, gathered 140 miles above the planet with multinational spacecraft, held press conferences with reporters from several nations on earth.

—Moon Shot, p. 357

Public support for the Apollo program had dropped severely by the early 1970s. According to historian Joan Hoff there were four other reasons behind the deceleration of the American space program: (1) unlike Presidents Kennedy and Johnson, Nixon was not interested in space exploration. When serving as the vice-president in the Eisenhower administration, great efforts had been made to limit the growth of NASA. Nixon's inclination to restrain NASA was compounded by the fact that none of his closest advisors favored a vigorous space program, (2) NASA officials failed to adjust to the new budgetary processes implemented by the Johnson, Nixon, and Carter administrations, and were arrogant, even hostile, to the suggestion that their cost-control measures were inadequate, (3) the cold-war competition between the United States and Soviet Union that had led to NASA's rapid early expansion had been replaced by an era of "detente" where cooperation was encouraged, and (4) costly domestic initiatives focusing on the environment, crime, and urban renewal drew resources away from the space program and into these areas.

With the early termination of the Moon program, several years existed in the mid-1970s in which no manned flights were scheduled. Skylab would be launched in May 1973, but this program would end in less than a year. The Space Shuttle had yet to be built and no manned test flights were expected before the end of the decade. In order to retain its astronaut corps and trained flight controllers, a manned mission was needed to bridge the gap between Skylab and the first shuttle flight.

NASA's search for an interim mission happened to coincide with a lull in the cold war between the United States and Soviet Union. Agreements were signed between the two super powers on space law, the nonproliferation of nuclear weapons, and on the reduction of strategic armaments (SALT-1).

Building upon this momentum, President Nixon and Soviet Premier Kosygin signed a protocol called the "Agreement Concerning Cooperation in the Exploration and Use of Outer Space for Peaceful Purposes" on May 24, 1972. Under the terms of this accord, the United States and Soviet Union were to "develop cooperation in the fields of space meteorol-

ogy, study of the natural environment, exploration of near Earth space, the Moon and planets, and space biology and medicine." As a part of this agreement, a detailed book, *Principles of Space Biology and Medicine,* would be written jointly by American and Soviet doctors.

Much to NASA's delight, the protocol also called for the launch of a joint American–Soviet space mission. Such a mission was possible because it was in the political interests of both countries. In the early 1970s the Soviet Union encountered a series of space setbacks, including the deaths of the *Soyuz 11* crew in mid-1971. This disaster, coupled with their failure to place a man on the Moon, had tarnished the Soviet Union's image as a world power. By engaging in a space mission as an equal partner to the United States, they hoped to restore some of their lost prestige. Soviet space officials also hoped to learn more about the management and quality-control techniques America had successfully employed in the Apollo program. Russian scientists enthusiastically supported the Nixon/Kosygin agreement because it increased the resources available for space research. Boris Petrov, chairman of the USSR Interkosmos Council stated,

it is difficult to overestimate the significance of the agreement which has been concluded . . . outer space is becoming, in all aspects, an arena for broad international cooperation and demands joint efforts of many countries, especially those countries which already have made considerable achievements in this matter.

The United States saw at least five benefits by agreeing to cooperate in a joint space mission: (1) it would showcase the American skills developed in the lunar program before a worldwide audience, (2) it would demonstrate America's willingness to reduce cold-war tensions, (3) it would help repair the damage to America's "superpower status" after its defeat in Vietnam, (4) it would provide an opportunity for American scientists and engineers to gain firsthand knowledge about the capabilities of the Soviet space hardware, much of which was used by the military, and (5) it would meet NASA's need to bridge the gap between Skylab and the Shuttle in a relatively inexpensive manner. Little additional hardware would have to be built since equipment already purchased for the now completed Apollo program could be employed.

Even the *Wall Street Journal*, an ideological foe of communism, saw Apollo-Soyuz as a positive step in East–West relations. In a May 16, 1972 article, the space initiative was praised, "after years of competition and years of almost fruitless talks about cooperation, both sides are moving toward collaboration on specific projects." A joint American–Soviet spaceflight was approved by Congress with little dissent and was hailed by leaders in both countries as the beginning of a new era of United States–Soviet cooperation. A mutually agreed upon launch date was set for July 15, 1975.

The Joint American–Soviet Space Mission

Initially, the American–Soviet expedition called for an Apollo and a Soyuz craft to dock at each end of a Salyut space station. The crews would then engage in a series of experiments. However, this ambitious scheme was cancelled when it became evident the space station would not be ready by the proposed launch date. A modified flight was therefore planned which included a smaller number of scientific experiments and the joint docking of an Apollo and a Soyuz craft. The mission was appropriately named the Apollo-Soyuz Test Project (ASTP).

Crew Announcements and Flight Protocol

Both nations broke with tradition by announcing their crew selections two years before the mission's scheduled liftoff. On January 30, 1973, it was announced that the American crew would consist of astronauts Tom Stafford, Donald Slayton, and Vance Brand. About four months later the Soviets announced their crew would consist of cosmonauts Alexei Leonov and Valery Kubasov. Given the political overtones of this mission, both countries were eager to be seen as coopera-

Figure 33-1. The American and Soviet Apollo-Soyuz Test Project team. From left to right, Donald Slayton, Thomas Stafford, Vance Brand, Alexei Leonov, and Valery Kubasov.

tive. To emphasize their coequal status in this endeavor, officials decided that the American crew would speak Russian and the Soviet cosmonauts would speak English during the flight.

The ASTP Crew Members

Thomas Stafford, the commander of the American crew, was the most experienced space veteran on either team, having flown on *Gemini 6, 9,* and *Apollo 10.* He held the rank of brigadier general in the air force and had coauthored two books while serving as a flight instructor at Edwards Air Force Base, *The Pilot's Handbook for Performance Flight Testing* and *The Aerodynamics Handbook for Performance Flight Testing.* During training sessions Leonov would often burst into laughter after listening to Stafford attempt to speak Russian with his heavy Oklahoma twang.

Apollo-Soyuz would be Slayton's first spaceflight. Sixteen years had passed since he had joined NASA as one of the original seven Mercury astronauts. At the age of 50, he would be the "old man" of the mission. Time had not changed Slayton much since his early days as a World War II fighter pilot. He rarely used diplomatic language when expressing his views. When asked if he looked forward to this flight, he replied, "Mister, I'd get on my hands and knees . . . to make this flight."

The last member of the American crew, Vance Brand, held a B.S. degree from the University of Colorado in Aeronautical Engineering and an M.B.A. from the University of California at Los Angeles. He had served as a jet pilot in the marine corps and then as a test pilot for Lockheed Aircraft Corporation. By the time of his selection as an astronaut in 1966, he had flown in thirty different types of aircraft and had logged over 8,700 hours of flying time. During the Apollo program he served on the support crew for *Apollo 8* and *13* and as backup command module pilot for *Apollo 15.* He would have flown to the Moon aboard *Apollo 18* if that mission had not been cancelled. While training for Apollo-Soyuz, he also served as the backup commander for the last two Skylab missions. Vance Brand would pilot the Apollo CSM. Although 44 years old, this would be his first spaceflight.

Alexei Leonov was known around the world for completing mankind's first spacewalk during the mission of *Voshkod 2.* He was not only physically fit, but was fun-loving as well. Slayton liked him right from the start,

> Alexei was something of a story book. The guy was a top artist . . . went to a couple of military flight schools . . . and then—which proves he'd stop at nothing to qualify himself in any way he could—he made a hundred jumps as a paratrooper. . . . I had a sort of strange association with Alexei. He'd been teamed up with Oleg Makarov, and these two guys were assigned to fly the first mission to circumnavigate the Moon. That was late 1968, or early '69. Well he didn't go, of course. The Russian lunar program came unglued. And I never went to the Moon, either, so we had sort of a buddy feeling between us.

Kubasov was the "perfect second man for *Soyuz 19*." He was a brilliant flight engineer and an excellent complement to Leonov, who would pilot their craft. During Kubasov's *Soyuz 6* mission, he had conducted several experiments involving space welding, experiment teardown and repair, and metal smelting. Soviet officials jokingly told the American astronauts that "if something goes wrong and equipment breaks down, you have Kubasov, who can weld together whatever has come apart."

Old Adversaries Learn to Work Together

Having engaged in a cold-war competition for over a quarter of a century, there was naturally some initial "political baggage" connected with the Apollo-Soyuz mission. The Soviets distrusted the United States and were deeply disappointed at having lost the race to the Moon. On the other hand, the United States still possessed bitter memories of the late 1950s when the Soviets possessed superior space hardware. During that time America was having trouble placing a grapefruit-sized object into orbit, while the Soviets were orbiting massive, automobile-sized satellites. The Soviets had not been reluctant to call attention to this fact and now that circumstances had changed, some people felt it was time for them to eat crow. Others worried that the Soviets would be given access to America's advanced technology.

Slayton was angered by this type of reasoning and did not hesitate to tell people who made these statements that they were way off-base. In his usual forthright manner, he expressed his view to an interviewer,

Figure 33-2. Deke Slayton was one of the original Mercury astronauts, but the ASTP would be his first spaceflight. Grounded because of an irregular heartbeat, he had waited sixteen years for this chance to fly in space.

Let's get rid of this crap once and for all that the Russians were stuffing our technological secrets under their sweaters. We did not transfer carloads of technical data to them. You had to understand that the engineering of their Soyuz was twenty years old. But you also had to keep in mind that Apollo was hardly new; hell, it was a ten-year-old ship by then. We had already developed entirely new technologies for our Space Shuttle. What the Russians learned from us—if they learned anything at all—it was our system of management.

Apollo-Soyuz was a new experience for both countries, and in a very real sense, it was a ground-breaking political experiment. Slayton understood this fact, but also realized that Apollo-Soyuz would probably be his last chance to get into space. Reflecting on the difficulties each nation faced and his own personal situation, he later wrote, "Both sides just had to learn. And as for me, I had one overriding interest. . . . Let's just do it! Apollo-Soyuz was the last train outa Dodge for ol' Deke."

The Apollo-Soyuz Docking Module

Before the Apollo and Soyuz spacecraft could be joined together, a new docking mechanism had to be built. Responsibility for this task was assigned to the United States, but it would be jointly tested by engineers from both countries. The docking mechanism would be launched along with the Apollo craft using a Saturn 1B. Once in orbit, the Apollo CSM would turn around, reenter the rocket's upper stage in the same manner as was done during the Apollo program, and retrieve it.

In previous multicraft missions one spacecraft was responsible for performing most of the maneuvers. The "stationary" craft acted as a target with the docking mechanism attached to it. In the ASTP either craft could initiate a docking.

The ASTP docking mechanism employed a series of hydraulic attenuators and latches that pulled the spacecraft together. Once joined, the astronauts could travel between the two craft through a pressurized passageway.

Technical Issues Associated with the Apollo-Soyuz Mission

Several technical issues had to be overcome before a Apollo-Soyuz docking and crew exchange could take place. One of the main obstacles dealt with the differing atmospheric compositions within the two spacecraft. When in space, the air in the Apollo craft was almost 100 percent oxygen at a pressure of 5 pounds per square inch (psi). On the other hand, the atmosphere within the Soyuz was a mixture of nitrogen and oxygen at 14.7 psi. If the cosmonauts were suddenly exposed to the atmosphere within the Apollo craft, they would become incapacitated for the same reason deep-sea divers become ill when they rise to the surface too quickly.

Figure 33-3. The ASTP docking unit shown above the Apollo CMS. This unit would provide an air tight connection between the Apollo command module (bottom of picture) and the Soviet Soyuz.

Air bubbles would form in their blood creating a condition known as the bends.

A solution to this problem was reached after compromises were made by both space programs. The Soviets agreed to decrease their cabin pressure to 11 psi and to increase its oxygen content to 40 percent. The Americans agreed to construct the docking module so that its air content could be adjusted to compensate for the high percentage of oxygen in the command module.

Other minor technical problems were also addressed by the two programs. Adjustments were made for incompatible search and rendezvous radars, different communication systems, interfering external fixtures, and incompatible beacons and optical markers.

The ASTP Mission

The launches of the Soyuz and Apollo spacecrafts were closely coordinated. The Soviet flight, called *Soyuz 19*, took off first. It blasted-off into a clear blue sky from Tyuratam within 10 seconds of its scheduled time on July 15, 1975. The Soyuz used the same launch pad employed in the flight of Yuri Gagarin. In a Soviet first, live television views of this lift-off were broadcast.

Once the Soyuz was safely in orbit, John Young was sent to awaken Stafford, Slayton, and Brand. He knocked on the door of their quarters and told the crew, "You're friends are upstairs, right on schedule." Slayton replied, "We're over hurdle number one, now all we have to do is get our [expletive deleted] up there."

On the way to the pad Slayton was amused by Stafford's indifference to the fact that they would shortly be sitting atop a fully fueled Saturn IB. Stafford seemed much more concerned with his Russian pronunciation and repeated various expressions over and over. At the pad, Slayton paused as the others walked ahead and gazed at the gleaming rocket. "It's beautiful," he thought. The crew then rode the gantry elevator to the awaiting capsule. Slayton recalled the moment, "I have to admit I felt pretty good walking across the swing arm to the spacecraft . . . it was only thirteen years overdue. I never

---

Focus Box: The Apollo-Soyuz Docking Module

The docking module had a length of 105 feet, a diameter of 4.6 feet, and a weight of 2,700 pounds. Along its exterior wall were three VHF/FM transceivers, a uv spectrometer, a docking target, canisters of oxygen and nitrogen, and a front coupling device. The interior of the unit contained a life-support system, oxygen masks, and a fire extinguisher. The flow of air from these tanks could be adjusted to produce different mixtures of nitrogen and oxygen.

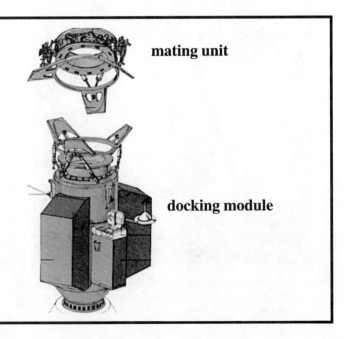

mating unit

docking module

Focus Box: The Apollo-Soyuz Mission

Both the Apollo and Soyuz craft would have to complete a number of maneuvers before a docking between the two craft could take place. These steps are shown in the NASA diagram to the right.

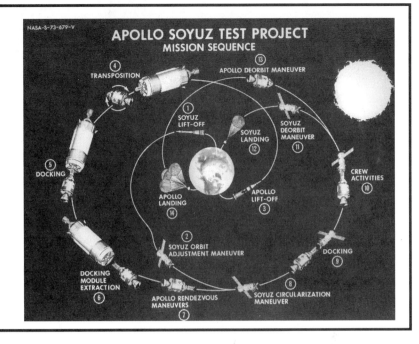

planned on being the world's oldest rookie astronaut but I wasn't going to complain." After a minor problem with an umbilical cord, the Saturn IB blasted off from Pad 39 within 4 seconds of its planned lift-off time. Slayton later wrote of their ascent,

> I'd debriefed every Gemini and Apollo crew, so I wasn't surprised by much that happened. The noise at lift-off was greater than I imagined . . . we had eight engines running back there, and they got even louder as we moved through Max Q, then things began to smooth out. Whoppo! Shutdown was pretty abrupt. You went from being pushed back in your couch to hanging in your straps. We were in zero-G.

The chase to catch *Soyuz 19* had begun.

Once Leonov and Kubasov were in their 133×112 mile orbit, they deployed the Soyuz solar panels and antenna and removed their spacesuits. They then checked the atmosphere control systems within the orbital module. Once docked with the Apollo CMS, the air in their cabin would be adjusted in the manner described earlier. Prior to launch, an on-board television system had failed, and no live images of the cosmonauts were available during the launch or rendezvous. When in orbit an attempt was made to repair this system, but it was only partly successful. As a result, nearly all pictures of the Apollo-Soyuz rendezvous and docking were from the Apollo craft.

When Stafford and his crewmates reached orbit, they were 4,100 miles behind *Soyuz 19*. As the two craft gradually approached each other, both crews were busy making preparations for the upcoming docking by fixing minor spacecraft problems. Docking was scheduled to occur on the 36th orbit of the Soyuz and everyone wanted it to take place without in-

cident and on time. On the morning of the seventeenth, the Soyuz was visible from the Apollo CSM and Slayton radioed the Russian crew. The efforts of both crews were shortly rewarded when a hard docking was achieved 6 minutes ahead of schedule, a few minutes before 11 a.m. Houston time. Soon after the spacecraft were securely fastened together, Soviet Premier Brezhnev issued the following statement,

> Speaking on behalf of the Soviet people, and for myself, I congratulate you on this memorable event. . . . The whole world is watching with rapt attention and admiration [of] your joint activities in fulfillment of the complicated program of scientific experiments.

After docking, the atmospheres in the docking unit and Soyuz were adjusted and the astronauts travelled down the tunnel to the Soviet craft. Stafford entered the Soyuz's orbital

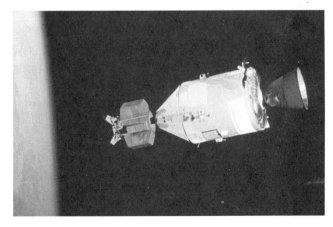

Figure 33-4. The Apollo CSM and docking module as seen from the Soyuz 19 *during the ASTP program.*

Figure 33-5. Soyuz 19 *as seen from the Apollo command module. Leonov would pilot the Soyuz when it docked with the CM. This version of the Soyuz had radar panels.*

module and was greeted by a smiling Leonov. As a worldwide television audience watched, Stafford and Leonov exchanged a genial handshake. From that point onward they would remain friends. Stafford and Slayton were then welcomed aboard the Soyuz by Kubasov.

As the American crew settled into the Soyuz, Slayton took a moment to remind Leonov of an earlier meeting between the two spacemen. During a dinner party in 1965, Leonov had told the audience that he looked forward to toasting an American astronaut on a future joint Soviet–American mission. Now aboard the Soyuz, Slayton reminded him of this statement. To his surprise, Leonov remembered the situation well. Indeed, prior to the flight, he had prepared tubes of food labeled "Russian Vodka" and "Old Vodka." He now offered one to Stafford and Slayton. Stafford was uneasy at breaking one of NASA's flight regulations regarding alcoholic beverages on a spacecraft. Leonov later recalled,

Tom [Stafford] said it was not possible; there were so many people watching on TV, including the President. I said it was a Russian tradition. Tom and Deke drank . . . but inside the tubes was soup. Deke complained, but I told him it was the thought that counts!

The Apollo and Soyuz craft remained connected for two days. During this time, four crew transfers were performed and five joint experiments were completed. Before closing the hatch between the craft, the two commanders exchanged boxes of tree seeds in an ancient Russian gesture of friendship.

On July 19 the Apollo and Soyuz crafts undocked and moved away from each other. They then engaged in a science experiment in which Slayton piloted the Apollo spacecraft so that it passed in front of the Sun to allow the cosmonauts to obtain measurements on the Sun's corona. The spaceships then docked again, this time with Leonov piloting the Soyuz into the docking mechanism. After completing a 3-hour evaluation of the docking mechanism and procedures, the two craft undocked for the last time.

On July 21, *Soyuz 19* began its preparations to return to Earth. Leonov and Kubasov concluded their experiments, put on their spacesuits, fired the Soyuz reentry rockets, and made a textbook perfect landing near the Soviet town of Arkalyk. After exiting the capsule, maintaining a longstanding postflight tradition, Leonov and Kubasov signed the capsule with chalk to commemorate the successful conclusion of their mission.

The Apollo astronauts returned to Earth about three days later. However, their landing was not at all routine. Following normal procedure the command module's excess fuel was dumped during descent, but as the fuel escaped toxic nitrogen tetroxide gas entered the capsule, causing Brand to lose consciousness. Fortunately, Stafford managed to put on an oxygen mask before being overcome by these fumes and then he aided his crewmates.

Focus Box: An Artist's Conception of the Apollo-Soyuz Docking

In the picture to the right, the Apollo command and service module is located at the bottom left and the Soyuz is located in the mid to upper right. Astronaut Stafford is shown in the docking module shaking hands with cosmonaut Alexei Leonov.

Figure 33-6. Alexei Leonov, one of the Soviet Union's best-known cosmonauts, possessed a good sense of humor and was a gifted artist.

After being plucked from the ocean and transported to the U.S.S. *New Orleans*, all three astronauts were hospitalized. Air containing 400 parts per million of nitrogen tetroxide is lethal. Doctors estimated the capsule air contained 300 parts per million of this toxic gas. In Slayton's words, it was a "pretty close" call. The incident was serious enough that the astronauts were not declared to be out of danger until July 29th. A short time later Stafford underwent surgery to remove a spot that had been found in his lung from x-rays taken during his recovery. If this spot had been detected prior to the flight, doctors would not have allowed him to fly. A biopsy later found this growth to be benign.

Science Results of the ASTP

Although the primary goals of ASTP were political, both the American and Soviet crews conducted scientific studies. The study of the solar corona has already been mentioned. The Apollo service module also carried a uv camera, a helium glow detector, and a soft x-ray telescope. Eleven other experiments were conducted by the Apollo crew in the field of materials processing. During the course of these experiments, magnets with properties superior to those manufactured on Earth were produced. A biology experiment was also completed to segregate kidney cells that produced significantly more urokinase (an anti-clotting factor) than on Earth. A series of Earth resource photographs were also acquired for later analysis by NASA scientists.

The cosmonauts also completed several experiments. These primarily dealt with the medical aspects of spaceflight and included the reaction of fungi in spaceflight, embryonic development, fish hatchings, and the growth of microorganisms.

Other Results of the ASTP

The ASTP was largely undertaken to improve the international images of the Soviet Union and the United States. Therefore, its aims were not purely scientific or engineering. In the political realm the ASTP was a great success. The world was pleased to see the two superpowers working together. The Soviets achieved all of their goals. Their space hardware was shown to be equal to that developed by the United States. Morale in the Soviet manned space program rose sharply and the lingering anguish at the loss of the Moon race was diminished. The Soviets also benefited by studying the management and quality-control practices of their American counterparts.

The United States was also pleased with the results of the ASTP. American engineers received technical data about the Soviet hardware they had only been able to guess at prior to this mission. NASA administrators were also given an unprecedented view of the inner workings of the Soviet program. This included information about operational and training procedures developed from years of flight experience. Finally, ASTP provided the bridge NASA needed between the Apollo program and the Space Shuttle era.

Political Aftermath of the Joint Apollo-Soyuz Program

As successful as the ASTP was, it did not lead to additional cooperative programs. The cold-war climate between the two superpowers deteriorated when the Soviet Union sent military forces into Afghanistan. Strong differences also arose over the political situation in Poland. President Carter expressed his displeasure at the way dissidents were treated by Soviet security forces by canceling planned scientific visits to Russia. This tense new period prevented further collaborations in space. Nevertheless, a precedent had been established on how to conduct mutually productive space activities. During the planning for ASTP, numerous technical exchanges had taken place. Forty-four group meetings and joint exercises had been completed and approximately 100 officials from each country had visited their counterparts' facilities. This pattern for cooperation would pay dividends during the Shuttle-Mir program.

The value and limitations of the Apollo-Soyuz flight were concisely described by Matthew von Bencke in his book *The Politics of Space,*

> When astronauts Stafford, Brand and Slayton and cosmonauts Leonov and Kubasov shook hands in space, they pushed back not the frontier of science but the political barriers which had previously prevented cooperation at this level. . . . It would have been unrealistic to have expected more: the facts remained in 1975 that the Soviet Union and the United States were superpowers competing around the world for influence. . . .

Apollo-Soyuz marked the end of an era in the American manned space program. ASTP was the last mission to use an Apollo craft and expendable boosters. After its conclusion, the remaining large pieces of Apollo hardware were donated to various museums and tourist centers in the United States and elsewhere in the world. NASA was now ready to move into a new period of manned exploration utilizing the Space Shuttle.

34 The Space Shuttle

We must sail sometimes with the wind and sometimes against it, but we must sail, and not drift, nor lie at anchor.

—Oliver Wendell Holmes

From the early 1900s explorers and science fiction writers have dreamed of a spaceship that could carry people to distant worlds. Konstantin Tsiolkovsky and Hermann Oberth had written about this possibility, and Goddard's lifelong goal had been to build a craft that could transport people to Mars. The pioneers envisioned ships containing crew quarters, living and work areas, and engine rooms. These imagined spacecraft were more akin to those seen in popular movies such as *Star Trek* and *Star Wars* than to the manned "capsules" flown by America and the Soviet Union.

In the 1930s and 1940s Eugen Sanger and Irene Bradt of Germany worked on the theory of a craft more like that envisioned by the early space pioneers. This totally reusable vehicle could operate within the Earth's atmosphere as well as at the threshold of space. The U.S. military constructed several of these high-altitude and high-speed aircraft after the second world war. The Bell X-1 flew faster than the speed of sound in 1949, and the X-15 operated at the fringe of space. Built in 1959 by North American Aviation, the X-15 provided unique data about high-performance rocket engines whose power could be adjusted during flight. As part of this program, special alloys were manufactured to withstand the high temperatures generated by atmospheric heating. The engineering data collected during the X-15 program served as an important resource for the next generation of high-altitude reuseable craft. NASA used this information in building the Space Shuttle. As Sanger and Bradt later wrote, "It was the only existing data base on winged manned reentry vehicles available when

Figure 34-1. Eugene Sanger was inspired by the work of Hermann Oberth to conduct theoretical studies on a spacecraft capable of operating in near-earth orbit and then landing like a normal aircraft.

the development of the Space Shuttle was begun in the 1970s."

The Roots of the Space Shuttle

In the 1950s, von Braun helped write a series of articles on manned spaceflight for *Collier's* magazine. On the front page of the March 1962 edition (see Figure 34-2), a sleek three-stage reusable ferry was depicted. These articles portrayed the spaceship in such a realistic manner that many people thought America was on the verge of making space travel a reality. Although it was fascinating to read about futuristic space journeys, von Braun knew our technology was far too primitive to construct these ships. Few rocket engineers seriously believed these craft could even be built in the foreseeable future. Indeed, ten years after the *Collier's*

Table 34-1. High-Performance Aircraft Leading to the Space Shuttle					
Craft	Weight (pds)	Speed (mph)	Year	Height (feet)	Year
Bell X-1	12,250	957	1948	69,000	1949
Bell X-1A	16,487	1,650	1953	90,440	1954
Bell X-2	24,910	2,094	1956	125,907	1956
X-15	31,275	4,520	1967	347,600	1963

Figure 34-2. *Von Braun wrote about a three-stage space ferry in* Collier's *magazine as part of his futuristic view of America's move into space.*

articles appeared, America was still struggling to place mul-timanned capsules into space.

While NASA was preoccupied with its Moon program, some preliminary work was completed by the military on the design of reusable spacecraft. Work on a winged, two-crew vehicle, called the X-20 Dyna-Soar, had progressed to the point where it was considered technically feasible, but this project was cancelled when a compelling military mission for it could not be found. NASA was interested in the X-20, but realized that as long as the Apollo program was active

Figure 34-3. *Walter Dornberger, von Braun's former military supervisor, worked for the American firm the Bell Aircraft Company after the war on winged craft that could reenter the atmosphere from space and land.*

the agency could only afford to conduct studies of the aerodynamic questions it posed.

The Political Debate over Whether to Build the Space Shuttle

By the middle of 1968 the last of the fifteen Saturn V boosters was built and their production line was shut down. NASA began looking beyond the Apollo era toward new programs that would keep it busy for at least the next twenty years. Among the projects under consideration was an Earth-orbiting space station, a Space Transportation System (STS) consisting of a fleet of reusable space shuttles and tugs, and a manned mission to Mars.

As the country entered the 1970s, NASA found there was little enthusiasm for an aggressive push into space. Still somewhat captivated by the success of the Apollo program, NASA Administrator Thomas Paine was surprised by the new hostile attitude he encountered when proposing new space initiatives. Liberal members of Congress decried the money allocated to NASA when in their view environmental and urban problems had reached a crisis level and crime was rising. NASA's budget requests were attacked as examples of "misplaced government priorities" and the Nixon administration failed to vigorously come to the agency's defense.

Unlike Presidents Johnson and Kennedy, Nixon was not a space enthusiast. As Eisenhower's former vice president he believed that the government should fund only a modest space effort. Even if Nixon had desired otherwise, the escalating costs of President Johnson's "Great Society" (a series of government social programs) and the Vietnam war made cutbacks in some programs unavoidable. Without the backing of the President or a large and vocal constituency, NASA was vulnerable to the budget ax. There was serious talk of canceling *Apollo 16* and *17* and eliminating or scaling back Skylab, the Space Shuttle, and several other NASA projects.

When NASA officials realized their original plan had little or no chance of being funded, the space tug idea was shelved and the agency agreed to postpone requests for a space station. Instead, they focused on the construction of the Space Shuttle, asserting it was the cornerstone of the nation's future manned space program.

Despite the emphasis NASA placed on the Shuttle, funding for it was not assured. NASA's new administrator, James Fletcher, tried to build political support for this craft by stating it would greatly reduce the cost of placing objects into space. One of the lessons of the Apollo program was that a program based on expendable vehicles was very expensive. During that era, space missions were likened to an airline that destroyed their planes after each flight and then built new ones. Congress saw this as a wasteful process and Fletcher made headway by emphasizing the savings a reusable space vehicle would make possible.

Figure 34-4. A seasoned politician, President Nixon realized that Congress was not in the mood to support another big and expensive space program like the Apollo program.

Figure 34-5. James Fletcher, the NASA chief administrator who succeeded Thomas Paine, was a strong supporter of a reusable spacecraft.

NASA had already begun design studies for a Space Shuttle in 1968–1969. These plans called for a vehicle that could place a 65,000-pound payload into orbit in a compartment that was 15 feet wide and 60 feet long. These dimensions were, in large part, dictated by the requirements of the Department of Defense, a major expected user of this craft. According to NASA's plan, the Shuttle would begin its journey on top of a jumbo-jet. Once at a high altitude, the shuttle would be detached as shown in Figure 34-6 and fire its engines to complete the journey into orbit. Planners estimated the shuttle would cost $10 to 15 billion to build this vehicle, a cost comparable to that of the entire Apollo program.

NASA officials soon learned this high price tag was far beyond what President Nixon was willing to support. The agency was directed to come up with a more modest proposal whose cost was around $5 billion. The Shuttle was downsized and reconfigured to use expendable boosters. Although this solution harkened back to the days of the Apollo program, the financial constraints NASA faced left it no other choice.

Three considerations came together in late 1971 that influenced President Nixon's decision to proceed with the construction of this scaled-down vehicle. The first was the strong endorsement of Casper Weinberger, deputy director of the powerful Office of Management and Budget. In an August 12, 1971 memorandum to Nixon he supported a vigorous space program, specifically mentioning the Space Shuttle and the NERVA, a nuclear-powered rocket. In his opinion these two projects had great commercial value. He also saw the space program as a way to improve America's post-Vietnam image as a superpower. He wrote,

There is real merit to the future of NASA, and to its proposed programs. The Space Shuttle and NERVA [the nuclear power rocket program] particularly offer the opportunity, among other things, to secure substantial scientific fallout for the civilian economy. . . . Most important is the fact that they give the American people a much needed lift in spirit

(and the people of the world an equally needed look at American superiority) . . . a cancellation now . . . would be confirming in some respects, a belief that I fear is gaining credence at home and abroad: that our best years are behind us, that we are turning inward, reducing our defense commitments, and voluntarily starting to give up our superpower status, and our desire to maintain world superiority.

This endorsement was followed by a memo from Fletcher listing five reasons why the Shuttle should be built: (1) having come this far, it was unthinkable for the United States to now forego manned space flight, (2) the Space Shut-

Figure 34-6. In early designs the Space Shuttle was envisoned as a two-part craft. A jumbo jet would lift the Orbiter to a high altitude and the shuttle would then use its own rockets to procede to space.

tle was the only meaningful manned space flight program that could be accomplished on a modest budget, (3) the Space Shuttle was essential for NASA to exploit the practical benefits of space, (4) the cost and complexity of the Shuttle had been reduced to the point where it was affordable, and (5) the Shuttle program would funnel significant resources into America's hard-pressed aerospace industry. These arguments, along with the political consideration that numerous jobs would be created in key states just prior to the 1972 national elections, persuaded Nixon to endorse the Shuttle's construction.

Naturally, the public announcement that America would build a space shuttle did not mention the political benefits Nixon envisioned, but pointed to possible advances in communications, weather forecasting, agriculture, pollution control, and other areas. The announcement noted,

> . . . all of these possibilities, and countless others with direct and dramatic bearing on human betterment, can never be more than fractionally realized so long as every single trip from Earth to orbit remains a matter of special effort and staggering expense. This is why commitment to the Space Shuttle program is the right next step for America to take, in moving out from our present beachhead in the sky to achieve a real working presence in space—because the Space Shuttle will give us routine access to space by sharply reducing costs in dollars and preparation time.

Components of the Space Shuttle

The final vehicle that emerged from the cost constraints imposed by President Nixon consisted of three components, (1) the Orbiter, (2) an expendable external fuel tank, and (3) two reusable solid rocket boosters. Each of these is described below:

The Orbiter: The Orbiter is about the size of a commercial DC-9 jet. It is 122 feet long, 78 feet wide, stands 57 feet high, has an unfueled weight of 160,000 pounds, and can carry a crew of up to eight people. This vehicle consists of two sections: a forward crew compartment and a cargo bay. The cargo bay is enclosed by two large doors. When in space, these doors are left open to prevent the craft from overheating (see Figure 34-7). The forward crew compartment consists of an upper and lower deck. The craft's flight control system and a computer used to manipulate equipment in the cargo bay are located in this section. This computer is used to operate a large robot arm, called the Remote Manipulator System (RMS), to lift objects out of the bay and into space. The lower portion of the crew compartment contains the astronauts' living and sleeping quarters and an airlock leading to the cargo bay.

The External Expendable Liquid Fuel Tank: The Shuttle is propelled by three powerful, permanent engines, each of which produces 375,000 pounds of thrust. They are fueled by liquid oxygen and hydrogen that is carried in a large external tank attached to the bottom of the Orbiter. This tank weighs

Figure 34-7. In this picture of the Challenger *its bay doors are open to space.*

67,000 pounds, has a diameter of 27.6 feet, and a length of 154 feet. When filled, the top portion of the tank contains 143,000 gallons of liquid oxygen weighing 1.4 million pounds at a temperature of −297°F. Below the oxygen tank is a 22.5-foot-long, 12,100-pound section. This part of the tank contains an apparatus for emptying the upper and lower tanks and an attachment for two external solid rockets. The bottom portion of the booster contains a 26,000-pound tank that holds 385,000 gallons of liquid hydrogen weighing 228,000 pounds at a temperature of −423°F. Just before the Shuttle reaches orbital speed the external tank is jettisoned and is then destroyed during atmospheric reentry.

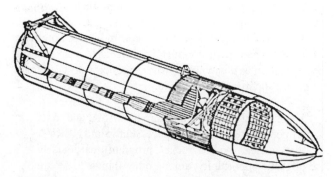

Figure 34-8. The large expendable liquid-fuel tank attached to the bottom of the Shuttle contains extremely cold liquid oxygen and hydrogen.

The Reusable Solid Rocket Boosters (SRBs): Attached to the Expendable Liquid Fuel Tank are two identical solid rocket boosters, the first such rockets used for manned flight and the most powerful solid rockets ever built. Each of these rockets is 149 feet long, has a diameter of 12 feet, and weighs nearly 1.3 million pounds. During flight they burn for a little more than 2 minutes and produce a combined 3.3 million pounds of thrust. Once their fuel is exhausted, they are jettisoned by eight small rockets at an altitude of about 40 miles. The boosters then fall sideways through the atmosphere (see Figure 34-10) until they drop to a height of 15,400 feet. At this point, a barometric switch opens a series of four parachutes and the booster lands in the ocean where it is retrieved by waiting ships. These rockets are then towed back to land and refurbished for future use (see Figure 34-11).

The Smaller Auxiliary Rockets: Once the Shuttle is in orbit, two additional engines located at the rear of the craft are employed to make major flight changes and to slow the craft for reentry. Each of these engines produces 6,700 pounds of thrust. Three sets of small rockets, one set located at the front of the Orbiter and two sets in the tail section, are used to make small course corrections and delicate maneuvers.

The Heat Shield

The thermal heat shield of the shuttle is one of its most impressive features. The nose-cone and leading surfaces of the shuttle are coated with a special nylon cloth impreg-

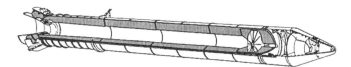

Figure 34-9. Each of the two solid booster rockets attached to the Space Shuttle is 149 feet long, 12 feet in diameter, weighs 1.3 million pounds, and produces 3.3 million pounds of thrust.

Figure 34-10. The Orbiter's two solid rocket boosters are jettisoned during flight and fall sidewards toward the Earth.

Figure 34-11. Once the solid rockets parachute into the ocean, they are retrieved by tugs and hauled back to land for later use.

nated with pyrolized carbon. This material can withstand temperatures of 3,000°F. The body of the craft is protected by special silica tiles that absorb the heat of reentry and radiate it back into space. This remarkable material can be heated in a furnace to a very high temperature and then safely placed in a person's hand a few seconds later. The black-coated tiles located on the belly of the shuttle can withstand reentry temperatures of 2,300°F while the white tiles on the shuttle's upper body can withstand temperatures of 1,200°F.

Spacelab

As the Apollo program was fulfilling its goal of placing a man on the Moon, NASA officials were looking for international international partners who could share the costs of the Shuttle missions. The formation of these collaborations was both fiscally sound and politically shrewd. Congress was reluctant to have the United States appear as an unreliable team player, so once a multinational effort was approved, continued support for the program was virtually guaranteed.

The European Space Agency (ESA) and Canada expressed a desire to join the Shuttle program. The ESA agreed to construct a reusable laboratory that would fit into the Orbiter's cargo bay if NASA agreed to pay the ESA to construct a second unit. This reusable laboratory was named Spacelab. The cost of the first Shuttle flight to carry this module was to be shared by NASA and the ESA. NASA would then be reimbursed for additional European Spacelab flights.

Spacelab modules can be pressurized and are equipped with scientific instruments. Each module is 13 feet in diameter and 8.9 feet long and draws power from the Shuttle. A U-shaped instrument pallet was also built for the cargo bay, and be oriented by a Shuttle astronaut from within the crew compartment or by ground controllers. Up to five pallets can be joined together or a combination of pressurized modules and pallets can be carried.

Shuttle Operations and Support Contractors

The lead NASA center for shuttle operations is the Johnson Space Flight Center in Houston. Launches were to take place from the Kennedy Space Center in Florida and from the Vandenberg Air Force Base in California. New facilities were built at the Kennedy Space Center to install shuttle payloads and to refurbish the craft between missions. A 3-mile-long runway was added to accommodate Shuttle landings.

The Kennedy Space Center is used for Shuttle missions whose orbit is inclined to the Earth's equator by less than 57 degrees. Flights requiring higher inclinations depart from Vandenberg. The Marshall Space Flight Center is responsible for work on the Shuttle engines, the large liquid fuel tank, and the booster rockets. All vibration tests of the Shuttle's integrity are conducted at Marshall.

Several different vendors were awarded major contracts for various components of the Shuttle. Among these companies are Rockwell International (forward and rear fuselage and cargo hold doors), General Dynamics Convair, Grumman Aerospace, and Republic Aircraft (central fuselage, wings, and vertical stabilizer), Martin Marietta (main external tanks and recovery systems), Thiokol (solid rocket engines), McDonnell Douglas (forward and rear skirts), and United Space Boosters (integration and assembly).

The Orbiter Fleet

The Space Shuttle fleet consists of five craft, each named for real or fictional vessels. One of the shuttles is employed for ground tests, while the other four are used in space. By popular request, the first Orbiter was named the *Enterprise* in honor of the spaceship in the television series *Star Trek*. Ironically, the Enterprise is the shuttle that is used only for ground tests. The *Columbia* is named after the first Navy frigate to sail around the world and the command module of *Apollo 11*. The third shuttle, *Discovery,* honors an exploration ship that sailed into Hudson's Bay in 1610–1611, while the fourth shuttle, *Atlantis,* bears the name of an oceanographic research vessel that sailed from 1930–1966. The namesake of the fifth shuttle, *Challenger,* comes from a ship that explored both the Atlantic and Pacific oceans from 1872–1876. This was also the name of the lunar module of *Apollo 17*. The *Challenger* was destroyed in a tragic accident on May 7, 1992, and was replaced by another vessel, the *Endeavour*. This craft was named in honor of the ship used by the famous explorer, Captain James Cook.

Reductions in the Proposed Shuttle Flight Schedule

One of the main arguments used to persuade Congress to build the Orbiter fleet was that it would greatly reduce the cost of placing objects in orbit. This argument was based on the assumption that the Shuttle would be the sole launch vehicle employed by the American space program. To meet the anticipated launch demand, NASA estimated 570 Shuttle missions would be flown between 1981 and 1991. However, once the Shuttle program began, unexpected difficulties arose and the number of flights was revised downward to 312. As more problems arose, this value was further reduced to 210 flights and the duration was increased to 1994. Even this estimate proved to be overly optimistic. Instead of the planned sixteen flights per year, only five to six launches actually took place. The greatest number of yearly launches achieved was nine and this was obtained only once in 1985.

By the end of the ten-year period between 1981 and 1991 only forty-two flights had taken place compared to the original estimate of 570. To be fair, the entire shuttle fleet was not always available during this time. The *Columbia* was grounded for a period of eighteen months while modifications were made to its systems and the *Challenger* disaster grounded the entire fleet for almost two years. In addition, the idea that the Shuttle would serve as America's sole launch vehicle was abandoned in the late 1980s when other vehicles were pressed into service. As a result of these decreased numbers, the average cost of each of the first 20 Shuttle flights was a rather hefty $257 million.

Shuttle Accomplishments

Despite NASA's failure to achieve its original launch rate, an impressive number of milestones were obtained. Several of these, listed by shuttle, are given below.

Columbia: The Columbia was the first Orbiter to reach space. It was launched on April 12, 1981, the twentieth anniversary of Yuri Gagarin's historic flight. Among the accomplishments of this craft are the first tests of the Shuttle's remote manipulator arm (*STS-2*), the first Shuttle to launch commercial satellites (*STS-5*), the first six-man crew (*STS-9*), and the first American spacecraft to carry a foreign visitor, Ulf Merbold of West Germany. Columbia also carried Spacelab into orbit on November 23, 1983, and Spacelab Life Sciences 1 in June of 1991.

Challenger: Challenger was the second Orbiter to fly in space. Launched on April 4, 1983, its accomplishments include: (1) the first seven- and eight-member crews placed in orbit, (2) the first Extra Vehicular Activity within the cargo bay (*STS-6*), (3) the first American spaceflight to carry a female astronaut, Dr. Sally Ride (*STS-7*), (4) the first night Orbiter launch and landing (*STS-8*), (5) the first untethered spacewalk from an Orbiter, and (6) the first capture, repair, and redeployment of a nonfunctioning satellite, the Solar Maximum satellite. Challenger was also frequently used for Spacelab flights.

Discovery: The first flight of Discovery took place on August 30, 1984. Its achievements include: the first Shuttle polar orbit flight, the first capture and return of a disabled satellite to Earth, the first orbital laser tracking system test, the first test of a 105-foot-tall solar array, and the launch of the famous Hubble Space Telescope (HST) in April 1990.

Atlantis: The first flight of Atlantis took place on October 3, 1985. This craft was equipped to handle cryogenic materials and is frequently employed for Department of Defense flights. Five of its ten flights between 1985 and 1991 were military missions. Three of NASA's premier scientific satellites, Magellan, Galileo, and the Compton Gamma-Ray Observatory, were launched from this craft. Magellan mapped the surface of Venus with unprecedented resolution, Galileo explored the atmosphere of Jupiter and its satellites, and the Compton Gamma-Ray Observatory (CGRO) gathered data about the most energetic astronomical objects in the Universe. The HST and CGRO belong to NASA's great observatory series which are intended to lead America's unmanned space program into the 21st century.

35   The Second-Generation Soviet Space Stations: *Salyuts 6* and *7*

Now we will have to embark on the next step—to put into orbit permanent orbital research stations with changing shifts of crews.

—President Leonid Brezhnev, May 1981

While the United States continued to launch expeditions to the Moon in the early 1970s, the Soviet Union pursued work on their space station. The first-generation Soviet space stations, *Salyuts 1* through *5*, were designed to accommodate a crew of two to three cosmonauts for a period of a few months.

A total of eleven Soyuz flights were launched to these stations, but as noted in the chapter, *The Soviet Salyut Space Stations*, a number of serious problems were encountered. *Soyuz 10,* the first manned flight to *Salyut 1,* docked but its crew was unable to open the airlock. The crew of *Soyuz 11* entered the station, but later died during a tragic reentry accident. *Soyuz 15* failed to dock with *Salyut 3,* and *Soyuz 18A* did not reach orbit because of a booster malfunction. The mission of *Soyuz 21* was prematurely ended when an acrid odor filled *Salyut 5,* and the automatic docking system malfunctioned on *Soyuz 23*. Only four of the eleven flights, *Soyuz 14, 17, 18B,* and *24,* could be considered successful. At the same time, the space stations had their own set of problems. Soon after lift-off, *Salyut 2* and *Cosmos 557* were beset by so many troubles that no attempt was made to occupy them.

The New Soviet Space Station

Naturally, the Soviets were concerned by these many failures, and after reviewing their causes, attempts were made to make the station more reliable. A redesigned station was produced that looked very similar to earlier Salyuts (see the Focus Box below). It consisted of successively larger cylindrical units joined by connecting fixtures called frustrums. The station's five main compartments were (1) a forward transfer compartment, (2) a work compartment, (3) a connecting frustrum, (4) a second work compartment, and (5) a rear service compartment. Information about each of these compartments is given in Table 35-1.

A new feature of the Salyut was the addition of a docking port to the rear of the station. A Soyuz craft could now be docked at the forward port while a supply vessel could be docked at the rear port. The station's propulsion system consisted of two rockets located on opposite sides of the outer rear rim. Groups of thrusters were also located along its exterior wall and provided the station with a pitch, roll, and yaw capability.

Scientific Instruments

The new generation of Salyuts contained three major scientific instruments designed to collect data in the areas of Earth resources, geography, meteorology, and astronomy. These instruments were:

The MKF-6M: This camera system simultaneously photographed the Earth in six different bandpasses. These images

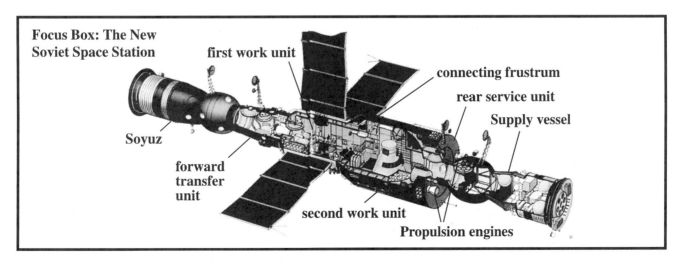

Focus Box: The New Soviet Space Station

first work unit

connecting frustrum

rear service unit

Supply vessel

Soyuz

forward transfer unit

second work unit

Propulsion engines

Table 35-1. Second-Generation *Salyut* Compartments			
Unit Name	**Length**	**Diameter**	**Components**
Forward transfer compartment	3.5m	2.0m	Docking drogue, rendezvous and search antenna, EVA hatch, astronomical camera
First work compartment	3.5m	2.9m	Solar panels, main control console, topological camera, biomedical equipment, storage units
Connecting frustrum	1.2m	>2.9m	Earth resources camera
Second work compartment	2.7m	4.2m	Submillimeter telescope, gamma-ray detector, treadmill, food lockers
Service compartment	2.2m	4.2m	Propulsion system and fuel tanks, transfer tunnel, docking port, trusters, rendezvous and search antenna, and a propellant transfer system

provided researchers with information about the health of crops, the presence of mineral deposits, and land uses. Each film cartridge contained 1,200 picture frames and weighed 26 pounds.

The KATE-140: This instrument consisted of a wide-angle, stereographic, topographic camera. Used to make contour maps, it could be operated by either ground controllers or by the space station's crew.

The BST-1M: This instrument obtained observations of the Earth's atmosphere from the ultraviolet to submillimeter wavelengths. Some bright astronomical objects, such as Venus, Mars, Jupiter, and Sirius, were also observed. This large instrument occupied a significant portion of the second work compartment. Because the BST-1M had to be cryogenically cooled and required a relatively large amount of power, it was used sparingly. The location of these three primary instruments along with several others are shown in the Focus Box on page 271.

The Launch of *Salyut 6* and *Soyuz 25*

The launch of the second-generation space station, *Salyut 6,* commemorated the 20th anniversary of *Sputnik 1.* Blasting-off from Tyuratam on September 29, 1977, it was placed into an orbit of 131 × 165 miles. Over the next week, ground controllers increased this height and made the orbit more circular. The first cosmonauts to travel to this new station, Vladimir Kovalyonok and Valery Ryumin, lifted-off a little over a week later on October 9 aboard *Soyuz 25.* Their mission was scheduled to last ninety days. Both Kovalyonok and Ryumin were space rookies. Kovalyonok was born on March 3, 1942, in the town of Beloye located near Minsk. From 1959 to 1963 he served in the Soviet Air Force. In 1965 he was placed in a reserve group of cosmonauts and later was appointed as the backup commander of *Soyuz 18A.* Ryumin was born on August 16, 1939, in far eastern Siberia. Trained in the army as a tank commander, he later returned to college and obtained a degree in computer science. Upon graduation

Figure 35-1. The crew of Soyuz 25: *Kovalyonok (left) and Ryumin (right).*

he worked as an engineer in Korolev's spacecraft design bureau and in 1971 was recruited as a Salyut cosmonaut.

The cosmonauts' launch and journey to the station went smoothly, but when they tried to dock, the connecting latches failed to close. Four more unsuccessful attempts were tried. Since the lifetime of the Soyuz batteries was only two days, the cosmonauts were forced to return to Earth. After landing they were publicly blamed for this failure, and in sharp contrast to the honor bestowed on other cosmonauts, they were not named as "Heros of the Soviet Union." After this episode, the commission overseeing spaceflights instituted a new policy requiring that at least one member of all future crews had to have spaceflight experience.

The failure of the *Soyuz 25* mission was a huge political embarrassment for the Soviets. By linking its launch to the

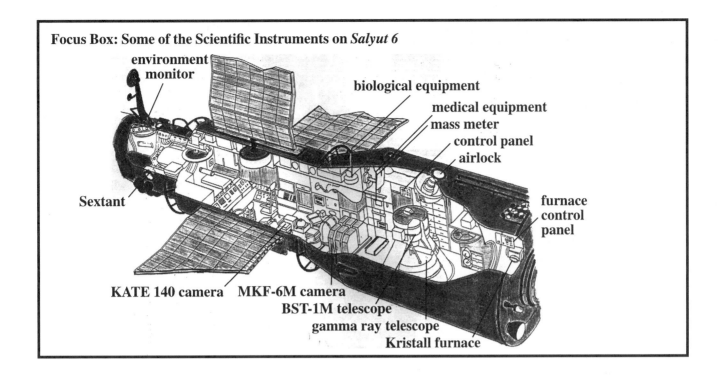

Focus Box: Some of the Scientific Instruments on *Salyut 6*

environment monitor

biological equipment

medical equipment

mass meter

control panel

airlock

Sextant

furnace control panel

KATE 140 camera MKF-6M camera

BST-1M telescope

gamma ray telescope

Kristall furnace

anniversary of *Sputnik 1*, they ensured worldwide coverage. However, to the engineers and scientists working on the *Soyuz 25/Salyut 6* project, this was the least of their concerns. They did not understand what had gone wrong! Was the Salyut's docking mechanism at fault? If so, the station might have to be abandoned.

The Flights of *Soyuz 26* and *27*

The only way to identify positively the cause of the docking failure was to send another flight to inspect the station's mechanism. On December 10, 1977, cosmonauts Georgi Grechko and Yuri Romanenko lifted-off from Tyuratam to complete this task.

The selection of Grechko and Romanenko for this mission was an excellent choice. Grechko had entered the cosmonaut corps in 1966 and was a veteran of the *Soyuz 17/Salyut 4* mission. He was also an experienced spacecraft designer. Romanenko had joined the Soviet space program in 1970 and was the backup crew commander for the prior Soyuz flight. He was very familiar with the Soyuz and Salyut station.

Rather than risk docking with the front port of *Salyut 6* as Kovalyonok and Ryumin had attempted, Grechko and Romanenko approached the rear port. As Romanenko piloted the Soyuz into the docking mechanism, controllers at the Russian control center held their breath. As the distance closed, Romanenko radioed, "Full pressure . . . full electrical coupling!" They had done it, they were docked!

Once joined to the Salyut, the cosmonauts entered the station and activated its systems. Their next task was to determine the condition of the front docking port. To accomplish this in-

Figure 35-2. The crew of Soyuz 26: *Romanenko (left) and Grechko (right).*

spection, Grechko would have to conduct an Extra Vehicular Activity (EVA). Eight years had passed since the last Soviet spacewalk, so this was not a trivial matter. Grechko exited through the forward hatch carrying a special tool to test the docking mechanism. Holding onto handrails along the hull, he slowly worked his way to the port. After performing several tests and finding nothing wrong, he radioed ground controllers, "The butt end is brand new—just as when it was machine tooled. There are no scratches or dents or traces. The cone is clear; not a scratch. The lamps, sockets and latches are in order."

Based on this good report, officials suspected the docking mishap had been caused by a problem in the Soyuz mech-

Figure 35-3. Cosmonaut Romanenko (Soyuz 26) *prepares to conduct an EVA to determine if the* Salyut 6 *docking mechanism is operative.*

anism. Unfortunately, there was no way to prove this because the mechanism had been jettisoned as a normal part of the craft's reentry.

While Grechko worked on the docking unit, Romanenko monitored his progress from inside the airlock. As Grechko finished and started back, Romanenko poked his head out and gave himself a slight shove upward to get a better view. Obeying Newton's law of action-reaction, his body continued outward toward open space. Caught off-guard, he quickly

Figure 35-4. The crew of Soyuz 27 *consisted of cosmonauts Dzhanibekov (left) and Makarov (right).*

tried to catch hold of the spacecraft, but to no avail. Luckily, Grechko saw his crewmate's distress and shouted, "Yuri, Yuri, where do you think you are going?" He then grabbed his crewmate and pulled him safely back into the craft. It was a close call.

Once back inside the airlock, a second life-threatening situation arose. When the cosmonauts tried to repressurize the unit, the airlock's instrument panel showed that a valve was stuck in an open position. If this was the case, the module would not be able to pressurize and cosmonauts would not be able to enter the station. If a remedy was not found, the outcome would be disastrous. When the oxygen inside their suits was consumed, they would die. After discussing this situation with ground controllers, it was decided to repressurize the module and hope the indicator was faulty. As air flowed into the module the cosmonauts were relieved to see the pressure was holding. They entered the Salyut, took off their space suits, and exhausted by these harrowing events, slept for ten hours.

The next day the cosmonauts settled into a routine work schedule. Their mission would eventually last ninety-six days and would break the eight-four-day space endurance record held by the American astronauts of *Skylab 4*.

On January 10, 1978, *Soyuz 27* was launched toward *Salyut 6* carrying cosmonauts Vladimir Dzhanibekov and Oleg Makarov. The primary objective of their flight was to test the station's front docking port. As their vessel approached the station, Romanenko and Grechko put on their spacesuits and entered their Soyuz craft. In the event of a docking accident this would be the safest place for them. Fortunately, no problems arose. The combined *Soyuz 27/Salyut 6/Soyuz 26* spacecraft established a new record—the first time three manned vessels had been joined together in space. The combined craft was 98 feet long, nearly one-third the length of a football field, and weighed approximately 32 tons.

After a five-day stay, Dzhanibekov and Makarov returned to Earth in the *Soyuz 26* craft. This swapping of Soyuz vehicles, where the fresher one was left attached to the station, would become a standard operating practice. Once these cosmonauts departed, the rear port of the station was available for a resupply mission.

Resupplying the Space Station

The Salyut space station revolved around the Earth at a height of about 168 miles. Even at this altitude, the tenuous atmosphere caused the station's orbit to decay slowly. To compensate for this, a reboost was necessary about every twenty days. Each time a reboost was undertaken, a considerable amount of fuel was consumed, reducing the number of remaining reboosts.

A second factor affecting the length of time the station could be occupied was the amount of available air, water, and food. During a one-year period a two-person crew would consume 10 tons of expendable supplies. In order to make longer missions possible, it was necessary to find an efficient way of resupplying the space station. Western experts suspected the

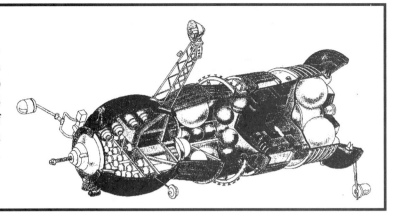

Soviets were working on a cargo vessel to fulfill this need and were not particularly surprised when *Progress 1*, the first such unmanned resupply vessel, was launched toward the station on January 20, 1978. This vessel is described in the Focus Box above.

Communist Solidarity: International Missions to *Salyut 6*

In November 1965 the Soviet Union joined eight other communist countries (Bulgaria, Cuba, Czechoslovakia, the German Democratic Republic, Hungary, Mongolia, Poland, and Romania) to create an organization called Intercosmos. The purpose of this organization was to plan and conduct joint space missions. Under its guidelines, Russia would provide launch facilities, "spaceships and stations," mission commanders, and training facilities. The other countries would provide people to be trained as flight engineers and/or research specialists. A follow-up agreement, signed in July 1976, called for the completion of experimental programs in astrophysics, space biology and medicine, communications, and meteorology. Joint missions were planned to take place from 1978 to 1983. Vietnam later joined this organization in 1979.

The first group of Intercosmos trainees arrived in Russia in December 1976. They were followed by a second contingent in March 1978. Although numerous experiments would be conducted under this program, particularly in the fields of space biology and medicine, the flights were largely a political statement. They stood in sharp contrast with the American effort in which few foreigners were allowed to participate. A list of the Intercosmos flights to *Salyut 6* is given in Table 35-2.

A New Shuttle Craft: The *Soyuz-T*

Following the loss of the crew of *Soyuz 11* only two cosmonauts were allowed to occupy a Soyuz craft. Although this was a wise safety measure, this decision also reduced the amount of work that could be completed during a flight. In 1976 the Soyuz was enlarged to accommodate a third crew member. This "new" craft was called the *Soyuz-T* (T standing for trioka, or three).

The Soyuz-T looked like a large version of the original Soyuz. It had an approximate weight of 7.6 tons, a length of 23 feet, and two large solar panels extended from its exterior wall. These panels provided sufficient power for a four-day flight. It carried 700 pounds of fuel, about 200 pounds more than its predecessor.

The first Soyuz-T flight to *Salyut 6* was launched on December 16, 1980. The unmanned craft automatically docked with the station and then remained attached for about three months to test the longevity of its systems. Once these tests were completed, the craft was instructed to undock and its descent module returned to Earth. With this trouble-free mission completed, cosmonauts Vladimir Aksyonov and Yuri Malyshev were launched on June 5, 1980. Aksyonov had flown earlier in *Soyuz 22* while Malyshev was a rookie. Their primary goal was to test the Soyuz-T hardware, and after docking with the station they returned to Earth after spending only four days in space.

Figure 35-5. The crew of Soyuz T-3 *was the first three-person crew since* Soyuz 11. *Left to right, Strekalov, Makarov, and Kizim.*

Table 35-2. The Intercosmos Flights to *Salyut 6*

Flight	Visiting Country	Launched	Crew (Visitor, Russian cosmonaut)
Soyuz 28	Czechoslovskia	March 2, 1978	V. Remek, A.N. Gubarev
Soyuz 30	Poland	June 27, 1978	M. Hermaszewsku, P.I. Kimuk
Soyuz 31	East Germany	August 26, 1978	S. Jahn, V.F. Bykovsky
Soyuz 33	Bulgaria	April 10, 1979	G. Ivanov, N.N. Rukavishnikov
Soyuz 36	Hungry	May 26, 1980	B. Farkas, V.N. Kubasov
Soyuz 37	Vietnam	July 23, 1980	P. Tuan, V.V. Gorbatko
Soyuz 38	Cuba	September 18, 1980	A. Tamaya-Mendez, Y.V. Romanenko
Soyuz 39	Mongolia	March 22, 1981	J. Gurragcha, V.A. Dzhanibekov
Soyuz 40	Romania	May 14, 1981	D. Prunariu, L.I. Popov

The next manned flight to *Salyut 6*, *Soyuz T-3*, was launched on November 27, 1980. For the first time in nine years, a crew of three cosmonauts (Oleg Makarov, Leonid Kizim, and Gennadi Strekalov) rode a Soviet craft into orbit. Makarov had flown in *Soyuz 12, 18A,* and *27* (although *18A* failed to reach orbit). This would be the first flight for Kizim and Strekalov. Like the previous *Soyuz-T* mission, this flight had limited objectives. The cosmonauts docked and entered the space station, checked its operational status, made several repairs, and completed experiments on the production of semiconductors, holographic photography, and on the growth of plants. Although this twelve-day mission was relatively brief, its primary objective of preparing the station for another series of flights was accomplished.

On January 24, 1981 a Progress cargo ship lifted-off and automatically docked with *Salyut 6*. Its engine was then used to place the combined craft into a higher orbit. The next *Soyuz T* flight, *Soyuz T-4*, was launched on March 12, 1981, and carried cosmonauts Vladimir Kovalyonok and Viktor Savinykh. After an uneventful trip, the crew entered the Salyut and quickly began work. They repaired a jammed solar panel, a malfunction that dropped the station's temperature to a chilly 50°F, unloaded the attached cargo vessel, and continued the experimental program begun by Makarov's crew. The Progress supply craft was then jettisoned and directed to reenter the atmosphere where it was destroyed. This would be the last supply vessel sent to *Salyut 6*.

In addition to the mission described previously, ten Intercosmos crews visited the station. The final two international crews were launched in 1981. *Soyuz 39* carried Mongolian cosmonaut Jugderdemidiyn Gurragcha and Soviet commander Vladimir Dzhanibekov. During their eight-day visit materials processing experiments were conducted and photographs of inaccessible regions of Mongolia were obtained. This flight was followed by *Soyuz 40*, the last mission to *Salyut 6*. It was manned by Romanian cosmonaut Dimitra Prunariu and Soviet commander Leonid Popov. During their eight-day flight they conducted medical and technology experiments. On May 26, 1981, Kovalyonok and Savinykh returned to Earth having completed a seventy-six-day mission.

Kovalyonok and Savinykh were the last cosmonauts to leave *Salyut 6*. By the end of their mission, the space station had been in orbit for nearly five years. Since lifting-off on September 29, 1977, sixteen Soyuz flights had transported thirty-three cosmonauts to the station. Three crews had conducted space walks to make repairs, matching the feat performed during *Skylab 2* and *3*. Not only was *Salyut 6* a technological success, it had also been scientifically productive. Data from over 1,300 experiments had been acquired in the fields of Earth resources, astronomy, space medicine, biology, and materials processing. In addition, more than 15,000 photographs had been collected for future study.

On April 25, 1981, a new and larger cargo vehicle called the Kosmos lifted-off from Tyuratam and later automatically docked with *Salyut 6*. The Kosmos engine was then used to propel the space station into a new orbit, but no subsequent manned missions were sent to this station. On July 29, 1982, *Salyut 6* reentered the atmosphere and was destroyed over the Pacific Ocean.

Salyut 7

The next space station placed into orbit, *Salyut 7*, looked nearly identical to *Salyut 6*, but it carried a different complement of instruments. *Salyut 7* was intended to support missions lasting longer than the 185-day period achieved by *Salyut 6*, so it contained more medical and exercise equipment. The BST-1M experiment located in the large work area was also replaced by an X-ray telescope and spectrometer.

Salyut 7 was launched from Tyuratam on April 19, 1982, approximately one year after the first manned flight of the American space shuttle and eleven years after the launch of *Salyut 1*. Its objectives included medical tests, astronomical studies, and technology experiments primarily dealing with materials processing. The first manned flight to *Salyut 7*, *Soyuz T-5*, lifted-off on May 13, 1982, with cosmonauts Anatoly Berezovoi (commander) and Valentin Lebedev (flight engineer) on-board. This was Berezovoi's first flight and Lebedev's second, having previously flown on *Soyuz 13* and

Focus Box: The *Soyuz-T*

The third-generation Soviet transport was the *Soyuz T*. The wing span of the two solar panels was nearly 35 feet. The first manned launch of this craft took place in 1980.

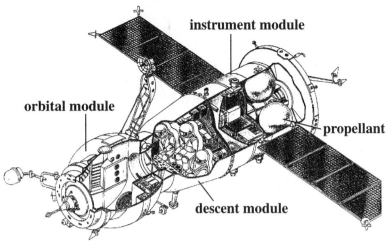

instrument module

orbital module

propellant

descent module

The Kosmos

The Kosmos was 42.6 feet long, had a maximum diameter of 13.8 feet, and had a mass of 22 tons. It was similar in size to the central section of the Salyut spacecraft. Two large solar panels supplied the Kosmos with power. Fuel tanks were mounted on its exterior wall, and two nozzles were attached to the back of the descent module. It could return 1,100 pounds of materials back to Earth in a pressurized descent module.

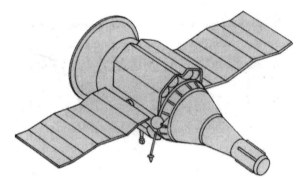

15. Both men were perfectionists, making it difficult initially for them to work together.

Berezovoi was born on August 11, 1942. After working as a lathe operator he joined the Army and was trained as a pilot. He spent two years as a flight instructor before joining the space program as a military cosmonaut on 1970. Lebedev was born on April 14, 1942, and had been fascinated with flight since his childhood. When attending the Moscow Aviation Institute, one of his instructors was Vasili Mishin, Korolev's deputy. Mishin recommended him for cosmonaut training, but Lebedev twice failed the required medical exams because of his high blood pressure. Finally, in 1971 he passed the exam and became a cosmonaut in 1972.

The scientific program of Berezovoi and Lebedev included studies in agriculture, biology, forestry, geology, and oceanography. After a routine launch and docking, the cosmonauts activated the station's systems and began their scheduled activities. A few days later they launched a small satellite, *Iskra-2*, through one of the station's airlocks.

A Welcome Break: Jean-Loup Chretien Visits *Salyut 7*

The first visiting crew to *Salyut 7* was launched on June 24, 1982, and consisted of cosmonauts Vladimir Dzhanibekov, Alexander Ivanchenkov, and a French "spationaut" named

Figure 35-6. The crew of Soyuz T-6: *left to right, J.-Loup Chretien, V. Dzhanibekov, and A. Lvanchenkov.*

Jean-Loup Chretien. At first Lebedev was concerned about these new people upsetting the good working relation he had finally established with Berezovoi. He wrote in his diary, "We're waiting for our guests nervously. Our relationship has settled. What impact will the new people have on us? The two of us have got used to each other and we're getting along well in our work. Now it's as though we'll have to start all over again."

These fears proved to be unfounded and Lebedev actually enjoyed the visit, especially the antics of Chretien. Lebedev wrote, "Jean is a funny man. He brought a quasimodo face mask up with him and when I approached the instrument panel a hairy image came out at me. I screamed! There was laughter all around."

The scientific program of this visiting crew emphasized medical experiments to measure the flow of blood in the brain, changes in acuity and depth perception, and the effectiveness of antibiotics in space. They also conducted investigations in materials processing and infrared astronomy. With the mission successfully completed, the visiting crew undocked and safely landed on July 2.

Svetlana Savitskaya: The Second Woman in Space

The next visiting crew, that of *Soyuz T-7,* was launched on August 19, 1982. It carried the second Soviet woman to fly in space, Svetlana Savitskaya. Nineteen years had passed since Valentina Tereshkova had flown aboard *Vostok 6,* so interest in this mission was high. Savitskaya was an exceptional person and better qualified for spaceflight than many of her male colleagues. Born in Moscow on August 8, 1948, her father, Yevgeny Savitskaya, was deputy chief of air defense forces. She held a degree in aeronautical engineering and in 1974 completed the Ministry of Aviation Production's civilian test pilot school. She became a superb pilot and an accomplished parachutist. At the age of 17 she held three world parachute records, and by the age of 24, was a world champion in all-around flying.

Savitskaya's participation in *Soyuz T-7* also met a political need. Sally Ride, America's first woman astronaut, was scheduled to fly aboard the Space Shuttle on June 18, 1983. The Soviet leadership wished to direct the world's attention away from that flight by placing its second woman cosmonaut in space. Once on-board *Salyut 7,* regular broadcasts featuring Savitskaya were transmitted and televised around the world.

Berezovoi and Lebedev Return from *Salyut 7*

After the departure of this visiting crew, Berezovoi and Lebedev continued their tasks. A second satellite, *Iskra-3,* was launched from the station on November 18. By this time the crew had been in space for over six months and were becoming exhausted. They asked if their mission could be ended two weeks early. Soviet doctors considered this request and agreed to move the return date up to December 10th. With their morale high, work on the station continued in a routine manner.

On December 10th Berezovoi and Lebedev departed *Salyut 7* and landed in weather so bad that the recovery teams could not immediately reach them. The first rescue helicopter crash-landed and its pilot had to "talk down" a second craft. Although Berezovoi and Lebedev needed help to walk, they recovered quickly and were able to spend New Year's Eve with their families.

Berezovoi and Ledebev took pride in their mission. They had established a new space endurance record of 211 days, had obtained 20,000 photographs of scientific interest, and had completed a host of experiments. The pictures they obtained were used to locate oil and gas fields in remote areas of the Soviet Union and to plan the route of the Soviet transcontinental gas pipeline and Baikal-Amur railway. The Soviet government claimed that the advance weather forecasting provided by the cosmonauts saved a billion rubles.

Near Disasters and Glorious Triumphs

With the return of *Soyuz T-7*, Vladimir Titov, Gennadi Strekalov, and Alexander Serebrov were launched on April 20, 1983 aboard *Soyuz T-8*. Once in orbit, their craft's rendezvous radar failed. A valiant attempt was made to rendezvous with the Salyut manually, but this effort consumed so much fuel officials were concerned there would not be

Figure 35-7. The crew of Soyuz T-7 carried the Soviet's second woman into space. The crew consisted of (left to right) A. Serebrov, L. Popov, and S. Savitskaya.

Figure 35-8. The crew of Soyuz T-8: (left to right) A. Serebrov, V. Titov, and G. Strekalov.

enough left to guarantee a safe return. After spending only two days in space, the cosmonauts returned without completing a docking.

Titov's flight marked the Soviet's fourth unsuccessful docking attempt. This was a frustrating situation because each failure seemed to have its own unique set of circumstances. How many more shakedown missions would be necessary before this problem was finally overcome?

Titov's backup crew, Vladimir Lyakhov and Alexander Alexandrov, were launched just a few months later on June 27, 1983 aboard *Soyuz T-9*. This time the flight proceeded normally and no docking problems were encountered. The cosmonauts unloaded 4 tons of supplies from a cargo vessel attached to the station and spent the next few months obtaining Earth resources photographs, locating schools of fish, conducting materials processing, and completing astronomical studies. During this time the space station was experiencing an increasing amount of trouble. One of its three solar panels failed, and during a resupply operation a fuel line ruptured. The interior temperature of the station dropped to a chilly 10°C (50°F) and the humidity rose to 100 percent.

In order to fix these problems, Vladimir Titov and Gennadi Strekalov were given a crash training course and were readied for flight. Both astronauts were anxious to redeem themselves after the docking incident on *Soyuz T-8*. Their flight would be hazardous, but both cosmonauts looked forward to it.

As Titov and Strekalov sat in their Soyuz capsule awaiting lift-off on September 27, 1983, preflight preparations proceeded normally. However 25 seconds before the R-7 was to blast-off, a fire broke out at the base of the launch pad. The situation quickly grew worse and their rocket escape system was fired, carrying the cosmonauts 1.5 miles from the launch site. Moments later the rocket exploded in an inferno of flames. Although the cosmonauts were bruised by their landing, they had no complaints. After viewing the devastated launch pad, they were glad to be alive.

Ground controllers notified Lyakhov and Alexandrov of the near disaster. The crew realized little could be done to help them until another flight was readied. Rather than simply wait in the cold, they decided to try to install two spare solar panels, already on-board, to the station's exterior wall. Neither cosmonaut had been trained for this task, so it was quite risky. Nevertheless permission was granted. Alexandrov described their spacewalks as "a tense, emotionally charged experience," but their efforts were successful. Shortly after verifying that the new panels were functioning properly, they undocked and returned to Earth on November 23, 1983. *Salyut 7* was now deserted.

The next crew to occupy the station was launched on February 8, 1984. It carried cosmonauts Leonid Kizim, Vladimir Solovyov, and Oleg Atkov aboard *Soyuz T-10*. This was Kizim's second flight, having served aboard *Salyut 6* and *Soyuz T-3*, but it was the first flight for Solovyev and Atkov. Their 237-day mission was intended to last twenty-six days longer than the record set on *Soyuz T-5*. During their time

Figure 35-9. The crew of Soyuz T-10 *(left to right, Atkov, Solovyov, and Kizim) set a new space endurance record of 237 days.*

aboard the station, Kizim and Solovyov conducted six EVAs, largely devoted to repairing a broken fuel line in the station.

Two other crews visited the station during their stay. The first carried visiting cosmonaut Rakesh Sharma, an officer in the Indian Air Force who had logged 1,600 hours of jet flight time. His mission was to use the MKF-6M and KATE-140 to search for water in arid regions of India as well as mineral and petroleum deposits. The second visiting flight (*Soyuz TM-12*) carried space veteran Svetlana Savitskaya. She would make history by completing the first spacewalk by a female. While outside the station she used a special tool to complete a series of cuttings, weldings, and soldering experiments (see Figure 35-10).

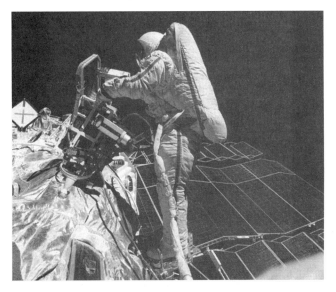

Figure 35-10. Svetlana Savitskaya became the first woman to perform a space walk, completing a welding experiment outside Salyut 7.

After this visiting crew departed, Kizim, Soloyvov, and Atkov completed their work and returned to Earth on October 2, 1984, leaving the station deserted. During their stay aboard *Salyut 7*, they had completed 600 scientific, technological, biological, and medical studies and obtained over 25,000 Earth resources photographs.

The Rescue of *Salyut 7*

After the return of Kizim and his crewmates, contact was unexpectedly lost with the unmanned *Salyut 7*. All attempts to reestablish contact were unsuccessful and officials were not certain what had gone wrong. On June 6, 1985, Vladimir Dzhanibekov and Viktor Savinykh were launched aboard *Soyuz T-13* to determine if the station could be salvaged. TASS had already prepared the Russian people for the worst by releasing a news bulletin stating that "the planned program of work aboard the *Salyut 7* orbital station has been fulfilled. . . ."

Dzhanibekov was one of the most experienced cosmonauts in the Soviet program, having completed successful dockings with *Salyut 6* (*Soyuz 27* and *29*) and *Salyut 7* (*Soyuz T-6* and *Soyuz T-12*). Savinykh was also a space veteran having served on *Salyut 6* for seventy-five days as a part of the *Soyuz T-4* crew.

When Dzhanibekov and Savinykh approached the nonresponsive station, the situation looked bleak. *Salyut 7* was slowly tumbling and its solar panels were not maintaining their proper orientation toward the sun. The motion of the station meant it would be very difficult to dock and the lack of Sun lock meant its interior would be very cold and dark. Undeterred, the cosmonauts approached the slowly tumbling station, matched this motion, and docked.

From this point onward, the situation steadily improved. The cosmonauts entered the Salyut and for ten days fought to get the station back into a working condition. They stabilized its motion and directed its solar panels toward the Sun. As

---

Table 35-3. Manned Flight to *Salyut 6* and *7*

Mission	Crew	Launched	Duration	Notes
Salyut 6/*Soyuz 26*	Romanenko, Grechko	Dec. 10, 1977	96 days	New endurance record
*Soyuz 27**	Dzhanibekov, Makarov	Jan. 10, 1978	6 days	First spacecraft exchange
*Soyuz 28**	Gubarev, Remek	Mar. 2, 1978	8 days	First Intercosmos flight
Soyuz 29	Kovalyonok, Ivanchenko	June 15, 1978	140 days	Spacewalk, endurance record
*Soyuz 30**	Klimuk, Hermaszewski	June 27, 1978	8 days	Photographs of Poland
*Soyuz 31**	Bykovsky, Jahn	Aug. 26, 1978	8 days	Materials processing
Soyuz 32	Lyakhov, Ryumin	Feb. 25, 1979	175 days	Station repairs, spacewalk
Soyuz 35	Popov, Lebedev	April 9, 1980	185 days	Live plants grown, atmosphere
*Soyuz 36**	Kabasov, Farkas	May 26, 1980	8 days	Materials science experiments
*Soyuz 37**	Gorbatko, Tuan	July 23, 1980	8 days	Materials science, agriculture
*Soyuz T-2**	Malyshev, Aksenov	June 5, 1980	4 days	*Soyuz-T* test flight
*Soyuz 38**	Romanenko, Tamayo-Mendez	Sept. 18, 1980	8 days	Organic crystal growth
Soyuz T-3	Kizim, Makarov, Strekalov	Nov. 27, 1980	13 days	Three-person mission
Soyuz T-4	Kovalyonok, Savinykh	Mar. 12, 1981	75 days	Scientific experiments
*Soyuz 39**	Dzhanibekov, Gurragcha	Mar. 22, 1981	8 days	Photographs of Mongolia
*Soyuz 40**	Popov, Prunariu	May 14, 1981	8 days	Last visit to *Salyut 6*
Salyut 7/*Soyuz T-5*	Berezovoi, Lebedev	May 13, 1982	211 days	Endurance record
*Soyuz T-6**	Dzhanibekov, Ivanchenkov Jean-Loup-Chretien	June 24, 1982	8 days	Medical experiments
*Soyuz T-7**	Popov, Savitskaya, Serebrov	Aug. 19, 1982	8 days	Second woman space
Soyuz T-9	Lyakhov, Alexandrov	June 27, 1983	149 days	Damaged fuel line
Soyuz T-10	Kizim, Solovyev, Atkov	Feb. 8, 1984	237 days	Endurance record
*Soyuz T-11**	Malyshev, Strekalov, Sharma	April 3, 1984	8 days	Yoga as cure for weightlessness
*Soyuz T-12**	Dzhanibekov, Savitskaya, Volk	July 17, 1984	12 days	First woman spacewalk
Soyuz T-13	Dzhanibekov, Savinykh	June 6, 1985	168 days	Space station repair
Soyuz T-14	Vasyutin, Grechko, Volkov	Sept. 17, 1985	65 days	Continued repairs on station
Soyuz T-15	Kizim, Solovyov	Mar. 13, 1986	125 days	Last time *Salyut 7* occupied
* *Intercosmos flight*				

Figure 35-11. Salyut 7 *as seen by the crew of* Soyuz T-13 *as they approached the station on their repair mission.*

power flowed into the nearly dead vehicle, its systems came back to life. The cosmonauts' efforts continued to bear fruit, and two supply vessels delivered needed materials. Steadily, the station was brought back to life.

The Soviet space officials seemed just as surprised as everyone that the rescue mission had succeeded. Now that *Salyut 7* was operational, they quickly moved to send a relief crew consisting of cosmonauts Georgi Grechko, Vladimir Vasyutin, and Alexander Volkov to board it. This mission, designated *Soyuz T-14*, was launched on September 17, 1985. Of this crew, Grechko was the only veteran, having spent thirty days in space aboard *Salyut 4* and ninety-six days aboard *Salyut 6*. Eight days after the relief crew arrived, Dzhanibedkov and Grechko returned to Earth.

A third Progress vessel was launched and docked with *Salyut 7* on October 2. It carried the normal supplies as well as a ton of scientific equipment, including a battery of astronomical telescopes. Shortly after unloading this equipment, cosmonaut Vasyutin developed a fever of 104°F. Soviet doctors thought it best to terminate the mission before his condition worsened and instructed the crew to return on November 21. The end of this flight must have been disappointing. According to Alexander Volkov, the original plan did not call for the crew to return until March 15, 1986. By that time, Savinykh would have established a new endurance record of 282 days. Upon his return, Vasyutin was hospitalized for about a month and recovered fully.

Conclusions

Salyut 6 and *7* demonstrated how far the Soviets had progressed towards their goal of building a permanent space station. Soviet manned flights were now becoming almost routine. The space endurance record had been extended to eight months and only the sickness of Vasyutin seemed to have prevented this record from reaching nine months. The list of Soviet accomplishments was impressive, two reliable resupply crafts, the Progress and Kosmos, had been constructed and thoroughly tested. Cosmonaut teams had conducted numerous engineering and scientific experiments, and large structures had been welded together in space. The space station had proven to be a reliable orbiting laboratory and the cosmonauts had demonstrated an impressive ability to modify and repair it under adverse circumstances.

By the mid-1980s, the Soviets had acquired a vast amount of medical data on the effects of prolonged space flight. Using this information they had developed exercise programs and food supplements which allowed their cosmonauts to readjust to Earth's gravity with a minimum number of complications. In this respect they were far ahead of their American counterparts who were preoccupied with their Space Shuttle program. As successful as the Shuttle would become, this craft was not intended to spend more than ten days in space. If astronauts wished to remain in space for longer periods of time, NASA had to gain much more experience with the problems posed by long periods of weightlessness.

36 The Soviet Super Station: The Mir Space Station

"The Mir complex is a tentative first step towards an orbital habitat. It is a tiny percursor, a technology demonstrator; the issue of scaling it up will come later, once the requisite technology is fully understood."

—**David Harland in the Mir Space Station**

Following the return of *Soyuz T-14*, a cargo craft (*Kosmos 1686*) was still attached to the unoccupied *Salyut 7*. Therefore, western observers were surprised when the core of a new space station, named Mir (Russian for "peace"), was placed into orbit on February 20, 1986. This date coincided with the opening of the Congress of the Communist Party of the Soviet Union, and in a break with tradition, news agencies were given videotapes of the station's launch just hours after it lifted-off. At first, observers thought this craft might join *Salyut 7* since they were in similar orbits. However, this theory was soon put to rest when Soviet officials announced that the module was the core of an entirely new space station.

The Mir Base Unit

The 22-ton Mir core module was similar in design to earlier Salyuts and had similar dimensions. The core module consisted of two cylindrical units, having a maximum diameter of 14 feet and a length of 43 feet. This made the base unit 5 feet shorter than the Salyut. Attached to the front of the module was a spherical 7-foot diameter unit called the forward docking compartment. Along its circumference were five docking ports and a radar, called the Igla, used for rendezvous (see #1 in the Focus Box below).

Once in the docking compartment, the cosmonauts entered the first work area by passing through a connecting airlock. The work area contained the station's eight control computers. Two solar panels were attached to its exterior. The forward work area opened into a larger 14-foot diameter compartment that contained the crew's exercise equipment, dinning area, and sleeping quarters.

Behind the second work compartment were the craft's propulsion system, a second set of rendezvous and docking radar, and a communication antenna. The propulsion system consisted of two main rockets and thirty-two attitude thrusters. As in prior Salyut's, the station could also be entered through a docking port at the end of this section.

The original Soviet plan called for five specialized research laboratories named Kvant 1, Kvant 2, Krystall, Spektr, and Priroda to be added over time to the Mir core unit. When completed, Mir would be the largest structure ever assembled in space. Because it would be very large and awkward to move, the approaching Soyuz or Progress vessel was expected to assume the primary role in docking.

Focus Box: The Mir Core Unit

A Soyuz TM craft is attached to the Mir core module in this drawing. The forward docking compartment is shown as (1), the first work area as (2), and the second work area as (3). The main control console is seen within the first work area. Attached to the exterior walls are two large solar panels that can be rotated to maximize their Sun exposure. The station's two main engines are seen at the rear of the craft (4). A sixth docking port is located between them (5).

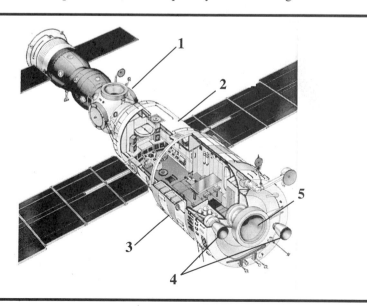

Initial Mir Missions

About a month after the launch of the Mir core module, cosmonauts Leonid Kizim and Vladimir Solovyov lifted-off aboard *Soyuz T-15*. Both cosmonauts were space veterans, having flown together on the record 237-day *Soyuz T-10/Salyut 7* mission.

On March 15, 1986, two days after their launch, Kizim and Solovyov docked and boarded Mir without incident. Mir's navigation and orientation computer was tested and its other systems brought on-line. Six days later a cargo vessel was launched from Tyuratam and automatically docked carrying 2 tons of supplies. The cosmonauts were kept busy unloading and inventorying these materials. On April 27 a second Progress delivered 2 additional tons of supplies and automatically began to transfer fuel to the station.

After verifying Mir was in working order, Kizim and Solovyov loaded 1,100 pounds of materials into their Soyuz and proceeded to *Salyut 7*. They completed a routine docking, entered the deserted station, and reactivated its systems.

Once *Salyut 7* was operational, Kizim and Solovyov conducted two spacewalks to retrieve materials left outside the craft to determine how their properties had been affected by prolonged exposure to space. They then erected a 50-foot aluminum tower to test construction techniques that would later be needed to build similar structures on Mir. While aboard the station they also collected observations of the area surrounding the Soviet Chernobyl nuclear reactor. A cloud of radioactive gas had been accidently released and had contaminated the region near the plant. Salyut photographs, along with data from Soviet Meteor satellites, aircraft, and ground personnel, would be used by scientists to plan cloud seeding activities to prevent rain from washing deadly material from the contaminated soil into the water supply of the surrounding Ukrainian cities.

While Kizim and Solovyov were busy with these activities, ground controllers continued to monitor Mir. A new, unmanned test version of the Soyuz, called the *Soyuz TM-1*, was launched from Tyuratam on May 21, 1986. This craft had the same dimensions as the *Soyuz T* but contained improved communication and computer systems, a slightly larger cargo area, and a new docking system called Kurs. After automatically docking with the Station and remaining attached for seven days it departed on May 29 and returned to Earth the next day. By the middle of June the attached Progress had finished refueling Mir. It was undocked on June 22, leaving both lateral ports unoccupied.

By late June Kizim and Solovyov had been on *Salyut 7* for fifty days and began making preparations to return to Mir. They removed twenty of the Salyut's instruments, storing them in their Soyuz, and departed on June 26. Ground controllers then fired the Salyut engines, placing the station in a higher 288-mile orbit. The crew's journey back to Mir was uneventful, and soon after reentering the station, they began their second experimental program. Biological experiments were conducted and photographs of the German Democratic Republic were obtained in a joint exercise with other satellites, aircraft, and ground teams. Finally, the cosmonauts installed a new computer system. With all of these tasks completed, Mir was put into an automatic mode, and the cosmonauts returned to Earth on July 16th. Their entire mission had lasted 125 days, during which time Kizim established a new cumulative space endurance record of 373 days.

The First Long-Duration Mir Mission

Cosmonauts Kizim and Solovyov had thoroughly tested Mir's systems, but delays in the construction of the Kvant 1 research module postponed the lift-off of the next crew. On January 16, 1987, the launch of a Progress supply ship signaled

Figure 36-1. Soviet cosmonauts Solovyov (left) and Kizim prior to their launch aboard Soyuz T-15 *from Tyuratam.*

Figure 36-2. Cosmonauts Vladimir Solovyov (left) and Leonid Kizim (right) after returning from Mir on July 16, 1986.

Figure 36-3. Cosmonauts Yuri Romanenko (bottom) and Alexander Laveikin prepare to board Soyuz TM-2 *for a flight to Mir.*

that the Soviets were now ready to send another crew to Mir. As expected, about three weeks later, *Soyuz TM-2* blasted-off to Mir. It carried cosmonauts Yuri Romanenko, commander of *Soyuz 26* and *38*, and Alexander Laveikin, a space rookie.

After docking with Mir, the cosmonauts activated the station's systems, and Romanenko began unloading supplies from the attached Progress, while Laveikin adjusted to his first experience with weightlessness. By February 17 the supplies had been stored and about one week later the Progress was jettisoned. About four weeks later a second Progress docked with additional supplies and equipment. Since each cosmonaut consumed about 22 pounds of food, water, and oxygen per day, the station had to be resupplied about every month and a half.

Romanenko and Laveikin were completely devoted to their mission and worked 15-hour days, starting at 8 a.m. and ending at about 11 p.m. They inventoried the station's consumable supplies, installed additional components into Mir's electrical and medical systems, and brought a new KATE-140 Earth resources mapping camera and Korund 1M furnace on-line.

Once these tasks were complete, they collected data on ocean currents and water runoff from the Caucasus Mountains. Medical experiments using the Gamma-1 and ultrasonic cardiograph were completed and forty-eight industrial Korund experiments employing various exotic materials such as cadmium, selenium, and indium antimoride were conducted.

Kvant 1: An Astrophysical Observatory

On March 31, 1987, about two months after the lift-off of Romanenko and Laveikin, Kvant 1 was launched. This 16-foot long, 14-foot diameter, 12-ton module contained a battery of astronomical telescopes and a package of biomedical experiments. The launch of Kvant 1 proceeded smoothly, and like the Progress craft, it was intended to dock automatically. However, when the module approached to within 700 feet of Mir, a collision was avoided by just 40 feet when its radar system failed.

After spending several days analyzing the reasons for this failure, engineers felt they understood what had gone wrong and another docking attempt was authorized. As Kvant 1 was maneuvered toward the rear port of Mir, Romanenko and Laveikin put on their space suits and entered their Soyuz craft as a safety measure. If a collision ruptured the station's wall causing the station's air to escape, they would be safe. This precaution proved unnecessary because a soft docking was achieved. The Kvant 1 was automatically pulled toward Mir to complete an airtight seal between the craft, but it then unexpectedly stopped just inches short of its locked position.

The Rescue of Kvant 1

Romanenko and Laveikin were puzzled by the Kvant 1 docking failure and looked out the rear Mir porthole to see what had gone wrong. Without an airtight connection, the crew could not enter the laboratory and it would be useless. Since no problem could be seen from inside Mir, the cosmonauts were asked to conduct an Extra Vehicular Activity (EVA). On April 11, both crewmen left the station through a forward docking port and proceeded down the station's 43-foot exterior wall, moving along prepositioned handholds. When Laveikin reached the observatory, he saw a white plastic bag was wedged into the docking mechanism and radioed this information to mission control. The 12-ton Kvant was instructed to fire its rockets and move slowly away from the station. When several inches away, Laveikin reached into the mechanism and removed the obstruction. It turned out to be refuse from the jettisoned Progress that had somehow escaped. Controllers then commanded the Kvant 1 to move back toward the station, and an airtight seal was achieved. Having saved the observatory, the cosmonauts retraced their path and reentered the station.

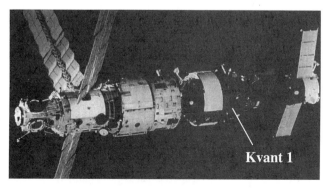

Figure 36-4. Kvant 1 was the first module added to the Mir core unit. It had a docking port attached to its rear that allowed resupply vessels or Soyuz craft to dock.

On the following day, Romanenko and Laveikin opened the hatch to the Kvant and began unloading its nearly 3 tons of cargo. They then jettisoned the module's propulsion unit, freeing its rear docking port. The cosmonauts then installed new computers in Mir and tested the linkages between the two craft.

On April 21, 1987, *Progress 29* was launched carrying food, water, fuel, and equipment. Once docked and unloaded, a second cargo vessel lifted-off with still more supplies. Having all the equipment they now needed, Romanenko and Laveikin conducted materials science studies, used the Kvant gyros to change Mir's orientation, and tested a high precision star tracker to correct slight drifts in the station's motion. On June 9, the Kvant x-ray telescopes were used to observe a newly detected supernova in the Large Magellanic Cloud, SN1987A. The cosmonauts also installed two Earth observation instruments and began plant growing experiments in the core module.

With the added power demands of Kvant 1, Mir's two solar panels were unable to provide enough power to satisfy all of the station's needs. The situation became so critical that the availability of power dictated the order in which experiments could be performed. Romanenko and Laveikin conducted two spacewalks in June to install a third solar panel which had been brought to the station aboard Kvant 1. Once in place, the station's power capacity was increased by 26 percent to 11.5 KW. The cosmonauts then returned to their experimental program of obtaining Earth resources photographs, making semiconductors, processing biological substances, and conducting plant growing studies.

Despite this heavy workload, Romanenko and Laveikin enjoyed being on Mir and for relaxation they sang while one of them played a guitar. Much to their surprise and slight embarrassment, they later learned Western monitors could hear their songs on the open radio frequency employed by Mir for ground communications.

International Flights

As was the case with the Salyut space stations, other countries, especially those politically aligned with the Soviet Union, were encouraged to participate in Mir's experimental program. The first "international" visitor was Mohammed Faris from Syria. His mission was to complete joint Syrian–Soviet experiments and obtain Earth resources photographs of Syria and the eastern Mediterranean. These tasks were similar to those performed on prior international flights where the visiting cosmonaut's country was surveyed from space. His Soviet crewmates were cosmonauts Alexander Viktorenko and Alexander Alexandrov. Their flight, designated *Soyuz TM-3*, was scheduled to blast-off from Tyuratam on July 22, 1987.

While preparations for this flight continued, medical personnel detected an irregularity in Laveikin's heartbeat from electrocardiograms transmitted from the station. Concerned about his condition, doctors instructed him to return along with the crew of the visiting flight. Cosmonaut Alexandrov, who had completed a 149-day mission on *Salyut 7,* was asked to remain on Mir with Romanenko.

The visiting mission proceeded in a routine manner and after eight days in space they returned on July 30, 1987. Laveikin was sent to the National Cardiology Research Unit in Moscow for tests, and after a thorough examination was declared fit for future missions. Faris returned to his duties with the Syrian Air Force.

An all-Soviet crew consisting of cosmonauts Vladimir Titov, Musa Manarov, and Anatoli Levchenko was the next to occupy Mir. This crew, seen in Figure 36-5, lifted-off on December 21, 1987. Titov and Manarov were scheduled to relieve Romanenko and Alexandrov, who would return with Levchenko. By this time Romanenko had established a new

---

Focus Box: The *Kvant 1*

The *Kvant 1* was 16 feet long, had a maximum diameter of 13.6 feet, a mass of 12 tons, and contains 1411 ft^3 of habital space. It contained two scientific instrument packets and six gyroscopes called Gyrodins that were used to control the space station's attitude.

One of the two instrument units contained an astronomical package called the Roentgen. This included the Soviet Pulsar-1 hard X-ray telescope/spectrometer, the English TTM wide-angle X-ray telescope, the European Space Agency Sirene 2 spectrometer, and the West German Foswich X-ray telescope. The second scientific package contained the Glazar telescope and the Svetlana automated electrophoresis unit that was used for biotechnology experiments.

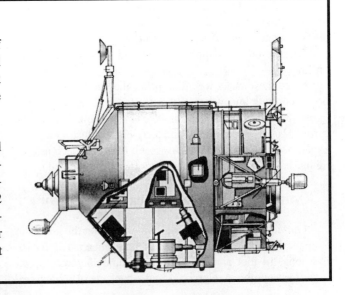

Figure 36-5. The crew of Soyuz TM-4: (top to bottom) Manarov, Levchenko, and Titov.

space endurance record of 326 days, three months longer than the previous record. Physically worn out, he was anxious to return home. Titov had commanded the *Soyuz T-8* flight that had failed to dock with *Salyut 7* in 1983 and had narrowly escaped with his life when his *Soyuz T-10* rocket exploded on the pad. Levchenko had been selected to pilot the new Soviet space shuttle, called Buran, and was added to the flight at the last moment to gain some experience with weightlessness. He later died of a brain tumor on August 10, 1988, about nine months after returning to Earth. The Buran backup pilot, Alexander Schukin, also died in a jet crash a few days after Levchenko's death. The deaths of these two pilots dealt a blow to the Soviet shuttle program since only a small number of pilots had been trained to fly this craft.

Figure 36-6. Cosmonauts Romanenko (left), Manarov (center), and Levchenko (right). Romanenko's space endurance record of 326 days aboard Mir would be broken by Manarov.

Table 36-1. Manned Flights to Mir I

Mission	Crew	Launch Date
Soyuz T-15	Kizim, Solovyov	Mar. 13, 1986
Soyuz TM-2	Romanenko, Laveikin	Feb. 6, 1987
Soyuz TM-3	Viktorenko, Alexandrov, Faris	July 22, 1987
Soyuz TM-4	Titov, Manarov, Levchenko	Dec. 21, 1987

Once on-board Mir, the cosmonauts continued the research program began by the prior crew. They collected astrophysical observations using the *Kvant 1* telescopes and carried out Earth resources, medical, and materials processing experiments. In order to prevent their muscles from deteriorating, the cosmonauts completed a daily set of exercises which included jogging for 3 miles on a treadmill and riding a stationary bicycle for 6 miles. At the end of each day, they relaxed by reading books or by watching videotapes.

Titov and Manarov were visited twice by international crews. The first visit took place in June 1988 when Soviet cosmonauts Anatoli Solovyov and Viktor Savinykh and Bulgarian Alexander Alexandrov docked aboard *Soyuz TM-5*. During their stay, they completed a research program consisting of forty-six experiments. The second visiting crew carried Afghan Abdul Mohmand and Soviet cosmonauts Vladimir Lyakhov and Valeri Poliakov. This mission was largely conducted for propaganda purposes, taking place just before the Soviet Army abandoned Afghanistan to the rebel mudjahaddin. However, the return flight of Lyakhov and Mohmand almost ended in a space disaster.

Another Close Call: The Return of Lyakhov and Mohmand

The procedure the Soviets had established for leaving the space station was for the departing crew to use the craft brought up by the prior crew. In accordance with this system, on September 5 cosmonaut Lyakhov and Mohmand undocked from Mir aboard *Soyuz TM-5*. Cosmonaut Poliakov, a crew member of *Soyuz TM-6* and a medical doctor, remained on the station to monitor the health of Titov and Manarov as they attempted to break Romanenko's 326-day space duration record.

As *Soyuz TM-5* prepared to reenter the Earth's atmosphere, its computer system failed to fire the retro-rockets at the proper time. Realizing this would cause them to overshoot their intended impact area, Lyakhov aborted the descent. Three hours later a second reentry failed when the computer instructed the craft to perform a brief preprogrammed rendezvous maneuver. Canceling this command, Lyakhov manually fired the reentry rockets, but after 60 seconds the computer overrode these instructions and turned the engines off. As their craft continued to spiral toward Earth, the situation looked grim. If control of their craft could not be regained, Lyakhov and Mohmand would be burned alive.

Fearing that the worst was about to happen, the Russian national news agency, *Izvestia,* issued a statement about the unfolding drama to prepare the public. However, Soviet ground controllers had carefully analyzed the combined effect of the prior capsule firings and told Lyakhov how to reprogram the Soyuz computer to account for them. He was instructed to attempt a landing on the following day. This time the retrorockets fired properly and the Soyuz descended safely to Earth.

After this harrowing flight, Lyakhov was flown to Moscow and Mohmand returned to the Afghan capital city of Kabul where his space suit was put on public display. However, his adventure had no effect on the military situation in Afghanistan. A short time later, the mudjahaddin captured Kabul and Mohmand fled to Germany. The victorious rebels seized his suit and sold it to raise money for their army.

New Visitors and a New Space Record

A third international flight was manned by Soviet cosmonauts Alexander Volkov and Sergei Krikalev and Frenchman Jean-Loup Chretien. This was Chretien's second flight, having participated in the earlier 1982 joint Soviet–French flight, *Soyuz T-6.* His principle tasks involved medical experiments and a spacewalk to erect a truss structure developed by the French space agency. He was assisted in this effort by Volkov. Chretien also tested an improved space suit jointly developed by Soviet and French engineers.

With the successful completion of this mission, Chretien returned to Earth with cosmonauts Titov and Manarov, both of whom had established a new endurance record, exceeding Romanenko's record of 326 days by forty days. After being retrieved by a Soviet recovery team, the cosmonauts were sent to Moscow for medical exams. Although the Titov and

Mission	Crew	Launch Date
Soyuz TM-5	Solovev, Savinykh, Alexandrov	June 7, 1988
Soyuz TM-6	Lyakhov, Poliakov, Mohmand	Aug. 29, 1988
Soyuz TM-7	Volkov, Krikalev, Chretien	Nov. 26, 1988

Table 36-2. Manned Mir Missions II

Manarov initially had difficulty balancing themselves in Earth's gravity, they were able to walk normally after just two days. Several weeks later they resumed their normal duties.

Mir was now occupied by three cosmonauts: Poliakov, Volkov, and Krikalev. Originally, the Soviets had planned to launch two additional laboratories while this crew was on Mir, however, the construction of these units was far behind schedule. It seemed unlikely they would be ready in the next few months so the crew was directed to return at the nominal end of their mission on April 27, leaving Mir temporarily unoccupied. The cosmonauts' return proceeded normally and Poliakov was in such good shape after his 240-day flight he was able to perform basic gymnastic exercises one day after landing.

Progress M and the Soviet Manned Maneuvering Unit

On August 23, 1989, an improved version of the Progress supply ship, Progress *M-1*, was launched. Unlike its predecessor which was totally destroyed during reentry, this craft carried a recoverable capsule called the Raduga. This 770-pound cone-shaped object had dimensions of 5 feet by 2.5 feet, giv-

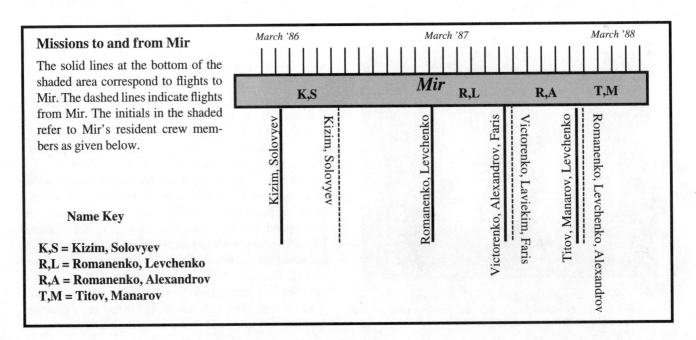

Missions to and from Mir

The solid lines at the bottom of the shaded area correspond to flights to Mir. The dashed lines indicate flights from Mir. The initials in the shaded refer to Mir's resident crew members as given below.

Name Key

K,S = Kizim, Solovyev
R,L = Romanenko, Levchenko
R,A = Romanenko, Alexandrov
T,M = Titov, Manarov

ing it a volume of 120 liters, comparable to the size of a tall chest of drawers. It could transport up to 330 pounds of materials from Mir back to the Earth.

On September 5, 1989, about two weeks after the launch of this cargo vessel, cosmonauts Alexander Viktorenko and Alexander Serebrov lifted-off on a mission to attach another of the long-delayed modules to Mir and to test the Russian equivalent of the American Astronaut Maneuvering Unit. This space suit possessed its own set of thrusters and permitted a cosmonaut to move through space. Other tasks assigned to this crew included routine maintenance of the Mir core module, Earth resources studies, and materials processing and astronomical experiments.

The Kvant 2: An Earth Resources Module

The next Mir module, Kvant 2, was launched on November 26, 1989. It had a mass of 20 tons, a length of 40.7 feet, and a maximum diameter of 14.3 feet. The pressurized volume of the station was increased by 47 percent to 6,719 ft^3 and Kvant 2's solar panels increased the station's power supply by 7 KWs.

Kvant 2 would serve as an orbiting multipurpose laboratory. It contained a second "Elektron" life-support system, a "Vika" oxygen production system, and biological and Earth resources experiments. On its exterior wall was a remotely controlled platform with three spectrometers and a large multispectral television camera. This module also contained a bigger airlock to allow the cosmonauts to make easier entrances and exits from the station. External fixtures allowed electronic components and building materials to be mounted to test how different kinds of materials behaved in a space environment.

Figure 36-7. The Kvant 2 module showing a Gamma-2 multispectral camera attached to its outside wall (top).

Figure 36-8. Kvant 2 extends toward the bottom part of this picture. The Soviet manned maneuvering unit is shown as an insert.

Once in orbit, one of the Kvant 2 solar panels failed to deploy. Fortunately, ground controllers soon devised an ingenious plan to deal with this problem. The module was "spun," creating a centrifugal force that pulled the panel outward from the module's body. Once this task was accomplished, Kvant 2 was brought to within 16 miles of Mir, but at this point its navigational system aborted the docking. Engineers studied the cause of this failure and devised a new docking procedure. Four days later a second docking attempt was successful. The manipulator arm of Kvant 2 was then used to move the module from Mir's front port to a side port. Once this operation was complete, Mir acquired an "L" shape appearance (Figure 36-8).

Kvant 2 also carried the Soviet manned maneuvering unit, new cameras, gyros, water regeneration equipment, and a shower. Cosmonaut Serebrov later tested the maneuvering unit by flying 149 feet from the station during a 4-hour spacewalk on February 1, 1990. Four days later Viktorenko again tested this unit during a 4-hour EVA. After completing these tests, the maneuvering unit was stored away for later use in the Soviet space shuttle program.

Krystall: A Materials Processing Laboratory

On February 11, 1990, cosmonauts Anatoli Solovyov and Alexander Balandin lifted-off in *Soyuz TM-9* to relieve Viktorenko and Serebrov. The main objective of their mission was to join Mir's third experimental module, Krystall, to Mir. This module contained five furnaces for materials processing experiments, a pair of high-resolution Earth observation cameras, an ultraviolet telescope, and a cosmic ray detector. Attached to its exterior wall were two 118-foot retractable solar panels.

Continuing the string of bad luck that had plagued the other modules when they reached orbit, Krystall's flight control system failed soon after launch. This caused its first docking attempt to fail, but four days later engineers used a backup program in the module's computer to complete this task. The module's manipulator arm was then used to move this unit to the port opposite Kvant 2, giving Mir the "T" shaped appearance seen in Figure 36-9.

In July Solovyov and Balandin conducted two EVAs to repair some thermal panels on their Soyuz craft. Western observers sensationalized this task by reporting that the two Russian cosmonauts had "no real possibility of returning to Earth," a charge that was dismissed as "groundless" and "incomprehensible" by Soviet Deputy Flight Director Viktor Blagov. He noted that, if needed, another Soyuz could always be sent to retrieve the crew. During Solovyov's and Balandin's first EVA, they accidently damaged the Kvant 2 airlock, forcing them to enter the station through another porthole.

Non-Communist Visitors to Mir

On August 1, 1990, a relief crew consisting of cosmonauts Gennadi Strekhalov and Gennadi Manakov began their journey to Mir. This was Strekhalov's fourth flight and Manakov's first. Their primary goals were to use the Krystall furnaces to make a variety of semiconductors and to repair the damaged airlock on Kvant 2. After this crew arrived, Solovyov and Balandin departed in a routine crew exchange.

Manakov and Strekhalov were visited on December 2, 1990, by a Japanese journalist, Toyohiro Akiyama. His fellow crewmen were cosmonauts Viktor Afanasayev and Musa Manarov. The Tokyo Broadcasting Service had paid the Soviets $12 million for the privilege of providing their viewers with telecasts from Mir. Although Akiyama had a difficult time adjusting to weightlessness, he managed to broadcast several live segments. At the end of this mission he departed with cosmonauts Manakov, and Strekhalov. Manarov and Afanasayev remained on Mir and continued the station's ex-

Figure 36-9. Between 1990 and 1995 Mir consisted of four modules: Kristall (1), the core module (2), Kvant 1 (3), and Kvant 2 (4).

Table 36-3. Manned Flights to Mir III

Mission	Crew	Launch Date
Soyuz TM-8	Viktorenko, Serebrov	Sept. 5, 1989
Soyuz TM-9	Solovov, Balandin	Feb. 11, 1990
Soyuz TM-10	Manakov, Strekalov	Aug. 1, 1990
Soyuz TM-11	Afanasayev, Manarov, Akiyama	Dec. 2, 1990
Soyuz TM-12	Artsebarsky, Krikalev, Sharman	May 18, 1991

perimental program, repaired the damaged airlock, and made other alterations to the exterior of the station. On January 16, *Progress M-6* docked with additional supplies and an extendable 40-foot crane. A week later Manarov and Afanasayev installed it on the Mir core module using the techniques employed by Kizim and Solovyov during their earlier *Soyuz T-15/Salyut 7* mission. The purpose of this apparatus, visible in Figure 36-9, was to move large packages on the outside of the station to different locations.

Helen Sharman, a British research technologist for the Mars company, was the next international visitor to Mir. She flew to the station aboard *Soyuz TM-12*. During her eight-day trip she conducted medical, plant, and chemical experiments, and spoke to British schoolchildren with the assistance of English amateur radio operators. Her Soviet crewmates, Anatoli Artsebarski and Sergi Krikalev, conducted five spacewalks to repair the radar antenna on Kvant 1 and to build a second 46-foot-high structure on Kvant 1. Attached to the top of this girder they placed a Soviet flag. On May 26, Afanasayev, Manarov, and Sharman returned to Earth in *Soyuz TM-11*.

A Troublesome Era for Mir

On August 19, 1991, a new and troubling era began for the Soviet space program. On this date, Soviet President Gorbachev signed a treaty granting member countries of the Soviet Union a greater degree of independence from Russia. However, this act was viewed with such alarm by hard-line communists they decided to stage a coup. Gorbachev was arrested and sent to his home in the Crimea. Unfortunately for the plotters, the takeover attempt lacked both widespread public support and the backing of the armed forces. Street fights broke out in Moscow, and in a memorable event, Boris Yeltsin, the president of Russia, gained international acclaim for his courage when he stood atop a defecting Soviet tank to rally the public against the conspirators. A few days after this heroic act the insurrection collapsed and the instigators were arrested. Despite the rapid turn of events, the coup revealed the unstable nature of the communist state, and almost overnight the Soviet Union began to break apart.

Table 36-4. The Mir Modules		
Kvant 1	March 31, 1987	Astrophysical Lab
Kvant 2	Nov 26, 1989	Earth Resources
Kristall	June 1, 1990	Materials Sciences

While these tumultuous events were unfolding, cosmonauts Artsebarski and Krikalev were still orbiting the Earth aboard Mir. In an effort to remain politically neutral, ground controllers provided them with news from Soviet central TV, which favored the coup, and Russian radio, which opposed it. Understandably, the cosmonauts were uncertain about how to proceed. Their training manuals stated that if the Soviet Union was to go to war, they were to return, but there was no plan if the government was overthrown! In the end they decided to wait for instructions from whichever faction prevailed.

The collapse of the Soviet Union left the Soviet space program in turmoil. Two international flights, one carrying Kazakh Takhtar Aubakirov and the other an Austrian, Franz Viehbock, were merged into a single mission called *Soyuz TM-13*. Since two seats in the return vessel would now be occupied by visitors, only one of the cosmonauts presently on Mir could be relieved. To accommodate this situation, Krikalev agreed to remain until the next flight. His crewmate, cosmonaut Artsebarski, returned to Earth on October 10 with the two international visitors. A replacement crew was launched on March 17, 1992, consisting of Russian cosmonauts Alexander Viktorenko and Alexander Kaleri and German Klaus-Dietrich Flade. The Russian crewmen would remain on Mir while Volkov and Krikalev returned with Flade.

Conditions in the former Soviet space program continued to deteriorate. In an attempt to restore order, Yeltsin created a new organization, the Russian Space Agency, to take charge of the booster rockets, spacecraft, and space station. Unfortunately, this action created further confusion when the other countries of the former Soviet Union followed his lead and established their own space agencies. Jurisdictional disputes developed over who controlled what, and agreements had to be reached over ownership of ground stations, tracking ships, as well as the launch complex at Tyuratam itself. As the Soviet empire collapsed, the Russian currency lost its value and runaway inflation forced many experienced workers to leave the space program to find higher paying jobs.

In this confused state, hopes for an improved space station, Mir 2, vanished. It was not even certain whether Mir's two remaining modules, Spektr and Priroda, would be finished and launched. The Mir core unit had a design life of five years, but had been in space for six years. The expected lifetime of Kvant 1 and 2 was only three years, but they had been in space for five and three years, respectively. Could the final Mir laboratories be launched before these modules became unusable? Was this work economically justifiable?

Despite these uncertainties, there was considerable international support for keeping Mir operational because it was the only orbiting spacecraft capable of supporting long-duration missions. Several nations expressed an interest in sending people to the station to conduct experiments and were willing to pay for this access. Once the visiting crew of *Soyuz TM-14* left, they were followed by another visiting flight carrying Frenchman Michel Tognini and cosmonauts Anatoli Solovyov, and Sergei Avdeyev. In October 1992 an agreement was reached between NASA and the Russian Space Agency to use the Space Shuttle to support Mir opera-

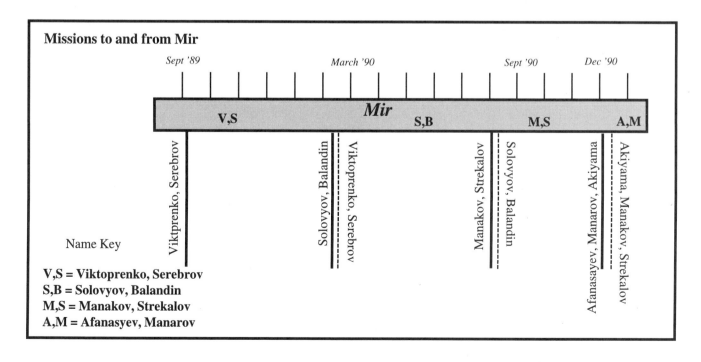

Missions to and from Mir

Name Key

V,S = Viktoprenko, Serebrov
S,B = Solovyov, Balandin
M,S = Manakov, Strekalov
A,M = Afanasyev, Manarov

Table 36-5. Manned Flights to Mir IV		
Mission	**Crew**	**Launch Date**
Soyuz TM-13	Volkov, Aubakirov	Mar. 12, 1992
Soyuz TM-14	Viktorenko, Kaleri, Flade	Mar. 17, 1992
Soyuz TM-15	Solovyov, Avdeyev, Tognini	July 27, 1992
Soyuz TM-16	Manakov, Poleshchuk	Jan. 24, 1993
Soyuz TM-17	Tsibliev, Serebrov, Haignere	July 1, 1993
Soyuz TM-18	Afanasayev, Usachyov, Poliakov	Jan. 8, 1994

tions and to allow American astronauts to board the station and conduct experiments. The European Space Agency also booked two flights to Mir in 1994 and 1997.

Impressed by this show of support, the Russian government authorized three additional flights, *Soyuz TM-16, 17, and 18,* to prepare the station for extended international missions. Ironically, it was the harsh economic conditions created by the fall of the Soviet Union rather than the acts of statesmen that finally opened the door for cooperative international space efforts. Although Tsiolkovsky, Goddard, and Oberth had long since died, they undoubtedly would have been pleased by the dawn of this new era.

Mir Perseveres through Troubling Times

One should not say that this second Russian revolution replaced Communism with capitalism. It is more appropriate to describe it as the overturn of an authoritarian system that maintained public order, in favor of a chaotic arrangement in which long-established practices could no longer hold.

—T. A. Heppenheimer in *Countdown*

With the collapse of the once powerful Soviet Union, the Russian economy was left in shambles. Soaring inflation destroyed the value of World War II pensions, and many veterans were forced to sell household goods and personal items in the city streets just to survive. Hardships abounded as school teachers, coal miners, and others went unpaid for months at a time. Public unrest grew and many longed for a return of the old communist system where freedoms had been restricted, but few went hungry.

The Russian space program was not spared during this crisis. The agency's 1992 budget was slashed by two-thirds, and military expenditures for space research were cut in half. Draftees were sent to maintain the launch facility at Baikonour, but living conditions were so poor they rioted. At least three people were killed and several buildings were burned to the ground. Circumstances at Plesetek, the Soviet's northern cosmodrome, were not any better. The major Russian newspaper, *Izvestiya*, reported, "the imprint of abject poverty of the soldiers can be seen everywhere." A cook told visitors, "I can no longer remember when there was fresh meat." Hardships were not limited to the common soldier, a general reported that in his command, "a total of 255 families of officers have no quarters." If the Russian space program and its crown jewel, the Mir space station were to survive, officials realized help had to come from outside their country. The government had so many urgent needs to deal with, it had little time to worry about space. Fortunately, assistance quickly arrived. The United States, France, and other countries signed contracts with Russia to send people to conduct long-term studies on Mir. In exchange for this privilege, the Russian government was paid tens of millions of dollars. The timely infusion of this cash not only saved the Soviet program, but helped prevent further economic chaos.

The Post–Soviet Union Space Program

In 1991 Soviet cosmonaut Sergei Krikalev found himself in a unique situation. Launched before the fall of the Soviet Union, he was still on Mir when an independent Russia was established. Because of his peculiar circumstance, Western news organizations referred to him as "the last Soviet citizen."

Even though conditions in Russia were becoming increasingly harsh, initially operational support for Mir continued in a surprisingly normal fashion. There were enough resources already in the pipeline to temporarily insulate Mir from the turmoil engulfing the country. Mission control was still manned by competent controllers and Krikalev collected data using the station's instruments. These results were then sent back to Earth in Progress Raduga capsules as though nothing had changed.

As the political structure of the Soviet Union underwent radical changes, Russia entered a new and uncertain era. *Soyuz TM-13, 14,* and *15* were launched as part of the Mir visitor program and to raise more cash. While the visiting cosmonauts from Austria, Kazakhstan, Germany, and France conducted research programs in Earth resources, medicine, and engineering, their Russian crewmates maintained and upgraded the station.

As part of the *Soyuz TM-15* flight, a propulsion unit, called the VDU, was attached to the top of the Mir Sofora crane to provide a more fuel efficient method of turning the station (see Figure 37-1). When in place it saved up to 85 percent of the fuel previously required to make maneuvers.

The Mir Experimental Program Continues

On January 24, 1993, *Soyuz TM-16* traveled to Mir carrying cosmonauts Gennadi Manakov and Alexander Poleshchuk. This flight was somewhat unusual in that it did not carry a paying international visitor and it docked at the station's Kristall port. This port would later be used by the American Space Shuttle, Atlantis, to bring crew members and supplies to the station.

Several experiments were completed during the *Soyuz TM-16* mission, including the placement of different types of building materials on the station's exterior wall to determine how their properties changed when exposed to the vacuum of space. This experiment and others like it were intended to provide essential data about the types of materials that could be used in the future construction of the International Space

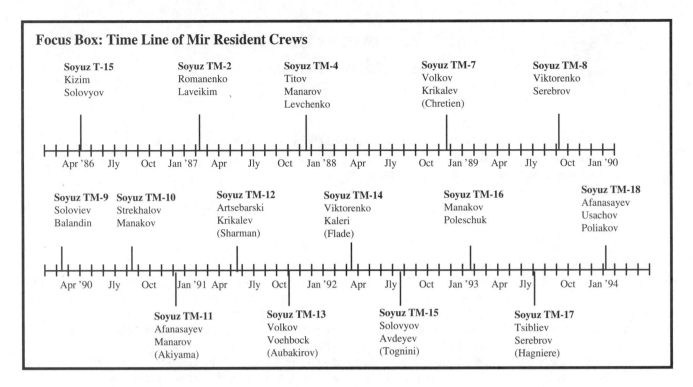
Station. A second experiment, funded by NASA, provided information about how fluids moved through granular material in a prototype device to grow food in space.

During this mission, a particularly memorable experiment, first suggested by Hermann Oberth in the 1920s, was conducted. After the cargo vessel, *Progress M-15*, was unloaded, this ship was moved away from the station and its jets fired to rotate the craft. The centrifugal force created by this spinning motion deployed a 66-foot diameter mirror. The goal of this exercise was to determine if reflected sunlight could be used to illuminate cities located in Earth's polar regions during their long dark winter. The mirror projected a 3-mile-wide bright spot onto the Earth and proved this concept was sound.

In an additional test, Manakov used a video camera on the Progress to maneuver this craft from the station. Due to the increasing number of solar panels, girders, and antenna on Mir, future dockings relying on preprogrammed instructions might not be possible. The goal of this test was to determine if a ship could be guided to a port by a cosmonaut inside the sta-

Figure 37-2. In the above picture of Mir, Kvant 1 is located at the bottom of the station and a Soyuz capsule is docked at the front port of the base module. Krystall is pointed toward the camera and Kvant 2 is located beneath it. The fixtures on the left and right edges of Kvant 1 are solar panel mounts.

Figure 37-1. The VDU, a propulsion unit attached to the top of the Sofora crane. The crane itself was attached to Kvant 1.

tion. Manakov and Poleschuk also installed two solar panel mounts on Kvant 1.

On July 1, 1993, cosmonauts Vasili Tsibliev and Alexander Serebov along with French cosmonaut Jean-Paul Haigniere lifted-off from Baikonour to relieve Manakov and Poleshchuk. Tsibliev and Serebrov later conducted several space-walks to erect a second but smaller girder on Kvant 1 and to examine whether the exterior of the station had been damaged when Mir passed through the Perseid meteor shower in early August. During their inspection, sixty-five pits and holes were found, the largest being a 4-inch diameter hole in one of the station's solar panels.

International Use of Mir Grows

In the fall of 1992 NASA agreed to a multifaceted cooperative program with the Russian Space Agency. During its first stage, a cosmonaut would fly aboard the Space Shuttle to familiarize the Russians with the capabilities of this craft. In the second phase, Shuttle flights would ferry supplies to Mir over a two-year period and leave astronaut researchers on the station. The goals of these flights were to allow the astronauts to test equipment that would be used on the International Space Station, to gather medical data on the long-term effects of weightlessness, and to conduct research projects that could not be completed in the relatively short flights of the Shuttle.

In order to dock with Mir, the Space Shuttle would use an apparatus built for the Russian shuttle Buran. This cylindrically shaped module was installed into the forward bay of the Atlantis during one of its scheduled maintenance periods and connected to the Shuttle's cabin by an airtight tunnel. In addition to the United States, the European Space Agency (ESA) paid Russia to transport its spacemen to Mir using a Russian Soyuz spacecraft. Realizing Mir was a source of much-needed cash, the Russian space officials announced they would continue to make the station available to visitors and would launch the long-delayed Spektr and Priroda modules.

Problems Begin to Develop on Mir

On January 10, 1994, Soyuz TM-18 docked at Mir's rear port carrying the relief crew of Viktor Afanasayev, Yuri Usachyov, and Valeri Poliakov. This was the first flight for Afanasayev and Usachyov, but Poliakov had previously flown on Soyuz TM-6. Their research program consisted of medical and technology experiments funded by the German government. About a week later, Tsibiliev and Serebrov undocked for their return journey to Earth. Once free of the station, they maneuvered their Soyuz to the Kristall port to obtain pictures of this area to aid the training of American Shuttle pilots who would dock at this port. While undertaking this task, their Soyuz accidentally bumped against the station, causing an undetermined amount of damage.

In fulfillment of the first phase of the U.S.–Russian cooperative program, Sergei Krikalev ("the last Soviet citizen")

lifted-off aboard STS-60 Discovery on February 3, 1994 as a mission specialist. This flight went smoothly and included the launching of a German satellite and Spacelab 2 activities.

On July 1, 1994, Yuri Malenchenko and Talgat Musabayev were sent to relieve Afanasayev and Usachyov. During their stay a series of troubling problems arose. On August 27, Progress M-24 failed to dock during a routine resupply mission and nearly collided with one of the station's solar panels. A second docking attempt also failed when the ferry rebounded from the docking collar. Malenchenko then used the remote control system tested by Manakov to ease the ship into port.

On September 9th, Malenchenko and Musabayev conducted an Extra Vehicular Activity (EVA) to examine whether Progress M-24 had damaged the docking collar during its impact. Their visual inspection revealed no harm had been done and they proceeded to the Kristall port to examine the area where Soyuz TM-17 had struck the station. A small tear in the thermal blanket was found and repaired.

Russian engineers had assumed the Progress M-24 docking failure was caused by an error in the ferry's computer system. Therefore, they were surprised when on October 6, 1994, the next Soyuz flight (Soyuz TM-20) experienced a similar failure. This craft was manned by cosmonauts Alexander Viktorenko, Yelena Kondakova, and ESA cosmonaut Ulf Merbold. Merbold's mission was the first of two long-term visits planned by the ESA. It included twenty-nine experiments, twenty-three of which dealt with the medical effects of weightlessness on the human body, four with materials science questions, and two with technology issues. As Viktorenko, Kondakova, and Merbold were about to dock, the Soyuz capsule suddenly jerked out of alignment. Viktorenko quickly took over manual control and piloted their craft safely into the port. Officials were now convinced the problem was

Figure 37-3. Dr. Valeri Poliakov (wearing glasses) had hoped to set a new endurance record of eighteen months, but delays at Tyuratam reduced his time in space to 437 days.

Figure 37-4. ESA cosmonaut Merbold poses with his Russian crewmates Musabayev (left) and Kondakova (right).

in Mir's systems, but what had caused this malfunction? Were some of Mir's older automatic systems finally beginning to fail?

Problems continued to plague the station. Another episode occurred when Mir's storage batteries were inadvertently drained by overuse. Activities within the station had to be curtailed for a few days as the batteries were recharged from the station's solar panels. A more serious incident occurred when a small fire broke out in one of the station's Elektron oxygen generation systems. By acting quickly, Poliakov was able to prevent this situation from becoming life-threatening, but this life-support unit was completely destroyed. With the loss of this unit, only one other generator remained to supply the crew with oxygen. On November 1, Merbold completed his work and departed along with some of his experimental results. His remaining data would be returned by the crew of the upcoming American Space Shuttle flight *STS-71*.

By January 1995, Russian engineers thought they understood why the automatic dockings of *Progress M-24* and *Soyuz TM-20* had failed. An incorrect value for the center of mass had been entered into the visiting ship's computer. This caused the craft to lose its alignment when minor adjustments were made just before a docking. In order to test this theory, Viktorenko and Kondakova changed the code in their *Soyuz TM-20* computer and then flew this craft several hundred feet away from the station. When they then attempted to rejoin Mir, automatically, the docking went perfectly. NASA officials were quite pleased to learn this problem had been solved. Both Russia and the United States looked forward to the upcoming Shuttle flight to Mir.

38   Flights of the Space Shuttle

It's all part of taking a chance and expanding man's horizons. The future doesn't belong to the fainthearted; it belongs to the brave.

—Ronald Reagan, Oval Office address to the nation, January 28, 1986

The Space Shuttle's first orbital flight, *STS-1*, occurred on April 12, 1981. By January 1, 1999 an additional ninety-four flights had been completed. Because of the large number of flights which have occurred, it is not practical to discuss each of them in this book. Therefore, this chapter will focus on three memorable missions: the first orbital Space Shuttle flight, *STS-1*, the Challenger disaster, *STS-25,* and the return of John Glenn to space aboard *STS-95*.

Preparations for *STS-1*

The test program adopted for the Space Shuttle was unique among NASA's manned space programs. Before astronauts undertook orbital flights in Projects Mercury, Gemini, or Apollo, the booster rockets and spacecraft were tested using unmanned capsules. Several such tests were conducted in Project Mercury, two were completed in Project Gemini, and six were conducted in the Apollo program. *STS-1* would be the first time NASA allowed astronauts to fly aboard the maiden orbital flight of a new craft.

Another new feature of *STS-1* was that the Shuttle employed two new solid rocket boosters, the first such rockets used in NASA's manned space program. These fifteen-story-high boosters provided 80 percent of the Shuttle's thrust at launch. There was no way to jettison these engines while they were still burning, and if they did not fire within one second of each other, the craft would become uncontrollable.

Part of the reason NASA adopted this different approach lay in the cost of the spacecraft. Individual capsules were expendable, whereas a Shuttle was intended to be reusable. Officials thought if a problem arose during a flight, the chances of recovering the craft were higher if an experienced test pilot were at the controls instead of a computer. This decision also had a considerable downside because it placed the astronauts in danger. Every component of the Shuttle had been subjected to extensive ground and low-altitude flight tests, but it was an exceedingly complex vehicle and unexpected problems were almost certain to arise. If the malfunction were serious, the astronauts could easily lose their lives.

The Shuttle is both a rocket and an aircraft. It is intended to operate in space and then glide through the atmosphere to a

Figure 38-1. The launch of America's first orbital flight of a Space Shuttle broke with tradition by carrying a crew of two astronauts.

predetermined landing site. However, during reentry the pilot does not have the use of engines to alter the Shuttle's flight path. If something goes wrong during its landing approach, there is no way to fly to an alternate site. The Shuttle will crash. This procedure is considerably more risky than that used in the Mercury, Gemini, or Apollo programs. In those efforts, the crew essentially rode their capsule down to Earth without attempting to alter its course.

April 10, 1981 was set as the launch date for *STS-1*. Despite the danger associated with this maiden flight, any number of astronauts would have volunteered to fly it. This was the type of mission test pilots lived for. To help ensure its success, NASA selected two of its ablest astronauts as crew members, John Young and Robert Crippen.

The Crew of *STS-1*

John Young was born on September 24, 1930, in San Francisco, California. Married to the former Susy Feldman of St. Louis, Missouri, they have two children and two grand-children. In 1952 he received a degree in aeronautical engineering from the Georgia Institute of Technology. Following graduation, he joined the U.S. Navy, was trained as a pilot, and was accepted as an astronaut in 1962. For relaxation, he enjoys wind surfing, bicycling, reading, and gardening.

At the time of the launch of *STS-1,* Young was a space veteran. The astronaut who had the closest personality to Young was Gus Grissom, his crewmate on *Gemini 3*. Both of these people were no-nonsense professionals and disliked media intrusions into their private lives. Michael Collins, *Apollo 11* command module pilot, wrote of Young, "John is witty, but rarely allows himself to be perceived as such, especially when outsiders are around. He prefers a cloak woven partially of engineering mumbo-jumbo and partially of aw shucks, t'ain't nothin." Following *Gemini 3*, Young served as the command pilot on *Gemini 10*, as the CSM pilot on *Apollo 10*, and as commander of the *Apollo 16* mission to the Descartes Highlands. Even at the age of 50, he still had not lost his enthusiasm for spaceflight.

STS-1 would be Robert Crippen's first spaceflight. Known as "Crip" to his fellow astronauts, he was born in Beaumont, Texas, on September 11, 1937. He is married to Virginia Hill of Corpus Christi and they have three daughters. In 1960 he received a bachelor's degree in aerospace engineering from the University of Texas.

Originally a navy pilot, Crippen flew in the air force X-20 program, and was chosen to participate in the U.S. Air Force Orbital Laboratory Program in June 1966. When this program was cancelled three years later, Crippen joined NASA as an astronaut. This move was made during a difficult period for NASA. Cutbacks in the Apollo program meant the astronauts would have to wait a long time before being assigned to a flight. Crippen realized it would be several years before he would fly in space. Young specifically requested Crippen as his crewmate on *STS-1* because of his extensive knowledge of the Shuttle's computer systems.

STS-1 Preflight Considerations

Tensions were high at the Kennedy Space Center as the April 10, 1981 launch date of *STS-1* approached. The Space Shuttle was the largest and most expensive NASA program since Apollo and hundreds of media representatives were watching. Officials were confident all would go well, but knew a failure would be a public relations disaster. Fortunately, Young and Crippen were able to keep themselves above the fray and even downplayed the significance of their flight in their prelaunch press conference. Crippen described it as routine,

> It is a test flight, and that basically is what this entire 54-hour 37-orbit mission consists of, to make sure we can get up into orbit properly, and don't have any problems, that we can make sure all the systems on the vehicle function as they should, and we'll go through a systematic checkout of basically everything that's aboard, and make sure that we can fly reentry like we planned. If we can just get up and get down, even if we had to do it all in one day, that would satisfy 95 percent of the objectives of the flight. . . . We'll also be testing out the environmental control systems, the cooling systems on board the spacecraft—we will even be doing mundane things like checking out the potty. . . ."

In his addresses to the press, Young spoke of his faith in the Shuttle and the extensive ground and flight test program the Columbia had passed. "If there is a vehicle we can have confidence in, I believe it's the Columbia."

A Failed Launch Attempt

On April 10 the astronauts were awakened at 2 a.m. to begin the now standard prelaunch preparations. By 5 a.m. they were aboard the Shuttle and the countdown was proceeding normally. Over one million people lined the streets leading to Cape Canaveral to see the launch, and the Queen Elizabeth II, which had made a special voyage to the Cape, was anchored a safe distance offshore to give its passengers a spectacular vantage point from which to watch this event.

Nine minutes before the end of the countdown a problem was discovered in the acid level of the Shuttle's number 3 fuel cell. The countdown was stopped to allow the seriousness of this problem to be discussed, and if needed, to replace the fuel

Figure 38-2. The crew of STS-1. *Seated, Robert Crippen, front, Robert Young.*

cell. While this was occurring a second problem arose in one of Columbia's five computers. After a delay of several hours, it was decided to scrub the mission and try again on Sunday, April 12. Upon hearing this news, Young replied, "What a pity, on such a lovely day."

Eight Minutes to Orbit

April 12, 1981 was the twentieth anniversary of Yuri Gagarin's historic flight. Through an accident of fate it now marked a second historic event, the launch of America's first Space Shuttle. The countdown procedure was repeated and things ran smoothly. The number of people present to see the launch was even larger than for the previous attempt. Well over one million people lined the shores and roads of the Cape with their eyes turned toward the Shuttle's launch site, Pad 39A. As the countdown reached zero, public speakers barked, "We've gone for main engine start. We have main engine start. . . . Lift-off! Lift-off of America's first Space Shuttle—the Shuttle has cleared the tower!"

As the Columbia rose into the sky, Crippen's heart rate increased to 130 beats per minute while that of the veteran Young reached only 85. Just 132 seconds after leaving the pad, the Shuttle was 27 miles high and its solid rocket boosters were jettisoned. People on the ground let out a loud cheer when they saw these engines separate from the Shuttle. A little over 8 minutes after leaving the pad, Young and Crippen were in orbit. Young radioed Houston Mission Control, "The view hasn't changed any—it's really something else." Crippen added, "I tell you, John has been telling me about it for years, but there is no way to describe it."

NASA officials were extremely pleased with how well everything had gone. Launch Director George Page told reporters, "It was a great job. We got off on time . . . had very

Figure 38-4. When Young viewed the back of the Shuttle through a rear window in the cabin, he was surprised to see that some of the heat resistant tiles had come off.

few problems . . . what more can I say? We are thrilled and proud today."

Once in orbit, Young and Crippen tested the Shuttle's cargo bay doors and found they could be opened and closed without any problem. Young reported some of the Shuttle's heat resistant tiles on the back engine casing were missing. NASA tried to de-emphasize the significance of this fact to the news media, but officials were worried. Some parts of the Shuttle would be heated to over 2,000°F during reentry, and if those areas were unprotected, the craft could be destroyed. The belly of the Shuttle would deflect most of the heat during reentry, so NASA requested that the American spy satellite KH-11 view this section of the craft. Much to their relief, the tiles on the underside of the Shuttle all appeared to be in place.

After their first meal in orbit, the crew used the TV cameras in the main cabin to send live pictures back to Earth. Young commented on the mission's progress, "We've done every test we're supposed to do, and we are up on the time-lines, and the vehicle is performing beautifully. . . . No systems are out of shape. . . . All the Reaction Control System (RCS) jets have been firing and the vehicle is just performing like a champ. Really beautiful, it's delightful up here!" Crippen echoed these sentiments and then signed off, ending the crew's first day of activities.

After a rest period in which neither astronaut was able to get much sleep, they were awakened by ground controllers. The schedule for day two was crammed with activities, including tests of the shuttle's RCS thrusters, the craft's attitude stability, and the cargo bay doors. Two TV broadcasts, one involving a congratulatory message from Vice President Bush, were also planned. All of these activities were successfully completed, and after a very long day the crew prepared the Shuttle for its return to Earth.

The landing conditions at Edwards Air Force Base were nearly perfect. The skies were clear and only a light breeze stirred the air. A large and joyful crowd, measured in the hundreds of thousands, was on hand to watch the landing. At 11:30 a.m. Young and Crippen closed the Shuttle's bay doors

Figure 38-3. Part of the crowd that gathered to see the launch of STS-1. *The first mission of a new program typically draws a large crowd of spectators.*

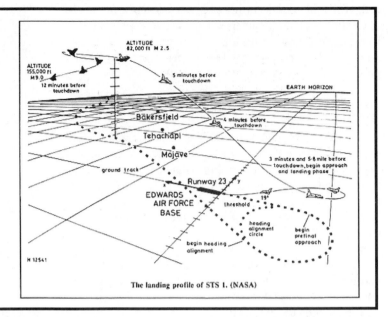
and put on their flight suits and helmets. At 2:21 p.m. the Shuttle's two OMS engines were fired for 160 seconds and the craft began its journey back to Earth. Houston Mission Control radioed the crew, "Nice and easy does it, John, we're all riding with you."

The first phase of the reentry was handled by the Shuttle's automatic systems. At a height of 115,000 feet Young took manual control of the craft and guided it toward Edwards. As they flew over the Pacific Ocean and approached the United States at six times the speed of sound, Young remarked, "What a way to come to California." The landing gear was lowered at a height of 400 feet while the shuttle was traveling at 211 mph. Moments later it touched down and Joe Allen of Mission Control radioed, "Welcome home, Columbia." In a nonchalant tone Young replied, "Do you want me to take it up to the hanger?"

During the next hour, the astronauts stayed on-board the Columbia turning off its systems. Young then emerged to inspect his craft. Delighted with the flight, he approached ground crew personnel to shake their hands.

The flight of *STS-1* was judged to be a success. All of its major objectives had been achieved. However, 300 tiles had been damaged during the flight. Most were dislodged during the launch and others had been damaged by debris picked up during landing. Further analyses found that the vibrations created by the ignition of the solid rockets had loosened many of the tiles. Others had been damaged by falling ice from the supercold external tank. NASA engineers made appropriate changes to address each of these problems. The only significant concern with the Shuttle's interior system was that the zero-g "potty" that Cernan had joked about before launch failed a few hours before the end

Figure 38-5. The landing speed of STS-1 was higher than NASA engineers had predicted. To allow for this possibility the landing site was on a long runway in a dry lake bed at Edwards Air Force Base.

Figure 38-6. After the landing was complete, it was clear that the exterior of Columbia had suffered significant damage. This was particularly obvious in the nose region of the craft.

of the flight. All in all, *STS-1* was a great start to a new era in space travel.

STS-25: The Challenger Disaster

NASA had built an impressive record of accomplishments since the start of its manned program in the 1960s. Four Mercury and ten Gemini flights as well as six missions to the Moon had been completed without the loss of a single life. These programs were followed by the successful Apollo-Soyuz and Skylab flights. Although astronauts Grissom, White, and Chaffee had died in a test pad fire during *Apollo 1*, this accident had occurred during a ground test. The closest NASA had come to losing a crew in space was during the nerve-racking flight of *Apollo 13*. As humbling as this experience should have been, its successful conclusion actually increased the public's feeling that NASA could handle whatever problems might arise during a spaceflight. Influenced by this sentiment, some of NASA's own managers began to believe the agency was infallible.

The Crew of *STS-25*

John Young, commander of *STS-1* and chief astronaut at the time of *STS-25,* knew the dangers spaceflight posed and tried to combat the growing attitude of complacency among NASA's personnel. He could enumerate an "awesome" number of failures that could lead to the loss of a Shuttle and its crew, but few people seemed to take these concerns seriously. Besides, the pressure to keep the program moving forward

Figure 38-7. Sharon Christa McAuliffe was selected from over 11,000 teacher applicants to fly aboard the Challenger as part of NASA's Teacher in Space Program.

Figure 38-8. The crew of STS-23 *(51-D). Back row (left to right), Onizuka, McAuliffe, Jarvis, Resnik. Front row (left to right), Smith, Scobee, McNair.*

was formidable. NASA had justified the Shuttle program on the basis that it would be more cost effective than using expendable vehicles. This promise had yet to be fulfilled because the Shuttle launch rate was too low. Nevertheless, things were improving. In 1984 five Shuttle flights were launched. In 1985 the number of flights reached nine, and in 1986 fifteen launches were planned. Safety issues were not forgotten, but they were dealt with in a piecewise fashion as time permitted. Blinded by its string of successful launches, NASA had unconsciously created an environment that almost guaranteed a disaster would occur.

On August 27, 1984, President Reagan announced NASA would include a "citizen" passenger on future Shuttle flights. The first such passenger would be Sharon Christa McAuliffe, a social studies teacher from Concord High School in New Hampshire. She would be followed in early 1986 by a journalist. Although McAuliffe and the yet to be named journalist would be the first "citizen" passengers in the Shuttle program, they would not be the first nonprofessional crewmen. Senator Jake Garn, chairman of the Senate committee that oversaw NASA's budget, had flown aboard *STS-16* on April 12, 1985, and Congressman C. William Nelson, a subcommittee chairman of the House Technology Committee had flown aboard *STS-24* on January 12, 1986.

The Flight Plan of *STS-25*

The flight of *STS-25* was to last 6 days and 34 minutes. It had three primary objectives, one in the area of technology, a second in science, and the third in public education. The first objective would be met by deploying a 19-ton satellite named the Tracking and Data Relay Satellite (TDRS). Once operational, it would improve communications between Earth-orbiting spacecraft and ground controllers. The crew would then deploy a second satellite, the Shuttle-Pointed Tool for Astronomy (SPARTAN-203). This satellite would observe the tail and coma of Halley's comet using two ultraviolet spectrometers and cameras to study the comet's chemistry and composition. At the end of the flight, this satellite was to be retrieved from orbit and returned to Earth. McAuliffe would conduct the educational aspect of *STS-25*. During the flight she would broadcast live television lessons from the Shuttle to classrooms on Earth. The first lesson was titled, "The Ultimate Field Trip," and the second, "Where We've Been, Where We're Going, Why?" She would also perform experiments to illustrate Newton's laws, magnetism, effervescence, and the operation of simple machines.

In addition to these primary objectives, several other experiments were scheduled for this mission, including the Fluid Dynamics Experiment, the Comet Halley Monitoring Program, the Phase Partitioning Experiment, and three Student Involvement Program experiments.

The crew of this flight consisted of Francis R. Scobee (commander), Michael J. Smith (pilot), Judith A. Resnik (mission specialist), Ellison S. Onizuka (mission specialist), Ronald McNair (mission specialist), Gregory B. Jarvis (payload specialist), and Sharon Christa McAuliffe (payload specialist). *STS-25* would be the second Shuttle flight for Scobee, Resnik, McNair, and Onizuko, and the first for Smith, Jarvis, and McAuliffe. Jarvis and McAuliffe were on temporary assignments to NASA and would return to their regular jobs after the mission was completed.

The Launch Decision

STS-25 was originally planned to blast-off from the Kennedy Space Center on January 22, 1986 but was delayed four times due to slippages in the launch of *STS-24*, technical problems, and bad weather. As a result of these postponements, NASA officials were anxious to launch *STS-25* to maintain their subsequent flight schedule. In May the Challenger was scheduled to launch two satellites, Ulysses, a European satellite to study the polar regions of the Sun, and Galileo, an American probe to study the giant planet Jupiter. If the launch window for this flight were missed, NASA would have to wait another thirteen months until the Earth and Jupiter were again in a proper alignment for a launch.

On January 27 an attempt was made to launch the Challenger, but during the countdown sensors indicated that the Challenger's hatch-locking mechanism had not been engaged. A repair team was sent to fix this problem, but by the time the repair was completed, winds at the center had risen beyond those permitted for lift-off. The flight was scrubbed at 12:35 p.m. and rescheduled for January 28 at 9:38 a.m.

Soon after the January 27 postponement, launch officials became concerned about the cold weather at the site. Larry Wear, solid rocket motor manager, called Morton Thiokol-Wasatch in Utah to ask if this would effect the operation of the solid rocket boosters. Thiokol convened a meeting to discuss this question and their engineers expressed concern that the O-rings located at the junction of the separate sections of the booster might lose their resiliency and fail to contain the hot gases created during ignition. At 5:45 p.m. on January 27th, a telephone conference was held between Thiokol, Mar-

shall Spaceflight Center, and the Kennedy Space Center to discuss this matter. Thiokol urged the launch be delayed until the afternoon of the 28th when the temperature at the site would be higher. Faxes supporting this recommendation were sent to Marshall for review. A second meeting was scheduled for 8:15 p.m later that evening.

During this evening meeting, Thiokol engineers used the faxed data to argue that the O-rings were unreliable at cold temperatures. They pointed out that on the January 5, 1985 flight of *STS-20,* the primary O-ring had failed to restrain the hot gases when the launch temperature was 53°F. This was much warmer than the expected temperature at the time of the launch of *STS-25*. The secondary ring had stopped the flow of hot gases, but if a more extensive flow had developed, the craft might have been in jeopardy. When NASA officials asked Thiokol management for its recommendation, they stated the launch should not be allowed to take place unless the temperature at the Cape was above 53°F.

Marshall participants were surprised by Thiokol's reply because it amounted to the institution of a new "Launch Commit Criteria," offered on the eve of a flight. Larry Mulloy, Marshall Solid Rocket Booster manager, responded, "My God, Thiokol, when do you want me to launch, next April?" He then went on to challenge the basis of their "no launch" recommendation, suggesting their data analysis was faulty. He pointed out that when all the data was examined, the presence of O-ring erosion or blow-bys was not correlated with temperature, and even if the cold weather allowed gases to blow-by the primary O-ring, the data showed that the gas would not get by the secondary ring. George Hardy, Marshall deputy director of Science and Engineering, was then asked for his opinion. He responded that he was, "appalled at the Thiokol recommendation," but then added he "would not recommend launch against the contractor's objection."

To several Thiokol engineers Hardy's use of the word "appalled" was very intimidating. They felt the NASA managers were requiring them to "prove that we should not launch" and were clearly unconvinced by the discussions. Thiokol's position was further weakened by the fact that many of the other Thiokol engineers present at the meeting agreed Mulloy's criticisms were valid. Thiokol engineer Brian Russell later said,

> The fact that our recommendations were questioned was not all that unusual. . . . In fact, it's more the rule. We are often times questioned on our rationale, which is the way I believe it should be. . . . We felt in our presentation of the data that we had to include all of the data that could possibly be relevant, even though not all of it tended to support our point of view. . . . So his [Mulloy's] argument was rationale and logical to my mind. I didn't agree with it necessarily.

Faced with the conflicting views of its own engineers and NASA's skepticism on the validity of their recommendation, Thiokol officials requested a 5-minute break that led to a 30-minute off-line discussion. During this session Thiokol Senior Vice President Jerry Mason asked his engineers to again explain their concerns. After this was done, he requested the attending managers to vote on a recommendation. Three voted in favor of letting the launch go forward, while one, Robert Lund, was opposed. Mason then asked Lund to "take off his engineering hat and put on his management hat." Thiokol's Jack Kapp recalled that moment,

> I certainly understood that Bob Lund was under a lot of pressure. You know, I don't care what anybody says, I felt pressured that night. These things were apparent to me. NASA had heard our arguments in detail and essentially had rejected them out of hand. You know, Larry Mulloy made the statement "I have a contrary opinion" right after hearing all of our arguments, so that was a factor. Mr. Hardy's statement that he was appalled at the recommendation was a factor providing pressure. These are knowledgeable, brilliant men. I have nothing but respect for Mr. Hardy and Mr. Mulloy. I worked with them for years. That was a factor. The fact . . . our general manager had voted that we go ahead and fire . . . just left the weight of the world on Bob Lund's shoulders.

After considering Mason's comment, Lund changed his vote to a go. He later spoke about his reasoning,

> Number one, it was pointed out that the conclusion we had drawn that low temperature was an overwhelming factor was really not true. It was not an overwhelming correlation that low temperature was causing blow-by. We have blow-by both at low temperatures and high temperatures. The second thing . . . we have also run lots of tests where we have put a lot of erosion on O-rings, and we could experience a factor of three times more erosion on the O-ring and still have it work just fine, so we had a very large [safety] margin on erosion. The last one was the Mc-Donald [point] that during that original pressurization, if it blows by there is only going to be so much blowing because that doggone thing [the secondary O-ring] is right there and seated. It is still going to seal, and you are not going to have a catastrophic set of circumstances occur.

When the teleconference was resumed, Thiokol presented its new recommendation favoring a launch to the NASA participants. The meeting broke up at 11:15 p.m.

The Morning of the Launch

On January 28, at 1:30 a.m. a NASA ice inspection team was sent for a 2-hour inspection of the launch pad and found ice forming. Rockwell International, the builder of the Shuttle, was contacted and asked if this condition endangered the

Figure 38-9. Ice on the launch pad of the Challenger. Never before had NASA attempted to launch a Shuttle under such conditions as these.

performance of the craft. A second inspection of the pad was started at 7 a.m., and this time 3-foot-long icicles were found to be hanging from the handrails and platform and sheets of ice covered the gantry where the crew would enter the Shuttle. Rockwell workers viewing the pad on closed-circuit TV described the icy scene as "something out of Dr. Zhivago." Based on these inspections, officials at the site decided to postpone the launch for a few hours to allow the ice to melt.

At 8:30 a.m. the Challenger crew entered the Shuttle and prepared for lift-off. Once aboard the Challenger, Mission Control radioed Scobee, "Let's hope we go today." He replied, "We'd like to do that." At 9 a.m. the Mission Management Team met for the last time to decide whether or not to allow a launch. Rockwell felt falling ice might damage the solid rocket boosters and advised against a launch. A few Thiokol representatives also urged that the launch be postponed. When asked for his opinion, Charlie Stevenson, head of the ice inspection team replied, "the only choice you've got today is not to go." Despite these opinions, when the vote was taken, the recommendation was to launch.

The Launch of *STS-25*

The temperature at Pad 39A at the time of lift-off was a frigid 36°F, seventeen degrees cooler than on any previous Shuttle launch. The crowd at the Cape to watch the launch of the Challenger had fallen compared to the previous day.

Thousands of people, including Smith's brother, Jarvis's mother, the governor of New Hampshire and his son, and many of McAuliffe's friends had been unable to remain an extra day. Seated in the V.I.P. bleachers were Christa's parents, sister, and brother. Her husband Steve and their two children watched from the roof of the Launch Control Center.

NASA had wanted to recapture public attention with its Teacher in Space Program and had succeeded. Students across the nation were glued to the television sets in their schools. At Concord High School, McAuliffe's school, students sat in the auditorium, blowing noisemakers and cheering as the moment for take-off approached. In the White House Nancy Reagan watched the prelaunch countdown on a television while her husband worked on his State of the Union address. In his book, *"I Touch the Future . . . ," The Story of Christa McAuliffe,* Robert Hohler described the interest this flight had generated,

> Not since the glories of the Moon landings had space lured so many. From the Virgin Islands to an Eskimo village in the Arctic Circle, they waited—two and a half million students and their teachers . . . all along the space coast people stopped what they were doing and turned to the sky.

The Tragedy Unfolds

As the three engines of *STS-25* burst to life, ground controllers heard Judith Resnik excitedly remark, "A-a-all riiight!" and Smith calmly radioed, "Here we go." Later a close examination of videotape of the Shuttle revealed an ominous sign. Just 0.68 seconds after ignition, a puff of black smoke escaped from the aft section of the right solid rocket booster (see Figure 38-10). The smoke revealed that the rubber O-rings in this section of the rocket were burning and that the joint was not sealed. The O-ring failure discussed by the Thiokol engineers the night before and dismissed by their own managers had occurred.

To observers on the ground the black puff of smoke was unnoticeable and the crowd in the V.I.P. bleachers cheered as the Shuttle continued upward. Even if observers had seen the black smoke, at this point nothing could have been done. Once the solid rocket boosters fire, there is no way to abort the flight. There was nothing the crew or the ground crews could have done to avoid the ensuing catastrophe.

About 58 seconds after lift-off, a small flame visible on enhanced frames of the launch videotape, broke through the wall of the lower right booster. Half a second later this flame had grown in size and was heating the strut joining the booster to the Shuttle's large external tank (see Figure 38-11). Sixty-five seconds after lift-off, this flame penetrated the external tank and hydrogen began to escape. At 72 seconds, the heated strut broke and the bottom of the solid rocket began to swing wildly. It hit one of the Shuttle wings and then the external tank. Weakened by the flames from the leak, the liquid hydrogen tank ruptured. When this occurred the top of the tank shot upward, breaking the liquid oxygen tank. Milliseconds later

Figure 38-10. Black smoke is seen escaping from the lower aft section of the right solid rocket booster in this videoframe of the launch.

the external tank exploded in a foaming white cloud of water vapor. As debris from this massive explosion spread across the sky, the solid rocket boosters sped out of the water vapor cloud in opposite directions. To avoid the possibility they might hit land and cause damage, the U.S. Air Force safety commander destroyed them by sending a self-destruct signal.

The crew was unaware of the solid rocket booster problem right up to the time of the explosion. Seven seconds before the explosion, Smith radioed a routine airspeed indicator check to Mission Control. A few seconds later Scobee increased the power to the Shuttle's main engines in response to a message from ground control, "Roger, go at throttle up." Three seconds later Smith was heard to say, "Uh-oh" and then silence.

The explosion of the Challenger caught NASA spokespeople by surprise. The public affairs officer continued to broadcast statistics about the flight even though the Challenger had been destroyed, "One minute and fifteen seconds, velocity 29 hundred feet per second, altitude 9 nautical miles, down range distance 7 nautical miles." People listening to this were confused and thought that the white cloud might be part of the booster separation process. Then there was a long pause and the announcer continued,

> Flight controllers looking very carefully at the situation. Obviously a major malfunction. We have no

Figure 38-11. Flames from the right aft solid rocket booster are seen in videoframes taken of the launch. The circled area shows the flame.

Figure 38-12. As the Challenger's external tank ruptured and the liquid hydrogen and oxygen mixed, a giant white cloud was created. It would take some parts of the craft 30 minutes to finally fall into the ocean.

downlink. We have a report from the flight dynamics officer that the vehicle has exploded. The flight director confirms that. We are looking at checking with the recovery forces to see what can be done at this point. Contingency procedures are in effect.

Crowds watching this scene from the ground and on televisions were horrified by the apparent disintegration of the Challenger. They were momentarily heartened, however, when one parachute was observed descending towards the Earth. Perhaps a crew member had survived? This hope soon faded when it was realized that the parachute belonged to one

of the destroyed solid rocket boosters. Nobody who saw the explosion then doubted that the entire crew had perished.

Reactions to the Challenger Disaster

The initial public reaction to the Challenger disaster was one of shock. Everyone expected the flight to proceed without incident, so its abrupt and tragic end came as a complete surprise. President Reagan postponed that night's State of the Union address and expressed the nation's grief in a speech broadcast from the White House,

> For the families of the seven, we cannot bear, as you do, the full impact of this tragedy. But we feel the loss, and we're thinking about you so very much. Your loved ones were daring and brave. . . . They wished to serve, and they did. They served us all.

The president's anguished feelings reflected those of the rest of the nation. The loss of McAuliffe, the popular teacher passenger, was especially hard to bear. Over the next few days pictures of the calamity were broadcast repeatedly. The entire nation felt as though they had just lost several close friends in a tragic accident.

The Rogers' Commission

Within a week of the Challenger disaster, President Reagan appointed a committee to determine the cause of this disaster and to make recommendations to prevent a similar event in the future. The House Committee on Science and Technology held its own investigation. The presidential commission was headed by William Rogers, former Secretary of State in the Nixon administration and included several well-known people, astronauts Neil Armstrong (retired) and Sally Ride, physics Nobel laureate Richard Feynman, air force Major General Donald Kutuna, executive vice-president of Hughs Aircraft Company, Albert Wheelon, retired air force Brigadier General Chuck Yeager, and the former editor-in-chief of *Aviation Week & Space Technology*, Robert Hotz.

The Rogers Commission examined nearly 122,000 pages of documents, hundreds of photographs, and interviewed more than 160 people. During the investigation they learned that the O-ring problem had been discussed by NASA and Thiokol. Postflight inspections of the primary and secondary O-rings on Shuttle flights *STS 2, 11, 13, 16, 20, 23, 24, 25, 27, 30, 31,* and *32* at least one of the rings had been eroded. The committee noted that blow-bys of the primary O-ring had occurred in eight of these flights. The commission was dismayed to learn that despite these warnings, neither NASA nor Thiokol had taken corrective action. In the words of the commission members, these organizations were playing a "kind of Russian roulette" and had failed to recognize the urgency of this problem. Because they had "got away with it last time" these organizations assumed it was safe to continue.

Four months after their work began, the committee issued a report. It concluded,

> The consensus of the Commission and participating investigative agencies is that the loss of the Space Shuttle Challenger was caused by a failure in the joint between the two lower segments of the right Solid Rocket Motor. The specific failure was the destruction of the seals that are intended to prevent hot gases from leaking through the joint during the propellant burn of the rocket motor.

The committee made nine recommendations, all of which were adopted by NASA:

1. The Solid Rocket Booster joint and seal would be redesigned.
2. The Space Shuttle management structure was to be reviewed, more astronauts were to be placed in management positions, and an independent flight safety panel was to be established.
3. A safety review of all critical items in the shuttle would be undertaken.
4. A new office, the Office of Safety, Reliability, and Quality Assurance would be created that reported directly to the Director.
5. Communications between the NASA Centers was to be improved.
6. The Orbiter's brakes and nosewheel steering would be improved.

Figure 38-13. The Rogers Commission recommended that an escape system be devised to allow the crew to abandon the Shuttle during the guide phase of its flight. The system selected for this purpose is shown above. Crew members slide down a pole-like structure to avoid being struck by the Shuttle's wing.

7. An escape system would be developed to allow the crew to exit the shuttle during controlled guided flight.
8. The Shuttle launch rate was to be tied to available resources and reliance on the Shuttle as the nation's only launch vehicle would be dropped.
9. Test and maintenance procedures would be improved.

The Aftermath of *STS-25*

A poll taken several months after the release of the Roger's Commission report found Americans blamed NASA for the loss of the Challenger and its crew. However, criticism of the agency was tempered by the realization that spaceflight is a risky business, that accidents are bound to occur. Isaac Asimov, a prominent science and fiction writer, expressed the change in peoples' attitude toward spaceflight the Challenger disaster caused,

> All of a sudden, space isn't friendly. All of a sudden, it's a place where people can die. . . . Many more people are going to die. But we can't explore space if the requirement is that there be no casualties; we can't do anything if the requirement is that there be no casualties.

Despite the loss of the Challenger crew, the same poll in which the public blamed NASA for this tragedy showed they still felt positive about the nation's space program. Support for Space Station Freedom and a manned mission to Mars, NASA's upcoming major initiatives, actually increased after the accident.

STS-95: John Glenn Returns to Space

On June 24, 1962, the editorial in the magazine *This Week* read, "Today as never before American's are hungry for heroes." Although the nation may have lacked an abundance of people to look up to, it certainly possessed one hero, John Glenn. Glenn's orbital flight of February 20, 1962 was a "first" for America and he was widely admired for the humility he displayed in the months leading up to this mission. In discussing his role in the space program, Glenn constantly used terms like "an honor," "duty to country," and "a privilege" when describing his role. These were not just slogans to Glenn, he deeply believed in these virtues. When asked why the Mercury 7 volunteered for the space program he said,

> I think we are very fortunate that we have been blessed with the talent that have been picked for something like this. I think we would almost be remiss in our duty if we didn't make full use of our talent. Every one of us would feel guilty, I think, if we didn't make the fullest use of our talent in vol-

unteering for something like this—that is as important as this is to our country and the world in general right now.

Although Glenn had some limited experience dealing with the public, he was still astonished at their attitude toward the astronauts. In his book, *John Glenn: A Memoir*, he wrote about the "tidal wave of attention" they received, and that, "People looked at us as if we had stepped out of the pages of science fiction or descended from another planet."

Glenn's Mercury Postflight Grounding

Much to his chagrin, Glenn's career as an active astronaut would be brief. After returning from his historic orbital flight aboard *Freedom 7*, he was puzzled to find himself grounded. Years later he discovered why this had occurred. Glenn wrote, ". . . I was told that President Kennedy told NASA that he preferred I not be on active flight status." Kennedy never told him the reason for this decision, but Glenn suspected that, "Maybe he was afraid of the political fallout if something happened to me." In fact, Glenn was too important to the country and too valuable a spokesperson for the space program to risk losing in another mission. Though this decision was understandable from a political point of view, Kennedy's directive did not sit well with Glenn,

> If the decision would have been left up to me, I would have preferred to remain in the active astronaut corps. Soon after my flight though, I was offered an astronaut training and management position at NASA Headquarters which I chose not to do. . . . Since I was not going to be on active flight status, I stayed on with NASA for a couple of years to plow my Mercury experience back into the program and then went on to other pursuits.

Realizing he would not fly again in space, Glenn resigned from NASA in January 1964. Shortly after leaving, Glenn announced he would run in the Ohio Democratic primary for a seat in the U.S. Senate. However, a month after beginning his campaign, he injured his inner ear in a fall and was left "virtually immobile." With an expected recovery time of eight to twelve months, he felt it would be impossible to mount an active campaign. On March 30 he called a news conference to announce he was withdrawing from the race. From his hospital bed he told reporters, "I do not want to run as just a well-known name. No man has a right to ask for a seat in either branch of Congress merely because of . . . orbiting the Earth in a spacecraft."

Glenn again attempted to win the Democratic nomination in 1970 but lost to Howard Metzenbaum, a lawyer and former Ohio state legislator. In 1974 another opportunity arose when one of Ohio's senators, William Saxbe, resigned to become the U.S. Attorney General. Glenn won the nomination for this seat and went on to win the general election over his Republican opponent by a wide margin. He was subsequently reelected to the Senate in 1980, 1986, and 1992.

During Glenn's years in the Senate he worked on several issues, including nuclear weapons treaties, fiscal management, foreign relations, and, of course, the space program. In 1995, while preparing for the debate over funding for the International Space Station, his attention was drawn to a chart in the book *Space Physiology and Medicine*. He was struck by the fact that many of the physical changes listed, such as bone loss, disturbed sleep patterns, loss of coordination, and balance disorders are similar to symptoms experienced during the aging process. Intrigued by these similarities he thought, "Here's something that ought to be looked into."

The Aging Process and Space

After conferring with doctors at the National Institute on Aging, Glenn thought the addition of an older person to a Space Shuttle crew might have real merit. He later wrote, "It became apparent to me that age might be an advantage instead of a disadvantage. There are about ten things in a human body that change in space that are like the frailties of old age."

NASA doctors were aware of the similarities between the aging process and the medical effects of weightlessness and were interested in studying this matter. Based on discussions with these doctors and physicians at the National Institute on Aging, Glenn approached NASA Administrator Donald Goldin with the idea of sending an older person on an upcoming Shuttle flight. "I really did go in and say somebody ought to do this. Whether it was me or not . . . But right along with it was, 'And look Dan, I'd like to be the guy that does this.' There wasn't any doubt about that."

At first Goldin did not think Glenn was serious about returning to space, but after relentless prodding, he finally decided to proceed providing that the value of flying an older person on the Shuttle could be confirmed by an outside panel of medical experts, and Glenn passed the same physical exam required of all crewmen. Goldin also stated he would not tolerate pressure on Glenn's behalf from the White House. Based on these criteria, Goldin thought the chances that Glenn would fly were slim, "First, I didn't think he was going to pass the physicals. . . . I knew it was going to be tough. And second, I thought it would be awfully tough to get some good peer reviewed science."

Events soon changed Goldin's mind. After meeting with the directors of the National Institute on Aging and the National Institute of Health, he found that both organizations thought such a mission would be scientifically valuable. Secondly, over the years Glenn had remained in good physical condition, never smoking, exercising regularly, and watching his weight. He passed the required physical exams without much difficulty.

Based on these findings, Goldin made his decision on January 15, 1998. "I threw everyone out of my room. And I sat in here, and I scripted all the issues. I had a little checklist that I made on a scrap of paper, and I went down the checklist to satisfy myself that everything was done right. . . . It was a very big decision that had to be made. You know, it was not a comfortable decision." After completing his analysis, he called Glenn, "You're the most persistent man I've ever met. You've passed all your physicals, the science is good, and we've called a news conference tomorrow to announce that John Glenn's going back into space." Glenn was overjoyed. He was assigned to *STS-95* which was expected to lift-off in October 1998.

The Crew of *STS-95*

STS-95 was manned by seven astronauts, Curtis L. Brown (commander), Steven W. Lindsey (pilot), Stephen K. Robinson (mission specialist), Scott E. Parazynski (mission specialist), Pedro Duque (mission specialist), Chiaki Mukai (payload specialist), and John Glenn (payload specialist). All of the crew members except Duque and Glenn had prior Shuttle experience.

Glenn was proud of his fellow crewmates and of his "science-rich" mission. He was annoyed that the press seemed to ignore the scientific objectives of *STS-95* and focused on him personally. At one session close to the launch date, Glenn scolded reporters,

> I was irritated at the press session . . . and I let it show. There were seventy-five to a hundred reporters and camera people there, more than had ever

appeared for tests like these before, and most of the questions were directed at me. History was repeating itself in that the questions were personal and had little to do with the scientific value of the mission. I said I wished the media would focus on the science and pay attention to my fellow astronauts and their accomplishments, but it did no good. I ended up doing most of the talking anyway.

The Scientific Goals of *STS-95*

The goals of *STS-95* extended beyond the medical issues Glenn would investigate. More than eighty experiments would be completed during the Shuttle's nine-day flight and a satellite, the *Spartan 201*, would be released to study the solar corona and the solar wind. Solar winds cause changes in the Earth's upper atmosphere which affect global radio communications. They also affect the operation of electronic equipment on manned and unmanned spacecraft and are thought to influence weather as well. After spending two days in space, the Spartan satellite would be retrieved and returned to Earth.

A second major experiment for *STS-95* involved the International Extreme Ultraviolet Hitchhiker Experiment. This apparatus, which is carried in the Shuttle's cargo bay, consists of six separate detectors that measure light in the extreme ultraviolet radiation (250-1,700 A). Objects to be examined included the Sun, Jupiter and its Moon Io, and hot stars. Astronomers used this data to construct models of planetary atmospheres, the heating of the solar system's Moons and comets, and stellar astrophysics.

A third major task of *STS-95* was to test the Hubble Space Telescope Optical Systems Test payload. This package contained a new cooling system for the telescope's near-

Figure 38-14. The crew of STS-15. *In the front row are Dr. Chiaka Mukai and mission commander Curtis Brown. Glenn is located behind Brown.*

Figure 38-15. John Glenn welcomed the opportunity to serve as a human test object to determine how older people would react to weightlessness.

infrared instrument (NICMOS), a faster telescope computer, a new engineering science solid state recorder, and a fiber optics communication system.

A fourth objective of this flight was to complete a set of thirty experiments located in a Spacehab module in the Shuttle's bay. Many of these were cooperative studies between NASA and the Canadian, European, and Japanese space agencies. They would collect data in the fields of materials science, plant growth, microgravity, protein crystallization, and medicine.

Glenn's Medical Program

Glenn's medical program consisted of two main experiments which would be completed in orbit and seven pre- and postflight tests. The two in-orbit experiments were designed to study sleep disorders and muscle atrophy. Astronauts do not sleep well in space, an affliction shared by many older people. Glenn and astronaut Dr. Chiaki Mukai would wear electronic sensors during four nights of the mission to gather data about their sleep patterns. Mukai would also take a drug, melatonin, to induce sleep. In the second experiment on muscle atrophy, Glenn and rookie astronaut Pedro Duque would take a pill containing alanine and an injection of histidine to determine if these proteins affected muscle production. David Liskowsky, NASA scientist in charge of life science experiments for Glenn's mission, explained,

> People thought if you exercised that would keep the muscles up . . . [But] even with exercise, though you can improve, or decrease the muscle atrophy, you still see some muscle atrophy in astronauts . . . We don't know whether they're not doing the right exercises, or maybe there's something else."

Figure 38-16. John Glenn adjusts his flight suit. Because these were orange colored the astronauts referred to them as "pumpkin suits."

The seven ground tests were intended to augment NASA's existing medical database. Liskowsky explained Glenn's contribution,

> These [medical tests] have been done on a number of other flights, and we have a pretty large database, for some of them as much as forty, fifty, sixty, astronauts. We have a pretty good idea of what a thirty-to fifty-five-year-old astronaut looks like on some of these measurements. Now we'll be getting that same data from Glenn.

Liskowsky continued,

> We know what he looks like before the flight and how he compares to the younger astronauts before the flight, and now we'll see what the changes are after the flight. Has he changed the same amount, or has he changed double the amount, or maybe less? We don't know, and these are the questions we're asking.

As the human body ages, bones become fragile and porous. This condition, known as osteoporosis, effects 25 million Americans, most over the age of 50. On Earth osteoporosis develops over many years, but in space similar changes occur in only a few months. Glenn's Shuttle flight would last nine days so the amount of bone loss was expected to be small. Nevertheless pre- and post-medical exams would be conducted to measure these changes.

Glenn took a humble but realistic view of his role on *STS-95*. In his memoir he wrote,

> Another research subject was a seventy-seven-year-old, 190-pound specimen who was known as payload specialist two. In many ways I would be as much of a guinea pig as I had been in 1962. Then, the other astronauts and I had been human subjects in the feasibility of space flight. In 1998 I would be a human subject seeking answers on the aging process.

The Launch and Flight of *STS-95*

STS-95 was scheduled for lift-off on October 29, 1998. It was a beautiful day, and after completing a traditional breakfast of steak and eggs, the crew suited-up, left their quarters, and headed to the launch pad. As they walked by the support people gathered to wish them a safe journey, Glenn looked at their faces and recalled his earlier trip, ". . . the expressions were different this time. When I took my walk from crew quarters on the day *Friendship 7* was finally launched, I was going solo and it was a first flight. There was more uncertainty on the faces then."

After a short ride to the launch pad, the astronauts rode the gantry elevator to the Shuttle's crew hatch. Before entering the craft, Glenn looked to the south where *Friendship 7* and

Figure 38-17. The launch of STS-95 *on October 29, 1998.*

chapter, and by the end of their mission all of these studies had been completed. At the time of this writing, the collected data is still being analyzed.

Conclusions

In this chapter just three of the Shuttle's many flights have been discussed. Since its first flight in 1981, the Space Shuttle has served as the workhorse of the United States manned space program. Intended to be the first part of a bigger program, the Space Shuttle will play a major role in the construction of the upcoming International Space Station discussed later in this book. However, the Shuttle's potential benefit in this area has already been demonstrated by its successful support of the Mir space station. The flight of *STS-1* ushered in a new era of manned spaceflight that we are now beginning to exploit.

STS-25 will always be known as the Challenger Disaster. It is the only American spaceflight in which people lost their lives. This tragedy had a predictable and chilling effect on America's space program. The "Teacher in Space" aspect of this mission ensured a large audience would be watching its lift-off. The explosion was a harsh reminder that machines built by humans are still fallible and failures in the realm of spaceflight are unforgiving.

John Glenn's October 1998 flight provided a much needed boost to the American spirit severely damaged by the Challenger Disaster. Although the primary aims of *STS-95* were technological and scientific, its unheralded benefit was to again point the nation's attention skyward and to rekindle its somewhat dormant sense of adventure. Glenn was a genuine American hero and his willingness to again fly in space, despite the dangers, was inspirational. Some attacked Glenn for seeming to use his position in the U.S. Senate to accomplish his long-held dream of returning to space. Despite this criticism, there is no evidence he acted improperly. If anything, his status as a national hero made his assignment to the flight by NASA Administrator Goldin more difficult.

The data returned by *STS-95* is still being studied, but it will undoubtedly prove useful to NASA's goal of reducing the long-term medical problems posed by spaceflight. In a broader sense, Glenn's mission has already proven to be of tremendous value. It reminded us of one of our most noble traits, our willingness to accept the grave risks inherent in exploring the unknown. History has shown that a nation's greatness can be measured by its willingness to support these types of expeditions. Glenn's flight renewed the commitment of Tsiolkovsky, Goddard, and Oberth to work toward the fulfillment of our destiny, to travel into the final frontier of space.

the other Mercury flights had lifted-off over thirty-five years ago. The launch pad was still there, but the gantry had long since been removed. The blockhouse used during those days was now a museum. Glenn recalled a statement he had been told earlier, "Somebody had pointed out that more time had passed between *Friendship 7* and this Discovery mission than had passed between Lindbergh's solo transatlantic flight and *Friendship 7*. It didn't seem that long to me, but that is the way lives pass when you look back on them, in the blink of an eye."

The astronauts were kept busy with a full schedule of activities beginning as soon as the Discovery reached orbit. While Brown fired the Shuttle engines to make their orbit circular, Glenn and Paraznski activated several biomedical experiments in the Spacehab module.

Like all missions, *STS-95* had its lighter moments. Because of the great public interest in this flight, NASA permitted the crew to participate in a live television segment with Jay Leno, the host of the *Tonight Show*. During this exchange, Commander Brown stole the show. In an unusually serious question, Leno asked him what he could see from orbit. Brown seized the moment and replied, "Well Jay, sometimes, if the lighting is good, we can see the Great Wall of China, but we just flew over the Hawaiian Islands and we saw that. And Baja California. You can see the pyramids from space, and sometimes rivers and big airports. And actually, Jay, every time we fly by California we can see your chin."

The crew was dedicated to the completion of the experimental program described in the previous sections of this

39   The Space Shuttle/Mir Era

> ... there was no better way for NASA to prepare for the future than to have a cadre of astronauts learn to live and work in space by spending tours on Mir, because everything from the exercise regime to the logistics system was new to the agency.

— David M. Harland in *The Mir Space Station*

In June 1994 NASA and the Russian Space Agency signed a three-part Memorandum of Understanding. The first part called for Russian cosmonauts to fly aboard the Space Shuttle to familiarize themselves with the capabilities of this craft. During the second phase, five (later increased to seven) astronauts would conduct long-duration missions on Mir. An important goal of these flights was to collect engineering data essential to the construction of the International Space Station. They would also permit American space officials to become familiar with Russian operating procedures and to develop working relations with their counterparts in the Russian space agencies. The third phase of the agreement involved the construction, launch, and joint operation of the International Space Station.

For the first phase of this plan, cosmonauts Sergei Krikalev, Vladimir Titov, and Anatoli Solovyov were selected to fly aboard the Shuttle. During the second phase, astronauts Thagard, Lucid, Blaha, Linenger, Foale, Wolf, and Thomas would conduct experiments on Mir.

Phase 1: The Cosmonaut Shuttle Flights

On February 3, 1994, the "last Soviet citizen," Sergei Krikalev, lifted-off aboard the space shuttle *Discovery* as a member of the crew of *STS-60*. Krikalev was the first foreigner trained by NASA as a mission specialist. All others had served as payload specialists. During this mission, Krikalev participated in seven experiments. Exactly one year later, on February 3, 1995, Vladimir Titov, who had trained with Krikalev, was launched aboard the Space Shuttle *Discovery* (*STS-63*). This flight would be a "near Mir" mission, that is, the Shuttle would maneuver close to the station and fly around it, but would not dock. The goals of this fly-by were to verify flight procedures, check communication and navigation systems, and to verify that no protruding objects from Mir would interfere with a docking. Russian cosmonaut Vladimir Titov, a veteran of several space missions, accompanied the astronauts as a mission specialist.

After *STS-63* reached orbit a systems check revealed two of the Shuttle's thrusters were leaking fuel. One of these leaks was stopped by heating the thruster with direct sunlight, but the other stubbornly continued to lose fuel. Officials feared

Table 39-1. Shuttle Flights to Mir		
Shuttle Flight	**Launch Date**	**Crew**
STS-63 Discovery	Feb. 3, 1995	Wetherbee, Collins, Voss, Foale, Harris, Titov
STS-71 Atlantis	June 27, 1995	Gibson, Precourt, Baker, Dunbar, Harbaugh, Solovyev, Budarin, Dezhurov, Strekalov, Thagard
STS-74 Atlantis	Nov. 12, 1995	Cameron, Halsell, Ross, McArthur, Jr., Hadfield
STS-76 Atlantis	Mar. 22, 1996	Chilton, Searfoss, Lucid, Godwin, Clifford, Sega
STS-79 Atlantis	Sep. 16, 1996	Readdy, Wilcutt, Akers, Blaha, Apt, Walz, Lucid
STS-81 Atlantis	Jan. 12, 1997	Baker, Jett, Grunsfeld, Ivins, Wisoff, Linenger, Blaha
STS-84 Atlantis	May 15, 1997	Precourt, Collins, Foale, Noriega, Lu, Clervoy, Kondakova, Linenger
STS-86 Atlantis	Sep. 25, 1997	Wetherbee, Bloomfield, Titov, Parazynski, Wolf, Chretien, Lawrence, Foale
STS-89 Endeavour	Jan. 22, 1998	Wilcutt, Edwards, Jr., Reilly, II, Anderson, Dunbar, Shapirov, Thomas, Wolf
STS-91 Discovery	June 2, 1998	Precourt, Gorie, Ryumin, Kavandi, Lawrence, Chang-Diaz, Thomas

that as the Shuttle approached Mir, the experiments placed on the station's exterior would be coated by propellant. The flight plan called for the Shuttle to come within 33 feet of Mir, but if the leak persisted, officials agreed this distance would be increased to 400 feet.

During the three-day rendezvous, the astronauts attacked the remaining thruster problem by closing the Shuttle's fuel line. This caused the flow of escaping propellant to stop and allowed the Shuttle to attempt the original planned close approach. As the Shuttle neared Mir, the Russian cosmonauts reported they could see Shuttle commander Wetherbee waving his hand at them. The distance between the craft slowly decreased, and when they were over the Pacific Ocean, the 33-foot goal was achieved. To commemorate this occasion Wetherbee radioed the Mir crew, "As we are bringing our space ships closer together, we are bringing our nations closer together. The next time we approach, we will shake your hand and together we will lead our world into the next millennium." He then repeated this greeting in broken Russian. The Mir commander, Alexander Viktorenko, responded, "We are all one! . . . This is almost like a fairy tale. It's too good to be true."

After completing the rendezvous, the Shuttle moved away and circled the station, photographing it with an IMAX motion picture camera. Wetherbee later told the cosmonauts he especially enjoyed this part of the flight, "It was like dancing in the cosmos."

The flight of *STS-63* lasted only a little more than eight days, but accomplished two important tasks. It confirmed the Shuttle would be able to dock at the Kristall port and that simultaneous communications among Houston, the Russia control center at Kaliningrad, and the two spacecraft were possible. An overall assessment of the mission was later given by Wetherbee, "When all was said and done, it turned out to be easy." NASA was delighted with this flight. With the exception of the leaking thrusters, no obstacles were encountered. The next flight would transfer an American astronaut to the station.

The First American Astronaut on Mir

Oddly, the first American astronaut to occupy Mir was not brought to the station by the Space Shuttle, but in a Russian craft, the *Soyuz TM-21*. This flight, launched on March

Figure 39-1. The crew of STS-63 *(clockwise from top left), Vladimir Titov, Michael Foale, Janice Voss, Bernard Harris, James Wetherbee, and Eileen Collins.*

Figure 39-2. Dr. Norm Thagard, the first American astronaut to board Mir.

14, 1995 from Baikonur (formerly called Tyuratam), carried Dr. Norman Thagard and cosmonauts Vladimir Dezhurov and Gennadi Strekalov. Since the Soyuz could only accommodate a three-person crew, most of Thagard's equipment was scheduled to be brought to Mir aboard *Progress M-28*.

Thagard was a veteran astronaut and a member of the original Shuttle class of 1978. As a boy he had enjoyed reading and writing science fiction books. Following graduation from high school in 1961, he studied engineering and medicine at Florida State University, receiving an M.S. degree in engineering in 1966. He then underwent pilot training in the marine corps reserve and was sent to Vietnam where he flew 163 combat missions. In 1971 he left the military and resumed his engineering and medical studies. When selected as an astronaut candidate, he was serving as an intern at the Medical University of South Carolina. At the age of 49, he had flown in four prior shuttle flights: *STS-7, 51B, 30,* and *42*. His remaining wish before retiring from NASA was to fly in the Shuttle/Mir program.

Thagard's backup was Bonnie Dunbar. Born in Yakima Valley in Central Washington in 1949, Dunbar enjoyed reading Jules Verne and H.G. Wells and was inspired to become an astronaut by the launch of Sputnik. She was often described using terms such as ambitious, naturally reserved, and extremely smart, but other people described her as prickly, easily insulted, and somewhat insecure. Her selection came as a surprise to many people within the space program because she knew very little Russian and did not appear to be compatible with Thagard. Bill Readdy, a Shuttle pilot and Thagard's original backup, described this pairing as "the biggest mistake we made in the program." NASA flight surgeon Mike Barratt described the pairing as "matter and antimatter."

As some had feared, problems soon arose between the two astronauts. While training at the cosmonaut center at Star

Figure 39-3. Bonnie Dunbar served as Thagard's backup during training at Star City.

Figure 39-4. Dr. Norman Thagard (left) was the first American to board the Mir space station. In this picture he is seen with fellow crewman Gennadi Strekalov (right). The instrument located between them is a large-format IMAX camera.

City, Thagard completely immersed himself into the Russian setting. He spoke Russian, read Russian, trained in Russian, and quickly adapted to the Russian training regimen. Dunbar, who was far less fluent in Russian and found the cultural change distressing. She abhorred the male chauvinistic attitude at Star City and spoke out about her treatment, "You just get to a point in life where you say, 'Why am I having to deal with this? . . . I just lose patience with it now."

Thagard saw Dunbar was having a difficult time adjusting, but viewed much of her trouble as a refusal to adjust to the Russian way of working. He later offered the assessment, "She wants to impose her own attitudes on the Russians, rather than understand theirs. Bonnie just brought over to Russia all this feminist baggage that no one could understand." Relations between the two astronauts became strained and finally broke down over a relatively small matter involving telephone privileges. From that point Thagard avoided her. Dunbar felt powerless to alter the situation, "Norm has a line, and once you cross that line, you cross it forever. . . . Norm was stressed; what could I do? Go to my management and say, 'Normie doesn't like me'? This is not something I wanted anyone to know about. I was personally very embarrassed. If it got out, I felt they would blame me." Although she would fly to Mir as a mission specialist on *STS-71* and *89*, she would not conduct a long-duration mission on the station.

Normally, three to four years are spent planning and training for an eight-day Shuttle flight. Thagard's mission to Mir was scheduled to last 115 days, but he had barely a year to prepare. As a medical doctor, most of his experiments would deal with space biology. When in orbit he would serve as a human guinea pig, collecting his own blood, urine, and saliva samples and recording his food and water intake to monitor the body's adjustment to weightlessness. Additional data on his cardiovascular, metabolic, and musculatory systems as well as the strength of his bones and his psychological state would also be obtained.

A severe blow to Thagard's science program occurred seven months before his scheduled lift-off. NASA was informed that the Russian Spektr module would not be launched until near the end of his flight. Several of Thagard's experiments were to take place in this module and it was to carry much of his equipment into space. As a result of this change, his scientific program was reduced from fifty to twenty-eight experiments.

Despite these setbacks, Thagard still anxiously looked forward to his flight. Two days after lifting-off from Baikonur on March 14, 1995, he entered the station and eagerly began his research program. An important part of his program relied on the use of a freezer left on Mir by Ulf Merbold of the ESA in 1994. All of his biological samples were to be stored in it. About a week into the flight, the temperature inside the freezer unexpectedly began to rise. The crew attempted to fix the unit, but all of their efforts were unsuccessful. A disgruntled Thagard had to place as many of his test tubes as possible into the Mir refrigerator and a small secondary freezer. About 80 percent of the samples he had painstakingly collected had to be discarded.

At the beginning of May Thagard received some additional bad news. After spending about a month on Mir and carefully following a rigid diet, he was told by doctors that he had lost 17 pounds. This weight loss made the interpretation of his bone loss measurements useless. There was no way doctors could determine how much of the loss was due to the effects of weightlessness and how much was due to his weight loss. Prior to the flight, Thagard had feared he would lose some weight, but he was angered NASA doctors had not alerted him to this developing problem. He later described his shock at this news, "My first thought was: Where have you people been for the last six weeks."

This series of setbacks essentially brought Thagard's science program to a halt. He searched for other ways to use his time productively and offered to help his Russian crewmates with their tasks, but they politely declined. NASA psychologists became concerned about his level of inactivity, "Norm had to twiddle his thumbs for a month, which is difficult for anyone to do, especially a high achieving astronaut. Work underload is a terrible thing. . . . [The Russians] trained him on systems but they wouldn't let him touch the systems."

While Thagard tried to fill his time with activities, Dezhurov and Strekalov prepared for two major tasks. The first was to move both of the solar panels from Kristall to Kvant 1, and the second was to prepare the station to accept the soon-to-be-launched Spektr module. During the retraction of one of Kristall's solar panels, it jammed while still 25 percent extended. Ground controllers decided to leave it alone until this failure could be studied. Its presence would not interfere with the upcoming Space Shuttle docking and further action might make matters worse. As a result, only one of Kristall's solar panels was installed on the exterior of Kvant 1.

The Launch of Spektr

Spektr was launched on May 20, 1995. This module, designed to collect data on the Earth's atmosphere and magnetic field, had a mass of 26 tons and an interior volume of 2223 ft^3. In preparation for a docking, the station's robotic arm was used to move Kristall to an adjacent port. As Spektr approached the station, officials could not help but remember the earlier docking troubles they had experienced with the other modules. However, a docking was achieved on the first attempt.

Once docked, two pairs of solar panels attached to Spektr's external wall were to unfold. The first pair deployed

Figure 39-5. The Mir space station with Spektr (lower middle) attached. Kvant 1, with one solar panel, is located at the left end of the station; Kristall is joined to the core module and extends to the right. Kvant 2 points to the upper right.

as planned, but a clamp holding one of the other panels refused to open. Since the loss of this extra power was not a critical matter, officials decided not to attempt a repair until a procedure could be tested on Earth. Once an airtight seal between Spektr and the station was achieved, the crew unloaded nearly a ton of equipment for Thagard, including a centrifuge, ergometer, several laptop computers, and a much needed freezer. With this new equipment on-board he quickly restarted his science program, completing as many experiments as he could within his remaining time on Mir.

The First Shuttle/Mir Docking

The first American Space Shuttle to dock with Mir was *STS-71*. It launched on June 27, 1995 and retrieved Thagard and replaced cosmonauts Dezhurov and Strekalov with Anatoli Solovyov and Nikolai Budarin. Fitted in the Shuttle cargo bay were the Buran docking unit and a Spacelab module equipped with life science instruments. Both modules were connected to the crew cabin by an airtight tunnel.

Astronaut Gibson slowly maneuvered the Shuttle toward Kristall's port by looking out its upper rear windows and by using a video monitor. Precisely on time, the Shuttle settled into Mir's docking mechanism and its capture latches closed. About 15 minutes later, he fired the Shuttle's bottom thrusters, driving the two craft firmly together and creating an airtight seal. An hour later Gibson and Dezhurov greeted each other in the narrow tunnel connecting the two craft (see Figure 39-6).

Once this historic meeting was complete and the obligatory crew pictures were taken, Dezhurov, Strekalov, and Thagard entered the Spacelab module for a complete physical exam. This was the first time such tests were administered to a long-duration crew while still in space. The Shuttle also transferred water and supplies to the station and provided bolt cutters to free the solar panel on Spektr. In preparation for departure, Thagard's experimental results and those left behind by European Space Agency (ESA) cosmonaut Merbold were stowed aboard the Atlantis. Some nonfunctioning Mir equipment was also taken to help in the design of improved models.

Figure 39-6. A short time after docking, Vladimir Dezhurov (left) welcomed Robert Gibson on-board Mir.

Figure 39-7. STS-71 *was the first Space Shuttle to dock with Mir. Here it is near Mir's Kristall port.*

During the shuttle undocking, Solovyov and Budarin flew their *Soyuz TM-21* several hundred feet away to take pictures of its departure. However, when Atlantis undocked, it imparted a slight motion to the station causing Mir's computer controlled orientation-control system to shut down. The station began to tumble, but the motion was not large enough to prevent the cosmonauts from redocking. The computer failure was later traced to a software oversight and was easily corrected.

Thagard had been in space for 115 days, far shorter than Valero Poliakov's record 438 days. Nevertheless his research program provided NASA with information about the body's reaction to weightlessness over a time period similar to the three-month duty cycle anticipated for the International Space Station. He reported that Mir was roomy but had the feel of a "locker room" which had been continuously lived in for years. Working with his Russian counterparts had been enjoyable, but the language barrier kept him culturally isolated. He also missed his family. Having completed his wish to fly in a Russian spacecraft, he retired from NASA and accepted a faculty position at his alma mater, Florida State University.

NASA sought ways to correct some of the defects uncovered by Thagard's mission. More attention would have to be given to the astronaut's science program and communications with the ground would have to be improved. In the view of Frank Culbertson, director of the American side of the Shuttle/Mir program, "We weren't in sync at all." To remedy this problem, experienced flight controllers were recruited from NASA's Missions Operations Directorate but with little initial success. Phil Engelauf, an official at that directorate recalled,

They said, "We want volunteers to go out to Star City for eighteen months." There was a deafening silence.

Most of us have two- and three-year-old kids and don't want to live in Russia; that wasn't in the oath of office when we joined NASA. Not only couldn't we find volunteers, we couldn't spare people of the caliber they wanted. [But] we were basically told by management, "Phase One is broken. Go fix it."

STS-74: A Supply Mission

The next Shuttle flight to Mir, *STS-74*, was intended to carry supplies to the station. Launched on November 12, 1995, it was commanded by Kenneth Cameron. The major objectives were to rendezvous with Mir and deliver a new Russian docking module. This 15-foot-long unit had a mass of 4.6 tons (too heavy for a Progress vessel), two solar panels, and was intended to make future dockings easier.

Once the Shuttle docked, the internationalization of Mir was evident by the diverse nationalities on board. These included German ESA cosmonaut Thomas who had come to Mir aboard *Soyuz TM-22,* Russian cosmonauts, and American and Canadian astronauts. After attaching the new docking module and unloading a ton of water, food, and research equipment, the Shuttle departed on November 18 with 800 pounds of cargo.

Long-Term American Visits to Mir

By the end of 1995, Mir had proved its usefulness as a laboratory for medical and materials processing studies. Based on requests from the international community, the Russian Space Agency decided to make the station available for additional missions until at least the year 2000. This announcement prompted further interest from Germany, Japan, and China.

On February 21, 1996, a few months after the departure of *STS-74*, Mir's crew was relieved by cosmonauts

Figure 39-8. The docking module was slowly eased into place towards the Kristall port (top part of the picture) from the shuttle bay (lower part of the picture).

Yuri Onufrienko and Yuri Usachyov aboard *Soyuz TM-23*. Gidzenko, Avdeyev, and Reiter returned to Earth nine days later on March 1, 1996 in *Soyuz TM-22*.

STS-76: Shannon Lucid Establishes a New American Space Endurance Record

The primary goals of the third Space Shuttle mission to Mir, *STS-76*, were (1) to transport another astronaut, Shannon Lucid, to Mir to continue NASA's experimental program, (2) to deliver nearly a ton of cargo to the station, and (3) to attach an experimental package consisting of four instruments called the Mir Environmental Effects Payload (MEEP) to the newly installed Mir Docking Module. Two of the MEEP experiments would gather data about the amount of debris the station encountered in orbit while the other two would test the ability of various construction materials to withstand the harsh environment of space.

Lucid's research program placed great emphasis on the collection of engineering data for the International Space Station. In addition, she would perform studies in Earth resources, materials processing, the behavior of flames and liquids, biomedicine, and climatology and geology. After Thagard's flight, NASA resolved to provide its astronauts with more assistance during their scientific training program at Star City. However, Lucid and her backup, John Blaha, found that few improvements had been made. Lucid recalled,

> I remember during Norm's [Thagard] training there were all of these science people around, and so during his flight John and I said, "Okay, they're going to help us now." Then Norm landed, and they all disappeared. We were alone. John and I had a lot of conversations, you know, where we said we were actually worse off than Norm [had been].

Lucid and Blaha had become victims of a NASA administrative breakdown. After Thagard's lift-off, NASA hired a new science support contractor. Unfortunately, the new contractor, Lockheed Martin, mistakenly assumed Lucid and Blaha would receive their training at Star City from the Russians. As time passed it was obvious to Lucid a misunderstanding existed. She e-mailed Houston for help, "I was asking them, 'Does anyone have a list of experiments for the flight?' No. No one had a clue."

After it was clear Houston could not easily rectify matters, Lucid resigned herself to the situation. She later wrote, "I had no idea how things were going to go in space. If things fell apart, and your career's over, that was fine. I've had four good [Shuttle] flights. I was going to have a good time. That was it." This type of placid response was in keeping with Lucid's approach to life. As Bill Gerstenmaier, her mission operations director, later said, "Shannon's personality is not to complain at all."

Shannon's patient, cheerful disposition coupled with a lot of hard work would prove to be vital to the success of her

mission. While she was in orbit Gerstenmaier sent detailed instructions on how to complete each experiment based on his own experience with these activities on Earth. To veterans of the Shuttle program this procedure was astonishing, but in the end it worked.

Lucid's Mir Studies

Lucid's Shuttle flight successfully lifted-off from the Kennedy Space Center on March 22, 1996. A few days later she joined her Russian crewmates and friends, Yuri Usachyov and Yuri Onufrienko, who had arrived at the station aboard *Soyuz TM-23* about a month earlier.

Although objects aboard Mir were weightless, small vibrations within the station could arise from airflows, the movement of crew members, and the station's orbital motion. These vibrations were too small for humans to sense but large enough to have a significant impact on delicate materials processing experiments. To help create a more stable environment, Lucid employed a Canadian-built unit called the Microgravity Isolation Mount (MIM). This unit, shown in Figure 39-9, relied on electromagnetic fields to shield objects from vibrations. Metallurgical samples were melted in a furnace, placed on the MIM and then allowed to cool. A major objective of this work was to determine how various compounds solidified in space. On Earth, the presence of gravity allows hot and cool gases to separate, but in space this process is slowed down considerably. Lucid also employed another apparatus called the "glove box" (see Figures 39-10 and 39-11) to study how fluids and flames behaved. This work would help engineers construct fire fighting equipment for the International Space Station.

The primary biomedical experiment conducted by Lucid dealt with the growth of embryos. Japanese quail eggs were placed in an incubator and removed, one by one, over a period

Figure 39-9. The Microgravity Isolation Mount located in the instrument rank to the left was used for sensitive materials processings experiments.

Figure 39-10. Shannon Lucid (left) and her two Russian crewmates, Yuri Usachyov (center) and Yuri Onufrienko. The "glove box" is in front of her.

Figure 39-11. Lucid conducted her experiment on the development of quail eggs using an apparatus that contained interlocking plastic bags.

of sixteen days. They were then fixed in a paraformaldehyde solution for later examination. Lucid later described the care used in constructing this apparatus and then conducting experiments with it,

> NASA and Russian safety rules called for three layers of containment for the fixative solution; if a drop escaped, it could float into a crew member's eye and cause severe burns. Engineers at the NASA Ames Research Center designed a system of interlocking clear bags for inserting the eggs into the fixative and cracking them open. In addition, the entire experiment was enclosed in a larger bag with gloves attached to its surface, which allowed me to reach inside the bag without opening it.

Once analyzed, the abnormality rate among the embryos was found to be four times higher than in a control sample on Earth. Researchers speculated that the increased radiation

level on Mir and the slightly warmer temperature of the incubator were responsible for this result.

Lucid also conducted a test to determine if plants could be grown on a spacecraft to provide food and oxygen. A fast growing strain of wheat was planted and monitored by computer controlled equipment. During the course of this study Lucid became excited when some of the plants appeared to have produced seeds,

> At selected times, we harvested a few plants and preserved them in a fixative solution for later analysis on the ground. One evening, after the plants had been growing for about forty days, I noticed seed heads on the tips of the stalks. I shouted excitedly to my crew-mates, who floated by to take a look.

This was a potentially significant result. If seeds could be produced, it might be possible to create a self-sustaining food supply in space. Unfortunately, subsequent analysis at Utah State University found the seed pods were empty. Investigators speculated that the low level of ethylene in the station's atmosphere had interfered with the pollination process.

Of the four areas Lucid worked in, her favorite was Earth photography. This program permitted her to utilize the geology training she had received prior to the flight. Unlike other experiments, where she served as a skilled technician, in this area she played the role of an investigator, selecting objects for study and making real time adjustments in the program. The numerous photographs she obtained from the Kvant 2 observation window covered a long enough period of time to document global seasonal changes. These images were subsequently provided to oceanographers, geologists, and climatologists for detailed analysis.

Because of financial cutbacks in the Russian Space Agency, Mir missions were often extended beyond their scheduled ending date. This decreased the program's cost by reducing the number of Soyuz capsules and launch rockets. Therefore, cosmonauts Onufrienko and Usachyov were not surprised when their mission was increased by forty days. Less than a month later, Lucid received news that her return Shuttle flight would also be delayed. After the mission of *STS-75,* burn holes had been discovered in a new sealant used in the Shuttle's solid rocket boosters. Officials decided to remove this material from all of the rockets as a precautionary measure and replace it with the former material. Lucid took this news well and used the extra time to set up the proton crystal and cell culture experiment her replacement and friend, John Blaha, would utilize during his upcoming mission on the station.

Priroda: A Second Earth-Monitoring Module

While Lucid was still on Mir, the fifth and last space station module was launched from Baikonur. This module, called Priroda (translated Nature), lifted-off on April 23, 1996 and was a complementary module to Spektr. The Priroda instruments were built to gather data on the spread of industrial pollution, thermal variations in the oceans, the temperature structure of clouds, multispectral analyses of land masses, the ozone content of the atmosphere, and changes in the mean sea level. Although one of the module's battery buses failed during rendezvous, a smooth docking was accomplished on the first attempt.

On August 19, Mir was visited by *Soyuz TM-24,* carrying cosmonauts Valeri Korzun, Alexander Kaleri and French cosmonaut Claudie Andre-Deshays. Deshays's specialty was space medicine, and during her visit she conducted experiments to learn more about how the body regulates blood flow and senses blood pressure as it adapts to weightlessness. After this mission's successful conclusion, Onurfrienko, Usachyov, and Andre-Deshays left Mir on September 2, 1996. French space officials were so impressed with Deshays's mission, they booked two additional flights for 1999.

On September 16, 1996, *STS-79* finally lifted-off toward Mir to replace Lucid with Blaha. Once Blaha was aboard, Lucid showed him where the various pieces of equipment were stored and discussed how best to work on the station. After setting a new American space duration record of 188 days, Lucid departed and landed on September 26. Lucid later reflected on the remarkable changes that had taken place in U.S./Russian relations,

I had spent my grade school years living in terror of the Soviet Union. We practiced bomb drills in our classes, all of us crouching under our desks, never questioning why. Similarly, Onufrienko and Usachyov had grown up with the knowledge that U.S. bombers or missiles might zero in on their villages. After talking about our childhoods some more, we marveled at what an unlikely scenario had unfolded. Here we were, from countries that were sworn enemies a few years earlier, living together on a space station in harmony and peace.

40 The Shuttle/Mir Program Reveals Unsettling Problems

We learned that we don't know everything about space flight. . . . We learned that long duration flight on a space station is quite a bit different than short flights on the shuttle. The shuttle is an airplane, and the station is a ship, and it's going to be at sea for a long, long time. It may be a long way from land, but you still got to keep it afloat.

—Frank Culbertson, director, Shuttle/Mir Program

Shannon Lucid's record-breaking flight provided NASA with a public relations bonanza. Upon her return she was awarded the Congressional Space Medal of Honor by President Clinton, the Order of Friendship Medal by Russian President Yeltsin, and was featured on the covers of *Newsweek* and *Scientific American*. It had been a long time since NASA had received such favorable treatment, and officials hoped the upcoming American flights to Mir would be equally rewarding.

The Learning Curve: Organizational Problems within NASA

The highly successful end of Shannon Lucid's Mir flight gave NASA officials a false sense that everything was functioning properly in the Shuttle/Mir program. However, John Blaha, a veteran of four prior Shuttle flights, knew better. Blaha was concerned about the organizational differences he saw between the Shuttle/Mir program and the Shuttle program. The Shuttle/Mir program seemed to have a much lower priority and substantially less information was provided to the astronauts. Nevertheless, like Lucid, he decided to soldier on in the hope that matters would improve.

Figure 40-1. Shannon Lucid established a new American endurance record while on Mir and received wide acclaim upon her return to Earth.

Blaha had four primary tasks to complete on Mir: (1) tissue growth studies, (2) experiments on the crystallization of alloys of colloids, (3) the testing of construction materials, and (4) the collection of engineering data. This last goal was particularly important. Measurements of the stresses experienced by Mir during maneuvers, dockings, and while orbiting the Earth would be used in designing the International Space Station.

Blaha's Preflight Training Program

To prepare for his mission to Mir, Blaha was first sent to the Defense Language Institute in Monterey, California, to learn Russian. Students traditionally required two years to complete this program, but because of Blaha's tight schedule, he was only allowed four months. When told the normal training period was six times longer than this, NASA officials seemed unconcerned, replying, "These are astronauts, they're really smart. They can do it quicker." This attitude reflected NASA's deep pride in the quality of the people selected to be astronauts. Time after time these individuals had proven themselves equal to any task assigned to them, but this attitude placed the astronauts under tremendous pressure.

Despite Blaha's best efforts, it took him one month just to learn the Russian alphabet. He began to worry. Would he be able to understand the highly technical instruction at Star City which would be given in Russian? This concern would turn out to be well founded. During his grueling eighteen-month program in Russia, he would often work 16-hour days. As his studies intensified, he grew increasingly unhappy with NASA's support system. Procedures for his Mir experiments arrived late, incomplete, or not at all. After complaining, he was assured matters were "well in hand" and that these problems were being addressed, but as the days passed little seemed to change. In addition to the lack of written documents, Blaha was also concerned about the on-site support. Issac (Cassi) Moore, the head of his Moscow-based ground team, arrived just three months before the scheduled launch and seemed to know little about the mission. Still struggling to keep up with his classes Blaha was baffled, "I thought it was just incredible. . . . Here I am, and I'm busier than heck, and I didn't have any time to spend with Cassi Moore. I mean I had none. Cassi had no concept of what was going on. He was joining the effort so, so late." Instead of developing a mutual feeling of trust, tensions between these two men grew.

More Training Problems

As bad as Blaha's relation with Moore was, the situation with his flight surgeon, Pat McGinnis, was even worse. Despite Blaha's attempts to get him involved, McGinnis seemed more interested in spending his time monitoring Lucid's flight or with the astronaut who would follow him, Jerry Linenger. An exasperated Blaha later complained "Pat avoided me!" They would spend little time working together until close to the launch.

After an exhausting year and a half of training, Blaha returned to Houston physically and emotionally drained. He was in no condition to handle the next shock that came his way. With his launch about a month away, both of his Russian crewmates were medically disqualified from the flight. Blaha would now have to go into space with two cosmonauts, Valery Korzun and Alexander Kaleri, he barely knew. A quick trip was arranged back to Star City to meet his new crewmen, and although the session went well, Blaha was still upset about their lack of joint training. Kaleri later admitted the cosmonauts also felt uneasy about this situation, but for a different reason, "We hadn't had a single training event with John. We could tell John was worried about this and we worried because John worried."

When Blaha returned to Houston, his frustration came to a head. He confronted Frank Culbertson, head of the American side of the Shuttle/Mir program, "Frank, I like you, but this is the single most screwed-up program I have ever been

associated with." He then went on to criticize NASA's failure to act on astronaut complaints and suggestions, "Norm Thagard told you all a lot of this stuff. Y'all wouldn't listen. You kept telling us how to fix it, rather than getting it fixed. Now eighteen months later we're having the same discussion." He complained that NASA's lack of oversight had placed an unreasonable burden on him and the American/Mir, astronauts, "We all launched tired, absolutely wiped out. What a way to go to a long duration spaceflight."

Blaha's Problems on Mir

Once aboard Mir the lack of joint training with his support teams began to cause problems. The time line controllers employed bore little resemblance to what Blaha actually needed to complete his tasks. To stay on schedule he was forced to work excessive hours, often sleeping only three hours per night. Realizing he was becoming exhausted, Blaha asked to be given fewer activities, but this request seemed to be ignored. Instead of his workload being reduced, more and more experiments were added.

One day while Blaha was communicating with his controllers, Korzun decided to take matters into his own hands. Breaking into Blaha's transmission and speaking directly to the Russian flight director, Korzun said, "I'm the commander of Mir, and I can tell you what they are doing with John Blaha over the last ten days has really been wrong. The Americans really need to get better organized." Blaha was pleased by this show of support, but was also saddened it was needed. He recalled, "Here I was, a veteran Shuttle pilot, a seasoned space flier, and I have been telling the ground the same things for ten days. They never heard me. But a Russian on his first spaceflight, they heard it. That's almost incredible." The hard

Figure 40-2. Frank Culbertson, head of the American side of the Shuttle/Mir program.

Figure 40-3. Blaha working in the Mir Spektr module.

lessons learned as a result of Skylab's long-duration missions had apparently been forgotten.

Embarrassed by Korzun's reprimand and by a subsequent scolding from Russian officials, Moore reduced Blaha's activities by 25 percent. This reduction was welcomed, but his work day still ran 14 hours or longer. Disappointed in NASA's response, Blaha would later state, "I was trying to run up a mountain, and the Russians were trying to help me, and the Americans were trying to bring me down." Only the combination of his reduced workload and the encouragement of his Russian crewmates allowed him to overcome his mounting frustration.

Linenger Replaces Blaha on Mir

STS-81 was launched from the Kennedy Space Center on January 12, 1997. On-board was Blaha's replacement, astronaut Jerry Linenger, a medical doctor with a specialty in epidemiology. Linenger's experimental program dealt with the physical problems posed by spaceflight, fluid physics, and materials science. He was also scheduled to conduct a spacewalk. This mission was expected to be fairly routine, but events would soon make it one of the most nerve-racking since *Apollo 13*.

Linenger arrived on Mir on January 15, 1997. During the next month he worked well with his Russian crewmates, Korzun and Kaleri, and spent much of his time in the Priroda and Spektr modules performing experiments on sleep, the behavior of an open flame, crystal growth, and Earth observations. His studies on crystal growth relied on a device called the

Figure 40-4. Jerry Linenger lifts-off aboard STS-81 *on his way to Mir.*

Figure 40-5. Astronaut John Blaha and his replacement, Jerry Linenger, share a happy moment during STS-81.

Liquid Metal Diffusion experiment, but soon after starting his work this unit malfunctioned. Hours were spent analyzing the system's computer software and resetting clocks controlling the operation of the experiment's oven, but to little avail. After struggling with this apparatus for several weeks, only three samples had been obtained. His schedule allocated fifteen days for the completion of this entire task. Linenger found his lack of progress discouraging.

The Celebration of "Army Day"

February 23 is celebrated in Russia as a national holiday, Army Day, a day to honor the Soviet armed forces. By this time, the number of people on Mir had grown to six. Korzun, Kaleri, and Linenger had been joined by Vasili Tsibliev, Alexander Lazutkin, and a visiting German cosmonaut, Reinhold Ewald. The Russian cosmonauts intended to celebrate this holiday by hosting a special dinner consisting of the station's best food. At the appointed time the entire crew assembled in the base module to share a meal consisting of red caviar, cheese, and sausages. Conversations were lighthearted, the food was good, and everyone enjoyed themselves. After chatting for a while, Linenger excused himself and returned to Spektr to begin a sleep experiment.

Fire!

As the remaining crew members sat in the core module talking and autographing posters and envelops for collectors back on Earth, Lazutkin floated into the Kvant to place a fresh canister into the "Vika" oxygen-generating system. With six people on the station, the Vika was used to increase the oxygen content of the air. Inserting this canister was a simple task and Lazutkin completed it with no trouble. All seemed well, but as he prepared to return to his crewmates he heard a soft SSHHHH . . . sound. Curious, he looked for the source of this noise, when suddenly sparks erupted from the newly installed

Figure 40-6. Valery Korzun would be the principle person fighting the fire in the Mir Kvant 1 module.

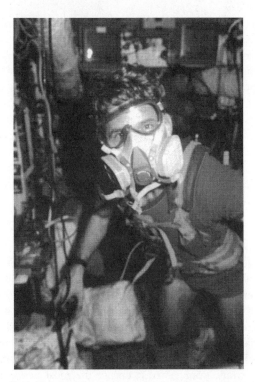

Figure 40-7. Concerned poisonous gases might have been released by the fire, Linenger (above) suggested the crew wear gas masks.

canister. As he watched, the sparks rapidly grew in number becoming a fierce 9- to 12-inch flame. Startled he thought, "This is unusual. It shouldn't be doing this. Why is it doing this?!" In a calm voice he called to his crewmates, "Guys we have a fire." Next to depressurization, a fire is the most feared event that can happen aboard a spaceship.

German cosmonaut Ewald was the next to see the flames in the Kvant and blurted out the Russian word for fire, "Pozhar." Tsibliev turned, saw what was happening, and shouted "Pozhar!" Korzun, seeing that Lazutkin was in trouble, quickly came to his aid but realized the flame was too large to be smothered. He called to his fellow crewmen, "Get the fire extinguishers!" An extinguisher from the base unit was handed to him, but by then the smoke was so thick and black it wasn't clear whether its chemical foam was even hitting the flame. Regardless, the fire was getting larger and it was approaching the opposite wall of the module. Korzun knew if a hole burned through this wall, the station's air would escape into space. He shouted, "Everyone to the oxygen masks!"

Thick smoke was now beginning to fill the base module and Lazutkin was ordered to prepare the forward Soyuz for an emergency evacuation. However, the path to the rear Soyuz was blocked by the fire. Without this craft, it would be impossible for everyone to leave the station. As the situation worsened, the fire alarm in the base unit went off and its sound reverberated throughout the station.

Linenger's Reaction to the Fire

While this frightening scene was taking place in the Kvant, Linenger was performing his sleep experiment in Spektr. He was unaware a problem existed until he heard the station's alarm. Propelling himself out of Spektr toward the base module he encountered Ewald and Tsibliev who told him there was a fire. When Linenger asked, "Is it serious?" he was told, "It's serious, it's serious!"

Together Linenger and Tsibliev quickly entered the Priroda module to retrieve its fire extinguishers, but found them securely attached to the wall. Rather than struggling to free these units, they entered Kvant 2. Linenger grabbed one of its extinguishers and hastened to the core module. When he entered this formerly peaceful module, he was shocked to see a wall of black smoke. Handing the extinguisher to Korzun he then headed off to Kristall to fetch another unit.

By this time Korzun had given up trying to douse the fire with foam and was shooting water directly onto it. After he had emptied three extinguishers the flame finally began to decrease in size. He shouted to Linenger, as the ship's only doctor, to check the medical supplies and see if any of the crew needed attention. Breathing heavily, Korzun turned back to the fire and with a sign of relief told Kaleri, "It's over, I think it's over."

Linenger Confronts Korzun

The fire on Mir caused the crew some anxious moments, but Linenger knew they were not out of danger. No one knew what chemicals might have been released into the air. As a doctor, he was particularly concerned about the possibility that the air might contain poisonous gases. He instructed the crew to wear gas masks until more information could be obtained from the ground. Fifteen hours after the fire, Korzun had still not informed controllers that a fire had taken place, so Linenger requested permission to speak directly to ground personnel. Korzun replied, "Jerry, just wait, we'll get to you." Fighting to restrain himself, Linenger told him, "You understand, Valery, we really need to do this now?" Despite this attempt at prodding Korzun still did not bring this matter up. Exasperated, Linenger realized ground controllers did not even know a fire had taken place. Impulsively, he interrupted his commander's ground transmission to inform them of the fire, but quickly stopped when he received an angry look from Korzun. After Korzun completed his transmission, he removed his headset and shouted, "Don't you *ever* do that again!" Annoyed by Korzun's lack of action, an unrepentant Linenger hotly responded, "Why are we not talking about this? This is important! You should know this. You're the commander!" As the two men argued, the discussion became heated. Linenger later recalled the scene, "We were really screaming at each other, real nose-to-nose stuff there for a minute or two."

Linenger would later claim the fire lasted 14 minutes, while Korzun estimated its duration was closer to 3 minutes. This time difference was actually an important matter. If the fire had lasted 14 minutes as Linenger claimed, then the station was in real danger and critics of the Shuttle/Mir program could have used it as a argument for canceling the remaining American flights to Mir. If the fire had lasted just a few minutes, as Korzun stated, it could be classified as a regrettable but minor incident and the program, largely funded by American dollars, would continue. When other crewmen were later asked about the fire's duration, they gave conflicting estimates. Ewald said he thought it lasted about 3 minutes, but Tsibliev estimated it lasted 9 minutes, but then added, "It's difficult to say. Nobody was watching the clocks."

A Russian committee was established to investigate the fire, and after a three-day inquiry concluded that "the cassette failure [was] an isolated event" probably caused by a manufacturing defect. NASA's safety engineers were unhappy with this analysis, pointing out that laboratory tests had not been able to duplicate the fire. However, as far as the Russian managers were concerned, the investigation was over. Based on this report, the Mir crew was instructed to use only recently manufactured canisters, but nobody was certain another fire would not occur.

Linenger Begins to Lose Faith in His Support Team

Linenger was deeply concerned about what he perceived as NASA's lack of concern over this incident. He transmitted a complete report of this episode, but five days went by before his flight surgeon, Tom Marshbarn, even received a copy. When told of this delay Linenger responded, "That is totally unacceptable. Understand, that thing came down on telemetry days ago, so it's been sitting in the system down there just rotting away for a few days. . . . What's going on?"

On Sunday, March 2, Korzun, Kaleri, and Ewald undocked from Mir and returned to Earth. Communication links between the ground and the station had been getting worse for some time and were now unreliable. Loud static often made transmissions impossible to understand. Tsibliev asked that

an alternative to this "ratty" communication process be found, but none were offered. Already upset with his controllers, Linenger's solution to this situation was to stop speaking directly with his support team and to communicate with them using e-mail.

Get Ready to Evacuate!

On March 4 Tsibliev conducted a test of a new instrument called the TORU. This unit was designed to allow a Mir crewman to dock an approaching Progress supply ship. After being assured by Russian officials that this procedure had been completed many times before, NASA officials offered no objection. According to the test plan, once the Progress was brought to within 4 to 5 miles of Mir, ground personnel would instruct the ship to head toward the station and then relinquish control to Tsibliev. A video camera on the Progress would come on and transmit pictures to Tsibliev's computer, and he would then guide the Progress to within a few hundred feet of the station. At this point controllers would reestablish contact and help dock the supply craft.

The TORU test initially proceeded as planned, but as the distance between the craft narrowed to 2 miles, the Progress camera failed to activate. As additional minutes passed, Tsibliev grew nervous because he had no way of knowing where the Progress was. He asked Lazutkin and Linenger to search for it through Mir's windows. As the 6-ton craft silently approached Mir, Tsibliev knew the crew was in danger, "It was the most uncomfortable [time]. I felt as if I'm sitting in a car, but I don't see anything from the car, and I know there is this huge truck out there bearing down on me. You don't know if it's going to hit you or miss you."

As the minutes passed, Tsibliev anxiously asked his crewmates, "Do you see it? Do you see it?" Finally Lazutkin spotted the craft, it was below Mir but appeared to be on a collision course! Fearing the worst, Tsibliev turned to Linenger and shouted, "Get in the spacecraft. Get ready to evacuate!"

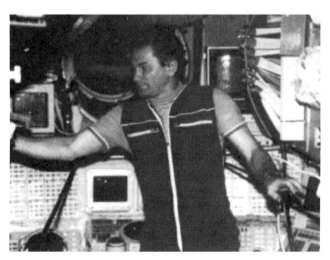

Figure 40-8. Cosmonaut Vasil V. Tsibliev worked exceptionally long hours to keep Mir operational.

Tsibliev was desperate. His computer screen was still blank. As Linenger looked out from the Soyuz he saw Tsibliev feverishly working the computer joystick while shouting to Lazutkin, "What's it doing!?"

Fifteen seconds before the expected impact Tsibliev's monitor suddenly flickered to life and he saw that the Progress would miss Mir. Lazutkin watched through a port-hole as the craft sailed by at a distance of only 200 meters. Shaken by this near collision, Tsibliev excitedly apologized to Linenger, "Jerry, what was I supposed to do? What could I do? The screen shows nothing! Nothing!" Collecting himself, he told his crewmates, "Guys, I never want to do that again."

Another Communication Breakdown

While the above events were unfolding American officials at Star City calmly went about their business unaware of the drama being played out in space. Russian officials later told them the docking had been unsuccessful, but left the impression this was not a serious concern. Unwittingly, Linenger helped keep NASA officials ignorant of what had happened by not mentioning this episode in his report. He recalled, "[I] assumed that they must know. How could they not know about it? Bad assumption I guess." Only after Linenger returned to Earth did officials learn of the near collision.

The Decision to Continue the Shuttle/Mir Program

Russian engineers were perplexed by the Progress docking failure and did not understand why its video camera on the Progress had failed. Perhaps the signal from the station's Kurs radar, a unit normally used for dockings, had interfered with its transmission. To test this hypothesis, they decided to ask Tsibliev to turn this radar off during a subsequent Progress flight. This request would have disastrous consequences.

By this time, Mir had been in space much longer than its design lifetime so its rotating cosmonaut crews were trained to handle minor failures and breakdowns. On the day after the TORU test, the station's only remaining Elektron oxygen regeneration system stopped working. The crew was reluctant to use the Vika unit located in Kvant 2 to produce oxygen because this unit was identical to the one that had earlier burst into flames. Despite their concern, they had to have oxygen if they were to remain on the station. Their apprehension was increased when the unit initially failed to work. After several attempts the Vika finally came on and operated properly.

In view of the series of problems Mir was having, NASA officials and Congress began to question if it was safe to send Linenger's replacement, Michael Foale, to the station. Wisconsin Congressman James Sensenbrenner reminded *Time* magazine, "We compromised safety before and we got the Challenger disaster." Culbertson admitted problems existed, but urged the program be continued, "It's very true the Russians are going through a difficult time right now. [But] just

as we expect them to stick with us on the International Space Station . . . I think it's important that we as partners stick with them during a difficult time with Mir." Blagov agreed, "If they [the U.S.] didn't want to send him, I would have understood. But the Russians would never do such a thing. If we are partners, then we should not forsake each other."

Foale Replaces Linenger

On April 8, the situation on Mir finally took a turn for the better when *Progress M-34* successfully docked with parts to fix the Elektron oxygen regeneration system, and 20 days worth of Vika canisters. Conditions steadily improved and Linenger was able to complete his experimental program and begin the tasks assigned to his successor.

Linenger's tour ended when *STS-84* arrived at Mir on May 17, 1997. Additional lithium hydroxide canisters and a new oxygen regeneration unit were transferred to the station. The old Elektron and burnt Kvant 1 Vika unit were disconnected and loaded for examination on Earth. Astronaut Foale boarded Mir, replacing Linenger. His experimental program consisted of studies on protein crystallization, Earth resources, space biology, materials processing, and engineering.

A Space Collision

Just as the prior unsettling episodes on Mir began to fade from the public's memory, disaster once again struck the station. On June 25, Tsibliev was engaged in a test with *Progress M-34* to determine why the previous docking attempt had failed. *Progress M-34* was moved away from the station and then maneuvered back toward its port. As the craft approached the station Tsibliev lost control and the Progress crashed into the Spektr module. Foale was unaware of the im-

Figure 40-9. Astronaut Michael Foale on the Mir space station.

pending collision until he heard a "loud bang" and felt the station shake. He then heard the unnerving hissing sound of air escaping from Spektr.

The crew quickly tried to seal off Spektr from the rest of the station before their entire air supply was gone. Unfortunately several cables snaked through Spektr's entry port and prevented its door from being closed. The fastest solution was to cut them, but even this process took several minutes to complete.

Cutting the cables saved the station's air supply, but this created another serious problem. These cables provided power from Spektr's four solar panels to the rest of the station. Since about half of the station's power was generated by these panels, their loss resulted in a severe power shortage.

In order to prevent the station's batteries from being drained, the crew rushed to turn off as many of Mir's systems as possible. The situation became so critical the crew was forced to turn off the gyrodynes controlling Mir's orientation and the station's communication equipment. Despite these efforts, the performance of many of the station's remaining systems dropped dramatically. With the orientation units offline, Mir began to drift. The remaining solar panels lost their sun orientation, reducing the available power even more. To restore the station's Sun alignment, Foale suggested the crew use their *Soyuz TM-25* to reorient the station. This was done and conditions finally began to improve.

To restore full power to Mir the Spektr solar panels had to be reconnected to the station's power grid. A procedure to complete this task was sent by ground controllers, but after some discussion they decided to leave this job to the replacement crew of Anatoli Solovyov and Pavel Vinogradov, who would be launched on *Soyuz TM-26*. These cosmonauts would have the benefit of receiving extensive training in the Russian underwater tank at the cosmonaut training center and would be better prepared. It was also hoped they would be able to locate and fix the rupture in Spektr.

Figure 40-11. Dave Wolf occupied Mir after the departure of Michael Foale.

Dave Wolf's Mir Mission

Foale was relieved by David Wolf when *STS-86* docked with Mir on September 28, 1997. Unlike the prior visiting astronauts, Foale had few regrets about leaving and later expressed the view that the station should be retired before another serious accident occurred. Wolf continued the experimental program of his predecessors, and despite the apprehensions raised by the numerous mishaps of the last mission, no serious problems arose during his stay.

The Final Shuttle Flights to Mir: *STS-89* and *91*

Compared to the prior missions, the last two Shuttle/Mir tours were tame. Both missions provided information for the construction of the International Space Station and continued the experimental program started by their predecessors. *STS-89* was launched to Mir on January 22, 1998, and carried approximately 6,000 pounds of cargo. This flight employed the Shuttle Endeavour and marked the first time a Shuttle other than Atlantis had docked with Mir.

Astronaut Andrew Thomas, the last American scheduled to complete a tour on Mir, relieved Dave Wolf, who had spent approximately four months on the station. During Thomas's stay, 27 studies, many of which built upon those previously undertaken in the areas of Earth resources, biology, microgravity, and technology, were completed.

Launched on June 2, 1998, *STS-91* was America's last mission to Mir. The duration of this flight was limited to that of a normal Shuttle mission, less than two weeks. Its experimental program was similar to other pre-Shuttle/Mir flights and included experiments to search for antimatter and dark

Figure 40-10. After bouncing off Spektr, Progress M-34 *struck and damaged one of the module's solar panels.*

matter in space. After retrieving astronaut Thomas, the Shuttle landed on June 12, 1998 at the Kennedy Space Center. This marked the end of the second phase of the American/Russian space collaboration. The next phase would involve the construction, launch, and operation of the International Space Station.

Reflections

Mir was the first space station to truly internationalize space research for peaceful purposes. In many respects it was also a human and engineering experiment to verify systems, suggest improvements, and test hardware for future stations. When Mir was launched, mankind's experience with orbiting space stations was still in its infancy. Just as the Wright brother's first airplane identified ways aircraft could be improved, Mir revealed ways in which future space stations could be made safer and more functional. The engineering data collected from Mir and the contributions it has made to the medical problems confronting long-duration spaceflight were more than adequate to justify the time, effort, and expense of this project.

The Shuttle/Mir program was the most extensive collaborative space effort ever undertaken. It dwarfed the Apollo-Soyuz program conducted in July 1975, the Salyut visitor program, and the other international visiting expeditions to Mir. The American Space Shuttle flights benefited both the United States and Russia. The United States was able to conduct long-duration crystal growth, biology, and technology studies that were not possible on the short Shuttle flights. Seven astronauts, beginning with Norm Thagard and ending with Andrew Thomas, provided NASA doctors and engineers with medical data about the human body's reaction to weightlessness, and engineering data essential to the construction of the International Space Station. The Russian government benefited from this collaboration through the infusion of critically needed cash at a time when the existence of its space program was threatened by economic turmoil. The Space Shuttle transported tons of equipment and supplies to Mir allowing the Russians to forego several costly Progress flights.

One of the most important results of the Shuttle/Mir program involved administrative issues. The lessons learned from this joint project would later pay dividends in the International Space Station program. After placing its astronauts aboard Mir, NASA gained some eye-opening experiences. The problems associated with long-duration space flights first encountered in the Skylab program had not been solved. Additional administration changes would have to be implemented before the International Space Station was launched.

41 The American Space Station Freedom Becomes the International Space Station

America has always been greatest when we dared to be great. We can reach for greatness again. We can follow our dreams to distant stars, living and working in space for peaceful, economic, and scientific gain.

—President Ronald Reagan, 1984 State of the Union Address

In his State of the Union address on January 25, 1984, President Ronald Reagan issued a similar challenge to the one made by President John F. Kennedy over twenty years earlier. Reagan announced, "Tonight, I am directing NASA to develop a permanent manned space station and to do it within a decade." Unlike Kennedy who saw the Moon race as a political contest between communism and democracy, Reagan assured his audience that the rationale for this new space effort extended beyond that of cold-war politics. He stated: "A space station will permit quantum leaps in our research in science, communications, in metals, and in lifesaving medicines which could be manufactured only in space."

President Reagan's interest in a space station was not new. On July 4, 1982, while welcoming the crew of Columbia back to Earth at Edward's Air Force Base, Reagan urged the nation, "to look aggressively to the future by demonstrating the potential of the Shuttle and establishing a more permanent

Figure 41-2. James Beggs was NASA's chief administrator under President Reagan and was a strong supporter of the space station.

presence in space." In 1983 he told a group of businessmen interested in pursuing commercial adventures in space, "I want a space station, too. I have wanted one for a long time." When elected president, Reagan's concern about the growing federal deficit moderated his support for the construction of a space station. David Stockman, the administration's budget director, had cautioned the president that the deficit would never go down if projects of this magnitude continued to be funded. These concerns were echoed by Presidential Science Advisor George Keyworth. Reagan took the warnings of these key advisors very seriously. As governor of California, he had gained a reputation as a fiscal conservative and during his presidential campaign he had pledged to control spending in Washington.

Reagan's reluctance to procede with a space station was finally overcome by three events. The first occurred when his friend and cabinet member, Attorney General William French Smith advised him not to be overly swayed by Stockman's concerns but to consider the historic significance of the

Figure 41-1. President Ronald Reagan formalized the American commitment to build a space station in his 1983 State of the Union address.

station. Smith noted, "I suspect the comptroller to King Ferdinand and Queen Isabella made the same pitch when Christopher Columbus came to court." The second event occurred when Stockman softened his stance and conceded that a small budget increase for NASA was possible. The third factor that influenced Reagan's decision was his desire to leave behind a favorable legacy. He stated, "I do not wish to be remembered only for El Salvador."

Reagan envisioned the space station, later named Space Station Freedom, as an international project rather than one financed solely by the United States. Nevertheless, the cold-war competition between the United States and the Soviet Union dictated which countries would be invited to participate. In his address to Congress, Reagan made this clear, "We want our friends to help us meet these challenges and share in their benefits. NASA will invite other countries to participate so we can strengthen peace, build prosperity, and expand freedom for all who share our goals."

This invitation was obviously meant to exclude the Soviet Union and its allies and to draw attention to the differences between the closed communist societies and the freedoms that existed in the United States.

In response to this call, Congress added $150 million to NASA's 1985 budget. However, the decision to include international partners, even those "friendly" to the United States, was controversial. Since America's space technology was significantly more advanced than that of other countries, government officials feared we would be "giving rival nations technical knowledge that only the United States held." Officials argued that technology was the key not only to military superiority, but to industrial success and that by sharing this information America might lose some of its commercial advantages. Despite these concerns, agreements were reached with Japan, Canada, and the European Space Agency (ESA) early in 1985. Brazil and Italy later signed more modest agreements to provide components to the station.

Criticisms about the loss of technology this collaboration might incur were lessened by the nature of the work the participating countries would complete. Japan was to construct a module for materials processing, life sciences, and technology experiments as well as transport vehicles. These projects required minimal assistance from the United States. Canada would build a 57-foot-long robotic arm for the station based on its prior Space Shuttle work. The European Space Agency would construct a science module modeled after the Spacelab units they had built for the Space Shuttle, as well as transport vehicles of their own design. Officals were relieved to learn that neither the Canadian nor European projects required the sharing of sensitive American technology.

Space Station Freedom's Difficult Start

Congressional meetings with NASA administrators to limit the cost of the space station were tense. Large overruns on the Space Shuttle program made Congress wary of NASA projections. One official described how congressional repre-

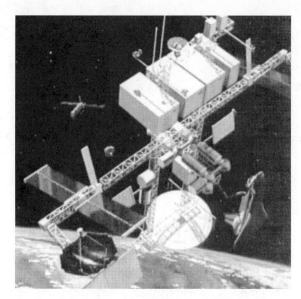

Figure 41-3. This version of Space Station Freedom was suggested in 1986. The Space Shuttle is seen with its robotic arm extended in the middle right of this drawing.

sentatives reacted to discussions about the station's cost, "It reached the scream level at about $9 billion." Realizing that the $9 billion figure was an upper limit to what Congress was willing to accept, NASA submitted a plan to build the space station for about $8 billion. This value was agreed to and start-up funds for the station were approved by large margins in both the House and Senate.

In March 1986, NASA reviewed eight different plans to build a station. The selected plan envisioned a "dual-keel" configuration like that shown in Figure 41- 3. It consisted of large metal trusses to which pressurized modules and large solar panels would be attached. The station would be manned by a crew of eight astronauts. As work on this facility began in 1988, NASA found it was unable to control the cost over-runs incurred by its contractors. Despite NASA's best efforts, the estimated completion cost of the station tripled during the next three years. As these costs continued to escalate, forces opposed to the construction of the station gained strength.

Support for Space Station Freedom Waivers

Unlike the Apollo program, support within the political and scientific communities for the space station was never very strong. Advocates saw it as a worthy scientific endeavor, while opponents viewed it as a waste of money. Senator William Proxmire, a fiscal conservative and station opponent, was particularly cynical in his judgement of NASA's reasons for proposing this project,

I am concerned that it will proceed regardless of the real need for such a program because your agency needs it more than the country needs it. I have long believed that your agency has a strong bias toward huge, very expensive projects because they keep your centers open and your people employed.

The National Academy of Science's Space Science Board, headed by Thomas Donahue, also took exception to the basic rationale offered by the Reagan administration for building the station. After a meeting on this subject, Donahue told reporters that his committee, "sees no scientific need for this space station during the next twenty years." He then went on to add fuel to the fire storm by adding, "If the decision to build a space station is political and social we have no problem with that . . . but don't call it a scientific program." Edward Keller, the father of the Atomic Bomb, expressed the view that the United States should use its resources to construct a Moon Base instead of a space station.

After President Reagan completed his second term in office, his former vice president, George Bush, was elected president in 1988. Bush was more than just an advocate of the station, he was a strong supporter of an enlarged space program. Just one year after becoming president, he urged Congress to fund a massive, new program called the Space Exploration Initiative (SEI). The SEI proposed to return people to the Moon by the year 2000, establish a Moon base, and place people on Mars by the year 2010. Its price tag, estimated to be a staggering $500 billion spread over two decades, scared even NASA's staunchest supporters.

The SEI made little progress in Congress with a huge thud. Congressional reaction was swift and negative. Barbara Mikulski, chair of the Senate Appropriations Subcommittee and a traditional ally of NASA, declared, "We're essentially not doing Moon-Mars." When funds for this program were virtually eliminated in the 1991 budget, both the White House and NASA realized SEI had no chance of being saved. In an admission of defeat, NASA's chief administrator, Donald Goldin, stated the space agency was, "putting it off until we're ready and the nation is able to afford it." The defeat of this initiative emboldened opponents of the space station and support in Congress mounted to "rein" in NASA spending.

By 1990 escalating costs threatened the continuation of the station. Vice President Quayle attempted to rally support by declaring that the station was a symbol of "the leadership in space of the United States." This type of appeal, focusing on the world's perception of the United States, had been successful in the past, but this time it failed to gain converts in Congress. Jim Sensenbrenner, chairman of the House Science Committee, expressed the view of many when he lamented, "We must get this program under control."

To help deal with an increasingly adversarial Congress, President Bush appointed the Advisory Committee on the Future of the United States Space Program. Headed by Norman Augustine, chief executive officer of Martin Marietta, the committee was charged with delineating NASA's main objectives and reviewing future programs. This panel agreed with critics of the station and recommended that it be redesigned to lessen its complexity and costs. NASA was faced

Figure 41-4. Donald Goldin, NASA chief administrator, successfully argued that plans for Space Station Freedom should be revised to include Russian participation.

with a conundrum. Congress compelled it to limit the scope of the station in order to save money but then criticized the agency for making deletions. *The New York Times* expressed the frustration of all parties when in an editorial about the station it concluded, "What a mess."

In view of these events, Space Station Freedom faced an uncertain future when Congress met to adopt the 1992 federal budget. Congressman Tim Roemer of Indiana energized opponents to the station by describing it as "a black hole quickly sucking away money from other promising scientific projects." Arguing that its costs were out of control and its technical feasibility doubtful, he urged that the station be canceled as quickly as possible. "The more daylight that shines on this program, the darker the outlook for its successful completion."

Fortunately for supporters of the station, the rationale for its construction was changing from a scientific to a social issue. The nation was in a recession and 75,000 jobs tied to station work were located in the powerful states of California, Texas, and Florida. The political risks of voting against this project now outweighed arguments about its need. When the matter came before Congress, the station survived by one vote.

NASA was pleased that the station had survived, but now found itself in a difficult position. Between 1991 and 1993 its inflation-adjusted budget had fallen by about 5 percent while the station's costs had skyrocketed. In the words of one NASA manager, the station was becoming "a self-eating watermelon." In order to continue work on this facility, NASA was forced to make painful cuts in its other programs as well as in its staffing levels. Even after this arduous process was complete, no one could predict what the attitude of the next Congress would be, but the signs were not good. The Congressional Budget Office again placed the space station at the top of its list of proposed government spending cuts. In March 1993, Congressman Alan Mollahan of West Virginia warned that, "NASA is not connecting with the American people, and the agency is losing its relevance." He went on to add that unless the station "contributes to U.S. economic competitiveness, politicians will view it as a cold-war anachronism."

The United States and Russia Agree to Build a Space Station Jointly

In March 1993 President Clinton, successor to President Bush, announced the ten-year budget for the station would be reduced by $18 billion. This decision was made to constrain a project whose costs seemed to continually rise. In the early 1980s NASA Administrator James Beggs had convinced President Reagan to support a space station by promising it would cost no more than $8 billion and because it was "the next logical step" for NASA. However, by the later 1980s its completion cost had escalated to $30 billion. When Clinton took office, NASA had already spent $8 billion of this money and had yet to launch a single module. Furthermore, it was unlikely a fiscally conservative Congress would keep funding this project. If the station was to survive a new approach was needed.

Along with the space station the Clinton adminsitration had inherited another problem, this one involving the proliferation of nuclear weapons. In November 1990 the Bush administration learned the Soviet Union intended to sell the technology to build sophisticated rocket engines to India. The United States objected to this plan fearing that India's long-time adversary and neighbor Pakistan would be forced to develop or acquire a similar technology. Although Yeltsin appeared to signal that this deal with India would be canceled, in the chaotic situation following the collapse of the Soviet Union no action was taken. Limited sanctions were therefore imposed on Russia by the United States. Seeing that this approach was having little effect, Anthony Lake, the new national security advisor for the Clinton Administration, sought an alternative way to have this deal cancelled. Building on an obvious Russian strength, Lake formed a committee to search for ways their rocket program could be used to raise cash by launching commercial satellites. Learning of this effort, Yuri Koptev, head of the Russian Space Agency, grabbed at this opportunity. Believing the United States would be willing to pay for Russian space expertise, in a meeting with NASA Administrator Goldin he proposed Russia join America in the construction of a joint space station. Goldin recalled this meeting,

I'll never forget when Yuri dropped the bomb that day. I was flabbergasted! I was worried we had no experience dealing with [a] station whatsoever— none. We had no major experience in logistics. We knew nothing about the hazards. For Apollo we had Mercury and Gemini. For a space station, which is more complicated, we had nothing.

Table 41-1. Assembly/Transportation Flights

Space Shuttle Flights (27)	
Assembly	21
Utilization and Outfitting	6
Russian Flights (44)	
Assembly	15
Crew Transport	10
Reboost	19
European Space Agency Flights (2)	
Assembly	1
Crew Transport Vehicle	

Following this discussion the Clinton administration put together a three-part package it hoped the Russians would accept in lieu of their arrangement with India: (1) the United States would not attempt to block the sale of completed engines to India, (2) Russia would be allowed to compete in the lucrative American satellite market, and (3) Russia would be invited to "participate" in the American space station effort. This plan was presented and tentatively agreed to at the Clinton and Yeltsin summit meeting held in early April 1993 in Vancouver. The term "participate" was left purposefully vague, but the Russians understood it would involve substantial purchases in Russia by the United States.

A few weeks later a Russian delegation arrived in Washington. They presented NASA officals with two offers: (1) Russia would build a space station for the United States for $7 billion, or (2) Russia would supply space station modules to NASA for $35 million each. Both of these suggestions were rejected as being too ambitious.

By early June the Vancouver accord seemed to be unraveling. Russia had not canceled its technology transfer agreement with India and the Clinton administration was threatening to impose additional sanctions by July 15. An American delegation was sent to Moscow on June 30 with a new offer to rescue matters. Russia would be given equal participation in a new International Space Station, paid several hundred million dollars to launch space station modules, and paid an additonal $400 million to allow American astronauts to conduct experiments on Mir between 1994 and 1997. NASA had three goals for its Mir program: to gain additional first-hand knowledge about the long-term effects of weightlessness, to test materials to be used in the construction of the International Space Station, and to allow it to develop an organizational structure for conducting large-scale collaborative space projects.

Russia had little choice but to agree to this plan if its space program was to survive. The financial situation in Russia was dire and substantial government cutbacks were inevitable. Yuri Koptev told his superiors that unless emergency measures were taken, the Russian space industry could collapse within a year. In this context the agreement proposed by Clinton was "heaven

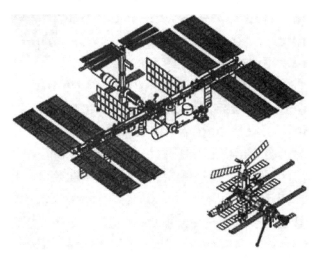

Figure 41-5. When it is complete, the size of the International Space Station (top) will dwarf that of Mir (lower right).

sent." Since both countries benefited from this agreement, little opposition arose, and in August 1993, Vice President Gore and Russian Prime Minister Chernomyrdin signed a document formalizing this arrangement. No one was more relieved at this signing than Koptev, who told a Russian reporter, "We're getting out of a tough spot today."

From the view of the United States, the financial chaos in Russia was an inducement to proceed with the space station rather than a reason to cancel it. In 1993 Clinton was perceived as a weak foreign policy leader and was widely criticized for his handling of events in Somalia, Haiti, and Yugoslavia. Now Russia seemed on the verge of collapse and Clinton's critics predicted this would be yet another foreign policy disaster. Former President Nixon urged the president to act and wrote in *The New York Times* that, "it would be tragic, if at this critical point, the United States fails to pro-

Figure 41-6. The Zarya (Sunrise) was the first module of the International Space Station to be launched. It lifted off from Tyuratam on a Proton rocket.

vide the leadership only it can provide." Clinton saw the space station as a way of funneling large sums of money to Russia and as a way of avoiding the charge that his inactivity had "lost Russia" back to communism.

Although $400 million is a substantial amount of money, it was not a particularly large sum relative to other government expenditures, for example, a single Space Shuttle launch costs $550 million. But with the collapse of the Russian currency, $400 million had an extraordinary purchasing power. The annual salary of a Russian scientist or engineer was about $500 and an entire research institute could function on about $100,000 per year. Even Clinton's political opponents had to concede that the potential benefits of this infusion of funds greatly outweighed other considerations.

The American/Soviet agreement also made it possible for the United States to reduce its space station expenditures. As part of this effort, Russia would construct the station's living quarters, two research modules, an array of solar panels, and transport vehicles. They would also supply Soyuz spacecraft to transport crew members to and from the station. Upon completion of the station Russia would pay 30 percent of its operating costs. If these commitments were honored, America's $400 million would be repaid several times over. The agreement appeared to be a great deal.

The nature of the changing arguments in support of the space station were noted by several people. John Pike, analyst with the Federation of American Scientists, contrasted the political motivations of the station, "For Reagan it was to whip the Evil Empire; for Clinton it is to be friends with the Russians."

Components of the International Space Station

The International Space Station will consist of at least seven major laboratories. It will be at least four times larger than Mir and have a mass of over 1 million pounds (see Figure 41-5). Orbiting the Earth every 90 minutes at an elevation of around 250 miles, it will be visible from the ground as one of the brightest star-like objects in the sky. The station will be placed in an orbit that will cause it to eventually pass over most of Earth's inhabited land.

The projected construction cost of the station is $65 billion, $25 billion of which will come from the United States, while its international partners will provide the balance. Thirty-seven flights, twenty-one by the United States, fifteen by Russia, and one by the ESA, will be required to assemble the station. The targeted completion date is 2004. Once functioning, operating costs will be shared in the following manner: 50 percent to the United States, 30 percent to Russia, and 20 percent to other international members.

During the first phase of its construction, eleven large structures will be joined together in a modular fashion. The order in which these parts will be added to the station is shown in Figure 41-7. The first module, the Zarya (1), was launched from Tyuratam on November 20, 1998. Built in

Figure 41-7. Phase 2 of the United States/Russian space collaboration will involve the initial construction of the International Space Station. The order in which modules will be added are shown by the numbers 1–11 in the drawing below.

1. Zarya
2. Unity
3. Russian service module
4. Universal docking module
5. Unity truss (Z1)
6. Solar power array
7. Science power platform
8. U.S. laboratory module
9. Space Station remote manipulator system
10. Airlock
11. Japanese experimental module

Russia but paid for by the United States, it has a mass of 20 tons. Attached to its outer wall are two solar panels that can generate 4 kW of power and fuel tanks that can hold over 6 tons of propellants. The Zarya possesses several engines for orbital maneuvers and orientation control during the early assembly phase. At its front end is an adapter with three docking ports.

A second module, an American-made compartment named Unity (2), was launched from the Kennedy Space Center aboard the American Space Shuttle Endeavour (*STS-88*) on December 15, 1998. As its name implies, the main purpose of this unit is to link other modules to the station. As seen in Figure 41-7, a large truss will later be attached. During *STS-88* the astronauts employed the Shuttle's robotic arm to reposition the Zarya and Unity modules. The Shuttle's thrusters were then used to push these two modules together, creating a 77-foot-long, 70,000-pound craft. Even at this stage the astronauts could sense the magnitude of the structure they were helping to build. Shuttle Commander Robert Cabana remarked, "This is going to be one heck of a space station when we get it done."

Once Zarya and Unity were locked together, three spacewalks were completed to connect their power and data cables. Sergi Krikalev and other crewmen then entered the complex and installed additional equipment. With these tasks complete, the Shuttle crew returned to Earth, leaving the station unoccupied.

The third module of the International Space Station, the Russian service module, will be the next compartment launched. This 20-ton unit (shown as number 3 in Figure 41-7) will be attached to the Zarya aft port and will provide living quarters for three people. Resembling the Mir core module, it

Figure 41-8. The Unity module was launched aboard STS-88 in December 1998. Here some of the crew of this flight inspect the module at the Kennedy Space Center.

Figure 41-9. Unity and Zarya were joined together during the mission of STS-88 using the Shuttle's remote manipulator arm and the Shuttle's thrusters.

also contains the station's control systems. Attached to the front of this module is a five-port docking adaptor. A large tower, with solar panels and thermal radiators, will later be attached to one of these ports.

After the Russian service module has been tested from Earth, a three-person crew (two Russians and an American) will travel to the station in a Soyuz craft. Their job will be to assist in the installation of the other modules. The Soyuz will

Figure 41-10. The service module while still under construction in Russia. Attached to its front is a five-port docking adaptor. A long truss structure is planned to extend upward from the top port.

Figure 41-11. The addition of the Japanese experimental module (11), the centrifuge accommodation module (12) and ESA Columbus Orbital Facility (13) will begin phase 3 construction.

serve as the crew's escape craft should an emergency arise aboard the station.

The next unit to be added to the station will be the universal docking module (4). It will be connected to a Zarya port and will allow later modules as well as an additional Soyuz craft to be docked. Following this module, the Unity truss structure (5) and its upper solar power array (6) will be attached to the station. The science power platform (7), will then be joined to another Zarya port. The last components to be attached during this part of the construction phase will be the U.S. laboratory module (8), the Canadian-built Space Station remote manipulator system (9) and an airlock (10).

The United States Laboratory will be 28 feet long and will carry instruments for medical, biological, and materials science experiments. The Canadian-built remote manipulator arm, a larger and more complex version of the one employed on the Space Shuttles, will be used to move objects about the station, to carry spacewalking astronauts, to capture and release satellites, and to lift large pieces of cargo. Once this unit is attached, the initial construction phase of the station will be complete.

Phase 3: The Completed International Space Station

During the final phase of the construction of the International Space Station additional modules from Russia, Japan, the ESA, and the United States will be attached. Russia currently plans to provide two research modules to the station, the first of which is scheduled to be launched at the beginning of this phase. The next unit to lift-off from Earth will be the Japanese experimental module (11). This module will consist of a pressurized compartment and an adjacent porch on which experiments requiring direct exposure to space can be placed. A small manipulator arm, located on the module, will be used to move items on the pallet. This module will be followed by the American-built centrifuge accommodation module (12) that will be joined to the U.S. lab. The ESA's Columbus or-

bital facility (13) will then be launched by an Ariane rocket and attached to a side port of the U.S. laboratory, opposite from the Japanese experimental module. It will house equipment for experimental programs in the life sciences, fluid physics, and materials science.

Figure 41-12. The first crew to occupy the International Station. From left to right, Sergei Krikalev, William M. Shepherd, and Yuri P. Gidzenko.

Figure 41-13: An artist's view of the completed International Space Station with the Space Shuttle docked.

The final module to be added to the station will be the habitation module. As its name implies, this compartment will be used to house station personnel. It will be made of aluminum or an inflatable but stronger than aluminum material called Kevlar.

Italy will provide the mini pressurized logistics module to the station. This 21-foot, pressurized module, has a storage capacity of about 10 tons and is intended to transport materials to the station in the Shuttle's bay. Italy will also provide a second, similar-sized module for life science experiments.

When the International Space Station is fully assembled there will be ample work and living space for a crew of six people. Its pressurized volume will be equivalent to that provided by two Boeing 747 jumbo jets. More than 850 hours of spacewalks will have been completed to construct the station, more than all of the previous Extra Vehicular Activities (EVAs) combined. The completion date for the station is estimated to be in the year 2005 but some slippage in this date can be expected.

Work of the Space Station

The International Space Station will serve as a laboratory for experiments in astronomy, medical science, biology, Earth resources, and materials sciences. Some of the areas to be studied and the reasons why this work must be conducted on the station are discussed below.

Protein Crystal Studies: Purer protein crystals can be produced in the weightless environment of space than on earth. It is hoped these samples can be used to better understand how enzymes and viruses function. Researchers hope these studies will lead to more effective drugs for the treatment of diseases such as cancer, diabetes, and emphysema.

Cell Development: Cultures grown aboard the station will provide new insights about the role that gravity plays in the operation of cells. This information has important implications for the establishment of human colonies in space.

The Medical Effects of Weightlessness: The long-term effects of gravity on the human body, especially on the body's muscles, heart, arteries, veins, and bone density, will be studied. This type of information is vital for the safety of crews sent on future missions to the planets.

Flames, Fluids, and Metals: All of these objects behave differently in space than on Earth. Combustion experiments are expected to provide new insights about how these processes operate. The convective motions that cause warm air to rise and cold air to fall on Earth are greatly reduced in a weightless environment. In a weightless environment materials do not separate as they do on Earth, so purer crystals and unusual compounds can be constructed. These purer compounds are expected to increase the efficiency of computer chips and allow advancements to be made in several other industries.

Materials Science: As noted in the last section, several studies on the International Space Station will be performed on the Japanese experimental platform. Experiments conducted in materials science will aid engineers in the design of sturdier and safer spacecraft.

Earth Observations: Large-scale, long-term changes on the Earth are difficult to observe from the ground. Changes in the oceans, forests, deserts, as well as large-scale weather formations such as hurricanes, typhoons, and extensive thunderstorms will be studied from the space station. The environmental damage caused by water and air pollution, deforestation, oil spills, etc., will be monitored from the station and the effectiveness of corrective actions will be determined.

Commercial Uses of Space: One of the goals of the International Space Station will be to develop new products and services that will improve living conditions on Earth. Studies in materials science, medicine, and Earth resources are expected to produce products of considerable economic value. Revenues from their sale will be used to defray the operational costs of the station.

Space Station Operations

The normal crew of the space station will consist of six people, a commander, a flight engineer, and four research scientists. However the station will be able to accommodate up to sixteen people for a period of a week. The commander and flight engineer will be responsible for the operation of the stations and will complete routine maintenance tasks. The role of the research scientists will be the equivalent of NASA's "payload specialists." Their primary responsibility will be to conduct the research programs on the station and to communicate these results to scientists on the ground using video and data links.

Crew will stay on the station for three to six months. Although the lifetime of the International Space Station will be at least ten years, officials are hopeful period this can be increased to twenty-five years by routinely completing preventative maintenance operations.

Conclusions

The justification for building a space station has changed several times since it was proposed in 1983. Originally, it was portrayed as a craft that would provide direct benefits to the ordinary man. Major advances in communications, medicine, meteorology, and manufacturing were promised. The station was then defended as a political response to the Soviet Union's Mir space station. It was spoken of as the symbol of the superiority of capitalism over communism. At its low point, the station was viewed simply as a way of creating aerospace jobs during a time of economic hardship. Other people defended the space station as the next logical step leading to America's permanent presence in space. The Space

Shuttle, they noted, was intended to service a space station. During the Clinton administration the American space station was canceled in favor of an international effort. At this time, the station was seen as an instrument of foreign policy to keep Russia from slipping back into a totalitarian state that would be hostile to American interests.

Ultimately, the best reason to build the International Space Station might not be in any of the areas in which its proponents sought support. Its greatest contribution may be its ability to satisfy the human need to explore and to push the boundaries of the unknown outward. While this is being accomplished the station will also teach the world how to work together toward a common goal.

Regardless of the reasons to build a station, it is now on an irreversible path toward completion. Andrew Lawler, a fellow of the Knight Science Journalism Program at the Massachusetts Institute of Technology, called attention to the social and historic nature of the station when he wrote, "It is a scientific United Nations and a milestone of scientific, technological, and political ingenuity of the highest order."

As laudable as the development of a new commercial product might be, projects with this goal have never been able to capture the imagination of a nation or inspire its people. Men like Tsiolkovsky, Oberth, and Goddard ultimately became world heros because of their adventurous spirit and ideas. They elevated human thinking by their willingness to challenge the restrictions that nature has imposed on us. As Robert Goddard wrote decades ago, the quest for new knowledge is part of mankind's very being. Once work on the station is complete, it seems certain that the next goal of humanity will be to return to the Moon, this time to stay. Plans for this mission are already being developed.

Moon Base: An Important Step into the Future

For ages humans have looked up at the Moon and dreamed of reaching it. Soon we may find ourselves gazing up from the Moon itself in order to explore the universe as never before.

—**Jack Burns, Observatories on the Moon,** *Scientific American,* **1990**

The early space pioneers could only imagine what it would be like to journey to the Moon and explore its mountains, maria, and hidden valleys. These adventurers devoted their lives to solving the problems of spaceflight, but never flew in a rocket. The fulfillment of their dream relied on the engineering and organizational skills of people like von Braun and Korolev and the courage of numerous astronauts and cosmonauts who tested the space vehicles and equipment.

Tsiolkovsky, Goddard, and Oberth would probably have been disappointed that the Apollo program did not result in the construction of a lunar base. Most likely, they would have been perplexed by this fact and shocked by the long seventeen year delay between the final Apollo flight and the start of discussions leading to a return to the Moon.

Man's Return to the Moon

On July 20, 1989, President George Bush commemorated the 20th anniversary of the Apollo 11 landing by announcing an ambitious, long-term plan for manned space exploration called the Space Science Initiative. In a speech announcing this program, Bush said, "I'm proposing a long-range continuing commitment. First, for the coming decade, for the 1990s, Space Station Freedom; and next, for the new century, back to the Moon, back to the future, and this time, back to stay. And then a journey to another planet, a manned mission to Mars." This multidecade effort was intended to begin America's irreversible advance into space. With this call to action, NASA began work on determining how these goals might be met.

Nearly everyone in the scientific community realized that establishing a permanent base on the Moon would be very difficult. Without a doubt, formidable problems would have to be overcome. As late as 1993, Don Wilhelms, a prominent Apollo geologist, gave the following pessimistic view:

Possibly the most important spin-off from Apollo is the concept of Spaceship Earth. Apollo may have been a first step into the cosmos, but further steps have not yet followed as we thought they would. Apollo taught that we *cannot* colonize space except

Figure 42-1. President George Bush took a leadership role by proposing that a moon base be built shortly after the year 2000.

on a very small scale. The astronauts had to bring absolutely everything with them to sustain their lives, and at the present rate the Earth will be worn out long before the knowledge to live cost-effectively on the Moon or another planet is developed. Earth is our home.

The lack of air and water on the Moon requires the presence of extensive life-support equipment if humans are to survive. The Moon's environment is often described by such terms as "extreme," "harsh," "hostile," and "dead." Even astronaut Edwin Aldrin when first setting foot on the lunar surface described the scene as "magnificent desolation." Why would we want to establish a base in such a forbidding place?

Advantages of a Lunar Base

Many of the conditions which make the Moon a difficult place to live, make it a nearly ideal site for other purposes. Possible uses of the Moon include (1) a site for an

astronomical observatory; (2) a launch site for solar system and deep space missions; (3) a mining site to materials for use on Earth, for space facilities, and for fuel; and (4) as a rest station for humans working in Earth orbit or in space. Each of these areas is discussed below.

The Moon as an Astronomical Base: The Moon offers several advantages for an astronomical observatory. The lack of an atmosphere allows all forms of light to reach its surface. Seismic activity is smaller by a factor of 10 million than on the Earth and plate tectonic motion is nonexistent. This allows kilometer size arrays of telescopes to be electronically combined into an instrument called an interferometer, an instrument that allows fine details to be seen in the structure of celestial objects.

In order for such an array to operate, the distance between the individual array elements must be maintained to one millionth of an inch. Seismic activity on Earth makes this criterion difficult to meet, but activity on the Moon is so small that distances between lunar objects remain essentially unchanged over long periods of time.

On the far side of the Moon, the night sky is extremely dark so optical telescopes would be able to observe exceedingly faint objects. Radio telescopes would also be able to detect weak celestial sources because their signal would not have to compete against human-created radio interference. The low background of radio noise makes it much easier to detect faint signals.

Astronomical detectors on Earth are often cooled to very low temperatures in order to increase their sensitivity. All electronic instruments, when activated, produce a small level of internal emission. This unwanted signal is referred to as

"noise." If placed at the bottom of deep craters near the Moon's poles, detectors would be continually shielded from the Sun's heating rays. This would produce a constant, very low temperature environment that would minimize instrumental noise while simultaneously eliminating the need to replace expendable coolants.

Finally, the surface of the Moon is rich in elements that are useful in constructing both scientific instruments and their supporting facilities. More will be said about this later.

The Moon as a Launch Site: The Moon's gravity is one-sixth that of the Earth making it easier to launch a spacecraft into space. The amount of gravity that a rocket must fight against also has a dramatic impact on the mass of rocket that can be launched. During the Apollo program, the bottom two stages of the enormous Saturn V rocket were needed just to place its SIV stage and CSM into Earth orbit. In contrast, when leaving the less massive Moon, only the engines of the lunar module were required for lift-off. In fact, the Moon's orbital velocity is so low, it is possible to place objects into orbit by using a magnetic or electrically powered catapult.

The Moon as a Resource Base: The Moon's highlands and maria are rich in aluminum and titanium, but neither of these elements is expensive enough to justify mining and transporting them to Earth. However, these elements and other minerals found on the Moon could play another important role. These materials could be used to construct modules that could then be joined together to build facilities on the Moon, in lunar orbit, or even in Earth orbit.

Table 42-1. A Comparison between the Moon and Earth		
Property	**Moon**	**Earth**
Mass	7.4×10^{22} kg	6.0×10^{24} kg
Radius	1,738 km	6,371 km
Surface area	3.8×10^{7} km^2	5.1×10^{8} km^2
Mean density	3.34 g/cm^3	5.5 g/cm^3
Gravity	1.62 m/sec^2	9.1 m/sec^2
Escape velocity	2.4 km/sec	11.2 km/sec
Rotation period	27.3 days	24.0 hours
Temperature (day)	107°C	22°C
Temperature (night)	−153°C	22°C
Atmosphere	10^4 mol/cm^3	2.5×10^{19} mol/cm^3
Seismic energy	2×10^{10} J/yr	5×10^{17} J/yr

Data taken from *Lunar Sourcebook*, p. 28.

The Skylab and Mir astronauts found it difficult to work in the weightless environment of space. Special tools and extensive training were required to perform even relatively simple tasks. The Moon's gravity would allow many of the building techniques developed on Earth to be used in the reduced gravitational field of the Moon. James Blacic of Los Alamos National Laboratory has found that the lack of water in the Moon's soil and rock makes it possible to construct super strong glasses that have the strength of steel. These glasses would be suitable for making the optical mirrors of astronomical telescopes as well as the structures to support them.

Oxygen is the most abundant element on the lunar surface. Once released from the soil, it can be employed in life-support systems, in the generation of electricity, and as a rocket propellant. As space travel becomes more common, the need for a relatively cheap and readily available source of liquid oxygen is likely to increase. Oxygen mined on the Moon could be used to supply these missions.

A final reason to mine the Moon is that its soil contains ^{3}He. This atom, which is rare on the Earth, is captured in the lunar soil from the solar wind. Scientists have suggested that ^{3}He might be more safely used to power fusion reactors on Earth rather than tritium (^{3}H).

The Moon as a Rest Station: Skylab and Mir programs have demonstrated that much remains to be learned about how the human body reacts to a reduced level of gravity. Cardiovascular troubles, muscular atrophy, and loss of bone calcium are among the problems known to arise from prolonged periods of weightlessness. The Moon could serve as a convenient location for space workers to recover from these effects. In addition to providing relief from these physiological problems, a Moon base could help alleviate the psychological problems that arise when humans are confined to cramped quarters for long periods of time. The Moon is, after all, a large, solid object, with plenty of room. This alone would make the Moon an attractive haven for people who work in space.

Some Disadvantages of Earth-Orbiting Observatories

The Moon offers several advantages over Earth-orbiting observatories. Some of these advantages are given below.

* The near Earth environment is cluttered with high velocity debris left over from forty years of rocket launches (see Figure 42-2). If hit by one of these particles, a satellite could be severely damaged or even destroyed.

* A significant amount of dust and gas is present at the height normal scientific satellites orbit the Earth. This dust scatters light and creates an illuminated background against which the faintest celestial objects cannot be seen. This high altitude gas also produces emission lines that contaminate optical spectra.

* When solar activity increases during the sun's normal eleven-year cycle, the Earth's atmosphere expands. The increased size of the atmosphere creates additional drag causing the satellite's orbit to decay at a faster rate, thereby reducing the satellite's lifetime.

* The Earth creates a significant amount of radio noise through the interaction of electrons in the solar wind with the Earth's magnetic field. Additional radio emissions arise from commercial broadcasts. The closer a satellite is to the Earth,

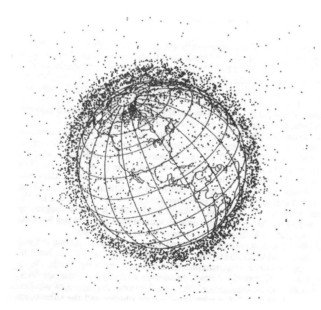

Figure 42-2. The amount of debris that is in orbit around the Earth is very extensive. This figure shows the distribution of only the larger size objects.

the greater the impact these emissions have on the detection of faint celestial radio sources.

* As structures orbit the Earth, their surfaces are alternately heated by the Sun and then cooled. This process causes thermal forces to arise that change the shape and dimensions of the structure. Gravitational gradients also distort large-sized structures.

Difficulties of a Moon Base

Despite the advantages a Moon base enjoys over a station placed in Earth orbit, such a facility does have disadvantages. Unlike the situation on Earth where meteorites are destroyed by frictional heating as they pass through the atmosphere, these particles hit the Moon's surface unimpeded. Materials brought back to Earth by the Apollo astronauts showed that exposed surfaces were pock-marked with tiny craters ranging in size from 1 to 10 microns. Over time these impacts will cause the optical surfaces of telescopes to deteriorate. Protective coverings will be required to reduce this damage.

A second problem a Moon base will have is associated with the lack of an atmosphere and the rotation of the Moon itself. As the Moon rotates, its surface enters periods of sunlight and darkness just as an Earth-orbiting satellite does. Because the Moon lacks a moderating atmosphere, temperatures drop quickly once the sun sets below the horizon. The day/night temperatures on the Moon differ by nearly 500°F. Special care will be required to insure that the thermal strain caused by this large temperature change does not damage the observatory's instruments or structures.

Water on the Moon

One of the surprising results of the Apollo program was that no water was found in any of the returned samples. No one expected to find liquid water on the Moon, but the absence of water within the chemical bonds in the soil or rocks was a surprise. This finding presented a serious problem for a manned Moon base. Currently, it costs about $40,000 to lift one pound of material from the Earth's surface to the Moon. Even if this cost could be decreased to $4,000, as NASA believes it can, the cost of supplying a Moon base with water would still be prohibitively high.

On May 5, 1998 a partial answer to this problem was provided by the announcement that the unmanned lunar probe, Lunar Prospector, had found water ice at the bottom of the perpetually shadowed craters at the Moon's poles. This result was not entirely unexpected. An earlier probe, the Clementine, had reported the possible detection of ice at the Moon's poles in November 1996.

The Moon's polar ice is believed to be mixed with the lunar soil, at a very low concentration of 0.3 to 1.0 percent. The total amount of water is estimated to be between 10 million to 1 billion metric tons. If collected together, this water would fill a fair size lake and would be sufficient to meet the needs of 600 people for a century.

The presence of water ice increases the possibility that a lunar base will eventually be constructed. According to William Feldman of Los Alamos National Laboratory, the Moon's water could be recovered in a fairly easy manner. Ice-laced lunar soil could be placed inside a closed chamber and heated using solar panels placed above the dark polar craters or by a nuclear power plant. "It'll be like making

---

Focus Box: The Clementine Images of the Lunar Poles

These pictures are composite images obtained by the Clementine spacecraft of the Moon's north (A) and south (B) poles. More than fifty separate images are summed together that span an entire lunar day. Bright areas show regions that are illuminated by sunlight during at least part of a lunar day. Dark regions are areas where sunlight does not hit the surface. Dark regions exist at both poles but more are present at the south pole. Ice might be present in some of these dark regions.

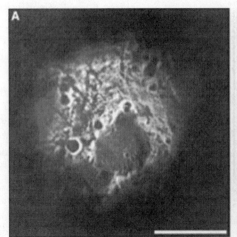 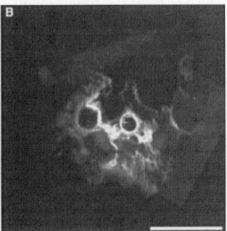

moonshine," Feldman said. The liquid water obtained from this process could then be transported to the lunar base by a variety of means. Once broken down into its component atoms, the lunar water could also be used to generate electricity, make breathable air, and serve as a rocket propellant. Feldman has also found that the left-over dirt from this process could then be used to make glass-like bricks which are strong enough to construct additional buildings for the Moon base.

What Does the Future Hold?

Until each of the options for a Moon base is fully explored, it is impossible to determine what this facility will look like or even where it will be located on the Moon. However, the case for an astronomical observatory seems particularly compelling. What might such an observatory look like?

The first lunar observatory will almost certainly consist of small or modest-sized instruments designed to study celestial objects at x-ray, ultraviolet, infrared, and radio wavelengths. Much of this light is shielded from astronomers by the Earth's atmosphere so the Moon serves as a natural platform from which to acquire these observations. Currently, studies at these wavelengths are conducted from spacecraft. X-ray and gamma-ray detectors are used to investigate exotic objects such as gamma-ray bursts, black holes, pulsars, and binary systems containing a neutron star and normal star. These objects emit prolific amounts of high-energy emission. By measuring the properties of this emission, astronomers can obtain information about these objects that cannot be acquired through any other means.

Initially, ultraviolet and infrared work at a lunar observatory might focus on completing a survey of the entire sky.

Figure 42-4. In this artist's drawing, minerals are being mined from the lunar surface for use in construction projects.

These studies might then be followed by investigations of interesting objects identified by that work. Next, long-term projects analyzing the light from variable stars or extragalactic objects might be undertaken. The long lunar night (fourteen Earth days) will make it possible to observe these sources for longer periods than are possible from a single location on Earth.

An important addition to a lunar observatory would be a low-frequency radio telescope. The Earth's atmosphere reflects these radio waves back into space. Therefore, astronomers have never been able to see what the universe looks like at these wavelengths. About 200 simple antennae could be spread over a 20-kilometer diameter circle and their signals could be electronically combined at a single location. Solar outbursts, the magnetic fields of the planets within the solar system, pulsars, the interstellar medium, and external galaxies, are likely to be strong low-frequency emitters.

After the lunar observatory has been functioning for many years, its instruments will undoubtedly be made larger and more sensitive. A 16-meter optical-infrared telescope capable of seeing objects forty times fainter than can be detected by the Hubble Space Telescope might be built. A 10-kilometer diameter array of forty-two ultraviolet telescopes, producing a resolution 100,000 times better than obtainable from a single instrument might also be built. This lunar telescope would be able to determine whether a coin located 250,000 miles away was a dime or a penny. Images of Jupiter and Saturn produced by this array would surpass even those obtained by the Voyager spacecraft that flew by these planets. Planets around nearby stars could be directly imaged and subtle motions in the atmosphere's of stars could be measured and analyzed.

The construction of a Moon base would constitute an important milestone in human exploration and today's technology is adequate to compete this task. Such a facility would be indisputable proof that the human race is willing to leave the cradle of Earth and begin its journey into space.

Figure 42-3. An artist's drawing of a lunar outpost. Work compartments are buried underground to protect their occupants from the hostile lunar environment.

43 Mission to Mars

If we have a spiritual need for a new frontier—and I believe we do—Mars is it. If there is a migratory drive with us—and I believe there is—it will lead us to Mars. If there is an extraterrestrial imperative, Mars is surely the next logical stepping stone on the endless journey to the stars.

—**Michael Collins in** *Mission to Mars*

The first realistic proposal to send an expedition to Mars dates back to a series of articles that appeared in Collier's magazine in 1952. The most detailed of these was "The Mars Project," written by Wernher von Braun. Despite the burst of excitement these articles created, the arms race with the Soviet Union left America with little time to tackle the technical problems such a journey presented. During the 1960s and 1970s NASA was preoccupied with winning the Moon race and constructing the Space Shuttle. Another decade would pass before an expedition to Mars was seriously reconsidered.

The Decision to Go to Mars

In 1985 President Reagan appointed a fifteen-person panel chaired by former NASA Administrator Thomas O.

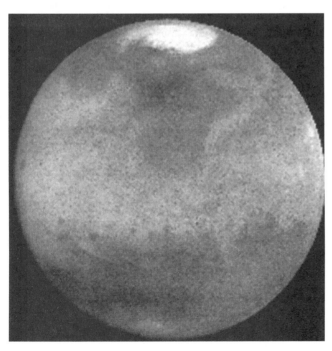

Figure 43-1. The planet Mars will be one of the first objects in the solar system to be explored by humans once they begin their journey into space.

Paine. Named the National Commission on Space, its goal was to recommend directions for America's space effort in the twenty-first century. The committee's findings were published in a 200-page report, "Pioneering the Space Frontier." It recommended projects in three broad areas: (1) satellite-based studies of the Earth, the solar system, and the universe; (2) manned expeditions to the Moon and the planets; and (3) commercial endeavors aimed at exploiting the unique environment of space. The committee concluded that one of NASA's long-term goals should be to build an outpost on Mars, and a target date of 2016 was given for the completion of this task. Shortly after the Paine report was released, an independent NASA advisory committee chaired by Daniel Fink, a former executive with General Electric, issued a separate paper stating the exploration of Mars should be the agency's "primary goal."

Progress toward a Mars mission was temporarily put on hold when NASA reconsidered its long-term objectives after the Shuttle Challenger disaster of January 28, 1986. A third committee, headed by former astronaut Sally Ride, was formed to review NASA's near and long-term missions. Using the Paine report as a starting point, the Ride Committee issued a report in August 1987 titled, "Leadership and America's Future in Space." Four programs were recommended: (1) Mission to Planet Earth; (2) Exploration of the Solar System; (3) Outpost on the Moon; and (4) Humans on Mars. The long-term objective of this program was to establish a Martian base. Under initiative 2 three probes would land on Mars and return soil samples to Earth. This information would then be used in initiative 4 to mount manned expeditions. In the words of the report, "We should adopt a strategy of natural progression which leads step by step, in an orderly, unhurried way, inexorable toward Mars."

Despite the endorsement of the Ride Committee, support in the space community for a future manned Mars expedition was not unanimous. Some members thought resources could be better spent investigating the Earth from space and that automated probes would be a more cost-efficient way of exploring Mars. Christopher Kraft, retired director of the Johnson Space Center, argued for a program centered on Earth observations and unmanned robotic planetary probes. When asked about a manned Mars mission he replied, "I believe our goals should be more modest and more aimed at solving the problems we have on Earth. The rate of progress in robotics and automation will make unmanned expeditions to the planets more productive, will provide strong technological stimulation, and will be considerably less expensive than sending humans along to do the job." John Logsdon, director of the Space Policy Institute at George Washington University, urged that a lunar base be built before proceeding to Mars. He wrote,

> Establishing a permanent base on the Moon is an essential stage of expansion beyond Earth. Human beings should go to Mars, but it does not make sense to bypass the Moon, a celestial island just offshore that offers significant knowledge for colonizing space. . . .

Advocates of a Mars mission disagreed with these positions. President Bush noted, "Men and women do not follow machines; they follow great men and women." Carl Sagan, a noted planetary astronomer, expressed the opinion that "a lunar base wouldn't be a detour on the road to Mars, but a trap. We would use up [financial] resources and indefinitely delay going to Mars. Mars is so much more exciting."

Despite these conflicting opinions, a consensus was forming within NASA that a Mars landing would be the agency's primary objective for the next century. Ironically, the Challenger disaster solidified support for this position. In 1988 Michael Duke, a NASA lunar and Mars geologist, wrote,

> Two years ago, you couldn't find anyone in authority who would talk about missions to Mars. That environment, a fear of ideas that might be considered too far-out, is definitely changing. After the Challenger accident, we got criticized for not being safe enough, but also for not being aggressive enough. We began to recognize that what we're doing is exploration, not simply transportation, and we ought to address ourselves to exploration goals.

In early 1989 President Bush signaled his support for an expanded space effort in an address to a joint session of Congress. He declared,

> In very basic ways, our exploration of space defines us as a people—our willingness to take great risks for great rewards, to challenge the unknown, to reach beyond ourselves, to strive for knowledge and innovation and growth. Our commitment to leadership in space is symbolic of the role we seek in the world.

To demonstrate his commitment to this vision he asked Congress to increase NASA's budget by 22 percent. Six months after this address, on July 20, 1989, the 20th anniversary of the Apollo 11 landing, Bush called upon the nation to accept ambitious new goals for its space program. In a speech at the Air and Space Museum in Washington, DC, he said,

Figure 43-2. The Sun on the Martian horizon as photographed by one of NASA's Viking landers.

> Today the United States is the richest nation on Earth—with the most powerful economy in the world. And our goal is nothing less than to establish the United States as the preeminent spacefaring nation. I'm not proposing a ten-year plan like Apollo. I'm proposing a long-range continuing commitment . . . first . . . for the 1990s . . . Space Station Freedom . . . next—for the next century—back to the Moon . . . back to stay. And then . . . a manned mission to Mars.

Differences in Direction between Apollo and the Proposed Mars Expedition

During the 1960s the Apollo program was portrayed as a race against the Soviet Union and then as an effort to collect scientific data about the origin and evolution of the Moon. Once the Moon race had been won and a substantial amount of lunar data had been collected, it was not evident what new results more flights would produce. NASA had never advocated building a lunar outpost, so Apollo was viewed as a completed program. *Apollo 11* astronaut Michael Collins verbalized America's lack of interest in sending additional expeditions to the Moon, "It was like the same Super Bowl being broadcast time after time. . . . When Apollo was over, it was over. It was not perceived as a gateway to the future but as an end in itself." In contrast to the Apollo program, NASA has gone to great lengths to emphasize that the goal of its Mars program is not just to visit Mars, but to establish a working colony.

Comparisons between the Earth and Mars

Mars and the Earth share many common features. Both of these planets possess a rocky surface, an atmosphere, and have similar densities. However, important differences exist. The atmosphere on Mars is quite thin, roughly equivalent to the density of air 24 miles above the Earth's surface. The

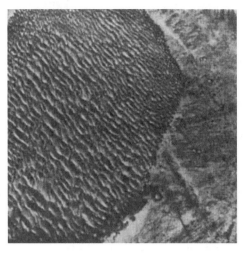

Figure 43-3. A field of sand dunes on the Martian surface.

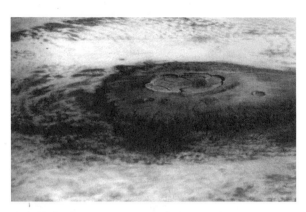

Figure 43-4. The Martian extinct volcano Olympus Mon. Light clouds are seen clinging to its sides.

Martian atmosphere also consists of 95 percent carbon dioxide, 2.7 percent nitrogen, 1.6 percent argon, and 0.15 percent oxygen. Conversely, the Earth's atmospheric composition is 78 percent nitrogen and 21 percent oxygen. The air of Mars is too thin and composed of the wrong gases to be breathable by humans. The atmospheric pressure on the surface of Mars is so low liquid water cannot exist on its surface—it would immediately boil into a vapor.

Mars is one-half the radius of the Earth, but because it lacks oceans has nearly the same amount of land area. Mars' gravity is about 0.38 that of the Earth. A person who could jump 2 feet on Earth could jump 5 feet on Mars. On average a day on Mars is 37 minutes longer than our day. The rotation axis of the Earth and Mars are tilted by similar amounts (25.2 versus 23.5 degrees) so both planets have seasons. Since a year on Mars is equivalent to 1.88 years on Earth, making the

Martian seasons are about twice as long as those on Earth. During summer, the highest temperature on the equator of Mars is a rather pleasant 72°F, but during the winter, temperatures of −150°F are common. Mars also has weather. When closest to the Sun, vast dust storms with wind speeds of 120 miles per hour can cover the planet. Over time the winds have produced huge fields of sand dunes (see Figure 43-3). Mars also possesses polar ice caps, mountainous regions, flat plains, valleys, and large canyons. The Martian ice caps contain both water ice and frozen carbon dioxide.

The largest known volcano in the solar system, Olympus Mons, is located on Mars (see Figure 43-4). This gently slopping, extinct volcano has a diameter of 340 miles and a height of 15 miles, roughly three times the diameter of Mauna Loa in Hawaii, and three times higher than Mount Everest. If placed in the United States, Olympus Mons would cover the state of

Focus Box: Properties of the Earth and Mars

Property	Earth	Mars
Distance from the Sun (Earth = 1)	1.00	1.52
Equatorial Diameter (miles)	7,926.0	4,221.0
Mass (Earth = 1)	1.000	0.107
Density (Water = 1)	5.52	3.93
Surface gravity (Earth = 1)	1.00	0.38
Length of a year (days)	365.25	686.98
Solar day (hours)	24.00	24.62
Escape velocity (Earth = 1)	1.00	0.45
Mean surface temp. (°F)	59.	−74.
Surface pressure (mbar)	1,000.	6.36

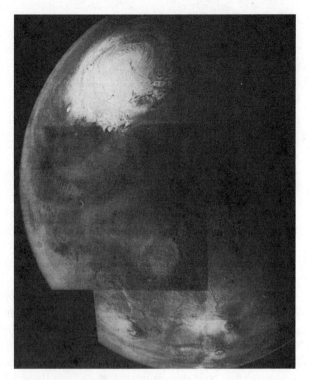

Figure 43-5. A mosiac picture highlighting the north polar cap and volcanos (lower center) on Mars. These views were obtained by U.S. spacecraft Mariner 9 *in 1973.*

Figure 43-6. The Mariner Valley as photographed by Mariner 9. *The valley is seen in the lower central portion of this picture.*

Arizona. Its summit caldera is 45 miles across, large enough to accommodate the entire state of Rhode Island.

A second equally impressive Martian feature is the Valles Marineris (Mariner Valley). This canyon is huge by Earth standards. It extends for 1,500 miles, has a maximum width of 400 miles, and a depth of 4 miles. If placed on the Earth, it would stretch from New York City to Los Angeles. Along its sides are numerous smaller canyons, comparable in size to the Grand Canyon in Arizona.

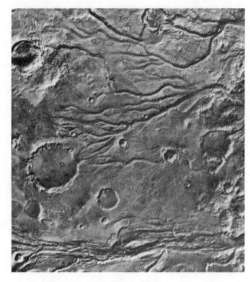

Figure 43-7. NASA scientists believe this series of dry riverbeds on Mars was formed by running water.

Many scientists believe that Mars was once a wetter place. Although no liquid water can currently exist on Mars, dry riverbeds, much like the arroyos in the American southwest, are seen on its surface (see Figure 43-7). The largest dry Martian riverbed is the 210 mile long and 60 mile wide Amazonis channel.

Why Go to Mars?

Several reasons why we should go to Mars are discussed below.

1. An expedition to Mars would provide a long-term investment in our technology. Profit has frequently been used to justify the costs of exploration. The first expedition authorized by the U.S. Congress, the Lewis and Clark Expedition, discovered new flora and fauna, but it did not produce immediate riches. The knowledge of the western geography that resulted benefited future generations. The advances made possible by a Martian expedition will also likely be of a long-term nature. In his book, *Mission to Mars*, Michael Collins supports a Mars expedition,

> I think the government has an obligation to its citizens not only to fight today's battles but to look to tomorrow, to assure that the country is headed in a direction that will create higher living standards. Government investment in space exploration is a stimulus to our high-tech economy. . . . America has been synonymous with opportunity, and the ability to go out into space for knowledge or profit continues that tradition.

2. It would provide new scientific knowledge about the development of life. Biologists are faced with a difficult situation when attempting to study processes that lead to the de-

velopment of life. Outwardly a tremendous variety of life exists on Earth, but at the molecular level all life is based on RNA and DNA. Because of this, it is nearly impossible to make general statements about how life might arise or adapt to various sets of conditions.

Two planetary probes, *Viking 1* and *2*, were sent to Mars in 1975 to conduct chemical and biological tests for the presence of life. After a journey of nearly a year, both landed successfully. Photographs from these probes were able to immediately rule out the possibility that larger forms of life exist on Mars. In addition no unusual compounds, perhaps liberated by life processes, were detected in the Martian atmosphere. Biology tests conducted at both landing sites, did not find conclusive evidence for microbial life. Despite these negative results, scientists were not dismayed. The Viking spacecraft sampled just two locations on Mars. Interesting places like the dry river beds or polar regions were not examined. Much work needs to be done before Mars can be declared to be a lifeless world.

3. It would fulfill the human need to explore. In *Mission to Mars* Collins argues a Martian expedition is a logical progression of the human desire to explore. He writes, "To me, the *why* of a Mars mission is rooted in the history of our planet and of this nation. . . . The urge to go, to see, to touch, to smell, to learn—that is the essence of it, not to mention the exhilarating possibility of encountering something totally unexpected."

Stephen Pyne, a historian at Arizona State University, also believes the urge to explore is a deeply rooted human trait. He argues that throughout history we have adapted new technologies to satisfy this need to explore, "We explore not because it is in our genetic makeup but because it is within our cultural heritage."

4. Exploration is a natural activity of a dynamic society. A nation's unwillingness to explore is often a sign of a stagnating or declining society. A century before Columbus set sail, China had the means of reaching America. Their ships had crossed the Indian Ocean and landed on the east coast of Africa. However, in 1435 China's new emperor decided to use the funds allocated for exploration to fight the Mongolian wars. In his book *The Discoverers*, Daniel Boorstein discusses how this decision stymied China's development as a force in the world, "Fully equipped with the technology, the intelligence, and the national resources to become discoverers, the Chinese doomed themselves to be the discovered."

5. It would deprovincialize our view of the Universe. In 140 A.D. the Greek astronomer Claudius Ptolemy published an Earth-centered model of the solar system in his book, the *Almagest*. In this model, the Sun and all of the planets revolved around the Earth. Today, we know the Earth revolves around the Sun, but we often act as though the latter were true. Our language is laced with expressions such as "sunrise" and "sunset." Frank White, a social scientist, writes that an entirely different view of the Universe might arise once we journey into space. In his book *Overview Effect*, he writes, "going into space is not about a technological achievement, but about the human spirit and our contribution to universal purpose. Space . . . is a metaphor of expansiveness, opportunity and freedom. More than a place or even an experience, it is a state of mind. It is a physical, mental, and spiritual dimension in which humanity can move beyond the current equilibrium point, begin to change, and eventually transform itself into something so extraordinary that we cannot even imagine it." By moving into space, humans will finally be able to see the Earth in its true context.

Focus Box: Trajectories to Reach Mars

Using the method shown below, a spacecraft is launched from Earth and later encounters Mars. This method, employed by the U. S. satellite *Mariner 4*, requires a travel time of about 9 months but is fuel efficient.

A second way to reach Mars is to follow the trajectory shown below. The craft arrives at Mars in 6 months. The crew would stay at Mars until the planet reaches portion (2) and then return to Earth.

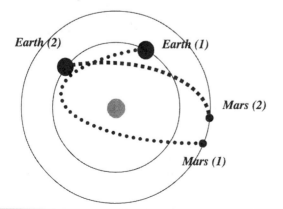

1.25 Year Round Trip

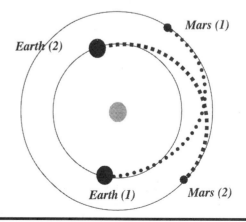

2.6 Year Round Trip

1960 Oct. 10	*Mars 1960A*	USSR	Launch failure
1960 Oct. 14	*Mars 1960B*	USSR	Launch failure
1962 Oct. 24	*Mars 1962A*	USSR	Fails to leave Earth orbit
1962 Nov. 1	*Mars 1*	USSR	Contact lost during transit
1962 Nov. 4	*Mars 1962B*	USSR	Fails to leave Earth orbit
1964 Nov. 5	*Mariner 3*	USA	Mechanical failure
1964 Nov. 28	*Mariner 4*	USA	Success, 22 pictures
1964 Nov. 30	*Zond 2*	USSR	Contact lost during transit
1965 July 18	*Zond 3*	USSR	Success, 25 pictures
1969 Feb. 25	*Mariner 6*	USA	Success, 75 pictures
1969 Mar. 27	*Mariner 7*	USA	Success, 126 pictures
1969 Mar. 27	*Mars 1969A*	USSR	Launch failure
1969 Apr. 2	*Mars 1969B*	USSR	Launch failure
1971 May 8	*Mariner 8*	USA	Launch failure
1971 May 10	*Cosmos 419*	USSR	Fails to leave Earth orbit
1971 May 19	*Mars 2*	USSR	Orbits Mars, lander crashes
1971 May 28	*Mars 3*	USSR	Orbits Mars, lander successful
1971 May 30	*Mariner 9*	USA	Success, 7,329 images
1973 July 21	*Mars 4*	USSR	Fails to orbit Mars
1973 July 25	*Mars 5*	USSR	Success, pictures/data
1973 Aug. 5	*Mars 6*	USSR	Orbits but lander fails
1973 Aug. 9	*Mars 7*	USSR	Flyby, lander failure
1975 Aug. 20	*Viking 1*	USA	Success, search for life
1975 Sept. 9	*Viking 2*	USA	Success, search for life
1988 July 7	*Phobos 1*	USSR	Flight failure
1988 July 12	*Phobos 2*	USSR	Flight failure
1992 Sept. 25	*Mars Obs.*	USA	Flight failure
1996 Nov. 16	*Mars 96*	Russia	Launch failure
1996 Dec. 4	*Mars Pathfinder*	USA	Success, deploys rover
1998 July 4	*Planet B*	Japan	Arrives 2003
1998 Dec. 10	*Mars Climate Orbiter*	USA	Fails to orbit Mars
1999 June 3	*Mars Polar Lander*	USA	Failed.

Getting to Mars

The Earth–Mars distance varies from 35 to 235 million miles. The amount of time needed to complete a journey to Mars will be a compromise between the velocity at which the craft travels and its fuel consumption. The most fuel-efficient path to Mars is one that smoothly leaves Earth's orbit and joins Mars' orbit (see Focus Box, p. 345). A craft following this trajectory would take about nine months to reach Mars. After arriving at the planet, the crew would then have to wait a year and a half until Mars and Earth were similarly aligned for the return trip. Using this method, the entire journey would take 2.6 years to complete.

A second way of reaching Mars would be to send a supply craft ahead of the manned vessel. Once the cargo craft arrives at Mars, the manned ship would then be sent along a fast, but fuel-inefficient trajectory, arriving in about seven months. If the crew then spent one month exploring the planet's surface and returned to Earth using fuel stored in the cargo ship, the entire mission would take about fifteen months. This plan greatly reduces the mission duration, but it possesses a major risk. When the expedition arrives at Mars, the ship must be refueled. If, for whatever reason, this cannot be done, the crew would perish.

A third way to travel to Mars would be to use the gravity of Venus to increase the ship's velocity. In this plan the expedition's ship would make a close approach to Venus to increase the ship's speed and then continue on to Mars. Al-

though this procedure seems rather complex, these types of "gravity boosts" are often employed by NASA for their planetary probes. If this trajectory was employed, the journey to Mars would take about eleven months. Assuming the crew spent two months exploring the planet and then returned to Earth using a minimum energy orbit, the entire mission could be completed in twenty-two months.

Crew Sustenance

While in space a person consumes 12 pounds of supplies per day—3 pounds of food, 7 pounds of water, and 2 pounds of oxygen. Assuming an eight-person crew for a Mars expedition and a twenty-two-month mission, 65,000 pounds of supplies, roughly equal to the weight of the Apollo command and service module, will be needed to sustain the crew. This estimate is a lower limit since it does not include water for such things as showers or laundry. The large quantity of needed supplies will insure that the crew size is small. Once at Mars the same procedure used during the Apollo program will probably be employed. Part of the crew will monitor the main ship while their crewmates journey to the surface.

Cost of the Mission

It is difficult to estimate costs for a mission whose planning and execution could take up to fifteen years. Students at the University of Michigan estimated the cost (in 1984 dollars) to be around $22 billion. The California-based Planetary Society commissioned a study that found this cost to be $40 billion. The Ride Committee arrived at a similar figure of $45 billion. Roald Sadeev, former head of the Soviet Institute for Space Research, puts the cost at $50 to $100 billion. Michael Collins thinks the cost might be as high as $200 billion if items such as a space station and Moon base are included as components of the Mars mission.

To put these numbers into perspective, in 1984 dollars, the entire Apollo program cost $74 billion and the military's B-2 bomber program cost $70 billion. A Mars expedition would be comparable to the cost of these programs. This is a considerable amount of money, but not out of line relative to other government programs. If spread over fifteen years, a Mars mission would cost less than the current U.S. per capita expenditure on cigarettes.

Conclusion

Predicting the future is often a futile exercise, but the assumption that a manned expedition to Mars will take place during the first half of the next century seems safe. NASA is scheduled to send an unmanned lander and orbiter to Mars in 2001. This craft will measure the depth of the Martian permafrost and attempt to produce fuel from the Martian soil. In collaboration with the European Space Agency (ESA), NASA will also launch a series of missions beginning in 2003

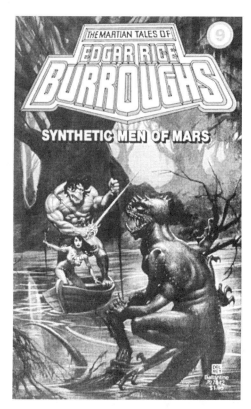

Figure 43-8. Edgar Rice Burroughs was able to capture the imagination of vast numbers of readers through his adventure-filled stories featuring strange Martian creatures.

or 2005 whose goal is to return material collected from Mars by the year 2008. The ESA also intends to send two of its own probes to Mars, the Mars Express and the Dynamo Micro-Orbiter, by 2005. This later vessel will gather data about the Martian magnetic and gravitational fields. Japan has already sent a probe, the Nozomi, toward Mars. It will go into orbit around that planet in 2003.

All of the above missions are a prelude to the main event, a manned mission. Favorable launch conditions exist in 2007, 2009, 2011, and 2014. Some scientists are cautiously optimistic that a mission will take place during one of those times. Raymond Ladbury, a contributing editor to *Physics Today*, expressed the following optimistic view in an October 1999 article, ". . . if the technologies now being evaluated prove viable and, most important, if the political will for such a mission holds, the first decade or so of the new millennium could see humans standing in the shadow of Olympus Mons."

At the present time it is not known what event or circumstances might cause us to make this type of commitment. Will it be driven by economic considerations that we are currently unable to foresee, or will it be driven by the "spirit of inquiry" Goddard wrote about in his youth? In *Mars Beckons*, John Wilford writes,

> . . . The first people who land on Mars will probably have no clearer idea of what will ultimately follow their achievement than did Columbus. . . . The first visitors to Mars will be constrained in their imagination by the technology at their command and the immediate motives that brought them there. . . . The vision that took [Columbus] across the sea into the unknown could not make him see that he had found a new world and understand where this would lead in a hundred years or 500 years. We should be excused, therefore, for not being able to imagine where Mars exploration will lead. . . .

Some journalists, like William Walter suggest the primary benefit of a Martian expedition will not be scientific but human. In *Space Age* he writes,

> Mars is not simply a planet, an astronomical curiosity, but a mystery and a beacon, a place as myth laden as Olympus or Valhalla or Shangri-la. The science fiction stories . . . made Mars a place where anything is possible, a locus of magic where we could play out our worst fears, greatest hopes, and most profound questions. Maybe that is why it seems inconceivable that we could ever forgo its exploration. We sense that in exploring Mars we may uncover more secrets about ourselves than anything else, and that it will lead, like the old myths themselves, to a better understanding of where we fit in the scheme of things.

Ultimately, it seems, Tsiolkovsky's prediction will be proven correct—mankind will not stay in the cradle forever.

Robert Goddard's dream of exploring strange new worlds will also be realized, perhaps sooner than many believe. The desire to travel into space is too deeply inbedded in our soul to be resisted for long. The technological barriers historically making space travel impossible have been removed. The most formidable remaining obstacle is our own hesitancy. Like a newborn bird, we are reluctant to leave the nest. Nevertheless, our inexorable journey into the final frontier has begun. Can the human departure from mother Earth be far behind?

Geological Results of the Apollo Program

The very success of the operations and the advances in knowledge gained tend to make us forget their earlier obscurity.

—Stuart Ross, *Lunar Science: A Post-Apollo View*

Between 1964 and 1968 America launched nine Rangers and five Lunar Orbiters to conduct a photographic reconnaissnace of the Moon. These craft mapped the Moon's entire surface with unprecedented resolution and strengthened the view that the Moon's surface was firm enough to support a manned landing craft. Five Surveyor missions soft-landed on the Moon's surface and provided additional evidence that a thick layer of lunar dust was not present. Based upon these findings NASA concluded that it was safe to send a manned expedition to explore the Moon. Over the next four years, six Apollo crews landed on the lunar maria and highlands. The names of many of these sites—Tranquility Base, Fra Mauro, and the Hadley Rille—are forever engraved in the memories of people who lived in that era.

The Apollo expeditions were impressive technological feats that also fulfilled the desired political goal of increasing the world's respect for America. But did these missions significantly increase our scientific knowledge about the Moon and its history?

Lunar Surface Features

Even with the unaided eye, it is obvious that the Moon's surface consists of three distinct regions, large dark areas called maria, bright mountainous regions called highlands, and craters. From our perspective on Earth, the maria resemble giant lunar parking lots paved with asphalt. Their circular shape was taken as evidence that they were created by enormous impact events. The Oceanus Procellarum and the Imbrium Basin have diameters of 1900 and 900 miles, respectively, and are larger than most American states. Using telescopes on Earth, astronomers see narrow crevasses called rilles along the boarders of the maria, but oddly few craters are seen on their surfaces.

The nature of the dark material that cover the maria was a matter of intense scientific debate. Some researchers thought it was acquired when the Moon orbited the Sun as a solitary object. According to this theory, the Moon was later captured by the Earth. Others felt the maria were covered by cooled molten rock. They hypothesized that early in the Moon's history, it was struck by an asteroid-sized object. This collision released

Table A1-1. Apollo Missions		
Mission	**Launch Date**	**Landing Site**
Apollo 11	July 16, 1969	Mare Tranquillitatis
Apollo 12	Nov. 14, 1969	Oceanus Procellarum
Apollo 14	Jan. 31, 1971	Fra Mauro
Apollo 15	Aug. 26, 1971	Hadley-Apennines
Apollo 16	Apr. 16, 1972	Descartes
Apollo 17	Dec. 7, 1972	Taurus-Littrow

such an enormous amount of energy that the surface rocks were melted. As the fluid-like material hardened it assumed the flattened surface we see today. Other scientists agreed that the maria were covered by cooled lava but thought that this material had been extruded onto the surface from the Moon's interior. This later idea was particularly attractive because, if true, the material covering the maria contained information about the Moon's interior composition.

In contrast to the maria, the lunar highlands are heavily cratered and appear lighter in color. Mountain ranges stretching for hundreds of miles and individual peaks reaching heights of 20,000 feet are found in these regions. Small analogues of the maria, called Cayley Formations, are also present. The highland regions were thought to be composed of pulverized material from the Moon's original crust.

Several prominent craters are located on the near side of the Moon, including Copernicus, Tycho, and Kepler. These craters have diameters of 55, 52, and 20 miles and depths of 2.0, 2.8, 1.4 miles, respectively (see the Focus Box, p. A-2). Bright streaks of material extend radially outward from the centers of these craters, presumably consisting of ejected material. Not all craters on the Moon are large. Telescopic observations show that the number of craters increases as their size decreases. This observation suggested the Moon's surface might be covered by innumerable small craters that could pose a danger during a manned landing.

Scientists had debated the origin of the lunar craters for over 300 years and in the early 1960s were still divided over

Maria
1. Mare Imbrium
2. Mare Serenitatis
3. Mare Trabquillitatis
4. Mare Crisium
5. Mare Fecunditatis
6. Mare Nectaris
7. Mare Nubium
8. Mare Humorum
9. Mare Cognitum
10. Oceanus Procellarum

Craters
a. Plato
b. Tycho
c. Copernicus
d. Kepler
e. Archimedes SR

whether these objects had been created by volcanic action or by the impact of meteorites.

The primary evidence supporting the volcanic origin of craters was their round shape. They look like the collapsed calderas of terrestrial volcanos. Advocates of this hypothe-

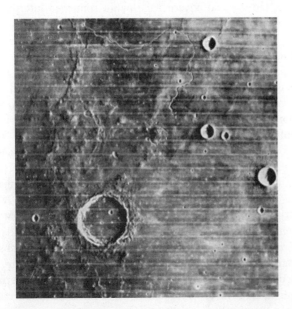

Figure A1-1. Nearly all of the craters on the moon have a very circular shape. Advocates of the volcanic origin of craters argued that impacts should create elongated craters.

sis argued that if the craters had been formed by infalling objects, then most would have elongated shapes, reflecting the wide range of angles at which the Moon was struck. Since few elongated craters are seen (see Figure A1-1), this observation seemed to pose a severe problem for the impact model.

Advocates of the impact origin of the lunar craters offered five pieces of evidence to support their view, most of which were incompatible with the volcanic hypothesis. First, the ejecta patterns associated with many of the large craters on the Moon resembled those created in laboratory high-velocity impact experiments. The collapse of a caldera, on the other hand, does not produce such a widespread deposit. Secondly, a direct relation is expected to exist between a crater's diameter and depth if it is created by impact events. Careful ground-based measurements, such as those plotted in Figure A1-2, show that the depth and diameter of the Moon's craters are closely related. No such relation is expected if the craters had a volcanic origin. The size of the ray patterns extending from several large craters provided a third piece of evidence supporting the impact model. During a collision sufficient energy is released to eject material over the vast distances on the Moon. However, if the craters were formed by volcanic processes, the energy released would not be adequate to produce these patterns. Fourth, in an impact event a central peak is often created at the center of the crater as the compacted material rebounds from the collision. Ad-

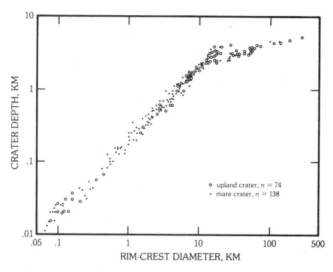

Figure A1-2. Craters on the moon follow a well-defined depth/diameter relation. This is expected if craters were formed from impact events.

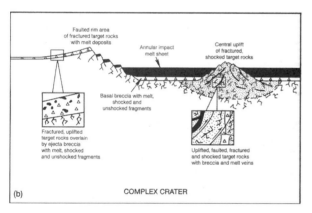

Figure A1-3. When an impact event creates a crater, a central peak and raised rim result. The energy released during the collision fractures the surface and lava floods the crater floor.

Figure A1-4. When an impact event occurs, the crater that is formed will often have a central peak or a ring of debris on its floor. As shown above, many lunar craters have these properties.

vocates of the impact hypothesis could show picture after picture of large lunar craters that possessed prominent central peaks. Additional experiments also demonstrated that the slumped walls and circular deposits seen on crater floors could also be explained by a high-velocity impact (see Figure A1-4). Neither of these features is predicted to exist in the volcanic hypothesis. Lastly, advocates of impact hypothesis found that circular craters, the primary evidence supporting the volcanic hypothesis, are produced in high-velocity impact tests *regardless* of the angle at which the projectile strikes the surface.

The strong evidence supporting the impact origin of lunar craters caused one prominent scientist in the mid-1960s to declare that the "battle is over." However, not everyone was willing to accept this theory. As we have seen, the search for volcanic craters played an important role in the selection of some Apollo landing sites, notably *Apollo 15* and *16*.

The Ages of Lunar Features

Before the mid-1960s, scientists estimated relative ages of regions on the Moon by using crater counts. If a region contained a large number of craters, it was considered to be old. Conversely, if few craters were seen, it was thought to be young. Crater densities in the highlands were higher by a factor of 32 when compared to the maria, indicating that the highlands were older. However, since the rate at which the Moon was hit by meteorites was unknown, this method only yielded *relative* age sequences. In other words, it was possible to estimate which features were older and which were younger, but it was not possible to assign them absolute ages.

A second method used to identify age sequences on the Moon was based on an examination of the lunar topology. The southeastern section of Mare Imbrium, shown in Figure A1-5, can be used to illustrate how this technique works. The Apennine Mountains, extending from the lower left to the

Figure A1-5. This view of the southeastern section of Mare Imbrium shows the Apennine Mountains (lower left to middle right), the lava-filled crater Archimedes (top center), and the craters Autolycus and Aristillus (darkened craters at top right).

middle right in this picture, form a portion of the Maria Imbrium's border. The order in which the identified features are believed to have formed is (1) the Imbrium Basin and the Apennine Mountains, (2) the crater Archimedes, (3) the Imbrium Basin lava fill, and lastly, (4) the crater Aristillus.

Scientists arrived at this sequence using the following logic. First, an asteroid-sized object strikes the Moon, blasting out a huge hole (the Imbrium Basin) and uplifting the Apennine Mountains along its rim. They hypothesize that the Moon's structure was deeply fractured by this impact and lava began to seep to the surface from the interior. After the basin was created, but before it filled with lava, a large meteor struck the basin, forming Archimedes. Lava emanating from deep basin fissures continued to flood this region, covering ejecta from Archimedes and seeping through the crater's rim. After the basin was filled and the lava solidified, an additional meteorite struck the surface, forming the crater Aristillus. Material ejected from this crater then fell on the hardened maria where it is still visible today.

Using this approach, researchers identified several formation sequences on the Moon. For instance, they deduced that the craters Tycho, Copernicus, and Kepler were created long after the maria and the highlands formed. However, as in the case of crater counts, this process does not allow absolute ages to be assigned; only relative formation sequences are obtained.

The Apollo Experimental Program

The Apollo missions were divided into three categories called G, H, and J, depending on their objectives. *Apollo 11* was the only G mission. Its goal was to collect a sample of lunar rocks and soil and deploy a small experimental package. *Apollo 12*, *13*, and *14* were H missions. They carried completely integrated ALSEPs and included two long Extra Vehicular Activities (EVAs). The three remaining flights, *Apollo 15*, *16*, and *17*, were J missions. They carried extensive experimental packages and three EVAs were conducted.

In order to increase our knowledge about the geology of the Moon, the astronauts landed on a variety of terrains. *Apollo 11* and *12* explored the lunar maria, *Apollo 14* investigated the ejecta blanket produced by the creation of the Imbrium Basin, *Apollo 15* landed near the Hadley Rille, *Apollo 16* set down in the lunar highlands, and the *Apollo 17* site contained both highland and maria material.

With each Apollo mission, the tools used to explore the Moon became more sophisticated. On *Apollo 11*, astronauts were only able to collect samples from within 100 feet of their landing site. On *Apollo 12*, the astronauts explored a larger region but their excursions were limited by the amount of oxygen in their backpacks. The astronauts used a hand-drawn cart on *Apollo 14* to transport equipment and samples to more distant locations. On *Apollo 15*, *16*, and *17*, the astronauts were able to travel several miles by using an electrically powered vehicle called the Lunar Rover. On these missions thorough geological investigations of the landing sites were performed.

While an astronaut team was busy completing tasks on the lunar surface, the third member of their crew obtained global observations of both the near and far side of the Moon from the orbiting Apollo spacecraft. Special instruments within the exposed side of the service module acquired high-resolution photographs that revealed objects as small as a desk while other sensors gathered mineralogical data and measured the Moon's gravitational field.

The 862 pounds of lunar rocks and soils returned from the Moon were divided into nearly 35,000 samples. Many were immediately sent to scientists around the world for study, but the bulk were kept in a special isolation chamber at the Johnson Manned Space Flight Center in Houston for later analysis. A list of the surface and orbital experiments carried out on the Apollo lunar missions is provided in the Focus Box on page A-5.

Scientific Results of the Apollo Program

The Apollo missions provided new data about the Moon's maria, highlands, craters, and interior structure. The insights gained in each of these areas is discussed below.

The Lunar Maria: Two Apollo missions, *Apollo 11* and *12*, visited lunar maria. Mare Tranquillitatis was chosen for the first manned landing because it was relatively flat and its location allowed the astronauts to be in constant communica-

Focus Box: Apollo Surface Experiments

Experiment	11	12	Mission 14	15	16	17
Passive seismic experiment	X	X	X	X	X	
Active seismic experiment			X		X	
Lunar surface magnetometer		X		X	X	
Solar wind spectrometer		X		X		
Suprathermal ion detector		X	X	X		
Heat flow experiment				X	X	
Charged particle lunar environment			X			
Cold cathode ion gauge		X	X	X		
Lunar field geology	X	X	X	X	X	X
Solar wind composition	X	X	X	X	X	
Laser ranging retroreflector	X		X	X		
Lunar dust detector		X	X	X		
Portable magnetometer			X		X	
Lunar gravity traverse						X
Soil mechanics			X	X	X	X
Far-UV camera/spectroscope			X			
Lunar ejecta and meteorites						X
Lunar seismic profiling						X
Surface electrical properties						X
Lunar atmospheric composition						X
Lunar surface gravimeter						X
Lunar neutron probe						X

tion with Earth. Officials thought that rocks collected from this site would provide data about the composition and age of the maria while minimizing the danger the crew faced.

If any concerns remained about the depth of the dust layer on the lunar surface, they were put to rest when *Apollo 11* astronaut Neil Armstrong set foot on the Moon. He radioed to Mission Control, "Yes, the surface is fine and powdery. I can kick it up loosely with my toe. It does adhere in fine layers like powered charcoal to the sole of my boots. I only go in a small fraction of an inch. . . ." He provided further evidence that the surface was firm when he reported, "The descent engine [of the lunar module] did not leave a crater of any size."

Core samples returned by the *Apollo 11* astronauts showed that the surface material was well mixed and homogeneous. The top layer was about 1 inch thick and consisted of loosely packed, porous material. Core samples from a depth of 12.5 inches showed that the density of the soil increased with depth, starting at a small value of 1.5 g/cm^3 and increasing to about 1.8 g/cm^3. For comparison, the density in the Earth's crust is about 2.7 g/cm^3.

The rocks returned by the astronauts were basalts, establishing beyond any doubt that the maria were hardened lava beds. But was this lava extruded from the Moon's interior or was it created from heat released during the impact of a mas-

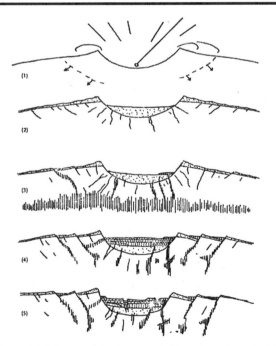

Figure A1-6. In panel (1) the maria is blasted out from the Moon's surface by the impact of a large asteroid-sized object and a shock wave propagates into the Moon; (2) debris fills the basin and the Moon's substructure is fractured; (3) subsurface magma flows into the basin; (4) major faulting produces concentric cliffs; (5) additional faulting and maria extrusions produce the present maria.

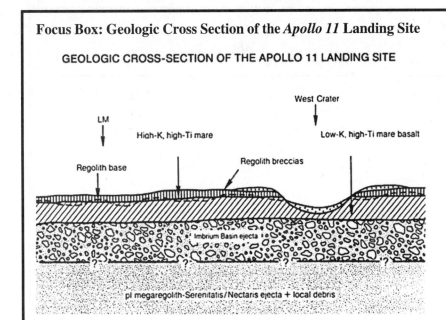

GEOLOGIC CROSS-SECTION OF THE APOLLO 11 LANDING SITE

Cross-section maps for each Apollo landing site were constructed by lunar geologists. This particular map of the *Apollo 11* landing site shows two different lava flows, labeled High-K, high-Ti mare and Low K, high-Ti mare basalt. Both flows cover ejecta from the Imbrium Basin. LM marks the landing position of the *Apollo 11* lunar module.

sive object? Lunar Orbiter photographs played an important role in answering this question by showing that none of the maria were covered by large amounts of debris. If the maria had been formed by impacts, some should have hardened before other basins were blasted out of the Moon. Ejecta from the newer basins would then have littered large portions of the older maria. The fact that no such deposits were found was disappointing to proponents of the impact/melt hypothsesis but they were not ready to give up the fight. They explained the lack of these features by postulating that all the basins were formed at essentially the same time. The material expelled by the basin-forming impacts simply sank into a Moon-wide sea of lava. This modified hypothesis naturally required that all of the maria have essentially the same age.

The alternative theory, that the basins were filled by lava long after the impact event, predicted that the subsurface lava flows had different ages. In this regard, the rocks returned from Oceanus Procellarum by *Apollo 12* proved decisive. These samples were several hundred *million* years younger than the rocks returned from Tranquility Base. The *Apollo 12* basalts also had a coarser texture, a much lower titanium abundance, and a redder color than those brought back by the crew of *Apollo 11*. This was conclusive evidence that all of the maria were not of the same age, different epochs of maria-filling had taken place on the Moon. Indeed chemical analysis of the *Apollo 11* samples suggested that at least two episodes of maria flooding had occurred within the Sea of Tranquility itself. As shown by the geologic cross section of this site in the Focus Box above, the first flow was low in potassium while the second was high in potassium. Apparently the maria did not condense from a single Moon-wide sea of lava as suggested by the advocates of the impact/melt hypothesis.

The relatively low quantity of volatile (easily vaporized) elements found in the *Apollo 11* and *12* basalts finally provided scientists with an explanation for the large size of the maria. The lack of these elements gave the lunar magma a low viscosity, permitting it to easily flow over extensive regions of the Moon's surface. This low viscosity also explains why few solidified flow fronts are seen in the Lunar Orbiter photographs.

Even though Tranquility Base appeared flat from Earth, it was pockmarked by craters having diameters of 1 to 66 feet. A 590-foot diameter crater was located just a quarter of a mile east of the landing site and was surrounded by a 820-foot wide field containing boulders as tall as 16 feet. The large number

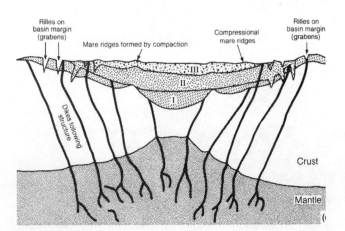

Figure A1-7. During the formation of a maria, multiple periods of lava extraction from the interior can occur.

Figure A1-8. This is a photograph of the Moon's southern highlands. Note that many of the larger crater floors have smaller craters within them.

Figure A1-9. A debris layer is seen covering the Fra Maurto region in this picture taken by Apollo 16.

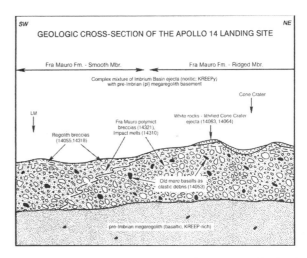

Figure A1-10. The landing site of Apollo 14 was covered by the thick layer of breccias.

of small craters confirmed the prediction based on telescopic observations that countless small craters would be found. The ground was also covered by a fine-grained soil that gave the area a subdued, soft appearance.

Based on crater counts and an assumed rate of meteor impacts, the maria were predicted to be young. Some parts of Oceanus Procellarum (the landing site for *Apollo12*), for instance, were thought to be only 3 to 6 million years old and maria age estimates of a few hundred million years were common. Few scientists believed the maria were significantly older than this. The rocks returned by the *Apollo 11* astronauts provided a rude awakening when they revealed that Mare Tranquilitatis was about 3.7 billion years old. Hardly a young feature! The lack of extensive cratering on the maria, like that seen in the highlands, suggested that the meteoritic impact rate had dropped precipitously by the time the flooded lunar basins hardened.

Although the nearest mountains were at least 24 miles away, rocks blasted from them were found among the *Apollo 11* samples. This proved impacts ejected material over considerable distances, a finding that was supported by the extensive ray patterns of material emanating from the craters Copernicus and Tycho.

The Maria/Highland Interface Region: The fact that the lunar highlands are heavily cratered is a testimony to the Moon's violent past. Shortly after its formation, the primitive crust was completely obliterated by collisions. Many of the basins and craters that already had been formed were destroyed during this period. For this reason, dating schemes based on the number of counted craters do not provide reliable age sequences in the highlands. The age of this region

would have to be determined from rocks returned by the astronauts.

Apollo 14, 15, and *17* visited the maria/highland interface region. The landing site of *Apollo 14,* north of the Fra Mauro Crater, has a crater density which is two to three times higher than that of the lunar maria and its surface was thought to be old. The Fra Mauro Formation is crossed by ridges 0.5 to 2.4 miles wide and tens of feet high (see Figure A1-9). Since these ridges radiate away from the Imbrium Basin, they were assumed to consist of material ejected from it. The largest crater within walking distance of the *Apollo 14* landing site is the 1,100-foot diameter Cone Crater.

Most of the rocks returned by the *Apollo 14* astronauts had a complex structure with breccia embedded within brec-

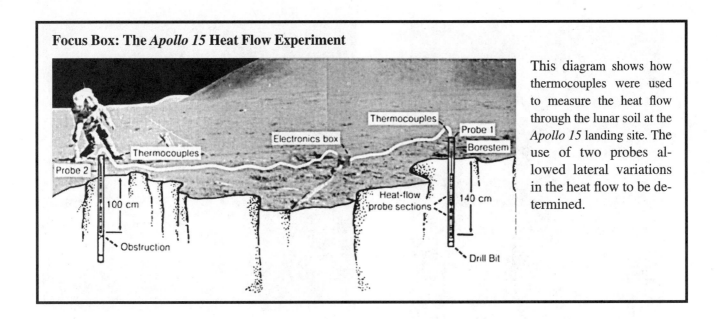

This diagram shows how thermocouples were used to measure the heat flow through the lunar soil at the *Apollo 15* landing site. The use of two probes allowed lateral variations in the heat flow to be determined.

cia. This suggests that they were produced in a violent impact event and supports the view that its material originated from the cataclysm that created the Imbrium Basin. The age of this material was measured to be 3.82 to 3.85 billion years old, a few hundred million years older than samples obtained from Mare Tranquillitatis. A geologic cross section of the *Apollo 14* landing site is shown in Figure A1-10.

Apollo 15 landed near the Hadley Rille and Apennine Mountains, a region where Mare Imbrium extends into the Apennine Mountains. There were four geology objectives to this mission: (1) to collect samples from the highland region that formed the rim of the Imbrium Basin, (2) to obtain samples from the mare floor, (3) to collect samples from the Hadley Rille, and (4) to measure the heat flow through the lunar surface.

The Apennine Mountains rise 1.2 to 3 miles above the mare surface and were thought to have been formed when

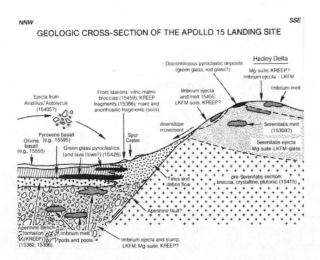

Figure A1-11. The landing site of Apollo 15, *near the Hadley Rille, was one of the most geologically diverse regions visited by the Apollo astronauts.*

the Imbrium Basin was blasted out of the Moon's surface. Therefore, these rocks could be used to confirm the formation time of the Imbrium Basin measured by *Apollo 14*. The Apennine samples yielded ages of 3.85 billion years, in agreement with those found by *Apollo 14*. The complex nature of this region was demonstrated by the discovery of the 4.5 billion-year-old "Genesis" rock, a remnant of the Moon's early crust. The basaltic lava samples obtained by the *Apollo 15* astronauts were found to be about 0.5 billion years younger than the Apennine samples. This age difference supported the hypothesis that the Imbrium Basin was flooded long after the initial impact event.

The Hadley Rille was certainly one of the most eye catching features visited by the astronauts, but when viewed from the Moon's surface it does not have the same crisp appearance that it has in telescopic views from Earth. Actually its walls slope downward at angles of 25 to 30 degrees and are covered by outcrops of large rocks. The average depth of the rille is about 1,000 feet and boulders are scattered on its floor.

Prior to the Apollo landings, scientists could only speculate about how the rilles were formed. Among the possibilities discussed were they were created from (1) pyroclastic flows, (2) running water, (3) collapsed lava tubes, or (4) by open lava flows. The first two possibilities were ruled out when no pyroclastic materials or hydrated minerals were found. Samples returned from the site suggested that the rille was formed from one of the last two mechanisms.

On the Earth, lava normally flows in open channels. As the molten rock splashes from side to side, high walls can be created along the edge of the stream. Under special circumstances, a top crust can form as the lava begins to cool, creating a lava tube. Prior to *Apollo 15* many scientists thought that the nearly mile wide width of the Hadley Rille made the formation of a stable roof unlikely. However, if the lava had a low viscosity, was very hot, and was rapidly extruded from beneath the surface, the Moon's low gravity made such a structure possible. The roof of the lava tube could then have been destroyed by meteor impacts. At the present time, the data is insufficient to determine whether the rille formed in this manner or as the result of an open lava flow.

Apollo 15 was the first mission to measure directly the heat flow from the Moon's interior. Because the Moon is a small body, its heat of formation should have been lost long ago. However, the decay of radioactive elements such as potassium, thorium, and uranium would continue to release heat. Therefore, heat flow measurements provided useful information about the Moon's interior composition and temperature. The value measured by the astronauts ($3.1 \pm 0.6 \times 10^{-6}$ W/cm^2) agreed with ground-based microwave measurements and was about half the value measured for the Earth. Apparently, the Moon possesses a fair quantity of radioactive elements and its interior is still quite hot.

Apollo 17 landed in the Taurus-Littrow Valley, at the interface between the highlands and Serenitatis Basin. The mountains surrounding the valley floor, in particular the North and South Massifs, were thought to be covered by rocks from deep within the Moon that were blasted to the surface by the creation of the Serenitatis and other basins. The valley floor was thought to be covered by mare basalts.

Samples from this region possessed an age of 3.7 to 3.8 billion years old, and like the soil on Mare Tranquillitatis, they were rich in titanium. As hoped, the North and South Massifs were found to consist of breccia. These rocks allowed researchers to determine that the Serenitatis Basin was formed about 3.87 billion years ago. The North and South Massifs had small differences in their composition. The South Massif contains slightly more calcium and aluminum, whereas the North Massif contains more magnesium and iron. This suggests that the material that covers these mountains came from different locations. The Sculptured Hills have a composition that is close to the average composition of all the areas sampled during the mission.

The landslide at the bottom of the South Massif is believed to have been created by ejecta from the event that created the crater Tycho. Samples from this feature suggest that Tycho was formed about 100 million years ago.

The Lunar Highlands

The highlands are the oldest exposed areas on the Moon's surface. Scientific opinion was divided on their composition, but a large segment of the community thought the highlands would consist of volcanic material. *Apollo 16* landed deep in the highlands in order to investigate this question. Its primary geologic objective was to collect samples from two different highland landforms, the Descartes Mountains and Cayley Plains (see Figure A1-12). Lunar geologists believed that both of these regions were formed by volcanic activity.

The Descartes Mountains resemble volcanic regions on Earth where the lava has a high viscosity. These types of flows create numerous mountain-like structures that can reach substantial heights. Samples from this region were ob-

Figure A1-12. The landing site of Apollo 16 *was deep in the lunar highlands.*

tained from the gentle slopes of the 1,800-foot-high Stone Mountain, located to the south of the landing site and from the much steeper Smoky Mountain. Transverse furrows seen on Stone Mountain looked similar to those observed on volcanic mountains on Earth, so this mountain was thought to have a volcanic origin.

The Cayley Plains look like a small lowland maria. They cover about 5 to 6 percent of the Moon's near side. Geologists thought they consisted of pryoclastic debris or low viscosity lava since few ridge lines are seen on their surfaces. To investigate this site, samples were collected near the South Ray and North Ray Craters. The South Ray Crater is located about 4 miles from the *Apollo 16* landing site, a short drive in the Lunar Rover. The North Ray Crater is located about 2.4 miles from the landing site. Samples collected from these regions yielded very young ages of 2 and 50 million years, respectively.

The rocks brought back by the crew of *Apollo 16* shocked mission scientists. The prevailing idea at that time was that the region would be covered with basalts, but all of the samples were breccias. The situation was pretty much summarized by Ken Mattingly, the pilot of the *Apollo 16* CSM. While still in orbit around the Moon, he was told by Mission Control that few, if any, basaltic rocks were being found by his crewmates. He replied, "Well, it's back to the drawing boards or wherever geologists go."

Rocks from the Descartes Mountains had an age of around 4.4 billion years, not much different from the age of the solar system itself. The Cayley Plains were found to be much younger, about 3.9 billion years old. As a result of this information, the Descartes Mountains are now believed to consist of ejecta from the formation of the Nectaus Basin and the Cayley Plains are thought to be covered by material from the Imbrium Basin.

Table A1-2. The Lunar Atmosphere

Atom	Number Density*	Atom	Number Density*
H	<17	Ar	4×10^4
He	$(2\text{-}40) \times 10^3$	K	16
Li	<0.01	Ca	<6
O	<500	Ti	<2
Na	70		

* atoms per cubic centimeter

The Lunar Craters: The Apollo missions removed any remaining doubt about the origin of the lunar craters. These objects were produced by meteor impacts. Even the glassy soil associated with some of the craters, like that found by Schmitt during *Apollo 17,* was determined to arise from impact melting, not volcanic activity.

Apollo Lunar Surfaces Experiment Package Results

The Apollo Lunar Surfaces Experiment Packages (ALSEPs) provided a large quantity of data about the longer term behavior of the Moon. Below are some of their major findings.

The Lunar Atmosphere: During the *Apollo 17* mission a mass spectrometer was deployed to collect data about the tenuous lunar atmosphere. A major source of these gases was thought to be volcanic eruptions, however, no gases expected to arise from this type of activity were detected. Rather the lunar atmosphere was found to consist of gases from the solar wind or from the radioactive decay of elements within the

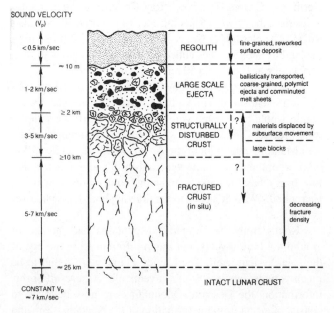

Figure A1-13. One of the major results of the Apollo program was a map of the internal structure of the Moon. This map shows the upper structure of the Moon's crust.

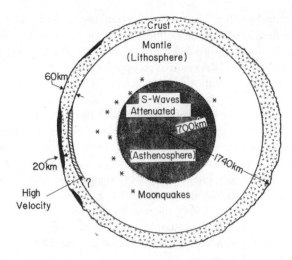

Figure A1-14. The structure of the Moon. The Earth is to the left in this figure. Stars represent the position of moonquakes. Note they are largely on the side facing the Earth. The region labeled High Velocity corresponds to a region where the velocity of waves dramatically increases. The dark areas on the surface are maria.

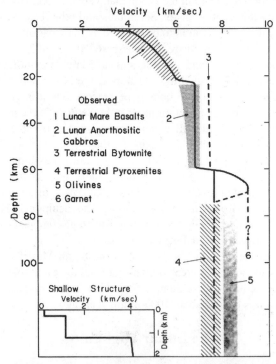

Figure A1-15. The velocity profile obtained by the Apollo seismographs revealed deep structures in the Moon's crust.

lunar soil. The quantity of gas discovered in the atmosphere was very small. Only one gas molecule exists for every 100 trillion present in our atmosphere. To put this into perspective, the amount of gas released during an Apollo landing was about the same as that present in the Moon's entire atmosphere. The Apollo landings increased the amount of gas in the Moon's atmosphere by a factor of about six!

The Moon's Structure: Data about the interior structure of the Moon was obtained by seismometers deployed at several of the Apollo landing sites. Between July 1969 and Sep-

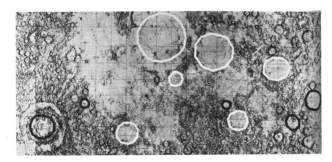

Figure A1-16. The location of several mascons are shown as circles in the above plot. Note how well the position of the mascons correlates with the position of maria.

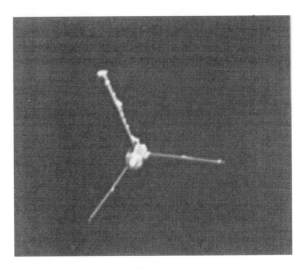

Figure A1-17. Two small satellites were placed in lunar orbit at the end of the Apollo 15 *and* 16 *missions. Their goal was to collect additional information about the Moon's gravitational field and solar activity.*

tember 1977, when the last station was turned off, 11,800 moonquakes, 118 of which were major events, were detected. They provided data about the structure of the Moon from a depth of a few miles to nearly 600 miles. Lunar seismic activity was found to be much weaker than on Earth. It was comparable to the weakest terrestrial earthquakes. The total energy released in a year is less than that given off by the explosion of a tenth of a pound of TNT. Most moonquakes occur deep within the Moon and appear to be triggered by the gravitational attraction of the Earth (see Figure A1-14)

The Lunar Seismic Profiling Experiment deployed on *Apollo 17* allowed the Moon's surface layer to be examined to a depth of a few miles. On July 17, 1972, a meteorite with a mass of about 1 ton struck the Moon's far side. The seismic waves produced by this impact were detected by the Apollo seismometers and revealed the deep structure of the Moon. Surprisingly, the core was found to be partially molten.

Lunar Mascons: The term mascon is an abbreviation for the words mass concentration. Mascons were detected by the additional gravitational force they produced on the orbiting Apollo CSM and on the subsatellites released by *Apollo 15* and *16*. As these vehicles passed above a mascon, their orbital velocity was changed. By noting when these perturbations occurred, the mascons were found to be associated with the large circular maria (see Figure A1-16). Two theories were developed to explain their presence. The first proposed that mascons were the embedded remnants of the object that blasted out the lunar basins, and the second suggested that they were caused by the vast amount of lava that filled the basins. The measured gravitational disturbances were not consistent with the presence of a single massive body but could be explained by a disk-shaped body of mass. Based on this finding, the mascons are attributed to thick beds of hardened lava.

The Moon's Magnetic Field: Samples returned by *Apollo 11* provided evidence that the Moon once possessed a global magnetic field. The subsatellites placed into orbit by *Apollo 15* and *16* also found magnetic anomalies over about 5 percent of the Moon's surface. Surface measurements obtained at the *Apollo 12, 14, 15,* and *16* sites similarly detected a weak magnetic field. The origin of this field and the reason for its disappearance are still debated by scientists. Several magnetic anomalies are located on opposite sides of the Moon from the younger maria basins, but it is not clear if these two phenomena are causally related.

Summary

During the Apollo program, astronauts returned over a third of a ton of lunar rocks and soil. These samples played a critical role in helping scientists unlock many of the Moon's long held secrets. They provided direct measurements of the Moon's composition, allowed the ages of many of its prominent surface features to be determined, and settled long-held disputes about the origin of the lunar basins and craters. In addition, the ALSEP experiments provided unique, long-term seismic data that allowed the Moon's interior structure to be determined. The most important scientific result of the Apollo program was that its results directly challenged existing theories about the Moon's origin. Based on the materials returned to Earth, scientists realized their models were in deep trouble and would have to be either drastically revised or completely discarded.

A2   The Origin and Evolution of the Moon

Somewhere near 4.5×10^9 years ago, we had to arrive at a volatile-and-iron-depleted moon near ten earth-radii, in prograde orbit with an orbital inclination to earth's equator of near 10 degrees. Almost everyone agrees such a stage occurred; there is less agreement on how it came about.

—Stan Peale, physicist, in preface to *Origin of the Moon*, 1984

In the previous chapter the geologic results from each of the Apollo missions were discussed. Since the last Apollo flight returned to Earth in December 1972, analyses of the lunar samples have revealed fascinating facts about the Moon's violent past, age, composition, internal structure, and current state. When combined, this new data revolutionized our thinking about the Moon's origin and permitted scientists to reconstruct the Moon's history from soon after its birth to the present time.

Pro and Con Arguments: Theories of the Moon's Origin

Prior to the Apollo program, scientists had developed three theories to explain the origin of the Moon: the capture theory, the coformation theory, and the fission theory. Each of these theories is discussed in this section.

The Capture Theory: According to this theory, the Moon formed beyond the orbit of Mars and was gravitationally captured by the Earth when its orbit decreased in size through gravitational interactions with other solar system objects. Advocates of this theory believed that the Moon formed far from the Earth because its low density of 3.3 g/cm^3 was incompatible with that of the Earth and the other planets of the inner solar system. These objects have densities between 3.7 and 5.4 g/cm^3. In the outer solar system the temperature is very low and the material found there consists of ices and other light compounds. An object that condensed out of this material would therefore have a low density.

Opponents of the capture theory were quick to identify two weaknesses in it. First, they pointed out that comparatively few objects in the solar system have a mass as great as that of the Moon. Considering the vast size of the solar system, it would therefore have been very unlikely for the Earth to have accidentally encountered such an object. Secondly, even if an encounter did occur, calculations showed that considerable dynamical difficulties exist with the capture of such a massive object. Much of the Moon's orbital energy would have to be converted into some other form of energy and a considerable amount of angular momentum would have to be lost. These requirements are difficult to fulfil. If the Moon did approach the Earth, it is much more likely to have collided with the Earth or have been swung back into space.

Advocates of the capture theory addressed the first of these problems by postulating that many other lunar-sized objects existed in the distant past. A chance collision between one of them and the Earth would therefore have been more

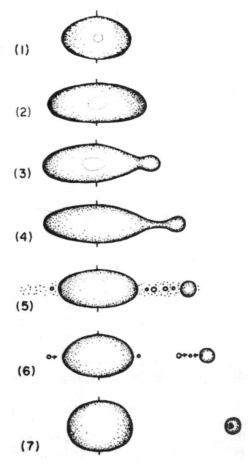

Figure A2-1. *In the fission theory a rapidly rotating, semifluid Earth ejects material from its equatorial region into orbit. This material was thought to reform into the Moon.*

likely than it is today. To booster this argument they pointed out that seven of the thirty-two major satellites in the solar system have a similar size to our Moon, and two of Jupiter's major moons, Io and Europa, have a similar density. They also noted that twelve of the planets' thirty-two major satellites have nonequatorial orbits, and six of them revolve in the opposite direction to the planet's rotation. These peculiar orbital parameters suggest they are captured objects. If other moon-like bodies existed and became attached to these planets, why couldn't the Earth have acquired its moon in a similar fashion?

The problems caused by the high energy and momentum associated with the Moon's motion were answered by assuming these values were decreased through tidal activity on the Earth and/or by collisions with smaller bodies already in orbit around the Earth. In most versions of the capture theory, the Moon is captured by the Earth as a cold object, and it remains cold throughout its lifetime. In this theory the lunar basins were formed by water, possibly acquired by the Moon during its gravitational interaction with the Earth.

The Fission Theory: The Fission Theory assumes the Moon formed from material ejected by a rapidly spinning Earth. As the Earth condensed out of the solar nebula and continued to contract, it spun faster and faster, eventually flinging material from its equatorial region into orbit. Figure A2-1 illustrates how the Moon is thought to have formed from this process.

George Howard Darwin, the second son of the famed naturalist Charles Darwin, was one of the principle advocates of the fission theory. He mathematically traced the motion of the Moon backward in time, estimating that 54 million years ago, it was only 6,000 miles above the Earth's surface. At that time, the rotation period of the Earth would have been a scant 5 hours and 36 minutes. Darwin wrote, "These results point strongly to the conclusion that if the Moon and Earth were ever molten viscous masses, then they once formed parts of a common mass." Although a 5-hour rotation period seems extraordinarily short, opponents to this theory argued it was still too long to result in the ejection of material. They presented calculations showing that a rotation period of less than 2.5 hours would be needed for a significant amount of mass to be expelled from the Earth. Proponents countered this objection by suggesting that tides raised on the Earth by the Sun worked in harmony with oscillations set up by the rapid rotation of the semifluid Earth to make it easier for material to be ejected. This argument temporarily satisfied critics.

Unlike the capture theory, which was forced to postulate that the Moon formed far from the Earth in order to explain its low density, the fission theory had no difficulty accounting for this fact. In this model, the Moon formed out of the Earth's crust and mantle where the density is very similar to that of the Moon.

The fission theory was also intriguing because it appeared to explain other facts about the Earth. In 1882, geologist Osmond Fisher suggested that the Pacific Ocean basin was the "hole" left in the Earth's surface by the ejected Moon. He further hypothesized that the separation of the American continent from Europe and Africa was initiated by this event. When the Moon's material was ejected, the Earth's crust broke into pieces and the land masses flowed over the mantle toward the hole.

By 1936 the fission theory was so widely accepted that its premise was incorporated into a radio program for children written by the U.S. Office of Education. A portion of the script read:

> Have you heard that the Moon once occupied the space now filled by the Pacific Ocean? Once upon a time—a billion or so years ago—when the Earth was still young—a remarkable romance developed between the Earth and the Sun—according to some of our ablest scientists.... The Sun's attraction raised great tides upon the Earth's surface ... the huge crest of a bulge broke away with such momentum that it could not return to the body of mother Earth. And this is the way the Moon was born!

Opponents of the fission theory were not convinced by the above arguments and pointed out that if oscillations developed from the Earth's rapid rotation, they would be quickly damped out by viscosity within the Earth's mantle. They also noted that the Moon did not orbit the Earth above its equator as would have occurred if the Moon was ejected from the Earth. An additional problem with the fission theory was that it did not explain why the Earth possessed an initial rapid rotation. If a planet forms by the accretion of modest-sized objects, models suggest that the resulting body would rotate very slowly. Some of the accreted material would increase the planet's rotational velocity while other material would cause it to decrease. There is no reason why the accreted material would impact the Earth so as to favor one direction over the other. Models of this process showed that the infalling material would form a body with a very slow rotation. Given this situation, the settling of dense materials from the crust to the core would be insufficient to increase the Earth's rotation rate to anywhere near that required to eject matter.

The Coformation Theory: The third theory, the coformation theory, states that the Earth and Moon formed together in space. In this theory the Earth gravitationally attracts material from its environment and a debris ring forms around the Earth. The Moon is thought to then condense out of this material. In this model, the compositions of the Moon and the Earth are expected to be identical.

This particular model was very attractive because it is essentially a scaled-down version of the widely accepted model used to explain the formation of the entire solar system. Its primary problems were that it had difficulty explaining why the Earth and Moon have different densities, why a circular ring of material formed around the Earth, and how the Earth acquired a significant initial rotational velocity.

Table A2-1

*TABLE 1. Compositional Similarities and Differences Between the Upper Mantle of the Earth and the Mantle of the Moon.**

		Earth	Moon
Oxygen isotopes		same	same
Uncompressed density, bulk planet (gcm^{-3})		4.45	3.34
Siderophile element depletions relative to CI			
	W	22 ± 10	22 ± 7
	Ga	4–7	20–40
	P	43 ± 10	115 ± 25
	Mo	44 ± 15	1200 ± 750
	Re	420	10^5
Refractory element abundances	U(ppb)	20	35†
Volatile/Refractory element ratios	K/U	10^4	2500
Volatile element ratio relative to CI	Cs/Rb	1/7	1/2
Mg/(Mg + Fe) ratio		0.89	0.80‡

*After Drake (1986).
†Cannot rigorously rule out a bulk concentration in the Moon as low as 20 ppb.
‡A few estimates based on the most magnesian terrae rocks approach 0.87.

* A siderophile element preferentially enters into a metallic phase when cooling.

The Apollo Data

Based upon an examination of the samples returned by the Apollo astronauts, several new facts emerged about the Moon. It was found that (1) refractory elements, those elements that condense at high temperatures, such as titanium and calcium are relatively abundant on the Moon; (2) volatile elements, elements that are vaporized at low temperatures such as bismuth, thallium, silver, and gold, are rare on the Moon; (3) the ratios of the isotopes of oxygen (^{16}O, ^{17}O, and ^{18}O) on the Moon are identical to those on the Earth; (4) no minerals brought back from the Moon contained water, indicating that the Moon never had any surface water; (5) unlike rocks on Earth that are either sedimentary, metamorphic, or igneous, all of the lunar samples were igneous; (6) the Moon was found to be a fairly homogeneous body, little element settling appears to have taken place; and (7) the Mg/(Mg+Fe) ratio on the Moon was measured to be different from that of Earth. These data are summarized in Table A2-1. The abbreviation, CI, refers to carbonaceous chrondrites, primitive meteorites whose composition is assumed to be the same as the nebula out of which the solar system formed.

Applying the Apollo Evidence

Prior to the Apollo era, there was no consensus about which, if any, of the theories dealing with the origin of the Moon was correct. A scientific stalemate had been reached. The new data brought back by the Apollo astronauts was able to provide a real breakthrough in this area because each of the prior models made testable predictions about the present-day composition of the Moon.

The coformation theory predicted that the abundances in the entire Earth and Moon should be nearly identical. The fission theory predicted that the composition of the Moon should closely resemble the Earth's crust and mantle. The capture theory predicted that few similarities would be found between the composition of the Moon and the Earth because

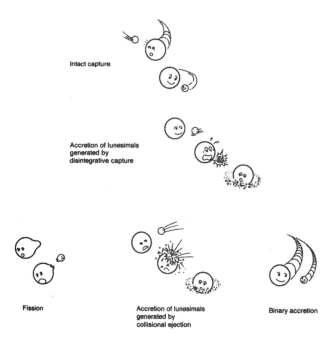

Figure A2-2. The above cartoon portrays several different hypotheses about the formation of the Moon.

these bodies would have formed under different circumstances and at different locations in the solar system.

The Confrontation between Theory and Observations

Comparisons between the elements found in the Moon and Earth are shown in Figures A2-3 and 4. According to the coformation theory, the Earth and Moon should have nearly identical compositions. However, Figure A2-2 shows that when compared to the Earth, the Moon is deficient in magnesium, sodium, and potassium and enriched in titanium, calcium, and aluminum. If the compositions of these two bodies were the same, all of the points in this figure should lie along the solid diagonal line. Advocates of coformation theory realized that there is no easy way to reconcile their predictions with this data.

According to the fission theory, the material in the Moon came from the Earth's crust and mantle. There should, therefore be a close relation between the compositions of the Earth's mantle and the Moon. Figure A2-3 already had shown that differences exist in the composition of the Earth's mantle and the Moon. Figure A2-4 also shows that the refractory elements, with the exception of rubidium (Rb), are uniformly en-

riched on the Moon. None of these enhancements are predicted by the fission theory. Finally, the Mg/(Mg+Fe) ratios given in Table A2-1 should be much closer if the fission theory is correct.

According to the capture theory, the Moon was formed in the outer regions of the solar system so its composition should bear little resemblance to that of the Earth. Unfortunately, as seen in Figure A2-4, a strong relation exists between several elements found in the Earth and the Moon. With the exception of rubidium (Rb), the abundances of these trace elements are highly correlated. If the Moon had been formed elsewhere, the points in this plot should be scattered *randomly* distributed.

A second Apollo finding that challenged the validity of the capture theory involved the Moon's isotopic oxygen abundance. It was known that the ratio of these isotopes varies with location in the solar system in a predictable fashion. If the Moon had formed far from the Earth, different isotopic ratios would therefore be expected. To the surprise of advocates of the capture theory, these values were found to be identical for the Earth and Moon. This meant that the Earth and Moon must have been formed at the same radial distance from the Sun, undermining one of the capture theory's main statements.

The data returned by the Apollo astronauts left scientists in a perplexed state. *None* of the theories developed to explain the origin of the Moon fit the data! Based on the composition of the lunar samples and the dynamical problems mentioned earlier, the capture theory, in particular, was declared to be "horrendously improbable." Advocates of the fission and coformation theories also found themselves in deep trouble and scrambled to find ways to alter their theories so conflicts with the Apollo data could be resolved.

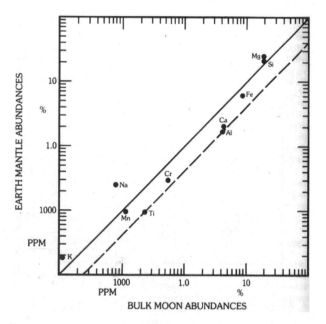

Figure A2-3. Elements in the Earth's mantle compared to those found in the Moon (number per million atoms).

A New Model for the Formation of the Moon

By 1982 scientists, led by William Hartmann, Jeff Taylor, and Roger Phillips, decided the time was ripe to convene a conference so researchers could share their latest ideas on the origin of the Moon. Scheduled for October 1984, it attracted a large number of participants. Anticipating that several new models for the Moon's formation would be presented, the conference organizers titled the last two sessions "My Model of Lunar Origin I" and "My Model of Lunar Origin II."

Despite the initial expectation that a consensus would not be reached, agreement rapidly developed over a "new" theory, that the Moon had formed as the result of the glancing impact of a Mars-sized object with the Earth. This was actually not a new idea at all, but had been suggested in 1946 by Harvard geologist Reginald Daly. Hartmann was simply unaware that it had already been proposed. The main premise of this model was that the Moon had two parents rather than one.

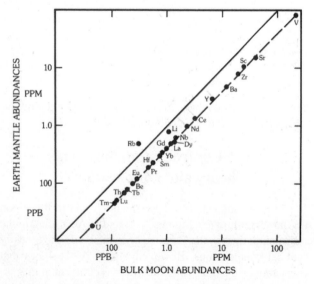

Figure A2-4. Rare earths found in the Earth's mantle compared to those found in the Moon (number per million atoms).

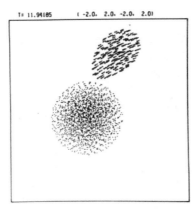

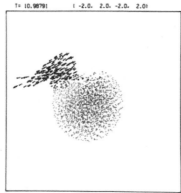

Figure A2-5. Three snapshots displaying the numerical results of a collision between the Earth and a Mars-like object. The Earth is the central, large object. The Moon is seen coming together in the lower frame.

The Giant Impact Model

The giant impact model assumes that the early solar system was populated by numerous large bodies and that one of these objects collided with the Earth at a glancing angle. In this collision, the impactor was completely destroyed and a significant amount of material from the Earth's crust and mantle was blasted into orbit. Much of this material was vaporized by the enormous amount of energy released in this collision. Once in space the gaseous component of this material dissipated leaving behind elements that have a high vaporization temperature. These eventually cooled and collected into the present-day Moon, as shown in Figure A2-5.

Unlike prior models of the origin of the Moon, the giant impact theory adequately accounts for all of the observed data. It explains why the Moon possesses few volatile elements while being enriched in refractory elements. The volatile atoms escaped when in their gaseous state, while the refractory elements remained behind and gradually gravitationally collected themselves in a new body. The identical Earth–Moon isotopic abundance of oxygen is accounted for because much of the Moon's material came from the Earth. The relatively low density of the Moon arises from the fact that it originated from matter in the Earth's crust and mantle which also has a low density. Finally, computations show that the iron in the impactor would have fallen back to Earth rather than collecting into the Moon. This explains why the Moon is globally deficient in iron.

As an additional benefit, the giant impact model provides an explanation for the Earth's rapid rotation. It is the result of the off-centered nature of the collision. Just as striking the rim of a spinning wheel causes the wheel to spin faster, the glancing impact of the Mars-like object (best illustrated by the middle box of Figure A2-5) caused the Earth to rotate faster.

The Evolution of the Moon's Surface

Once the Moon condensed out of the debris ejected by the devastating collision between the Earth and the Mars-sized object, its surface evolved through three separate stages.

Phase One: In the earliest stage, the Moon's surface was covered by a deep sea of liquid lava. The heat required to create this huge molten ocean could have come from the collision between the Earth and the Mars-like object or from the impacts of numerous but less massive objects. Low-density materials in the fluid Moon rose to the surface and hardened into a crust. During this era, the lunar surface was continuously bombarded. Some of the larger impacts blasted lunar basins out of the surface, while smaller ones created craters. The highland surface was pounded, mixed, and crushed into composite rocks, forming the breccias. This phase lasted for 0.6 billion years and at its conclusion the Moon was heavily pitted. Pools of lava, ejected from deep within the Moon, collected at the bottom of the lunar basins (see Figure A2-6).

Phase Two: The next stage in the Moon's evolution lasted for 2 billion years. During this time many craters were covered or partially filled with lava. The lava in the basins began to solidify giving the Moon its present distribution of light and dark areas (see Figure A2-7).

Phase Three: This phase of the Moon's evolution extended from about 2 billion years ago to the present time. The outer crust had now cooled and was rigid to a depth of about 600 miles. The mantle was firm enough to support the overlying solid magma oceans. Lava outflows from the interior ceased since the semifluid magma was now so deep that

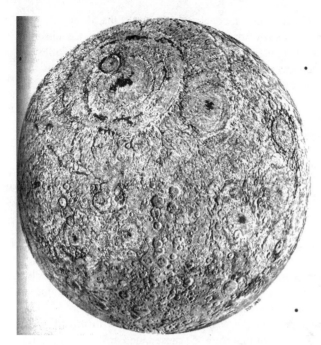

Figure A2-6. *In the first evolutionary phase of the Moon's surface, the Moon is scarred by infalling material, but the maria have not yet formed.*

Figure A2-8. *The face of the present Moon is not expected to change for billions of years.*

Figure A2-7. *The Moon as it appears after the lunar maria were filled but before the formation of Copernicus or Tycho.*

it was essentially impossible for it to reach the surface. Fragmented regions under the surface became locked in place by the surrounding solid structures and seismic activity stopped. Terrain-altering events are now rare and are caused by meteor impacts, such as the ones that created the craters Tycho and Copernicus. The face of the Moon became fixed in its present shape.

Contributions of the Apollo Program

The Apollo program made contributions to eleven different areas of lunar research. These are summarized in Table A2-2. As noted in this table, specific research areas were targeted by different missions. This was unavoidable since some flights landed in the lunar maria while others landed in the highland regions. When the Apollo results were combined with prior studies, scientists were able to construct a detailed history of the Moon. The ages of the Moon's geologic periods as well as other notable events are listed in Table A2-3.

Table A2-2

Appendix 3. Progress toward Scientific Objectives at Time of Apollo 17 Site Selection (between Apollos 15 and 16).

Objective	Apollo					
	11	12	14	15	16	17
Early lunar history	—	m	m	M?	?	E
Old crustal and interior materials	—	—	—	M?	?	E
Major basins (>250 km) and mascons	m	m	M	M	—	E
Highland crustal evolution	—	—	m	M	M	E
Mare fillings	M	M	—	M	—	D
Large craters (>40 km) and their products	—	m	—	—	—	E
Postmare internal history	m	M	—	M	?	E
Regolith evolution	M	M	m	M	M?	D
Regolith interactions with extralunar environment	M	m	m	M	M?	D
Present interior, physical, and chemical state	m	m	M	M	M?	E
Lunar heterogeneity	—	m	M	m	?	E

Source: From memorandum prepared for Noel W. Hinners by William R. Muehlberger and Leon T. Silver, dated 30 November 1971.
Abbreviations: M = major contribution; m = significant but limited contribution; E = essential; D = desirable but less urgent.

Abbreviations: M= major contribution, m= significant but limited contribution. Adapted from *To a Rocky Moon* by Wilhelms, p. 366.

Table A2-3

Appendix 4. Geologic Periods and Notable Events in Lunar History

Approximate time (aeons ago)	Event
4.5	Accretion of Moon in Earth orbit.
4.5–4.2 (?)	Differentiation of crust and mantle; plutonism, volcanism, and impact mixing and melting.
4.2(?)	Crustal solidification and formation of oldest preserved impact basins.
4.2–3.92	Formation of at least 30 pre-Nectarian basins.
3.92	Nectaris basin impact, beginning Nectarian Period.
3.92–3.84	Formation of 10 more Nectarian basins, including Serenitatis and Crisium.
3.84	Imbrium basin impact, marking Nectarian-Imbrian period boundary; eruption of oldest dated intact mare lava flows.
3.8	Formation of last large basin (Orientale), marking Early Imbrian-Late Imbrian epoch boundary.
3.8–3.2	Eruption of most voluminous mare lavas and pyroclastics; continued though diminished impact cratering.
3.2	Imbrian-Eratosthenian period boundary.
3.2–1.1	Continued mare volcanism and impact cratering.
1.1	Eratosthenian-Copernican period boundary.
0.81	Copernicus impact; approximate time of last mare eruptions.
0.11	Tycho impact.

Taken from *To a Rocky Moon* by Wilhelms, p. 367.

An Overview

As recently as 1967, a U.S. Geological report gave the following dismal appraisal of our understanding of the Moon's origin and evolution, ". . . no conclusions can be drawn other than that the interpretation of the existing data leads to many ambiguities." There is no doubt that this sit-uation was dramatically improved by the Apollo program. Based on information collected during this program we have now developed a model of the Moon's origin that explains all of the available data and provides dates for the Moon's major geologic epochs.

It is often asked if these same results could have been obtained at far less cost and danger if automated probes had been sent to explore the Moon. Most scientists respond that this is not the case, that the astronauts played a crucial role. Stuart Taylor, a noted lunar scientist who participated in the analysis of the lunar samples, wrote, ". . . there are many facets of lunar exploration and establishment of 'ground truth' that could only have been accomplished by the astronauts. The obvious ability to collect documented and oriented rock samples which are vital for so many studies, is one such [example]." He suggests that if only unmanned probes had been employed, "perhaps a decade, perhaps more, of patient work would have been needed to get the story straight and to uncover the long sequence of lunar events."

Technology has advanced considerably since the early 1970s. In the future, robotic geophysical stations or specialized probes may be sent to the Moon to unravel other mysteries about our nearest neighbor. Eventually, however, humans will return to the Moon, but this time to stay. Plans are now under-way for the construction of a permanent lunar base that will be built and occupied during the next century. The type of work that might be performed at this base and the advantages it offers have been discussed in Chapter 42, Moon Base: An Important Step into the Future.

The Science of Skylab

Scientifically, Skylab was a bold beginning. It was the first long-term venture into the exploration of the universe from near-Earth space. It looked inward to Earth, outward to our Sun, and scanned the universe.

—From *Skylab, Our First Space Station*

The astronauts aboard Skylab completed experiments in the areas of astronomy, the biomedical sciences, Earth applications, and space applications. The objectives in each of these areas were as follows:

Astronomy: to gather ultraviolet and x-ray observations of the Sun, to collect data in the fields of stellar and galactic astronomy, to measure and identify the source of the diffuse galactic x-ray background.

Biomedical Science: to study how living organisms adapt to a weightless environment, to determine the physiological effects of prolonged weightlessness on the human body, and to test procedures to counteract these negative effects.

Earth Applications: to gather data on air and water pollution and to obtain images of the Earth for scientists in the fields of agriculture, forestry, geography, oceanography, meteorology, and geology.

Space Applications: to evaluate the habitability of the space station and its systems, to discover ways that the productivity of future stations could be increased and its environment made more livable. An additional goal was to examine novel manufacturing processes to develop compounds that might have commercial value.

Astronomy from Skylab

The primary new astronomical data collected by Skylab was in the ultraviolet and x-ray regions of the spectrum. Ultraviolet wavelengths extend from about 1200A to 3500A, whereas the higher energy x-ray wavelengths extend from about 1A to 100A (1A = 10^{-8} cm = 0.4×10^{-8} inches). Both of these forms of light cannot be observed from the ground because they are absorbed by molecules of oxygen and water vapor in the atmosphere. The type and quantity of light emitted by a celestial object provides essential clues about the physical processes that power these sources. Without access to this light, the nature of some of the most energetic objects in the Universe would remain a mystery.

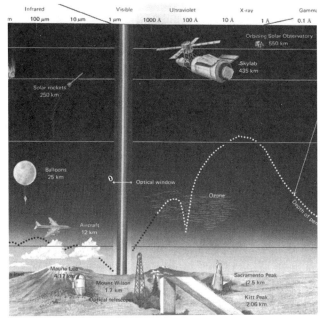

Figure A3-1. Not all forms of light reach the Earth's surface. The dotted line shows the depth that the wavelengths of light given at the top of the picture can penetrate the atmosphere.

Skylab's astronomical telescopes were clustered together in the Apollo Telescope Mount (ATM). This battery of telescopes, shown in Figure A3-2, was attached to the top of the orbital work station and consisted of eight separate instruments: two x-ray telescopes, an extreme ultraviolet spectroheliograph, an ultraviolet spectroheliometer, an ultraviolet spectrograph, a visible light coronograph, and two H-alpha telescopes. Together they allowed the emission from a celestial object to be measured from the x-ray to visible regions of the spectrum (i.e., from 2–7000A).

Prior to launch, all of the Skylab astronauts were trained to use the ATM. They took special classes in solar physics and were given wide discretion to alter an observing program to study quickly changing phenomena. ATM observations were controlled from within the multiple docking adapter where the astronauts could see an image of the Sun on a television screen. Several of Skylab's astronomical findings are discussed below.

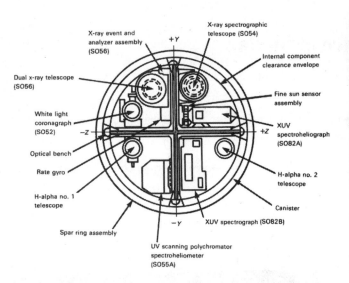

Figure A3-2. (left) Looking down at the telescopes in the Apollo Telescope Mount. In this view, all eight telescopes can be seen. They operated at x-ray to visible wavelengths. (right) Identification of the instruments in the left side panel.

Skylab's Stellar Research Program

Hot stars emit the majority of their light below 3000A and most of this light is absorbed by our atmosphere. Prior to 1973 sounding rockets had been employed to obtain ultraviolet spectra for about thirty stars. Astronomers use these spectra to determine a star's surface temperature and gravity. During its lifetime, Skylab would acquire nearly 400 ultraviolet spectra, increasing the available ultraviolet data on stars by over a factor of ten.

Astronomers use a short-hand notation, called spectral type, when referring to a star's surface temperature. A spectral type consists of one of the letters O, B, A, F, G, K, M followed by a number between 0 and 9. O stars have surface temperatures higher than 25,000°K, whereas M stars have temperatures of around 2,500°K (°K = 273+ °C). Examples of stars with spectral types in between these two extremes are B5, G2, and K9. The surface gravity of a star is designated by a Roman numeral between I and V, which is attached to the spectral type. Large stars which have a luminosity class of I

Focus Box: The Surface Gravity and Temperature of Stars

Skylab ultraviolet spectra included a range of spectral type and luminosity class stars. The strength of the CIV line at 1549A is used as a temperature indicator since its strength dramatically increases when moving from B5 to O6 stars. This same trend is also evident for other luminosity class stars. The luminosity class of the star can be determined from the Si IV lines located at 1394 and 1403A. The strength of these lines increases in moving from luminosity class V to I. The presence of an emission line just to the right of the CIV line indicates that mass is being ejected from the star. Based on the strength of this emission feature, O9.5 II-I stars are losing mass at a rapid rate.

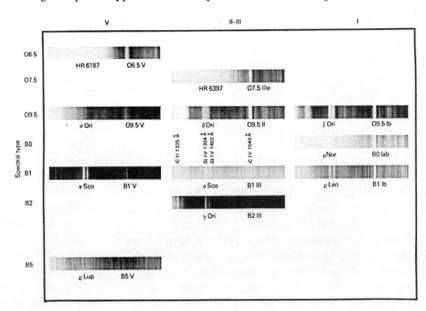

Taken from *Skylab's Astronomy and Space Sciences*.

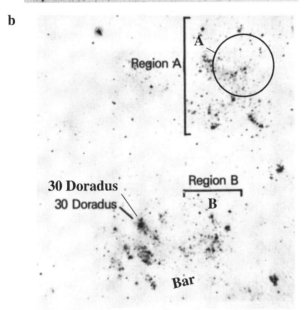

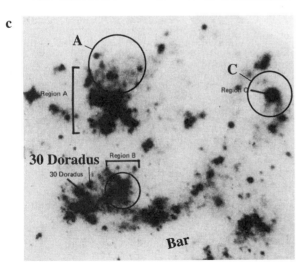

Figure A3-3. (a) An optical image of the Large Magellanic Cloud. (b) An image taken at 2,500A. (c) The same object seen in light with a wavelength of 1,400A.

have low surface gravities. A F2V star has a higher surface gravity than a F2I star but both of these stars have the same surface temperature since their spectral types are F2.

Both the surface temperature and gravity can be determined by the strength of absorption lines in a star's spectrum. By providing ultraviolet spectra, Skylab therefore allowed astronomers to vastly increase the sample for which these values were known (see the Focus Box on page A-22).

Skylab's Galactic Research Program

Galaxies are huge conglomerations of stars. They can be tens of thousands of parsecs in diameter and can contain hundreds of billions of stars. Billions of galaxies exist in our Universe but they have three basic shapes: spiral, elliptical, and irregular. The Large Magellanic Cloud (LMC) is one of the nearest galaxies to the Milky Way. Located at a distance of about 53,000 parsecs (1 parsec = 1.8×10^{13} miles), it contains billions of stars distributed in a chaotic fashion. Gas and dust clouds comprise about 40 percent of its mass and stars are being continuously formed within it. Although the majority of galaxies have either a spiral shape (like our own Milky Way) or an elliptical shape, approximately 25 percent of all galaxies do not have a definable shape, and like the LMC, they are classified as irregular galaxies.

Figure A3-3a shows an image of the LMC taken through a red filter by Dr. Karl Henize. A bar-like structure composed of a red star and a large region of glowing hydrogen gas called 30 Doradus dominates this view. Two faint glowing hydrogen gas clouds (Regions A and C) are located above and to the right of the bar. Figure A3-3b shows a Skylab ultraviolet photograph of the LMC. The emission from Region A remains strong, but the bar-like structure is absent. Figure A3-3c shows a far ultraviolet picture taken from the Moon during the *Apollo 16* mission. Region A is the most prominent feature even though it was barely visible in Figure A3-3a, and a new emission site, Region B, is visible. It is clear that these type of multiwavelength photographs provide dramatically different views that are important for learning about the structure and energetics of galaxies.

The Glowing X-ray Sky

X-rays created in space by energetic objects are absorbed when passing through the Earth's atmosphere. Therefore, before the launch of the Skylab, the x-ray sky had only been explored by unmanned probes. By 1973 about 100 x-ray sources had been discovered, but they possessed a bewildering range of properties. Some appeared to be associated with stars while others covered an extended region of the sky. Some were located within our galaxy, while others were beyond it. Some emitted pulsed x-ray radiation, while others emitted this light in a continuous fashion. In addition to this perplexing picture, the sky itself seemed to be uniformly glowing in soft x-rays.

An important goal of Skylab was to identify the source of this diffuse x-ray emission. The instrument used for this purpose, the Galactic Soft X-ray Experiment, was actually not aboard Skylab. It was carried in the instrument unit of the *Saturn 1B* that transported the third crew to Skylab. This detector continuously observed the sky as it trailed behind and below Skylab. Surprisingly, its data revealed that the widespread x-ray emission does not arise from the combined emission of thousands of weak, unresolved, high-energy sources as astronomers had believed. Rather it arose from previously undetected clouds of hot gas located between the stars.

Comet Kohoutek

Comets are mountain-sized objects composed of frozen ices and dust particles. Created billions of years ago when the solar system was forming, their composition reflects the nebula from which the Sun and planets formed.

Although millions of comets exist, their combined mass is less than that of our Moon. They are generally small objects, having irregular shapes and a diameter of less than 9 miles. During their journey around the Sun, their appearance changes. When far from the Sun, a comet is similar to a frozen mountain and is visible only because it reflects sunlight. As it travels nearer to the Sun, some of its material evaporates, forming a large cloud which surrounds the comet's nucleus. As the comet approaches the Sun this gaseous material is swept away by the Sun's light and forms a tail which can have a length equal to the distance between the Earth and the Sun (i.e., one astronomical unit, or A.U.).

Comet Kohoutek was discovered by Lubos Kohoutek of the Hamburg Observatory in March 1973. Early observations suggested it would provide a spectacular celestial display. Newspapers referred to it as "the comet of the century" and suggested it would be bright enough to be seen during the day. Some astronomers, however, were more cautious, noting that comet brightnesses were "notoriously unpredictable." Nevertheless, the view expressed by associate administrator for Manned Space Flight Dale Myers prevailed, "comets of this size come along once in a century. It really looks like the kind of thing you can't pass up." On August 16, 1973, NASA delayed the launch of Skylab's third crew so that they would be able to observe Kohoutek as it swung around the Sun. During this period, the comet is hidden from ground based observations studied by the bright daytime sky. Since Skylab was above the Earth's atmosphere, the sky appeared dark and the comet's behavior could be studied. Without a doubt, Comet Kohoutek was one of Skylab's most publicized targets.

From November 27, 1973 to December 26, 1973, Skylab instruments measured the size and brightness of Comet Kohoutek. Data collected during this period (plotted in Figure A3-4) show that as the comet neared the Sun, its size (top plot) and brightness (bottom plot) changed in a peculiar manner. Initially it increased in brightness in a steady fashion, but around December 12th Kohoutek began to fade. The brightness reached a minimum value around December 16, and then rapidly increased as it came to within 0.2 A.U. of the

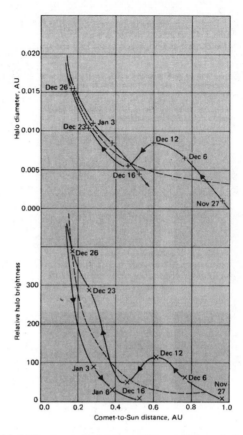

Figure A3-4. The size and brightness of Comet Kohoutek changed as it neared the Sun. An unexpected drop in its brightness and size occurred around December 16.

Sun. This behavior was thought to be caused by the evaporation of a layer of volatile material that boiled off from the nucleus just before December 12. To astronomers these plots supported the model that comets are composed of a mixture of ices and solid material, a viewpoint that described comets as giant "dirty snowballs."

Kohoutek produced a wealth of scientific data about comets, but from the public's perspective it was a dismal failure. Contrary to early predictions, Kohoutek never became a visually bright object. At its maximum, it was comparable in brightness to an average nighttime star. Newspapers expressed the public's disappointment in headlines across the country, "Kohoutek, The Flop of the Century," "The 'Comet of the Century' Went Phzzzt," and "The Cosmic Flopperoo."

Skylab's Observations of the Sun

The Skylab astronauts collected over 900 hours of data on the Sun. It was, by far, the most intensely studied astronomical object by Skylab. The Sun is not a static, unchanging object. It goes through both energetic and quiescent phases. During its active phase the Sun's surface is littered with dark spots called sunspots. Other features such as giant arcs of gas, called quiescent prominences, can be seen rising along the limb only to dissipate or to fall back to the Sun's surface.

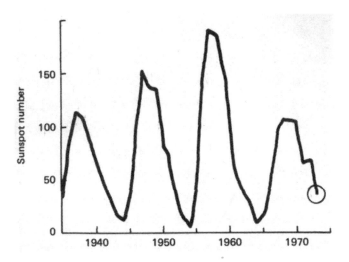

Figure A3-5. The number of sunspots varied significantly during Skylab's three manned missions. The number of spots on the Sun was similar during Skylab 2 and 3 but dramatically increased in the August-September time period during Skylab 3. The number of sunspots is an indicator of the level of solar activity.

Sudden, intense brightening, called flares, are also seen on the Sun's disk.

Based on years of study, astronomers realized that many of these phenomena are interrelated and intimately associated with the Sun's magnetic field. For instance, sunspots are formed in regions of intense magnetic fields and solar flares always occur near sunspots. Gas in the giant arch-shaped prominences flows along the Sun's magnetic field lines and fast-developing eruptive prominences occur in regions where the field is strong. Activity in the Sun's outer atmosphere (the corona) is also correlated with the number of sunspots.

The number of spots seen on the Sun's surface changes over a period of eleven years, reaching a maximum of about 200 spots and a minimum of near zero (see Figure A3-5). When Skylab was launched in May 1973, the Sun was entering its quiescent phase and few sunspots were present. However, the number of sunspots during the Skylab program varied from 0 to nearly 125. It was therefore possible to collect data about the Sun while it was in both its energetic and quiescent states. The following sections describe some of Skylab's discoveries about the Sun.

The Discovery of Super-Spicules

Spicules are large, fingerlike columns of gas in the Sun's lower atmosphere, the chromosphere. These gas columns rise and fall over time intervals of 5 to 10 minutes and can be seen projected against the blackness of space at the Sun's limb (see Figure A3-6). Skylab's ultraviolet images of the Sun's polar regions revealed a new class of extraordinarily large spicules. These gas columns rise 24,000 miles above the Sun's surface and last for up to 45 minutes. Despite numerous studies, twenty-five years after their discovery, astronomers still do not understand the origin of these objects.

Coronal Holes

In order to emit x-rays, a gas must have a temperature of at least a million degrees Kelvin. These high temperatures were known to be present in the Sun's outer atmosphere, called the corona. However, little else was known about this region. Using its two high resolution x-ray telescopes, Skylab astronauts discovered a new phenomenon in this region called coronal holes.

Coronal holes are regions in the Sun's atmosphere where the temperature and density are unusually low and where the magnetic field lines extend radially outward. These regions can cover much of the Sun's surface and appear black in x-ray pictures like that shown in Figure A3-7. Through these "holes," high-energy electrons, protons, and alpha particles leave the Sun, forming a solar wind. When these particles impact the Earth's ionosphere, radio communications are disrupted. During these times, the far northern and southern night skies are covered by shimmering, multicolored curtains of light called auroras.

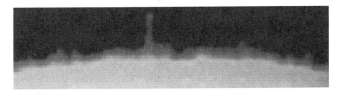

Figure A3-6. Giant fingerlike columns of gas were discovered to be present in the Sun's polar regions in Skylab ultraviolet photographs. Twice the diameter of the Earth and reaching heights of 40,000 kilometers, they are much larger than similar structures found elsewhere on the Sun.

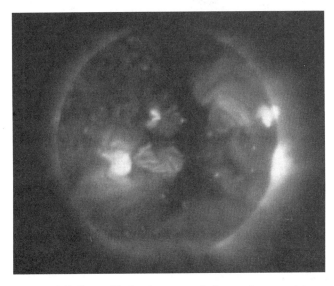

Figure A3-7. Coronal holes show up as dark areas in x-ray pictures of the Sun. They are always present at the Sun's pole. The solar wind escapes through these regions into space.

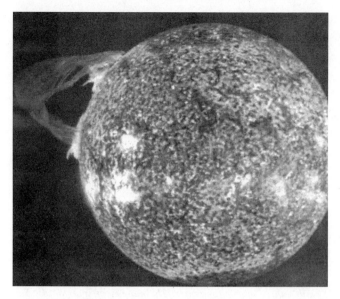

Figure A3-9. This picture shows an extremely large prominence. Skylab astronauts were able to observe its beginning, development, and end.

Prominences

Prominences are huge gaseous structures that extend up to 60,000 miles above the surface of the Sun. They are present during both the Sun's quiescent and active phases. Using Skylab's ultraviolet and x-ray telescopes, the astronauts were able to measure the temperatures within these structures. An example of how this was done is shown in Figure A3-8. Figure A3-8a shows a picture taken in light emitted from a gas at 20,000°K. Here the prominence is barely distinguishable from the surrounding region. This indicates that the prominence and surrounding gas have nearly the same temperature. In images that isolate features whose temperature is about 150,000°K, the prominence appears dark because it does not radiate much light at that temperature (see Figure A3-8b). At temperatures of $1.4 \times 10^{6°}$K (Figure A3-8c), the prominence is even darker. Based on pictures like these, it was determined that this prominence had a temperature of about 20,000°K.

The Outer Corona

The Sun's outer corona extends tens of millions of miles beyond the visible disk of the Sun. Nevertheless, it is difficult to observe from the ground because it is hidden by the bright daytime sky. Scientists lacked even basic information about the corona. How often do features in the corona change? Does the corona respond to violent events that occur on the surface of the Sun? Does gas ejected by the Sun reach the Earth?

The Sun's outer corona was extensively observed by the Skylab astraonauts. They found that the corona is dynamic and changes over time periods of days to months. When the Sun was in an active state, the astronauts detected over a hundred large-scale coronal disturbances. Some were related to solar flares,

Figure A3-8. Three views of a prominence in light originating from gas of temperatures of 20,000, 150,000, and 1,400,000°K. In the top view, the prominence is almost indistinguishable from the surrounding 20,000°K gas, indicating they have a similar temperature.

while others were associated with prominences. Eruptive prominences were also shown to create coronal disturbances that traveled outward at speeds of 2.2 to 4.3 million miles per hour. Some of this material reached the Earth and caused disturbances in the ionosphere.

Skylab's Biomedical Science Results

Based on postflight medical exams of the astronauts in the Gemini and Apollo programs, doctors had constructed a model of how the human body would react to prolonged weightlessness. This model predicted that (1) fluids normally located in the lower portion of the body would migrate upward, producing puffy eyelids, distended veins in the neck and forehead, and feelings of nasal congestion, (2) fewer red and white blood cells would be produced, (3) the potassium level in the body would be lowered, (4) the body would lose more fluids than on Earth, (5) the amount of bone calcium and muscle mass would decrease with time, and (6) after returning from space, there would be a short-term reduction in the ability of the astronaut's blood vessels to distribute nutrients to various parts of the body.

The biomedical experiments on Skylab were designed to provide data about an astronaut's changing cardiovascular, musculoskeletal, hematologic, vestibular, metabolic, and endocrine systems. Doctors sought answers to questions such as, Could nutritional changes lessen the loss of important chemical constituents in the body? How was the flow of blood in the body and the heart affected by exposure to zero gravity? Could adjustments in an astronaut's work schedule affect the body's circulatory system? How are blood cells and various body fluids affected by zero gravity? Did the rapid change from day to night in Skylab affect the natural human cycle of wakefulness and sleep?

In order to insure that the crew was in good health prior to flight, each crewman was given a thorough physical exam 30 days before lift-off. They were then placed in semi-isolation to prevent them from becoming infected by other people. During this time the astronauts only ate foods that would be available on Skylab to allow their bodies to adjust to a space diet.

Upon returning to Earth, the astronauts were given extensive postflight medical exams and were isolated for seven days or until their bodies' responses returned to their preflight values. Although American astronauts had never been in space for longer than a few weeks, NASA doctors felt adequate measures had been taken to minimize any adverse long-term health effects. Henry Cooper, in *A House in Space*, reported that the "redistribution of fluids was responsible for more of the trends the flight surgeons had noticed, and had a greater impact physically on the astronauts than any other single effect of weightlessness." Here are some of the medical findings of the Skylab mission.

Motion Sickness: Contrary to some scary predictions made before Skylab, after a short period of time, the astronauts rapidly adjusted to weightlessness and actually enjoyed it.

They did not become excessively tired and were able to work at an efficiency level that surpassed expectations.

Five of the nine Skylab astronauts had bouts with motion sickness early in their flight, but this condition quickly passed. Medical readings, routinely obtained to measure the astronauts' sensitivity to rapid motions, found that the astronauts' tolerance was actually greater in space than on Earth. Doctors were unable to predict whether an astronaut would experience motion sickness or not. Drugs could lessen the severity of space sickness but could not prevent it.

The Loss of Bone Material: All of the astronauts lost calcium and hydroxyproline, an amino acid associated with the rate at which bone mass is built while in space. Although the amount of calcium lost after eighty-four days (the longest of the three Skylab missions) was not serious, the depletion rate was large enough to cause doctors concern about the safety of future missions lasting more than a year. This result had unsettling implications for future missions to the nearby planets. Doctors also found that the astronauts' urine contained high levels of calcium. This was a concern because it could lead to the formation of painful kidney stones.

The Cardiovascular System: Data about the cardiovascular system and heart were provided by two experiments aboard Skylab, the bicycle ergometer (see Figure A3-10) and the metabolic analyzer. A surprising result of these experiments was that the astronauts' tolerance for exercise did not diminish with time. Measurements taken when the astronauts returned indicated that their bodies had begun to adjust to weightlessness.

In order to stay in good physical health the astronauts exercised daily. Rather than seeing this as a burden, they enjoyed these periods and took pride in maintaining the excellent physical condition their bodies were in prior to the flight. This became such an obsession that it caused one doctor to remark,

I have found myself asking, repeatedly, why there is this quite extraordinary emphasis on physical

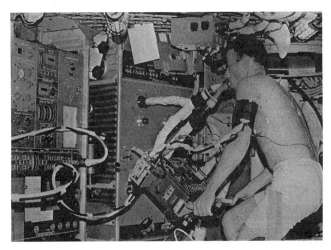

Figure A3-10. Astronaut Paul Weitz of Skylab 2 *exercises on the bicycle ergometer as part of his mission's metabolic test program.*

fitness for function in a weightless environment. Great muscular strength and endurance, which have obvious survival value in the jungle, are all but redundant in a zero-gravity environment. . . . We can select individuals already adapted to something closer to zero gravity. Here, I refer to sedentary, skinny, small individuals who would be better suited than these athletes.

Earth Resources

The Skylab Earth resources experiments provided scientists with new prospectives, but they also presented a challenge. As Cooper wrote in *A House in Space*, "The study of the Earth from space was so new that the scientists were still learning the capacities of their instruments, which were far more sensitive than the ones flown in unmanned satellites, with film that had much higher resolution." Multicolor photographs allowed scientists to classify large tracks of land into categories ranging from commercial areas to grasslands. These maps would be used by government officials in planning city growth and in determining how to meet their community's future water needs. Microwave measurements from the station proved useful in measuring the moisture content of soil, and discovering regions where a significant amount of pollution and erosion had occurred. The shape of the Earth was also measured using a radar altimeter and little-explored regions were mapped.

The astronauts never grew tired of observing the Earth. Pogue was impressed with how quickly things changed,

> Every pass was different. It was never the same from orbit to orbit. The clouds were always different, the light was different. The Earth was dynamic; snow would fall, rain would fall—you could never depend on freezing any image on your mind.

Carr was astonished at how little of the Earth's surface had been tamed by man. As he viewed the Earth from Skylab, he told controllers,

> There is nothing but a great big nothing out here now. Northern China, Outer Mongolia, and . . . the Golbi Desert. . . . Not much of the Earth is hospitable to man. We don't occupy much of our world. We're crowded into small areas.

Months after the last crew returned from Skylab, scientists were still studying the 46,000 pictures the astronauts had obtained. These frames contained a wealth of information about such diverse subjects as ice movements, volcanic eruptions, melting snow, floods, droughts, geothermal sites, sand dunes, ocean currents, and fish migrations. For the first time, these images allowed scientists to view the global impact of events on Earth.

Space Science

Each Skylab crew member made in-flight evaluations of the space station. Movies and videotapes of the crew were also scrutinized by engineers to identify areas where improvements could be made. Upon returning to Earth, each astronaut was also asked to provide suggestions about how the station could be improved. The areas identified included a less disorienting arrangement of instruments at workstations, the removal of breakable objects near module entrances, and the need for a workbench, and the desirability of minimizing the number of small objects on the station.

Skylab had fourteen experiments in its materials processing program. Crystals were formed from vapors, diffusion in liquid metals, and from the solidification of molten metals. These crystals could be made ten times larger than on Earth and they were essentially free of defects. This latter property made them particularly attractive to high-tech industries because defects degrade the electrical conductivity of the material.

Compounds could be constructed in space that were also ten times purer than on Earth. The weightless environment of Skylab also made it possible to create new components. When materials are mixed on Earth, gravity causes the elements to separate before the compound solidifies. In the weightlessness of space, the elements stay mixed during the cooling process.

Summary

Skylab was America's first space station and as such it served as a test bed for future stations while allowing the crew to conduct a full set of scientific experiments. Skylab proved to be a remarkably functional craft. The three crews spent a total of 171.5 days in space and completed nearly 300 experiments. The total cost of the program was about $3 billion. The planned versus completed number of activities in the areas of astronomy, biology, Earth, and space applications are given in Table A3-1.

The Skylab astronomy experiments provided unique data about the ultraviolet and x-ray emission of stars, particu-

Table A3-1. Skylab's Experimental Success Rate

Area	Planned	Actual	Difference
Astronomy studies	168	345	105%
Biomedical studies	701	922	32%
Earth applications studies	62	99	60%
Space application studies	274	277	1%

Adapted from *Living and Working in Space.*

larly from our Sun. Skylab greatly increased the amount of ultraviolet data astronomers possessed about stars and allowed the surface temperatures and gravities of the very hot O and B stars to be much more accurately determined. The ultraviolet observations of Skylab also revealed surprising large differences in the morphology of galaxies when compared to their visual appearance. Observations of Comet Kohoutek provided supporting evidence for the astronomical model that a comet is a large "dirty snowball." Prior spacecraft had hinted that new solar phenomena would be observable at x-ray and ultraviolet wavelengths and Skylab provided an unprecedented opportunity to search for them. As a result, coronal holes were discovered and a new class of spicules were found. Skylab data on the eruption of prominences and the changes they produced in the Sun's corona would take astronomers several years to analyze.

The biomedical results of Skylab provided both reassuring and troubling results. The model doctors had devised of the body's reaction to long-duration spaceflight was found to be fairly accurate. However, doctors were unable to prevent the onset of space sickness even though drugs were found which limited its severity. The medical feasibility of long-duration planetary missions was also called into question by the Skylab results.

Results in the areas of Earth resources, materials processing, and space science were more limited. Multicolor pictures obtained from Skylab were useful in categorizing land usage and for planning the future growth of cities. A number of different materials processing methods were also tested and several new compounds and crystals with commercially valuable properties were produced.

The United States had succeeded in building a very good space station on its first attempt, but it was also clear this knowledge could not be put to immediate use. America lacked the needed national commitment to proceed with a second-generation space station. In a repeat of the Apollo program, the hard-won advances of a major space program were not exploited. NASA was unable to persuade the political leadership to make the construction of a second space station a national priority.

Acknowledgments

Every effort was made to give credit to the sources of the figures, tables, and photographs used in this text. Chapter 14 was first printed in the July 2000 *Griffith Observer* and is reproduced with the permission of the Griffith Observatory. Acknowledgements for the other materials are given below.

Section 1-1 NASA AS8-8-14-2383, Focus Box (p. 2) (left) Konstantin E. Tsiolkovsky State Museum of the History of Cosmonautics, (center) Goddard Museum of Clark University, (right) Hermann Oberth Raumfahrt Museum, Figure 1-1 Tsiolkovsky Museum, F1-2 Centre de Documentation Jules Verne (Amiens), F1-3 Edgar Fahs Smith Collection/Univ. of Penn. Library, F1-4 and 5 Konstantin E. Tsiolkovsky State Museum of the History of Cosmonautics, F1-6, Ordway Collection U.S. Space and Rocket Center, F1-7 NASM 75-11483, F1-8 Konstantin E. Tsiolkovsky State Museum of the History of Cosmonautics, F1-9 Nicholas Fasciano in *Man in Space*, F1-10 James Harford. F2-1 to F2-10 Goddard Museum of Clark University, F2-11 Ordway Collection U.S. Space and Rocket Center, F3-2 Hermann Oberth Raumfahrt Museum, F3-3 Savajevo plate 19, F3-4 to F3-9 Hermann Oberth Raumfahrt Museum, Section II Figure 2-1 (left) The Works of H. G. Wells (right) Centre de Documentation Jules Verne (Amiens), F2-2 NASA, F2-3 *War of the Worlds*, F2-4 World Book Encyclopedia, F4-1 (a) Konstantin E. Tsiolkovsky State Museum of the History of Cosmonautics (b) Hermann Oberth Raumfahrt Museum (c) Goddard Museum of Clark University, F4-2 to F4-8 Ordway Collection U.S. Space and Rocket Center, F4-9 World Book Encyclopedia, F5-1 and 2 McGraw-Hill Companies, F5-3 NASM 76-7559, F5-4 Ordway Collection U.S. Space and Rocket Center, F5-5 *Space Age* by Walters, F5-6 Ordway Collection U.S. Space and Rocket Center, F5-7 *Space Age* by Walters, F5-8 Deutsches Museum Bildstelle (Munich), F5-9 Ordway Collection U.S. Space and Rocket Center, Focus Box p. 36 NASA MSFC 8915531, F5-10 National Archives NWDNS-111-SC-231809, F5-11 McGraw-Hill Companies, F5-12 NASA MFL-59-1414, F6-1 U.S. Navy, F6-2 NASA History Office, Focus Box p. 41 Sky and Telescope Nov 1957 and Dec 1957, F6-3 and 4 Washington Post, F6-5 NASA History Office, F6-6 Paul Donaldson Sky and Telescope Jan 1958, F6-7 NASA, F6-8 National Archives NWDNS-255-PV-CC, F6-9 Ordway Collection U.S. Space and Rocket Center, F6-10 NASA MSFC 5663627, F7-2 NASA 58-ADM-1, F7-3 NASA, F7-4 and 5 Ordway Collection U.S. Space and Rocket Center, F7-6 Novosti, F8-1 to 4 NASA *This New Ocean*, F8-3, NASA Project Mercury, Focus Box p. 53 NASA Project Mercury, F8-4 NASA Project Mercury, F8-5 NASA S64-19600, F8-6 NASA G60-2552, F-7 NASA 63-19199, F8-8 NASA *This New Ocean*, F8-9 S63-18198, F9-1 to 8 Novosti, Focus Box p. 64 Novosti, Section 3-1 NASA 58-ADM-1, S3-3 to 4 Novosti, S3-5 NASA, S3.6 NASA 70-18218. F10-1 NASA S62-3984, F10-2 NASA S63-23617, F10-3 Washington Post, F10-4 NASA G60-2742, F10-5 S63-22731, F10-6 NASA, F10-7 S67-19572, F10-8 NASA MSFC M61-1642-11, F10-9 NASA S61-02821

F11-1 to 3 NASA *This New Ocean*, F11-4 NASA S62-363, F11-5 NASA S62-06040, F11-6 NASA 62-MA6-178, F11-7 S62-02849, F11-8 S62-05458, F11-9 and 10 NASA *This New Ocean*, F12-1 NASA AS11-44-6667, F12-2 to 17 NASA Lunar Impact, F13-1 NASA, F13-2 NASA History Office, F13-3 to 6 NASA, Focus Box p. 99 NASA SP-200, Focus Box p. 101 Lick Observatory and NASA S66-68571, F13-7 NASA 66-H-476, F13-8 to 12 NASA, F14-1 Novosti, Focus Box p. 106. NASA History Office, F14-2 NASA History Office, F14-3 International Women's Air and Space Museum, F14-4 NASA History Office, F14-5 NASA G60-2739, F14-6 S64-34357, F15-1 NASA S65-54354, F15-2 NASA S65-04364, Focus Box p. 114. NASA S65-14257, F15-3 NASA, F15-4 NASA S66-54830, F15-5 S64-21560, F15-6 NASA Project Gemini, F15-7 NASA S65-H-1019, F15-18 NASA S65-28691, F16-1 NASA S65-H-1944, F16-2 NASA S65-63188, F16-3 NASA S65-66718, F16-4 NASA S66-24482, F16-5 NASA S66-32550, F16-6 NASA S66-37966, F16-7 NASA S66-34844, F16-8 NASA S66-54677, F16-9 NASA S66-54454, F16-10 NASA S66-59917, F17-1

NASA S79-35681, F17-2 NASA AS14-66-9344, F17-3 NASA S65-04549. F17-4 NASA S69-34072, F17-5 NASA, F17-6 NASA AS9-21-3212, F17-7 NASA SS66-23614, F17-8 NASA SS66-20416, F17-9 NASA S66-22932, F17-10 NASA Apollo Expeditions, F17-11 NASA S66-32074, F17-12 NASA S64-22331, F17-13 NASA SS66-23998, F18-1 NASA S66-30236. F18-2 NASA S67-21294, F18-3 NASA, F18-4 NASA S67-21295, F19-1 NASA S67-50433, F19-2 NASA S68-33119, F19-3 and 4 Apollo Expeditions, F19-5 NASA S68-50713, F19-6 NASA, F19-7 NASA Man in Space, F19-8 NASA S68-50265, F19-9 NASA AS8-14-2383, F19-10 NASA S69-17590, F19-11 NASA AS9-20-3064, F19-12 NASA AS9-21-3212, F19-13 NASA S69-34329, F19-14 NASA AS10-27-3873, F19-15 NASA Apollo Expeditions, F20-1 NASA S69-31739, Focus Box p. 146 NASA S69-34875, F20-2 NASA AS11-44-6574, F20-5 NASA AS11-37-5437, F20-6 NASA AS11-37-5458, Focus Box p. 150 NASA AS11-40-5866, AS11-40-5868, AS11-40-5869, F20-7 NASA AS11-40-5878. F20-8 NASA AS11-40-5942, F20-9 NASA AS11-40-5875. F20-10 NASA AS11-44-6642, F20-11 NASA S69-21365, Focus Box p. 154 Apollo Expeditions F21-1 NASA S69-38852, F21-2 NASA AS11-44-6667, F21-3 NASA AS12-46-6729, F21-4 NASA Apollo Expeditions, Focus Box p. 158 NASA AS11-40-5492, F21-5 NASA S69-59538, F21-6 NASA AS12-48-7133, F21-7 NASA S70-36485, F21-8 NASA S70-17696, F21-9 S70-35368, Focus Box p. 161 NASA S71-16745, F21-10 NASA AS13-062-8929, F21-11 NASA, F21-12 NASA AS13-59-8501, F21-13 NASA S70-35614, F22-1 NASA, F22-2 NASA S70-55387, F22-3 NASA AS11-44-6667, F22-4 NASA LOIV-120H3, F22-5 NASA AS14-66-9405, Focus Box p. 169 NASA S70-50764, F22-6 NASA AS14-68-9405, F22-7 NASA AS14-64-9089, F22-8 NASA S71-20784, F22-9 NASA S71-20374, F23-1 NASA S71-37963, F23-2 NASA LOIV-102H3, F23-3 NASA AS15-85-11425, Focus Box, p. 175 NASA AS15-88-11901, F23-4 AS15-85-11425, F23-5 NASA Where No Man Has Gone Before, Focus Box p. 176 Yerkes Observatory, F23-6 NASA S71-44150, F23-7 NASA S72-16660, F23-8 NASA Apollo Expeditions, F23-9 NASA AS16-114-18388, Focus Box p. 180 S72-147-V F23-10 NASA AS17-147-22465, F23-11 NASA S72-50438, Focus Box p. 182 NASA S72-3145-V, F23-12 NASA AS176-145-22165, F23-13 NASA AS17-140-21496, F23-14 NASA S72-55417, F23-15 NASA S72-55421, F23-16 NASA S72-55976, SIV-1 Goddard Museum of Clark University, SIV-2 Novosti, SIV-4 N. Korolev, F24-1 to 3 N. Korolev, F24-4 and 6 Novosti, F24-7 NASM NASM 96-16119, F24-8 and 9 N. Korolev, F24-10 NASM 80-19211, F24-11 to 14 Novosti, F25-4 NASA Landsat, F25-5 Novosti, F25-7 and 8 Soviet Manned Space Program, Focus Box p. 201. Almanac of Soviet Manned Space Flight (Gulf Publishing Company), Focus Box p. 204 Novosti, F26-1 Soviet Manned Space Program, Focus Box p. 205 Soviet Manned Space Program, F26-2 Novosti. F26-3 Soviet Manned Space Program, F26-4 to 7 Novosti, Focus Box p. 208 Novosti, F26-8 Soviet Manned Space Program, F26-9 and 10 Novosti, F26-11 James Harford, F26-12 Novosti, Focus Box p. 211 Handbook of Soviet Manned Space Flight (American Astronautical Society), F27-1 NASA, Focus Box p. 214 American Astronautical Society, F27-2 Novosti, F27-3 Novosti, F27-4 Soviet Manned Space Program, F27-5 to 7 Novosti, F27-8 NASA, F28-1 to 7 Novosti, Focus Box p. 223 Soviet Manned Space Program, F29-1 Novosti, F29-2 Novosti, F29-3 Novosti, F30-1 and 2 Novosti, Focus Box p. 231 Soviet Manned Space Program, F30-3 to 5 Novosti, F30-6 Soviet Manned Space Program, F30-7 and 8 Novosti

F31-1 NASA, F31-2 NASA S73-15240, F31-3 NASA, Focus Box p. 242 Soviet Manned Space Program, Focus Box p. 243 NASA S73-679-V, F31-4 NASA AST-32-2695, F31-5 AST-01-056, Focus Box p. 244 NASA S75-27290, F31-6 Novosti, F32-1 NASA S73-25654, F32-2 NASA S73-23952, Focus Box p. 248. NASA, Skylab Our First Space Station, F32-3 NASA, Focus Box p. 250 NASA, F32-4 NASA SL2-01-107, F32-5 NASA SL3-114-1660, F33-1 NASA S72-17512, F33-2 NASA S73-28419, F33-4 NASA, Skylab Our First Space Station, F33-5 NASA SL3-108-1282, F33-6 NASA S74-17306, F33-7 NASA SL2-02-161, F34-1 NASA, F34-2 Ordway Collection U.S. Space and Rocket Center, F34-3 Bell Aircraft Company, F34-4 NASA, F34-5 and 6 NASA, F34-7 NASA S83-35799, F34-8 and 9 NASA, F34-10 NASA S81-33179, F34-11 NASA S81-30954, Focus Box p. 269 Soviet Manned Space Program, F35-1 Novosti, Focus Box p. 271 Soviet Manned Space Program, F35-2 to 4 Novosti, Focus Box p. 273. Soviet Manned Space Program, F35-5 Novosti, Focus Box p. 275 Soviet Manned Space Program, F35-5 to 11 Novosti, Focus Box p. 281 Soviet Manned Space Program, F36-1 to 4 Novosti, Focus Box p. 284 Soviet Manned Space Program, F35-5 to 37-4 Novosti, F38-1 NASA S81-30492, F38-2 and 3 NASA, F38-4 NASA STS1-12-332, Focus Box p. 298 NASA, F38-5 NASA S81-30425, F38-6 NASA, F38-7 NASA S85-41239, F38-8 NASA S85-44253, F38-9 NASA 51L-10151, F38-10 NASA 51L-10098, F38-11 51L-10089, F38-12 NASA S86-38989, F38-13 NASA S88-29251, F38-14 NASA KSC-98PC-1440, F38-15 NASA, F38-16 NASA S98-04614, F38-17 NASA, F39-1 to 11 NASA/RSA, F40-1 NASA KSC-96PC-1113, F40-2 NASA KSC-96PC1007, F40-3 to 10 NASA/RSA, F40-11 NASA

F41-1 Ronald Reagan Presidential Library, F41-2 to 4 NASA, F41-5 Countdown, John Wiley & Sons, F41-6 NASA LG-1999-08-001-JSC, F41-7 NASA HQL-425, F41-8 NASA S97-17671, F41-9 NASA STS088-365-020, F41-10 NASA 97-E-04077, F41-11 NASA 97-10687, F41-12 NASA, F41-13 NASA HQL-426, F42-1 George Bush Presidential Library, Focus Box p. 336 NASA, F42-2 NASA Orbital Debris from Upper-Stage Breakup, Focus Box p. 338 NASA, F42-3 and 4 NASA, F43-1 Ordway Collection U.S. Space and Rocket Center, F43-2 NASA, F43-3 and 4 NASA/USGS, Focus Box p. 343 NASA. USGS, F43-5 to 7 NASA/USGS

Focus Box p. A-2 NASA AS11-44-6667, A1-1 Lick Observatory, A1-2 Lunar and Planetary Institute (Houston), A1-3 Lunar SourceBook, Cambridge University Press, A1-4 NASA, A1-5 Lick Observatory, A1-6 NASA, Focus Box p. A-6 and 7, Lunar SourceBook, Cambridge University Press, A1-8 and 9 NASA, A1-10 Lunar SourceBook, Cambridge University Press, Focus Box p. A-8 NASA, A1-11 Lunar Source Book, Cambridge University Press, A1-12 NASA, A1-13 Lunar Source Book, Cambridge University Press, A1-14 and 15 Lunar and Planetary Institute (Houston), A1-16 and 17 NASA, A2-1 and 2 NASA, A2-3 and 4 Lunar and Planetary Institute (Houston), A2-5 NASA, A2-6 to 8 (1971) Icarus 15, 368 Academic Press, Appendix Tables A2-1 Lunar and Planetary Institute (Houston), A2-2 and A2-3 *To a Rocky Moon* (Univ. of Arizona Press), A3-1 to 10 NASA, Focus Box p. A-22 NASA

References

Section 1: The Early Pioneers of Space Travel

Congressional Recognition of Goddard Rocket and Space Museum with Tributes to Dr. Robert H. Goddard, 1970, U.S. Senate Document No. 91-71, U.S. Government Printing Office, Washington, DC.

Manned Spaceflight Log by Tim Furniss, 1986, Jane's Publishing Company, London, England.

Men of Space by Shirley Thomas, 1960, Chilton Company, Philadelphia, PA.

Space Travel: A History by Wernher von Braun and Frederick I. Ordway III, 1985, Harper & Row, Philadelphia, PA.

The New Russian Space Programme by Brian Harvey, 1996, John Wiley & Sons, Inc., New York, NY.

Chapter 1: Konstantin Tsiolkovsky (1857–1935)

Behind the Sputniks: A Survey of Soviet Space Science by F. J. Krieger, 1958, Public Affairs Press, Washington, DC.

Jules Verne by Jean-Jules Verne, 1976, Taplinger Publishing Co. Inc., New York, NY.

Konstantin Tsiolkovsky and the Present Times by A. A. Kosmodemjansky in *History of Rocketry and Astronautics*, Roger D. Launius, editor, 1994, AAA History Series, Vol. 11, AAA Publications Office, San Diego, CA.

Korolev by James Harford, 1997, John Wiley & Sons, New York, NY.

Men of Space by Shirley Thomas, 1960, Chilton Company, Philadelphia, PA.

Space Exploration by J. K. Davies, 1992, Chambers Encyclopedic Guides, Edinburgh, New York.

NASA Technical Translations TT F-243, M. K. Tikhonravov, chief editor, G. I. Sedlenek, editor, T. N. Trofimova, technical editor, 1965, U.S. Government Printing Office, Washington, DC.

Space Age by William J. Walter, 1992, Random House, New York, NY.

Space Travel: A History by Wernher von Braun, Frederick I. Ordway III, and Dave Dooling, 1975, Harper & Row, New York, NY.

The New Russian Space Programme: From Competition to Collaboration by Brian Harvey, 1996, John Wiley & Sons, New York, NY.

Chapter 2: Robert Goddard (1882–1945)

Congressional Recognition of Goddard Rocket and Space Museum, 1970, U.S. Senate Document No. 91-71, U.S. Government Printing Office, Washington, DC.

Robert H. Goddard by Milton Lehman, 1988, A Da Capo Press, Inc., New York, NY.

Space Travel: A History by Wernher von Braun and Frederick I. Ordway III, 1985, Harper & Row, Philadelphia, PA.

The Papers of Robert H. Goddard, Volume I, 1898–1924, Esther C. Goddard, editor, G. Edward Pendray, assoc. editor, 1970, McGraw-Hill Book Company, New York, NY.

The Papers of Robert H. Goddard, Volume II, 1924–1945, Esther C. Goddard, editor, G. Edward Pendray, assoc. editor, 1970, McGraw-Hill Book Company, New York, NY.

This High Man by Milton Lehman, 1963, Farrar, Straus and Company, New York, NY.

Chapter 3: Hermann Oberth (1894–1989)

Hermann Oberth: Father of Space Travel by Helen B. Walters, 1962, The MacMillan Company, NY.

Hermann Oberth, The Father of Space Flight by Boris V. Rauschenbach, 1994, edited by B. John Zavrel, translated from German by Lynne Kvinnesland, West-Art, New York.

Sarajevo by Joachim Remak, 1959, Criterion Book, Inc., New York, NY.

Space Travel: A History by Wernher von Braun, Frederick I. Ordway III, and Dave Dooling, 1975, Harper & Row, New York, NY.

Section II: Beyond the Space Pioneers

Jules Verne: An Exploratory Biography by Herbert R. Lottman, 1996, St. Martin's Press, New York, NY.

Space Age by William Walter, 1992, Random House, New York.

Space Travel: A History by Wernher von Braun and Frederick I. Ordway III, 1975, Harper & Row, New York, NY.

Space Travel: A History by Wernher von Braun and Frederick I. Ordway III, 1985, Harper & Row, Philadelphia, PA.

The Politics of the Space by Matthew J. Von Bencke, 1997, Westview Press, A Division of Harper and Collins Publishers, Inc., Boulder, CO.

The War of the Worlds by H.G. Wells, 1964, The George Macy Companies, Inc., New York, NY.

The Works of H. G. Wells Volume I, 1924, Charles Scribner's Sons, New York, NY.

The World Book Encyclopedia (Volume 20), 1965, Field Enterprises Educational Corporation, Chicago, IL.

Chapter 4: World-Wide Interest in Rockets

Hermann Oberth: The Father of Space Flight by Boris B. V. Rauschenbach, 1994, West-Art, New York.

Korolev by James Harford, 1997, John Wiley & Sons, New York, NY.

Space Age by William Walter, 1992, Random House, New York.

Space Travel: A History by Wernher von Braun, Frederick I. Ordway III, and Dave Dooling, 1975, Harper & Row, New York, NY.

Chapter 5: Wernher von Braun (1912–1977)

Astronautical Engineering and Science, Ernst Stuhlinger, Fredrick Ordway, III, Jerry C. McCall, George C. Bucher, editors, 1963, McGraw-Hill Book Company, Inc., New York, NY. From an article by James L. Daniels, Jr., P366–375.

Hermann Oberth: Father of Space Flight by Helen B. Walters, 1962, The MacMillan Company, New York, NY.

Rockets and Missiles: Past and Future by Martin Caidin, The McBride Company, New York, NY.

Space Age by William J. Walter, 1992, Random House, Inc., New York, NY.

Space Exploration by J. K. Davies, 1992, W&R Chambers Ltd. New York, NY.

Space Travel: A History by Wernher von Braun, Frederick I. Ordway III, and Dave Dooling, 1975, Harper & Row, New York, NY.

The Men Behind the Space Rockets by Heinz Gartmann, 1956, David McKay Comp. Inc., New York.

The Peenumunde Wind Tunnels by Peter P. Wegener, 1996, Yale University Press, New Haven, CT.

The Rockets' Red Glare by Wernher von Braun and Frederick Ordway III, 1976, Anchor Press/Doubleday Garden City, New York.

Wernher von Braun by Ernst Stuhlinger, 1994, Krieger Publishing Co., Malabar, FL.

Wernher von Braun: The Man Who Sold the Moon by Dennis Piszkiewicz, 1998, Praeger Publishers, Westport, CT.

Chapter 6: Sputnik and the American Response

Between Sputnik and the Shuttle: New Perspectives on American Astronautics, Frederick C. Durant III, editor, 1981, AAS Publications Office, San Diego, CA.

Exploring Space: Voyages in the Solar System and Beyond by William E. Burrows, 1990, Random House, New York.

Ike: His Life and Times by Piers Brendon, 1986, Harper & Row, New York.

Russians in Space by Evgeny Riabchikov, 1971, Doubleday & Company, Inc., Garden City, NY.

The Politics of Space by Matthew J. Von Bencke, 1997, Westview Press, Boulder, CO.

The Sputnik Challenge by Robert A. Devine, 1993, Oxford University Press, New York, NY.

Wernher von Braun by Ernst Stuhlinger, 1994, Krieger Publishing Company, Malabar, FL.

Chapter 7: The Space Race Begins in Earnest

Exploring Space: Voyages in the Solar System and Beyond by William E. Burrows, 1990, Random House, New York.

Ike: His Life and Times by Piers Brendon, 1986, Harper & Row, New York.

NASA, The First 25 Years, 1958–1983, 1983, EP-182, Superintendent of Documents, Government Printing Office, Washington, DC.

Orders of Magnitude: A History of the NACA and NASA, 1915–1990 by Roger E. Bilstein, 1989, SP-4406, U.S. Government Printing Office, Washington, DC.

Space Travel: A History by Wernher von Braun, Frederick I. Ordway III, and Dave Dooling, 1975, Harper & Row, New York, NY.

Space Age by William Walter, 1992, Random House, New York.

The End of an Era in Space Exploration by J. C. D. Blaine, 1976, distributed by Univelt Inc., San Diego, CA.

Chapter 8: The Early Stages of Project Mercury

A Picture History: Rockets and Rocketry by David S. Akens, 1964, The Strode Publishers, Huntsville, AL.

Apollo Expeditions to the Moon, Edgar M. Cortright, editor, 1975, NASA SP-350, Superintendent of Documents, Washington, DC.

Between Sputnik and the Shuttle, Frederick C. Durant III, editor, 1981, AAS History Series Vol. 3, American Astronautical Society, Univelt Inc., San Diego, CA.

Handbook of Soviet Manned Spaceflight by Nicholas L. Johnson, 1980, Vol. 48, Science and Technology Series, American Astronautical Society, Univelt, Inc., San Diego, CA.

Project Mercury: A Chronology by James M. Grimwood, 1963, NASA SP-4001, U.S. Government Printing Office, Washington, DC.

Space Almanac, Second Edition by Anthony R. Curtis, 1992, Gulf Publishing Company, Houston, TX.

Space Travel: A History by Wernher von Braun, Frederick I. Ordway III, and Dave Dooling, 1975, Harper & Row, New York, NY.

This New Ocean: A History of Project Mercury by Lloyd Swenson, Jr., James Grimwood, and Charles Alexander, 1966, NASA SP-4201, U.S. Government Printing Office, Washington, DC.

The Origin of NASA Names by Helen T. Wells, Susan H. Whiteley, and Carrie E. Karegeannes, 1976, NASA SP-4402, Superintendent of Documents, Washington, DC.

This New Ocean: A History of Project Mercury by Lloyd S. Swenson, Jr., James M. Grimwood, and Charles C. Alexander, 1966, NASA SP-4201, Superintendent of Documents, Washington, DC.

We Seven by the Astronauts Themselves, 1962, Simon and Schuster, New York, NY.

Chapter 9: Yuri Gagarin—The First Man in Space

Between Sputnik and the Shuttle, Frederick C. Durant III, editor, 1981, AAS History Series, Vol. 3, American Astronautical Society, Univelt Inc., San Diego, CA.

Handbook of Soviet Manned Spaceflight by Nicholas L. Johnson, 1980, Vol. 48, Science and Technology Series, American Astronautical Society, Univelt, Inc., San Diego, CA.

Korolev by James Harford, 1997, John Wiley & Sons, New York, NY.

Life—The First Fifty Years: 1936–1986, Philip B. Kunhardt, Jr., editor, 1986, Time, Inc., Boston, MA.

Russians in Space by Evgeny Riabchikov, edited by Colonel General Nikolai P. Kamanin, translated by Gut Daniels, Doubleday and Company, Inc., Garden City, NY.

Starman: The Truth Behind the Legend of Yuri Gagarin by Jamie Doran and Piers Bozony, 1998, Bloomsbury Publishing, London, England.

The Soviet Manned Space Program by Phillip Clark, 1988, Salamander Book, New York, NY.

Section III: America Plans to Place a Man on the Moon

Ike: His Life and Times by Piers Brendon, 1986, Harper & Row, New York.

Man in Space: An Illustrated History of Space Flight, H.J.P. Arnold, editor, 1993, SMITHMARK Publishers Inc., New York, NY.

NASA: A History of the U.S. Civil Space Program by Roger D. Launius, 1994, Krieger Publishing Company, Malabar, FL.

NASA, The First 25 Years, 1958–1983, EP-182, 1983, Superintendent of Documents, Government Printing Office, Washington, DC.

Orders of Magnitude: A History of the NACA and NASA, 1915–1990 by Roger E. Bilstein, 1989, SP-4406, U.S. Government Printing Office, Washington, DC.

Space Age by William Walter, 1992, Random House, New York.

Spaceflight and the Myth of Presidential Leadership, Roger D. Launius and Howard E. McCurdy, editors, 1997, University of Illinois Press, Urbana, IL.

Space Travel: A History by Wernher von Braun and Frederick Ordway III, 1985, Harper & Row, New York.

The End of an Era in Space Exploration by J. C. D. Blaine, 1976, distributed by Univelt Inc., San Diego, CA.

The Politics of Space by Matthew J. Von Bencke, 1997, Westview Press, Oxford, England.

Chapter 10: Project Mercury's Suborbital Flights

A Picture History: Rockets and Rocketry by David S. Akens, 1964, The Strode Publishers, Huntsville, AL.

Apollo Expeditions to the Moon, Edgar M. Cortright, editor, 1975, NASA SP-350, Superintendent of Documents, Washington, DC.

Between Sputnik and the Shuttle by Frederick C. Durant III, editor, 1981, AAS History Series, Vol. 3, American Astronautical Society, Univelt Inc., San Diego, CA.

Exploring the Unknown: Selected Documents in the History of the U.S. Civil Space Program, Volume I: Organizing for Exploration, John Logsdon, editor, 1995, NASA SP-4218, The NASA History Series, U.S. Government Printing Office, Washington, DC.

Handbook of Soviet Manned Spaceflight by Nicholas L. Johnson, 1980, Vol. 48, Science and Technology Series, American Astronautical Society, Univelt, Inc., San Diego, CA.

Project Mercury: A Chronology by James M. Grimwood, 1963, NASA SP-4001, U.S. Government Printing Office, Washington, DC.

Space Travel: A History by Wernher von Braun, Frederick I. Ordway III, and Dave Dooling, 1975, Harper & Row, New York, NY.

The Origin of NASA Names by Helen T. Wells, Susan H. Whiteley, and Carrie E. Karegeannes, 1976, NASA SP-4402, Superintendent of Documents, Washington, DC.

The World Book Encyclopedia, 1965, Field Enterprises Educational Corporation, Chicago, IL.

This New Ocean: A History of Project Mercury by Lloyd Swenson, Jr., James Grimwood, and Charles Alexander, 1966, NASA SP-4201, U.S. Government Printing Office, Washington, DC.

We Seven by the Astronauts Themselves, 1962, Simon and Schuster, New York, NY.

Chapter 11: Project Mercury's Manned Orbital Flights

A Picture History: Rockets and Rocketry by David S. Akens, 1964, The Strode Publishers, Huntsville, AL.

Apollo Expeditions to the Moon, Edgar M. Cortright, editor, 1975, NASA SP-350, Superintendent of Documents, Washington, DC.

Between Sputnik and the Shuttle, Frederick C. Durant III, editor, 1981, AAS History Series, Vol. 3, American Astronautical Society, Univelt Inc., San Diego, CA.

Exploring the Unknown: Selected Documents in the History of the U.S. Civil Space Program, Volume I: Organizing for Exploration, John Logsdon, editor, 1995, NASA SP-4218, The NASA History Series, U.S. Government Printing Office, Washington, DC.

Handbook of Soviet Manned Spaceflight by Nicholas L. Johnson, 1980, Vol. 48, Science and Technology Series, American Astronautical Society, Univelt, Inc., San Diego, CA.

Project Mercury: A Chronology by James M. Grimwood, 1963, NASA SP-4001, U.S. Government Printing Office, Washington, DC.

Space Travel: A History by Wernher von Braun, Frederick I. Ordway III, and Dave Dooling, 1975, Harper & Row, New York, NY.

The Origin of NASA Names by Helen T. Wells, Susan H. Whiteley, and Carrie E. Karegeannes, 1976, NASA SP-4402, Superintendent of Documents, Washington, DC.

The World Book Encyclopedia, 1965, Field Enterprises Educational Corporation, Chicago, IL.

This New Ocean: A History of Project Mercury by Lloyd Swenson, Jr., James Grimwood, and Charles Alexander, 1966, NASA SP-4201, U.S. Government Printing Office, Washington, DC.

We Seven by the Astronauts Themselves, 1962, Simon and Schuster, New York, NY.

Chapter 12: Unmanned Lunar Reconnaissance Missions Part 1—Project Ranger

Countdown: A History of Space Flight by T. A. Heppenheimer, 1997, John Wiley & Sons, Inc., New York, NY.

Exploring Space: Voyages in the Solar System and Beyond by William E. Burrows, 1990, Random House, New York, NY.

Lunar Impact: A History of Project Ranger by R. Cargill Hall, 1977, NASA SP-4210, U.S. Government Printing Office, Washington, DC.

Project Ranger: A Chronology by R. Cargill Hall, 1971, JPL/HR-2, U.S. Government Printing Office, Washington, DC.

Space Exploration, 1992, Chambers Encyclopedic Guides, Chambers, NY.

Exploring Space: Voyages in the Solar System and Beyond by William E. Burrows, 1990, Random House, New York.

The Nature of the Lunar Surface, Proc. of the IUA-NASA Symposium April 15–16, 1965, Wilmot N. Hess, Donald H. Menzel, and John A. O'Keefe, editors, John Hopkins Press, Baltimore, MD.

The Nature of the Lunar Surface, C. T. Leondes and R. W. Vance, editors, 1966, John Hopkins Press, Baltimore, MD.

Where No Man Has Gone Before: A History of Apollo Lunar Exploration Missions by William David Compton, 1989, NASA SP-4214, U.S. Government Printing Office, Washington, DC.

Chapter 13: Unmanned Lunar Reconnaissance Missions Part 2—Projects Lunar Orbiter and Surveyor

Exploring Space: Voyages in the Solar System and Beyond by William E. Burrows, 1990, Random House, New York, NY.

Lunar Orbiter Photographic Atlas of the Moon by David E. Bowker and J. Kenrick Hughes, 1971, NASA SP-206, U.S. Government Printing Office, Washington, DC.

Moon Missions: Mankind's First Voyages to Another World by William F. Mellberg, 1997, Plymouth Press, Ltd., Plymouth, MI.

The Moon as Viewed by Lunar Orbiter by L. J. Kosofsky and Farouk El-Baz, 1970, NASA SP-200, 1970, U.S. Government Printing Office, Washington, DC.

The Nature of the Lunar Surface, Proc. of the IUA-NASA Symposium April 15–16, 1965, Wilmot N. Hess, Donald H. Menzel, and John A. O'Keefe, editors, John Hopkins Press, Baltimore, MD.

The Nature of the Lunar Surface, C. T. Leondes and R. W. Vance, editors, 1965, U.S. Government Printing Office, Washington, DC.

Space Exploration, 1992, Chambers Encyclopedic Guides, Chambers, NY.

Unmanned Space Project Management: Surveyor and Lunar Orbiter by Erasmus H. Kloman, 1972, NASA SP-4901, Superintendent of Documents, Washington, DC.

Voyages in the Solar System and Beyond by William E. Burrows, 1990, Random House, New York.

Where No Man Has Gone Before: A History of Apollo Lunar Exploration Missions by William David Compton, 1989, NASA SP-4214, U.S. Government Printing Office, Washington, DC.

Chapter 14: The Mercury 13 and the Selection of Women Astronauts

Hearing Before the Special Subcommittee on the Selection of Astronauts, July 17 and 18, 1962, Eighty-Seventh Congress, U.S. Government Printing Office, Washington, DC.

Jerrie Cobb: Solo Pilot by Jerrie Cobb, 1997, Jerrie Cobb Foundation, Inc., Sun City, FL.

Chapters 15 and 16: Project Gemini: Parts I and II

Countdown by Frank Borman with Richard J. Serling, 1988, Silver Arrow Books, New York, NY.

Gemini by Virgil Grissom, 1968, The MacMillan Company, New York.

Jane's Spaceflight Directory 1986, Reginald Turnill, editor, Jane's Publishing Inc, New York, NY.

Liftoff: The Story of America's Adventure in Space by Michael Collins, 1988, Grove Press, New York, NY.

Manned Spaceflight Log by Tim Furniss, 1986, Jane's Publishing Company, Ltd., London, England.

On the Shoulders of Titans: A History of Project Gemini, by Barton C. Hacker and James M. Grimwood, 1977, NASA SP-4203, U.S. Government Printing Office, Washington, DC.

Project Gemini, A Chronology by James M. Grimwood and Barton Hacker with Peter J. Vorzimmer, 1969, NASA Historical Series SP-4002, U.S. Government Printing Office, Washington, DC.

Project Gemini, The New Space Encyclopedia, E. P. Dutton and Co., New York.

Schirra's Space by Wally Schirra with Richard N. Billings, 1988, Bluejacket Books, Naval Institute Press, Annapolis, MD.

Space Exploration by J. K. Davies, 1992, Chambers Encyclopedic Guides, Chambers, Ltd., New York.

The End of an Era in Space Exploration by B. C. D. Blaine, An American Astronautical Society Publication, 1976, Univelt, Inc., San Diego, CA.

The Early Years: Mercury to Apollo-Soyuz, NASA Information Summaries PMS 001-A (KSC) May 1987.

The Moon Explorers by Tony Simon, 1970, Scholastic Book Services, New York.

Chapter 17: The Flying Machines of Apollo

Apollo and Universe, S. T. Butler, editor, 1968, Pergamon Press, Toronto.

Apollo: Expeditions to the Moon, Edgar M. Cortright, editor, 1975, National Aeronautics and Space Administration, Washington, DC.

Countdown: A History of Space Flight by T. A. Heppenheimer, 1997, John Wiley & Sons, New York, NY.

Moon Missions By William F. Mellberg, 1997, Plymouth Press, Ltd., Plymouth, MI.

NASA: A History of the U.S. Civil Space Program by Roger D. Launius, 1994, Krieger Publishing Company, Malabar, FL.

NASA Historical Data Book Volume II: Programs and Projects 1958–1968 by Linda Neuman Ezell, NASA SP-4012, 1988, U.S. Government Printing Office, Washington, DC.

Project Apollo: The Way to the Moon by P. J. Booker, 1969, American Elsevier Publishing Company, Inc., New York.

Space Travel: A History by Wernher von Braun, Frederick I. Ordway III, and Dave Dooling, 1975, Harper & Row, New York, NY.

The Apollo Spacecraft: A Chronology, Volume II by Mary Louise Morse and Jean Kernahan Bays, 1976, NASA Historical Series, SP-4009, Superintendent of Documents, Washington, DC.

The McGraw-Hill Encyclopedia of Space, 1967, McGraw-Hill Book Company, New York, NY.

Chapter 18: Apollo 1—The High Cost of Bravery

A Man on the Moon: The Voyages of the Apollo Astronauts by Andrew Chaikin, 1994, Penguin Books, New York, NY.

Apollo Expedition to the Moon, Edgar M. Cortright, editor, 1975, NASA SP-350, Superintendent of Documents, Washington, DC.

Chariots for Apollo: A History of the Manned Lunar Spacecraft by Courtney G. Brooks, James M. Grimwood, and Lloyd S. Swenson, Jr., 1979, NASA SP-4205, Superintendent of Documents, Washington, DC.

Countdown by Frank Borman and Robert Serling, 1988, Silver Arrow Books, William Morrow, New York.

End of an Era in Space Exploration by B. C. D. Blaine, An American Astronautical Society Publication, 1976, Univelt, Inc., San Diego, CA.

Orders of Magnitude: A History of NACA and NASA, 1915–1990 by Roger Bilstein, 1989, NASA SP-4406, U.S. Government Printing Office, Washington, DC.

Space Travel: A History by Wernher von Braun, Frederick I. Ordway III, and Dave Dooling, 1975, Harper & Row, New York, NY.

The McGraw-Hill Encyclopedia of Space, 1967, McGraw-Hill Book Company, New York, NY.

Chapter 19: Apollo Missions 4–10—From the Earth to Lunar Orbit

Apollo Expeditions to the Moon, Edgar M. Cortright, editor, 1975, NASA SP-350, Superintendent of Documents, Washington, DC.

Chariots for Apollo: A History of the Manned Lunar Spacecraft by Courtney G. Brooks, James M. Grimwood, and Lloyd S. Swenson, Jr., 1979, NASA SP-4205, Superintendent of Documents, Washington, DC.

Countdown by Frank Borman and Robert J. Serling, 1988, Silver Arrow Books, William Morrow, New York.

Countdown: A History of Space Flight by T. A. Heppenheimer, 1997, John Wiley & Sons, Inc., New York, NY.

End of an Era in Space Exploration by B. C. D. Blaine, An American Astronautical Society Publication, 1976, Univelt, Inc., San Diego, CA.

Jane's Spaceflight Directory, 1986, Reginald Turnill, editor, Jane's Publishing House, London, England.

Men from Earth by Buzz Aldrin and Malcolm McConnell, 1989, Bantam Books, New York, NY.

Orders of Magnitude: A History of NACA and NASA, 1915–1990 by Roger Bilstein, 1989, NASA SP-4406, U.S. Government Printing Office, Washington, DC.

Space Almanac, Anthony R. Curtis, editor, 1990, Gulf Publishing Company, Houston, TX.

Space Travel: A History by Wernher von Braun, Frederick I. Ordway III, and Dave Dooling, 1975, Harper & Row, New York, NY.

This New Ocean: A Story of the First Space Age by William E. Burrows, 1998, Random House, Inc., New York, NY.

Chapter 20: Apollo 11—The First Landing of Humans on the Moon

A Man on the Moon by Andrew Chaikin, 1994, Penguin Books, New York, NY.

Apollo Expeditions to the Moon, Edgar M. Cortright, editor, 1975, NASA SP-350, Superintendent of Documents, Washington, DC.

Chariots for Apollo, A History of the Manned Lunar Spacecraft by Courtney G. Brooks, James M. Grimwood, and Lloyd S. Swenson, Jr., 1979, The NASA History Series, SP-4205, Superintendent of Documents, Washington, DC.

First on the Moon by Neil Armstrong, Michael Collins, and Edwin E. Aldrin, Jr. written with Gene Farmer and Dora Jane Hamblin, 1970, Little, Brown and Company, Boston, MA.

Liftoff by Michael Collins, 1988, Grove Press Publication, New York, NY.

Lunar Sourcebook, G. H. Heiken, D. T. Vaniman, and B. French, editors, 1995, Cambridge University Press, New York, NY.

Men from Earth by Buzz Aldrin and Malcolm McConnell, 1989, Bantam Doubleday Dell Publishing Group, New York, NY.

Space Travel, A History by Wernher von Braun, Frederick I. Ordway III, and Dave Dooling, 1985, Harper & Row Publishers, New York, NY.

The Illustrated History of NASA by Robin Kerrod, 1986, Gallery Books, New York, NY.

Their Own Great Stories, *Life*, August 22, 1969, pp. 24–29.

This New Ocean: The Story of the First Space Age by William E. Burrows, 1998, Random House, Inc., New York, NY.

Three Men Bound for the Moon, *Life*, July 4, 1969, pp. 16–28.

Chapter 21: Apollo Missions 12 and 13—A Rough Start to Lunar Exploration

Apollo 13 by Jim Lovell and Jeffrey Kluger, 1994, Pocket Books, (Simon & Schuster Inc.,) New York, NY.

Apollo Expeditions to the Moon, Edgar M. Cortright, editor, 1975, NASA SP-350, Superintendent of Documents, Washington, DC.

Moon Mission: Mankind's First Voyages to Another World by William F. Mellberg, Plymouth Press, Ltd., Plymouth, MI.

Space Almanac, Anthony R. Curtis, editor, 1992, Gulf Publishing Company, Houston, TX.

The Man on the Moon: The Voyages of the Apollo Astronauts by Andrew Chaikin, 1994, Penguin Books, New York, NY.

Where No Man Has Gone Before: A History of Apollo Lunar Exploration Missions by William David Compton, 1989, NASA SP-4214, Superintendent of Documents, Washington, DC.

Chapter 22: Apollo 14—Alan Shepard Returns to Space

A Man on the Moon by Andrew Chaikin, 1994, Penguin Books, New York, NY.

Apollo Expeditions to the Moon, Edgar M. Cortright, editor, 1975, NASA SP-350, Superintendent of Documents, Washington, DC.

Lunar Sourcebook, G. H. Heiken, D. T. Vaniman, and B. M. French, editors, 1995, Cambridge University Press, New York, NY.

Manned Spaceflight Log by Tim Furniss, 1986, Jane's Publishing Company, Ltd., London, England.

Moon Missions by William F. Mellberg, 1997, Plymouth Press, Ltd., Plymouth, MI.

Moon Shot by Alan Shepard and Deke Slayton, 1994, Turner Publishing, Inc., Atlanta, GA.

Where No Man Has Gone Before: A History of Apollo Lunar Exploration Missions by William David Compton, 1989, NASA SP-4214, Superintendent of Documents, Washington, DC.

Chapter 23: Apollo 15–17—Lunar Exploration takes Priority—the J Missions

A Man on the Moon: The Voyages of the Apollo Astronauts by Andrew Chaikin, 1994, Penguin Book, New York, NY.

Apollo Expeditions to the Moon, Edgar M. Cortright, editor, 1975, NASA SP-350, Superintendent of Documents, Washington, DC.

Dynamic Astronomy by Robert Dixon, 1989, Prentice Hall, Englewood Cliffs, NJ.

For All Mankind by Harry Hurt III, 1988, The Atlantic Monthly Press, New York, NY.

Jane's Spaceflight Directory 1986, Reginald Turnill, editor, 1986, Jane's Publishing Company, Ltd., London, England.

Lunar Sourcebook: A User's Guide to the Moon by Grant Heiken, David Vaniman, and Bevan French, 1991, Cambridge University Press, New York, NY.

Moon Missions: Mankind's First Voyages to Another World by William F. Mellberg, 1997, Plymouth Press Ltd., Plymouth, MI.

The Illustrated History of NASA by Robin Kerrod, 1986, Gallery Books, New York, NY.

Where No Man Has Gone Before: A History of Apollo Lunar Exploration Missions by William David Compton, 1989, NASA SP-4214, Superintendent of Documents, Washington, DC.

Section IV: The Early Soviet Manned Space Program

Korolev by James Harford, 1997, John Wiley & Sons, Inc., New York, NY.

Hermann Oberth: The Father of Space Flight by Boris Rauschenbach, 1994, West-Art, Clarence, NY.

Mayday: Eisenhower, Khrushchev and the U-2 Affair by Michael Beschloss, 1986, Harper & Row, Publishers, New York, NY.

Robert Goddard: Pioneer of Space Research by Milton Lehman, 1988, Da Capo Press, Inc., New York, NY.

Russians in Space by Evgeny Riabchikov, 1971, Doubleday & Company, Garden City, NY.

Soviet Space Exploration: The First Decade by William Shelton, 1968, Washington Square Press, New York, NY.

The Crisis Years: Kennedy and Khrushchev 1960–1963 by Michael Beschloss, 1991, Harper & Row, Publishers, New York, NY.

Chapter 24: Sergei Pavovich Korolev (1907–1966)

Autospy for an Empire by Dmitri Volkogonov, 1998, Simon & Schuster Inc., New York, NY.

Korolev by James Harford, 1997, John Wiley & Sons, Inc., New York, NY.

Man in Space: An Illustrated History of Space Flight, H.J.P. Arnold, editor, 1993, CLB Publishing Ltd., Surrey, England.

Russians in Space by Evgeny Riabchikov, 1971, Doubleday & Company, Inc., Garden City, NY.

Since Stalin by Boris Shub and Bernard Quint, 1951, SWEN Publications, New York.

Space Age by William J. Walter, 1992, Random House, Inc, New York, NY.

The New Russian Space Programme by Brian Harvey, 1996, Praxis Publishing, Chichester, England.

The Stalinist Terror in the Thirties by Borys Levytsky, 1974, Hoover Institution Press, Stanford University.

Chapter 25: The Soviet Launch Facilities and Vehicles

Almanac of Soviet Manned Space Flight by Dennis Newkirk, 1990, Gulf Publishing Company, Houston, TX.

Handbook of Soviet Manned Space Flight by Nicholas L. Johnson, 1980, Vol. 48, Science and Technology Series, American Astronautical Society, Univelt, Inc, San Diego, CA.

Jane's Spaceflight Directory 1986, Reginald Turnill, editor, Jane's Publishing Company, London, England.

Korolev by James Harford, 1997, John Wiley & Sons, Inc., New York, NY.

Rand McNally Goode's World Atlas, Edward B. Espenshade, Jr., editor, 1987, Rand McNally & Company, New York, NY.

Rockets of the World by Peter Alway, 1999, Saturn Press, Ann Arbor, MI.

Russians in Space by Evgeny Riabchikov, 1972, Doubleday & Company, Inc., Garden City, NY.

Space Age by William J. Walter, 1992, QED Communications (Random House Inc).

Space Exploration by J. K. Davies, 1992, W & R Chambers, Ltd., Edinburgh, England.

The Mir Space Station: A Precursor to Space Colonization by David M. Harland, 1997, John Wiley & Sons, New York, NY.

The New Russian Space Programme by Brian Harvey, 1996, Praxis Publishing Ltd., West Sussex, England.

The Soviet Manned Space Program by Phillip Clark, 1988, Salamander Book, Ltd. a division of Crown Publishers, Inc., New York, NY.

Chapter 26: Vostok and Voskhod—The Soviet Union's Early Manned Missions

Handbook of Soviet Manned Space Flight by Nicolas L. Johnson, 1980, Science and Technology Series, American Astronautical Society, Vol. 48, Univelt Inc., San Diego, CA.

Jane's Spaceflight Directory 1986, Reginald Turnill, editor, Jane's Publishing Company, Limited, London, England.

Korolev by James Harford, 1997, John Wiley & Sons, Inc., New York, NY.

Russians in Space by Evgeny Riabchikov, 1971, Doubleday & Company, Garden City, NY.

Space Travel: A History by Wernher von Braun and Frederick I. Ordway III, 1985, Harper & Row, New York, NY.

Soviets in Space by Peter L. Smolders, 1971, Taplinger Publishing Co. Inc., New York, NY.

The New Russian Space Programme by Brian Harvey, 1996, John Wiley & Sons, New York, NY.

The Soviet Manned Space Program, Phillip Clark, 1986, Salamander Books Limited, New York, NY.

Chapter 27: The End of the Moon Race—A Deadly Dash to the Finish

Jane's Spaceflight Directory 1986, Reginald Turnill, editor, Jane's Publishing Company, Limited, London, England.

Korolev by James Harford, 1997, John Wiley & Sons, New York, NY.

Manned Spaceflight Log by Tim Furniss, 1986, Jane's Publishing Company, Ltd., London, England.

Soviet Space Exploration: The First Decade by William Shelton, 1968, Washington Square Press, New York, NY.

Soviets in Space by Peter L. Smolders, 1974, Taplinger Publishing Co., Inc., New York, NY.

Space Travel: A History by Wernher von Braun and Frederick I. Ordway III, 1985, Harper & Row, New York, NY.

The Illustrated History of NASA by Robin Kerrod, 1986, Gallery Books, New York, NY.

The New Russian Space Programme by Brian Harvey, 1996, John Wiley & Sons, New York, NY.

The Soviet Manned Space Program by Phillip Clark, 1986, Salamander Books Limited, New York, NY.

Chapters 28–30: The Soviet Space Station Program

Countdown, A History of Space Flight by T. A. Heppenheimer, 1997, John Wiley & Sons, Inc., New York, NY.

Jane's Spaceflight Directory 1986, Reginald Turnill, editor, Jane's Publishing Company, Limited, London, England.

Manned Spaceflight Log by Tim Furniss, 1986, Jane's Publishing Company, Limited, London, England.

Russians in Space by Evgeny Riabchikov, 1971, Doubleday & Company, Garden City, NY.

Soviets in Space by Peter Smolders, 1973, Taplinger Publishing Co. Inc., New York, NY.

Space Almanac by Anthony R. Curtis, 1992, Gulf Publishing Company, Houston, TX.

Space Travel: A History by Wernher von Braun, Frederick I. Ordway III, and Dave Dooling, 1975, Harper & Row, New York, NY.

The New Russian Space Programme by Brian Harvey, 1996, John Wiley & Sons, New York, NY.

The Soviet Manned Space Program by Phillip Clark, 1986, Salamander Books Limited, New York, NY.

Chapters 31 and 32: Skylab

A House in Space by Henry S. F. Cooper, Jr., 1976, Holt, Reinehart, and Winston, New York, NY.

A New Sun: The Solar Results from Skylab by John A. Eddy, 1979, NASA SP-402.

Apollo Expeditions to the Moon, Edgar M. Cortright, editor, 1975, NASA SP-350, Superintendent of Documents, Washington, DC.

"I Touch the Future . . .", The Story of Christa McAuliff, by Robert T. Hohler, 1986, Random House, New York, NY.

Jane's Spaceflight Directory 1986, Reginald Turnill, editor, Jane's Publishing Company, Limited., London.

Living and Working in Space, A History of Skylab by W. David Compton and Charles D. Benson, 1983, The NASA History Series, NASA SP-4208, Superintendent of Documents, U.S. Government Printing Office, Washington, DC.

NASA, The First 25 Years, 1958–1983, NASA EP-182, 1983, U.S. Government Printing Office, Washington, DC.

Skylab, Our First Space Station, Leland F. Belew, 1977, NASA SP-400, U.S. Government Printing Office, Washington, DC.

Skylab Science Experiments, George Morgenthaler and Gerald E. Simonson, editors, 1975, Volume 38, Science and Technology Series of the American Astronautical Society, Tarzana, CA.

Skylab's Astronomy and Space Sciences, Charles A. Lundquist, editor, 1979, NASA SP-404, U.S. Government Printing Office, Washington, DC.

Space Exploration by J. K. Davies, 1992, Chambers Encyclopedic Guides, Edinburgh, NY.

The Illustrated History of NASA by Robin Kerrod, 1986, Gallery Books, New York, NY.

Chapter 33: The Apollo-Soyuz Test Program

Deke! by Donald K. Slayton and Michael Cassutt, 1994, Tom Doherty Associates Inc., New York, NY.

Jane's Spaceflight Directory, Reginald Turnill, editor, 1986, Jane's Publishing Company, London.

Manned Spaceflight Log by Tim Furniss, 1986, Biddles Limited, Guildford, Surrey, England.

Moon Shot by Alan Shepard and Deke Slayton, 1994, Turner Publishing, Inc., Atlanta, GA.

Soviet Manned Space Flight by Nicholas L. Johnson, 1988, Vol. 48, Space and Technology Series of the American Astronautical Society, Univelt, Inc., San Diego, CA.

Space Exploration by J. K. Davies, 1993, W & R Chambers Ltd., New York, NY.

Space Travel: A History by Wernher von Braun, Frederick I. Ordway III, and Dave Dooling, 1975, Harper & Row, New York, NY.

The Illustrated History of NASA by Robert Kerrod, 1986, Gallery Book, New York, NY.

The New Russian Space Programme by Brian Harvey, 1996, Praxis Publishing Ltd., West Sussex, England.

The Politics of Space by Matthew J. Von Bencke, 1997, Westview Press (a Division of Harper Collins Publishers), Boulder, CO.

The Soviet Manned Space Program by Phillip Clark, 1988, Orion Books, New York, NY.

Who's Who in Space by Michael Cassutt, 1999, MacMillan Library Reference, New York, NY.

Chapter 34: The Space Shuttle

Manned Spaceflight Log by Tim Furniss, 1986, Jane's Publishing Inc., New York, NY.

NASA: A History of the U.S. Civil Space Program by Roger D. Launius, 1994, Krieger Publishing Company, Malabar, FL.

Nixon: The Triumph of a Politician 1961–1972 by Stephen E. Ambrose, 1989, Simon and Schuster, New York, NY.

Space Exploration by J. K. Davies, 1992, W & R Chambers Ltd., New York, NY.

Space Shuttle by Melvyn Smith, 1985, Haynes Publishing, Inc., Newbury Park, CA.

Space Shuttle: The History of Developing the National Space Transportation System by Dennis R. Jenkins, 1996, Walsworth Publishing Company, Marceline, MO.

Space Travel: A History by Wernher von Braun, Frederick I. Ordway III, and Dave Dooling, 1975, Harper & Row, New York, NY.

Spaceflight and the Myth of Presidential Leadership, Roger D. Launius and Howard E. McCurdy, editors, 1997, University of Illinois Press, Urbana, IL.

The Space Shuttle Story by Andrew Wilson, 1986, Hamlyn Publishing Group, New York, NY.

Chapter 35: The Second Generation Soviet Space Stations—Salyuts 6 and 7

Handbook of Soviet Manned Space Flight by Nicholas L. Johnson, 1980, Volume 48, Science and Technology Series, American Astronautical Society, Univelt Inc., San Diego, CA.

Jane's Spaceflight Directory 1986, Reginald Turnill, editor, Jane's Publishing Company Ltd., New York, NY.

Space Almanac by Anthony R. Curtis, 1992, Gulf Publishing Company, Houston, TX.

Space Travel: A History by Wernher von Braun, Frederick I. Ordway III, and Dave Dooling, 1985, Harper & Row, New York, NY.

The New Russian Space Programme by Brian Harvey, 1996, Praxis Publishing Ltd., West Sussex, England.

The Soviet Manned Space Program by Phillip Clark, 1988, Salamander Books Limited, New York, NY.

Chapter 36: The Soviet Super Station—the Mir Space Station

Almanac of Soviet Manned Space Flight by Dennis Newkirk, 1990, Gulf Publishing Company, Houston, TX.

Reaching for the Stars by Peter Bond, 1997, Cassell Publishers, Ltd., London, England.

The Mir Space Station by David M. Harland, 1997, John Wiley & Sons Publishers in association with Praxis Publishing Ltd., West Sussex, England.

The New Russian Space Programme: From Competition to Collaboration by Brian Harvey, 1996, Praxis Publishing Ltd. in association with John Wiley & Sons Ltd., Chichester, West Sussex, England

The Soviet Manned Space Program by Phillip Clark, 1988, Orion Books, New York, NY.

Chapter 37: Mir Perseveres through Troubling Times

Countdown: A History of Space Flight by T. A. Heppenheimer, 1997, John Wiley & Sons, Inc., New York, NY.

The Mir Space Station: A Precursor to Space Colonization by David M. Harland, 1997, John Wiley & Sons, New York, NY.

The New Russian Space Programme: From Competition to Collaboration by Brian Harvey, 1996, John Wiley & Sons, New York, NY.

Chapter 38: Flights of the Space Shuttle

Anti-Aging in the Space Age by Bob Delmonteque, N.D., 1998, *Journal of Longevity*, Vol. 4, No. 9, p. 12.

Jane's Spaceflight Directory 1986, Reginald Turnill, editor, Jane's Publishing Company, Ltd., London, England.

Liftoff: The Story of America's Adventure in Space by Michael Collins, 1988, Grove Press, New York, NY.

Man in Space: An Illustrated History of Space Flight, H.J.P. Arnold, editor, 1993, CLB Publishing Ltd., Surrey, England.

Manned Spaceflight Log by Tim Furniss, 1986, Jane's Publishing Company, Ltd., New York, NY.

Space Almanac, Anthony R. Curtis, editor, 1992, Gulf Publishing Company, Houston, TX.

Space Shuttle Story by Andrew Wilson, 1986, Hamlyn Publishing Group, Middlesex, England.

Space Shuttle: The History of Developing the National Space Transportation System by Dennis R. Jenkins, 1996, Wadsworth Publishing Company, Maeceline, MO.

Space Shuttle: U.S. Winged Aircraft: X-15 to Orbiter by Melvyn Smith, 1985, Haynes Publications Inc., Newbury Park, CA.

The Illustrated History of NASA by Robin Kerrod, 1986, Gallery Book, New York, NY.

The Last Heroes, John Glenn Flies Back to Space by Richard B. Stolley/Photography by Ralph Morse, *Life*, October 1998, p. 50.

Why We Must Venture into Space by Sen. John Glenn, *Parade Magazine*, October 25, 1998, p. 4.

Chapter 39: The Shuttle/Mir Era

Countdown: A History of Space Flight by T. A. Heppenheimer, 1997, John Wiley & Sons, New York, NY.

Dragonfly: NASA and the Crisis Aboard Mir by Bryan Burrough, 1998, HarperCollins Publishers, New York, NY.

Six Months on Mir by Shannon W. Lucid, May 1998, *Scientific American*, Vol. 278, Number 5, p. 46.

The Mir Space Station: A Precursor to Space Colonization by David M. Harland, 1997, John Wiley & Sons, New York, NY.

The New Russian Space Programme: From Competition to Collaboration, by Brian Harvey, 1996, John Wiley & Sons, New York, NY.

Who's Who in Space by Michael Cassutt, 1999, MacMillan Library Reference, New York, NY.

Chapter 40: The Shuttle/Mir Program Reveals Unsettling Problems

Countdown: A History of Space Flight by T. A. Heppenheimer, 1997, John Wiley & Sons, New York, NY.

Dragonfly: NASA and the Crisis Aboard Mir by Bryan Burrough, 1998, HarperCollins Publishers, New York, NY.

Six Months on Mir by Shannon W. Lucid, May 1998, *Scientific American*, Vol. 278, Number 5, p. 46.

The Mir Space Station: A Precursor to Space Colonization by David M. Harland, 1997, John Wiley & Sons, New York, NY.

The New Russian Space Programme: From Competition to Collaboration, by Brian Harvey, 1996, John Wiley & Sons, New York, NY.

Chapter 41: The American Space Station Freedom Becomes the International Space Station

Countdown, A History of Space Flight by T. A. Heppenheimer, 1997, John Wiley & Sons, Inc., New York, NY.

Dragonfly: NASA and the Crisis Aboard Mir by Bryan Burrough, 1998, HarperCollins Publishers, Inc., New York, NY.

NASA: A History of the U.S. Civil Space Program by Roger D. Launius, 1994, Krieger Publishing Company, Malabar, FL.

Onward to Space by Andrew Lawler, *Astronomy*, December 1998, Vol. 26, Number 12, p. 42.

The $48 Billion International Space Station Assembly Begins, *U.S. News and World Report*, December 7, 1998, Vol. 125, Number 22, p. 60.

The Mir Space Station: A Precursor to Space Colonization by David M. Harland, 1997, John Wiley & Sons, New York, NY.

The Politics of Space: A History of U.S.–Soviet/Russian Competition and Cooperation by Matthew J. Von Bencke, 1997, Westview Press, Boulder, CO.

The White House: An Historic Guide, 1995, National Geographic Society, Washington, DC.

This New Ocean: The Story of the First Space Age by William E. Burrows, 1998, Random House, New York, NY.

Chapter 42: Moon Base—An Important Human Step into the Future

George Herbert Wallace Bush by David Valdez, 1997, Texas A&M Press, College Station, Texas.

Lunar Sourcebook, G. H. Heiken, D. T. Vaniman, and B. M. French, editors, 1991, Cambridge University Press, Cambridge, England.

Observatories on the Moon by Jack O. Burns, Nebojsa Duric, G. Jeffery Taylor, and Stewart W. Johnson, 1990, *Scientific American*, 262, Number 3, p. 42.

Orbital Debris from Upper-Stage Breakup, J. P. Loftus Jr., editor, 1989, Progress in Astronautics and Aeronautics, Vol. 121, American Institute of Aeronautics and Astronautics, Inc., Washington, DC.

Space Exploration by J. K. Davies, 1992, W & R Chambers, Ltd., New York, NY.

The Clementine Bistatic Radar Experiment by S. Nozette et al., 1996, *Science*, 274, p. 1495.

Chapter 43: Mission to Mars

Dynamic Astronomy by Robert T. Dixon, 1989, Prentice-Hall, Inc., Englewood Cliffs, NJ.

Explorations: An Introduction to Astronomy by Thomas T. Arny, 1998, McGraw-Hill, New York, NY.

Frontiers of Astronomy by Michael A. Seeds, 1990, Wadsworth, Inc., Belmont, CA.

Journey into Space by Bruce Murray, 1989, W. W. Norton & Company, Inc., New York, NY.

Mars Beckons by John Noble Wilford, 1990, Alfred A. Knopp Press, Inc., New York, NY.

Mission to Mars by Michael Collins, 1990, Grove Press, Inc., New York, NY.

Rediscovering Mars by Raymond Ladbury, 1999, *Physics Today*, Vol. 52, Number 10, p. 33.

Space Age by William J. Walter, 1992, Random House, Inc., New York, NY.

The Discoverers by Daniel Boorstin, 1983, Random House, New York, NY.

The Mir Space Station by David M. Harland, 1997, John Wiley & Sons, New York, NY.

The Search for Life in the Universe by Donald Goldsmith and Tobias Owen, 1980, Benjamin/Cummings Publishing Company, Inc., Menlo Park, CA.

The Third Great Age of Discovery by Stephen Pyne, December 1987, Unpublished lecture at George Washington University.

Appendix 1: Geological Results of the Apollo Program

Apollo 15 Preliminary Science Report, 1972, NASA SP-289, Superintendent of Documents, Washington, DC.

Dynamic Astronomy by Robert T. Dixon, 1975, Prentice-Hall, Englewood Cliffs, NJ.

Lunar Science: A Post-Apollo View by Stuart R. Taylor, 1975, Pergamon Press, Inc., New York, NY.

Lunar Sourcebook, G. H. Heiken, D. T. Vaniman, and B. M. French, editors, 1995, Cambridge University Press, New York, NY.

Planetary Science: A Lunar Perspective by Stuart R. Taylor, 1982, The Lunar and Planetary Institute, Houston, TX.

The Lunar Rocks by Brian Mason and William G. Melson, 1970, Wiley-Interscience, New York, NY.

The Planetary Scientist's Companion by Katharina Lodders and Bruce Fegley, Jr., 1998, Oxford University Press, New York, NY.

Appendix 2: The Origin and Evolution of the Moon

Chemical Evolution and the Origin of Life by Richard Dickerson in *The Quest for Extraterrestrial Life*, Donald Goldsmith, editor, University Science Book, Mill Valley, CA.

Lunar Science: A Post-Apollo View by Stuart Ross Taylor, 1975, Pergamon Press, Inc., New York, NY.

Lunar Sourcebook, G. Heiken, D. Vaniman, and B. French, editors, 1991, Cambridge University Press, Cambridge, England.

Nickel for Your Thoughts: Urey and the Origin of the Moon by Stephen G. Brush, 1982, *Science*, 217, p. 891.

Origin of the Moon, W. K. Hartmann, R. J. Phillips, and C. J. Taylor, editors, 1986, Lunar and Planetary Institute, Houston TX.

Planetary Science: A Lunar Perspective by S. R. Taylor, 1982, Lunar and Planetary Institute, Houston, TX.

The Moon in *The Solar System: A Scientific American Book*, 1975, W. H. Freeman and Company, San Francisco, CA.

To a Rocky Moon by D. E. Wilhelms, 1993, University of Arizona Press, Tucson, AZ.

Appendix 3: The Science of Skylab

A New Sun: The Solar Results from SkyLab by John A. Eddy, 1979, NASA SP-402, U.S. Government Printing Office, Washington, DC.

Biomedical Results from Skylab, Richard S. Johnston and Lawrence F. Dietlein, editors, 1997, NASA SP-377, Superintendent of Documents, Washington, DC.

Skylab, Our First Space Station, Leland F. Belew, editor, 1977, NASA SP-400.

Jane's Spaceflight Directory 1986, Reginald Turnill, editor, Jane's Publishing Company, Limited., London.

Living and Working in Space, A History of Skylab by W. David Compton and Charles D. Benson, 1983, The NASA History Series, NASA SP-4208, Superintendent of Documents, U.S. Government Printing Office, Washington, DC.

NASA, The First 25 Years, 1958–1983, 1983, NASA EP-182, U.S. Government Printing Office, Washington, DC.

Skylab Science Experiments, George Morgenthaler and Gerald E. Simonson, editors, 1975, Volume 38, Science and Technology Series of the American Astronautical Society, Tarzana, CA.

Skylab EREP Investigations Summary, 1978, NASA SP-399, Superintendent of Documents, Washington, DC.

Skylab Explores the Earth, 1977, NASA SP-380, Superintendent of Documents, Washington, DC.

Skylab's Astronomy and Space Sciences, Charles A. Lundquist, editor, 1979, NASA SP-404, U.S. Government Printing Office, Washington, DC.

Space Exploration by J. K. Davies, 1992, Chamber Encyclopedia Guides, Edinburgh, NY.

Index

Atomic Energy Commission, 47, 48
Aubakirov, Kazakh Takhtar, 289, 290
Augustine, Norman, 327
Aurora 7, 83–84, 85, 210
Avdeyev, Sergei, 289
Averkov, S., 194

Babbitt, Donald, 133–134
Baikonour, collapse of Soviet Union and
    expenditures at, 291
Balandin, Alexander, 287
Balanin, Grigory, 188
Baldwin, Hanson, 66
Bales, Steve, 147
Barr, James, 56
Barratt, Mike, 310
Bassett, Charles, II, 121–122, 133
Bean, Alan
    Apollo 12 and, 155–159
    on Skylab, 241
    Skylab 3 and, 245, 246, 249
Becker, Karl, 32–33
Beggs, James, 328
Belka (dog), launch of, 56
Bell X-1, 261
Belotserkovsky, Sergei, 194
Belyayev, Pavel I., 203, 204
    Voshkod 2 and, 118
    Voskhod 2 and, 210–212
Belyoyev, 212
Beregovoi, Georgi, 216
Berezovoi, Anatoly, 274–275, 278
Berkner, Lloyd, 39
Big Joe program, 54–55
Big Muley, 179
Biomedical science knowledge from
    Skylab, A21, A27-A28
Blaha, John, 314, 315, 316
    preflight training program for, 317–318
    problems on Mir and, 318–319
Bondorenko, V. V., 203
Bone materials, loss of, A27
Boorstein, Daniel, 345
Borman, Frank, 67, 115, 122, 126
    Apollo 1 accident and, 134, 135
    Apollo 8 and, 139, 140–142, 217
    Gemini 7 and, 119–120, 130
Borman, Susan, 135
Bradt, Irene, 261
Brand, Vance, as member of ASTP,
    240–244
Brezhnev, Leonid, space race and, 201,
    210, 232, 243
British Explosives Act (1875), 28
British Interplanetary Society, 25–26, 27,
    28

British Interplanetary Society Journal, 28
Brown, Curtis L., 308
    STS-95 and, 306
Brucker, Wilber, 43–44
BST-1M, 270
Budarin, Nikolai, 312–313
Bumper rocket, 36
Buran, 285
Bush, George, space race and, 327, 335,
    342
Bykovsky, Valery F., 203, 204, 278
    Vostok 5 and, 86, 105, 111, 207–208

Cagle, K., 106
Calladan, David, 65
Cameron, Alastair, A16
Cameron, Kenneth, 313
Cape Canaveral, 197
Capture theory, A13-A14
Carbon dioxide problem, on *Apollo 13,*
    163
Cardiovascular system, A27-A28
Carpenter, Malcolm Scott, Jr., 115, 207
    Aurora 7 and, 210
    as backup on *MA-5,* 80
    MA-7 and, 82–84
    selection of, as one of first astronauts, 53
    selection of women astronauts and,
        109–111
Carr, Gerald, A28
    Skylab 4 and, 245, 246, 247, 248–249
Carter, Jimmy, space race and, 253, 259
Centaur rocket, 48
Centrifugal force, 4
Cernan, Eugene, 115
    Apollo 10 and, 143–144, 218
    Apollo 17 and, 181
    death of, 122
Chaffee, Roger, death of, on *Apollo 1,*
    133–135, 137, 299
Chaikin, Andrew, 174–175
Challenger, 266
    accomplishments of, 266
Challenger disaster. *See STS-25*
Chelomei, Vladimir, 200, 201, 219
Chernomyrdin, 329
Chernushka (dog), Soviet launch of, 58
Chertok, Boris, 198
Chretien, Jean-Loup, 275–276, 278, 286
Clinton, William Jefferson (Bill), space
    race and, 317, 328–329
Cobb, Jerrie, 106, 107–108, 111–112
Cochran, Jacqueline, 106, 108–109, 112
Coformation theory, A14
Cold war, rockets as measuring stick in,
    45–46

Collins, Eileen, 112
Collins, Michael, 115, 126, 296, 342,
    344–345, 347
    Apollo 11 and, 145–154
    Gemini 10 and, 123
Columbia, 266
Comet Kohoutek, A24
Command module (CM), 128
Compton Gamma-Ray Observatory, 267
Conrad, Charles, Jr., 115
    Apollo 12 and, 155–159
    Gemini 5 and, 118
    Gemini 11 and, 123–124
    repair of Skylab and, 241–242, 243
    Skylab 2 and, 245, 246
Cooper, Henry, 247, 248, A27, A28
Cooper, L. Gordon, Jr., 85–86, 115, 118
    selection of, as one of first astronauts, 53
Cosmonauts. *See also* Astronauts; specific
    original Soviet, 203
    on shuttle flights, 309–310
    Vostok, 203–204
    women, 86, 207–208
Cosmos 112, 199
Cosmos 133, 213
Cosmos 146, 215
Cosmos 154, 215–216
Cosmos 186, 216
Cosmos 188, 216
Cosmos 212, 216
Cosmos 213, 216
Cosmos 238, 216
Cosmos 557, 228, 269
Cosmos 573, 229
Cosmos 656, 230
Cosmos 772, 233
Crippen, Robert, 296, 297–298
Cronkite, Walter, 132
Culbertson, Frank, 313, 318
Cunningham, Newton, 92
Cunningham, Walter, *Apollo 7* and, 131,
    138–139

Daly, Reginald, A16
Darwin, George Howard, A14
Demin, Lev, 230
DeVoe, Barbara, 18
Dezhurov, Vladimir, 310, 312–313
Dietrich, Jan, 106
Dietrich, Marion, 106
Diffuse x-ray emission, A23-A24
Discovery
    accomplishments of, 266–267
    STS-63, 309–310
Dobrovolsky, Georgi, 225–228
Donahue, Thomas, 327